Geographies of Tourism and Global Change

Now accepted by Scopus! Content available on the Scopus site in the second half of 2023.

In a geographical tradition and using an integrated approach, this book series acknowledges the interrelationship of tourism to wider processes within society and environment. This is done at local, regional, national, and global scales demonstrating links between these scales as well as outcomes of global change for individuals, communities, and societies. Local and regional factors will also be considered as mediators of global change in tourism geographies affecting communities and environments. Thus *Geographies of Tourism and Global Change* applies a truly global perspective highlighting development in different parts of the world and acknowledges tourism as a formative cause for societal and environmental change in an increasingly interconnected world.

The scope of the series is broad and preference will be given to crisp and highly impactful work. Authors and Editors of monographs and edited volumes from across the globe are welcome to submit proposals. The series insists on a thorough and scholarly perspective, in addition authors are encouraged to consider practical relevance and matters of subject specific importance. All titles are thoroughly reviewed prior to acceptance and publication, ensuring a respectable and high quality collection of publications.

Asunción Blanco-Romero • Macià Blázquez-Salom
Editors

Spanish Tourism Geographies

Territorial Diversity and Different Approaches

Springer

Editors
Asunción Blanco-Romero
Department of Geography
Universitat Autònoma de Barcelona
Cerdanyola del Vallès, Spain

Macià Blázquez-Salom
Department of Geography
Universitat de les Illes Balears
Palma, Baleares, Spain

ISSN 2366-5610 ISSN 2366-5629 (electronic)
Geographies of Tourism and Global Change
ISBN 978-3-031-39779-0 ISBN 978-3-031-39780-6 (eBook)
https://doi.org/10.1007/978-3-031-39780-6

This Springer imprint is published by the registered company Springer Nature Switzerland AG
The registered company address is: Gewerbestrasse 11, 6330 Cham, Switzerland

Paper in this product is recyclable.

Foreword

Territorial diversity and different approaches presents a selection of recent research developments in the Spanish tourism geography. The editors, Asunción Blanco-Romero and Macià Blázquez-Salom, have compiled the contributions of the participants in the Colloquium of the Working Group on Tourism of the Spanish Association of Geographers jointly organized with the Commission on Tourism, Leisure and Global Change of the International Geographical Union in Maó (Balearic Islands) in October 2020. Paraphrasing Gibson (2021), this is a clearly place-dependent book written from the tourism destinations where authors are living in and/or researching.

As is well known, until the COVID-19 disruption, factors such as the intensification of economic globalization, market deregulation and financialization of economies, the rise of low-cost travel, short-term rental digital platforms, and the extensive adoption of tourism as a strategy for economic growth were drivers of increasing tourism activity everywhere. These global industry evolution factors also created many co-evolving local tensions and conflicts (social, environmental, and political) that were amplified during the COVID-19 pandemic (Milano & Koens, 2021).

Even before the pandemic, political initiatives such as the *Berlin Declaration on Transforming Tourism* (Transforming Tourism Initiative, 2017) or the European Travel Commission's *Tourism Manifesto* (European Travel Commission, 2020), representing totally diverse interests, critically re-evaluated the conventional growth trajectories of tourism. Thus, after COVID-19, the claim for (at least) the generalization of the adoption of policies, criteria, and practices derived from the UN's 2030 Agenda for Sustainable Development (even though it only mentions tourism a few times) throughout the Sustainable Development Goals (SDGs) have been more and more accepted, whereas de-growth and post-growth concepts, strategies, and practices have started to be increasingly discussed (Cheer et al., 2021; Prideaux et al., 2020; Sharpley, 2022).

From the academic perspective, the post-pandemic situation has offered a chance to critically interpret the multiple social, economic, and environmental consequences of gazing, visiting, and entertaining in a transition era where availability of free time is already in more countries a central component of contemporary life. In addition, large transportation, tourism, entertainment, media, and cultural corporations with enormous financial and technological capacities are taking advantage of it and are shaping places, tastes, and cultures. This is the general scenario that contextualizes most of the particular and detailed analysis included in this book from the perspective of the evolution, dynamics, and challenges of the tourism activity in the case of Spain after the COVID-19 pandemic. Hence, aside from introducing themes, issues, and challenges of tourism in Spain, one of the main contributions of the book is knowing how the understanding of the particular dynamics that emerge in Spanish destinations is evolving and which related knowledge-building processes and results can be used to develop or reshape theories, concepts, methods, and research procedures on a global scale. In this sense, this book provides an opportunity to facilitate knowledge transfer from the particular Spanish research tradition to the global circuits of tourism geography dissemination and practice.

The Spanish geography of tourism has been previously reviewed, and papers summarizing its scientific production from the beginning of the 1960s to 2022 have already been published in English. This is the case, for example, of the two reports about the Spanish tourism geography contributions made on the occasions of the 27th and 35th International Geographical Union conferences (Valenzuela et al., 1992; Blanco-Romero & Mínguez, 2022) or the revision of themes, topics, and approaches published in 1996 in *The Tourist Review* (Anton Clavé et al., 1996). An analysis of the Spanish tourism geographies as a "country" academic tradition legacy was also included in the review of the European research perspectives in geographies of tourism published by Emerald Group (Anton Clavé, 2013). What can be observed is that while well-established research topics in the Spanish tourism geography tradition exist, emerging tourism geography issues are continuously surfacing in the academic panorama. In this sense, the current book is an excellent compilation as it affords a clear view over many of them.

In the review of the Spanish geography of tourism up to the 1990s produced by Anton Clavé et al. (1996), various recurring research themes were identified. These included the transformation of coastal environments; the urban development processes associated with the spread of residential second homes; the analysis of tourism supply and demand; the measurement of economic and local impacts of tourism; the valorization of natural and cultural heritage for tourism; the development of tourism in non-coastal areas; the emergence of the historic city as a tourism product; and issues concerning tourism policy analysis. Interestingly, even though they take a renovated perspective – generally linked to the current global challenges previously mentioned – most of them are also addressed in this compilation. For instance, Chap. 2 discusses tourism in the less populated regions of Spain, Chap. 7 is about touristification and urban heritage in historic Spanish city centres, the subject of Chap. 14 is the role of landscape as a factor for tourism attractiveness, Chap. 15 examines tourism-related urban growth during the period 1990–2018, and Chap. 23's subject is package holidays and charter companies in Spain (2004–2021).

In general, positivist approaches, reflective critical analysis, and applied planning vocation can be attributed from the start to the tourism geography research practices in Spain. As noted in a review published in Spanish by Fernández Tabales et al. (2010), what changed at the end of the 1990s and the beginning of the 2000s was the development of new approaches and topics. These innovations included spatial modelling; economic development research; human agency analysis; environmental impact analysis; new destination evolutionary approaches; mobility studies; performativity; visitors' spatial behaviour; authenticity studies; and destination vulnerability, among others. Additionally, during this period, initial interest increased regarding destinations perception, identity, image, and the subjective discourse of the agents of tourism activity. Some of the chapters of the current book clearly relate to one or more of these topics. Chapter 4, for example, discusses nature conservation policies and tourism, Chap. 6 speaks of the fragility of successful tourism destinations linked to cultural itineraries, Chap. 10 is about the social production of Spain's coastal destinations, Chap. 12 examines beach erosion, and Chap. 21 expounds on how the impact of COVID-19 depends on the level of touristification and inequality of the space.

Finally, the book collects research focused on some of the currently more debated contemporary topics not only within the Spanish academic tourism geographies research tradition but also in the international academic arena. This positions the current Spanish tourism geographers at the forefront of the discussion of some relevant global issues of widespread interest. The first group of them are about current concerns related to our contemporary environmental, societal, and economic transition, such as climate emergence (Chap. 13), urban transformations (Chap. 16), inequality (Chap. 18), social vulnerability (Chap. 19), working conditions (Chap. 20), or social movements in tourism destinations (Chap. 22). The remaining contributions develop other topics related to tourism planning, management, or governance of destinations. This is the case of the analysis of the digital transformation (Chap. 5), undertourism (Chap. 3), touristification (Chap. 8), overtourism (Chap. 9), smart destination development (Chap. 11), platform economy (Chap. 17), and accessible tourism (Chap. 24).

Interestingly, most of the chapters are connected with internationally developed critical approaches and contribute to the ongoing global theoretical, epistemological, and conceptual debates from the particular perspective of the Spanish tourism geography research tradition. In so doing, the book meets the objective of the Springer Geographies of Tourism and Global Change Series by acknowledging the interrelationship of regional traditions of knowledge with wider processes within society and the environment and in this case, accordingly, between the particular development of a Spanish research tradition in tourism geography and the global concerns related to the geographical implications of tourism development. Reinforcing this, one of the book's added values is the bibliographies of each chapter, which contain important and relevant tourism geography- and tourism-related references lying outside of the English language circuits. Without any doubt, these bibliographies are a valuable tool for researchers in the tourism geography field and represent a precious legacy that allows those able to read them to gain insights about a partially different new landscape of research.

Another strong point of the book is the range, profile, and excellence of its contributors. Authors from 16 of the most highly reputed tourism geography research universities in Spain (Alacant, Autònoma de Barcelona, Autónoma de Madrid, Barcelona, Complutense de Madrid, Granada, Illes Balears, León, Málaga, Rovira i Virgili, Salamanca, Santiago de Compostela, Sevilla, Valencia, Vigo, Zaragoza) plus authors affiliated with other foreign universities and public institutions as well as independent researchers have participated in this joint overview. Thus, the book's chapters illustrate researches situated in time, space, culture, languages, and society and represent the diversity and complexity of an academic tradition with specific interests and problematizations that range, among other dimensions of geographical analysis, from the saturation of coastal spaces to the emptiness of rural areas. Moreover, the list of authors also includes multiple positions and research practices when approaching the actual evidence that tourism is a central iconic feature of neoliberal and global capitalism.

To summarize, in a context where English language dominance in disseminating tourism geography knowledge through academic publishing generates a kind of Anglo-American hegemony in theory building, concept formation, and selection of research cases (Wilson & Anton Clavé, 2013), this book introduces significant conceptual developments, discusses relevant research topics, explores meaningful case studies, and draws some key components for a future research agenda in tourism geography from the perspective of the actual dimensions of tourism in Spain, its main problems and challenges, and also, the type of approaches designed by the Spanish scholars devoted to analyse it. Throughout this sample of research contributions, it is possible to understand the interest and the potential role of the tourism geography academy in Spain and how it is committed to contributing to the social, economic, and environmental reconfiguration of tourism destinations under the current climate emergency, digital transformation, and social transition context.

Therefore, this book offers an approach of the current Spanish tourism geography as a contributing social science and also a contributing social actor aiming to participate in the discussion of a range of global issues from place transformation, sustainability, limits of growth, attractiveness, visitor behaviour, centrality, mobility to otherness. Even though this book does not include absolutely every approach, topic, and contribution made by tourism geographers in Spain, it stands as a clear illustration of the diversity and relevance of the research and knowledge dissemination of the Spanish tourism geography, the need to connect regional academic traditions with the hegemonic and dominant circuits of academic publishing, and the challenge to contribute to the search for solutions for the common good. Additionally, it reinforces the value of the analysis of economic, social, and political structures, strategies, and values in tourism research on the regional and local scales, and critically understands the uneven spatialities of tourism and the role of personal, cultural, social, economic, political, cultural, and environmental processes co-implicated with one another in the dynamics of tourism change and transitions.

Department of Geography, Universitat Rovira i Virgili, Tarragona, Spain

Salvador Anton Clavé

References

Anton Clavé, S. (2013). The geography of tourism in Spain: Institutionalization and internationalization. In J. Wilson & S. Anton Clavé (Eds.) *Geographies of tourism: European research perspectives* (pp. 151–177). Emerald Group.

Anton Clavé, S., López Palomeque, F., Marchena Gómez, M. J., & Vera Rebollo, J. F. (1996). Tourism research in Spain: The contribution of geography (1960–1995). *The Tourist Review, 51*(1), 46–64.

Blanco-Romero, A., & Mínguez, C. (2022). Research in tourism geography conducted by the tourism group of AGE (2012–2020). In *Spanish contribution to 35th IGC. Paris 2022. Time for Geographers* (pp. 265–286). Asociación Española de Geografía.

Cheer, J. M., Lapointe, D., Mostafanezhad, M., & Jamal, T. (2021). Global tourism in crisis: conceptual frameworks for research and practice. *Journal of Tourism Futures, 7*(3): 278–294. https://doi.org/10.1108/JTF-09-2021-227

European Travel Commission. (2020). *European tourism manifesto. Priorities.* https://etc-corporate.org/tourism-manifesto

Fernández Tabales, A., García Hernández, M., & Ivars Baidal, J. A. (Coords.). (2010). *La investigación de la geografía del turismo en las comunidades autónomas españolas: orígenes, desarrollo y perspectivas de una disciplina en el horizonte de la Geografía.* Asociación de Geógrafos Españoles.

Gibson, C. (2021). Theorising tourism in crisis: Writing and relating in place. *Tourist Studies, 21*(1), 84–95. https://doi.org/10.1177/1468797621989218

Milano, C., & Koens, K. (2021). The paradox of tourism extremes. Excesses and restraints in times of COVID-19. *Current Issues in Tourism, 25*(2), 219–231. https://doi.org/10.1080/13683500.2021.1908967

Prideaux, B., Thompson, M., & Pabel, A. (2020). Lessons from COVID-19 can prepare global tourism for the economic transformation needed to combat climate change. *Tourism Geographies, 22*(3), 667–678. https://doi.org/10.1080/14616688.2020.1762117

Sharpley, R. (2022). Tourism and development theory: Which way now? *Tourism Planning & Development, 19*(1), 1–12. https://doi.org/10.1080/21568316.2021.2021475

Transforming Tourism Initiative. (2017). *Berlin declaration on transforming tourism.* http://www.transforming-tourism.org/berlin-declaration-on-transforming-tourism.html

Valenzuela Rubio, M., López Palomeque, F., Marchena Gómez, M. J., & Vera Rebollo, J. F. (1992) Geografía del turismo y del ocio. In *La Geografía en España (1970–1990). Aportación española al XXVII Congreso Internacional de la UGI* (pp. 203–212). Fundación Banco Bilbao Vizcaya.

Wilson, J., & Anton Clavé, S. (2013). (Eds.) *Geographies of tourism: European research perspectives.* Emerald Group.

Contents

Chapter 1
Territorial Diversity and Different Approaches to Tourism from Geography in Spain

Asunción Blanco-Romero and Macià Blázquez-Salom

1.1 Introduction: Background and Academic and Social Context

The anglophone hegemony in the scientific sphere sidelines a large number of studies published in many other languages and from other cultural perspectives. This under-representation impoverishes the scientific debate, necessitating approaches that incorporate both of these aspects. Many academic studies in the field of tourism geography conducted by Spanish scientists are difficult to disseminate in non-Spanish-speaking international environments. The domestic perspective, or that of *insiders*, of these scientists residing in Spain bestows them with special codes for conducting interpretations and analyses based on their everyday proximity to a territory characterised by its intense touristification. The tourism and real estate specialisation that Spanish society, together with its territory and institutions, have forged since the beginning of "developmentalism" permeates this scientific analysis.

Our proposal in this book is to avoid academic colonialist bias with contributions to the knowledge of those who know the direct and everyday reality of the place. Similarly, we support studies developed by local scientists in the Latin American context and not the explanation of its reality from Spain (Blázquez-Salom & Cañada,

A. Blanco-Romero (✉)
Department of Geography, Universitat Autònoma de Barcelona, Cerdanyola del Vallès, Spain
e-mail: asuncion.blanco@uab.cat

M. Blázquez-Salom
Department of Geography, Universitat de les Illes Balears, Palma, Spain
e-mail: mblazquez@uib.cat

A. Blanco-Romero, M. Blázquez-Salom (eds.), *Spanish Tourism Geographies*, Geographies of Tourism and Global Change,
https://doi.org/10.1007/978-3-031-39780-6_1

2011). Tourism geography in Spain explains some unique features worthy of attention, such as: the afore mentioned "developmentalism",[1] the evidence of tourism saturation on the one hand and the depopulation of the interior[2] on the other, or the consequences of the different crises, such as the financial crisis in 2008, the health crisis of 2020 or the housing crisis due to its use as tourist accommodation. With this objective, this book presents a compilation and translation into English of the many different strands of knowledge to encourage mutual approaches based on insights and perspectives.

A highly prominent group in the Spanish academic contribution to tourism geography is the Working Group in this subject matter of the Spanish Geographical Association (Blanco-Romero & Mínguez, 2022). This association was founded in 1977 and its publications are mostly in the Spanish language. The international projection of its activity has been supported by the Tourism Commission of the International Geographical Union. Of its many activities, we can highlight the first joint Colloquium that took place in Maó during the de-escalation process following the COVID-19 pandemic in October 2020. The project for publishing this book emerged from this Colloquium.

Geographical research focused on tourism has acquired greater prominence in Spanish academia as an autonomous field over the last few decades, particularly in the regions in which the incidence of the industry is highest. The contribution of university academic research in Spain to improve the understanding of the tourism phenomenon is notable, rigorously analysing the characteristics and trends of its dialectic with the territory, on both a basic and applied level (López Palomeque et al., 2014). A reflection of this prominence is the emergence of scientific tourism geography journals promoted by teaching and research staff of public universities (*Cuadernos de Turismo, Investigaciones Turísticas, Papers de Turisme…*) and the significant number of papers written by Spanish researchers as well as the increase in the number of articles on tourism published in the scientific journals specialised in geography. The group of researchers engaged in the field of Spanish geography also promote transdisciplinary practices, for example, publishing in other journals specialised in tourism in collaboration with other social sciences: *PASOS. Revista de Turismo y Patrimonio Cultural, Journal of Tourism Analysis: Revista de Análisis Turístico* or *Enlightening tourism. A pathmaking journal.*

In parallel with the research developed in the universities, the importance of the Spanish tourism industry has aroused the interest of independent researchers, some of whom are committed to its critical analysis. Furthermore, autonomous research platforms have been developed for analysing global touristification in defence of the commons, dignified work or the rights to well-being, nature, etc. This is the case of social movements such as Alba Sud, a research centre that produces a vast amount of

[1] Period of economic and urban growth in the second stage of the Franco regime, based on the incorporation of Spain into Fordism, which was characteristic of the "Glorious Thirty".

[2] Referring to the territories in the centre of the peninsula affected by the demographic exodus and the lack of industrial economic development, today called "empty dying Spain".

material and organises training activities and debates. The contributions to the critical analysis of tourism from these independent fields include benchmark publications (Gascón & Cañada, 2005; Buades, 2006; Murray, 2015; Cañada & Murray, 2019, 2021; Cañada & Izcara, 2021; Alonso-Monasterio Fernandez, 2020; Schenkel, 2019; Teberga, 2021; Abellán et al., 2021; Rosa-Jiménez & España, 2023, etc.). Their contributions can help to expand the theoretical and methodological approaches to tourism studies. If we focus on these contributions, we may be able to address the deficits detected by Anton Clavé et al. (1996) in tourism studies, which have a thematic specificity but not a methodological one. Other geography specialities, such as economic or urban geography and other social sciences have been developed based on new paradigms and approaches, drawing from contributions in non-academic fields (Murray, 2013).

The first analyses carried out on the studies of tourism activity from a geographical perspective considered that the scientific production was scarce in proportion to the importance of the tourism phenomenon in Spain (Anton Clavé et al., 1996). Nevertheless, pioneer studies have been conducted since the 1960s, such as those by J. Vilà Valentí (1962) "El valor económico del turismo", by B. Barceló Pons (1964) "El turisme a les Illes Balears", or the studies by E. García Manrique (1968) and (1969) on the implications of tourism for the regional economy of Mallorca and on Spanish tourism. Furthermore, several degree dissertations were presented in the universities of Murcia (Sarrión Martínez, 1962; Melendreras Gimeno, 1964; Rodríguez De La Torre, 1965), Zaragoza (Pons Granja, 1967) and Barcelona (Rexac Cirici, 1967).

Currently, according to the Informe sobre la investigación geográfica en España: 2013–2020 (Report on geographical research in Spain: 2013–2020), there are 17 research groups dedicated to tourism from a geographical perspective. This is the most prolific line of research of Spanish geography in terms of the number of publications, research projects and doctoral theses (Fernández-Mayoralas, 2021). Spanish tourism geography has also produced an interesting number of reviews and analyses of its scientific production, covering from the beginning of the 1960s to 2022 and analysing different sources: publications of the AGE (Blanco-Romero & Mínguez, 2022), publications in national journals such as *Estudios Turísticos* (Anton Clavé et al., 1996), *Agricultura y Sociedad* (López Ontiveros & Mulero Mendigorri, 1997), *Anales de Geografía de la Universidad Complutense* (García Hernández & De la Calle Vaquero, 2004) or the *Boletín de la Asociación de Geógrafos Españoles* (Corral & de San Eugenio, 2013), and publications in Spanish editorials, such as *Antrophos* (Luis Gómez, 1988) and international publications, contemplating the analysis from a European perspective, such as *Emerald* (Wilson & Anton Clavé, 2013).

As shown by the research over the last few years, the changes taking place in tourism supply and demand have led to the emergence of a new tourism scenario, with the incorporation of new places and products. Furthermore, the hegemony of the conventional model of traditional tourist destinations has been questioned. The cases of touristification of the city or the restructuring of sun and beach tourist destinations particularly stand out. The research conducted by geographers on these transformations taking place in Spain has increased its collaboration with other disciplines, such as the humanities, anthropology, sociology or environmental sciences. Evidence of this boom can be seen in the activity reflected by the Tourism Working Group of the AGE, but also the participation of geography in academic events such as SOCANTUR (International Sociology & Anthropology of Tourism Congress) or the AECIT Congress (Spanish Association of Scientists in Tourism).

Among the wide diversity and complexity of the issues addressed in recent years, pressing questions continue to arise in tourism geography research, such as: the changes in the tourism-territory dialectic, the role played by territory in the tourism change processes, management in times of crisis or the roles of the tourism agents in new scenarios, among many others. Of these, it is clear that public planning and management continues to constitute a principal line of research, similarly to the frequent use of case studies as a way to approach the theoretical concepts and unique historical features of Spanish geography. Although the research topics are conditioned to follow lines of finance and interest in accordance with tourism policy, biased by the free market mechanisms, tourism geography is becoming more enriched with the emergence of critical studies. Critical tourism geography in Spain is characterised by not necessarily taking the side of this industry, responding more independently and rigorously to the social challenges and demands. Among the topics incorporated by autonomous studies carried out by independent researchers or NGOs, those focused on the conflicts and contradictions in the environment of the tourism industry particularly stand out: working conditions, socio-economic inequality due to gender, generation, origin…, unequal geographical development; commercial exploitation of housing, the city and the territory, gentrification, social reactions to touristification, the contribution of tourism to widening the metabolic rift or the climate emergency, etc.

This book seeks to present some of the principal current contributions of Spanish academia to the knowledge of tourism geography, classified into two groups: geographical areas affected by touristification and the variables of analysis of this industry. This categorisation shows a clear bias for domestic zeal in the repercussions of the tourism industry on Spain as a tourism destination, a recipient of tourist flows and investment capital. Both groups reflect a methodological specificity of Spanish geography, highly oriented towards the analysis of public policies and even the proposal of new planning and methodology formulas that go beyond diagnostic studies.

Below, we offer our own definition of the two groups into which we characterise the contributions to the book, and which correlate with one another, and which we consider represent the tourism geography studies in Spain at present (Sects. 1.2 and 1.3 of this introduction). The final section of this introductory chapter describes the aspects that, in our opinion, should contribute to the design of a future research agenda.

1.2 The Regional and Local Approach

The intense touristification process in Spain is embodied in the total reshaping of the territory. On the one hand, through the construction of accommodation establishments, retail outlets, facilities, infrastructure, etc.; and, on the other hand, with the dependency on other geographical areas destined to be their "back office". This process has materialised in the form of resorts and spaces created with this principal function. The expansion of the tourism business frontier has also materialised in the form of new functions of the city or rural environment and of the more natural space. This definition of geographical spaces guides a good part of the Spanish tourism geography studies. At the same time, the areas unaffected by tourism are defined, by defect, as operational spaces, as they are on the side-lines of the process and have, therefore, become suppliers of natural resources (food, minerals, energy...), labour or even constitute new opportunities for expanding the tourism business frontier.

The regional tradition of geographical studies in Spain is reflected in the analysis of tourism based on the characteristics of the geographical substrate. Therefore, the first four chapters of the book address the casuistry typical of the less touristified spaces, labelled as "undertourism" spaces: Spain's dying interior (understood as those territories that suffer from serious depopulation problems), interior territories, rural territories and natural spaces that have a higher or lower degree of protection.

Despite being one of the principal tourist destinations on a global level, Spain displays a major territorial imbalance in terms of the distribution of tourist flows. In order to mitigate this, the support of structural funds from the European Union has promoted nature tourism and rural tourism in interior territories as an economic solution to replace the agricultural and cattle farming activities and to attempt to slow down the depopulation of these areas. Addressing this problem is the objective of Chap. 2, "Tourism in empty Spain". Its authors, Xosé Somoza Medina and Marta Somoza Medina, argue that, despite the efforts made, the data show that many of these spaces arouse very little interest among the potential visitors and that the return on investment is, at the very least, questionable. The analysis of tourism supply and demand, sustainability and the territorial resilience of these places reveals, on the other hand, the maintenance of strong ties of the former inhabitants, embodied in what is known as emigrant tourism.

Some of the transformation processes of these spaces into tourist destinations include innovative initiatives within the so-called "proximity tourism", which is the predominant type of tourism arising from the pandemic in 2020. Chapter 3, written by Inmaculada Diaz-Soria, Asunción Blanco-Romero & Gemma Cánoves Valiente, offers an epistemological analysis focusing on the evolution of interior tourism in Spain. It contemplates a whole range of aspects from concepts such as rural tourism or endogenous development to more recent phenomena such as slow or proximity tourism, considering them as post-capitalist proposals. According to the authors, the principles of the "local turn" in tourism represent a development opportunity so as to benefit from proximity tourism, learning from the experiences of overtourism of coastal and urban destinations and based on the use of planning and management tools for a tourism adapted to each interior territory.

In Chap. 4, Nora Müller and Macià Blázquez-Salom analyse the trend towards the privatisation of the management of the public use of natural spaces. This trend has arisen from the advancement in neoliberal regulations that favour the interests of private property and the extension of the tourism business frontier. Their results associate this process with the commodification of nature and the enclosure of land for its conservation and tourism, in terms of environmental gentrification. The authors present alternative proposals of convivial conservation, the defence of the commons and socially-just degrowth.

The study of "back office" territories concludes with the proposal to extend the scope of applying the technology to the rural spaces, based on digitalisation, which constitutes one of the strategic investment lines of the recovery and resilience plans in Spain following the pandemic of 2020. To this end, Antonio Martínez-Puche, Luis Alfonso Hortelano Mínguez and Xavier Amat-Montesinos present the results of the European project SmartRural in Chap. 5. The application of the Smart concept to the creation of new networks and improved services in rural spaces and training in smart rural tourism are considered to be of major benefit to the inhabitants and companies of rural communities.

The COVID-19 pandemic has shown how, within a context of restricted mobility, tourism is a highly vulnerable economic sector as are those countries such as Spain whose economies depend heavily on it. In spite of all of this, thanks to proximity tourism, the majority of the interior spaces have exhibited a high level of resilience. Others, which have solid cultural, landscape and heritage products, but are highly dependent on international tourism, such as the "Camino de Santiago" (considered as a European Cultural Route) have seen their survival at risk in the health crisis. This is analysed in Chap. 6, written by Lucrezia Lopez, Rubén Camilo Lois González and Rossella Moscarelli.

In parallel to the tourism reality of the rural and natural territories, Spain has a wide variety of realities. Therefore, it is possible to find spaces that have been affected by major touristification process of different types. From this starting point, the subsequent contributions study the geographical areas that have experienced a tourism transformation that has been progressively more intense: historical quarters, small and medium-sized interior towns, the north Atlantic coast, the sun and beach tourist cities and tourist destinations that adopt digitalisation. In this way, in Chap. 7, Manuel de la Calle-Vaquero and María García-Hernández analyse the state of the historical quarters in Spanish cities where two recent major cycles can be identified: the first, from the 1990s to the systemic crisis of 2008 and the second from this crisis to the present day. Both cycles have their own characteristics in terms of tourism and the city but also in the way in which the discipline of Geography has analysed them.

Next, following the direction marked by the intensification of touristification, its effects on a sample of small and medium-sized interior towns are analysed by establishing sociodemographic, tourism or cultural variables in Chap. 8, written by Carmen Hidalgo-Giralt, Antonio Palacios-García, Diego A. Barrado-Timón and

Francisca Cea-D'Ancona. Their contribution corroborates the threat that the tourism-related commodification of the city represents due to its alienation from the local population, the overloading of the infrastructures or the deterioration of its cultural heritage.

Some Atlantic coastal cities also carry out tourism functions, with social and environmental characteristics threatened by the maturity of the supply and the tourist saturation. This stage of the evolution after touristification is analysed by Carlos Alberto Patiño-Romarís, Rubén Camilo Lois-González and Breixo Martins Rodal in Chap. 9, in order to establish the challenges of tourism planning and management in such a way as to avoid saturation and also capitalise on the best territorial components of these spaces. Their proposal is aimed at developing tourism governance measures based on interviews with representatives of interested social groups.

Mass tourism has given rise to coastal spaces designed to attract real estate investment that supports the sun and beach industry. This tourist city format is originated in the 1960s in response to the interests of the ruling social classes, particularly corporate entrepreneurs and high-ranking officials of the public administrations. Critical geography analyses this social production of the space and its continuous restructuring, in search of a reactivation of the hotel, real estate and financial profits. In Chap. 10, Macià Blázquez-Salom and Ivan Murray characterise the social role of some of the people and institutions within the framework of the theories of their agency interrelated with the structure, responsible for the creation and mutations of these sun and beach tourist cities.

Currently, in Spain, the policies of the public administrations seek to overcome the challenges of the post pandemic crises (particularly those related to climate and energy) through the digitalisation of social and environmental management processes. Chapter 11, written by Josep A. Ivars-Baidal, Francisco Femenia-Serra, Marco A. Celdrán-Bernabeu and David Giner-Sánchez contextualises this new paradigm applied to the Spanish tourism destinations.

1.3 The Thematic Approach Based on the Study of Variables of Analysis

Tourism geography benefits from the contributions of other related disciplines in the study of this industry. This second way of addressing tourism geography that has been developed in Spain is based on the analysis of variables and processes that particularly characterise the transformation generated by this industry. In this respect, the second section of this book compiles contributions referring to the study, on the one hand, of environmental issues: the beaches, water, landscape, land uses, sustainability indices in cities; and, on the other hand, of social, demographic and economic issues: gentrification, migrations, social vulnerability, working conditions, socio-economic inequality, social reactions, tourist packages and charter flights and accessible tourism.

The most successful tourism model in Spain depends on the availability of beaches and the abundance of sunshine, provided that climatic comfort is maintained. The coastline of the peninsula and the Balearic and Canary archipelagos have the climatic advantages of sunny and dry summers. The Mediterranean coast is also characterised by not having high tides or heavy swells or dangerous storms like those afflicting other tropical seas. The case of the Balearic beaches stands out due to their bioclastic composition, which makes them uniquely attractive due to the clarity of the water. Climate change threatens this resource due to the reduction in the area of the beaches. In Chap. 12, Guillem X. Pons, José Ángel Martín-Prieto and David Carreras-Martí quantitatively establish the incidence of erosive processes on the most popular beaches located in protected spaces of the Biosphere Reserve of Menorca by studying the evolution of its coastline and its use between 2001 and 2015.

Another characteristic climate aspect of the peninsula is drought, which reduces the availability of water and limits the time-old human activities, mainly in the Mediterranean region. Given that summer is the peak tourist season in this region, the traditional water use for irrigation has been transferred to supply tourist demands. This has been achieved, particularly, through the incorporation of non-conventional sources, such as desalination or the adoption of efficiency and savings measures in water consumption. In Chap. 13, Rubén Villar, Carlos Baños, Jorge Olcina and María Hernández analyse the rebound effect of these measures that favour urban growth through the relaxation of the environmental requirements for territorial planning, for example, due to the crisis generated by the COVID-19 pandemic. Ultimately, the recommendation has been to abandon the growth path in order to combat the climate emergency.

The quality of the landscape is another variable for evaluating the environment so as to establish the tourist offering of a territory. Tourist attractiveness depends, among other characteristics, on the identity of the landscape, which is the main attraction of the hinterland in the peninsula. In Chap. 14, Paloma Ibarra, Isabel Rabanaque, Elena De Uña and Montserrat Villarino establish a recent state of the question of tourism geography studies (2020–2022) related to the landscape attractions of Spain. The studies analysed reveal an interest predominantly in cultural, winemaking, mining and industrial and mountain landscapes. The authors examine in depth the threat that the industrial wind farms pose for the landscape value that attracts tourism, particularly in interior territories of the peninsula, which carry out energy supply functions from "empty" Spain. The use of opinion survey methods reveals the defence of the landscape integrity, due to their heritage value and their tourism profitability, against the threat of industrial wind turbine projects promoted by financial and energy groups.

The study of the change in land occupation and use reveals an abundance of information to explain the social production of the tourism space in Spain. The cartography of the artificialisation of the land, arising from urbanisation and the construction of infrastructure, has become an essential tool for analysing the historical periods of "developmentalism", with the incorporation of Spain into the growth period or the Glorious Thirty or the property bubble, which is closely linked to tourism in Spain. In Chap. 15, María-Dolores Pitarch and Carmen Zornoza describe the urban expansion of the latter in this last period (1990–2018) in some of the regions whose coast has been artificialised: Alicante, Mallorca, Murcia and Málaga.

The debate on the non-sustainability of the unrestrained urban growth and tourism model has been particularly harsh in Spain due to its dependence on these industries. Spanish geography has enriched the quality of this discussion through the establishment of indicator systems to contribute to well-informed decision-making. Anna Torres, Aurélie Cerdan and Pol Pareto identify both international and Spanish components and experiences of sustainability indicators for urban environments in Chap. 16. Their study is conducted within the recent phenomenon of touristification in cities which began to intensify in Spain after the 2008 crisis.

One of the aspects associated with this intensification of the tourism function in Spanish cities is gentrification, particularly after the legalisation of property rental for short-stay tourist accommodation. The emergence of rental platforms and the financialisation of real estate investment have intensified this process. However, the evaluation of the consequences of the COVID-19 pandemic on this socio-spatial segregation process is still tentative. Gustavo Javier Macías and Antonio Paolo Russo review the state of the question in the Spanish bibliography in Chap. 17, differentiating between before and after the pandemic. They go a step further and study the urban policies aimed at alleviating the impacts of gentrification.

In addition to gentrification, the mutations of the demographic landscape of the cities associated with touristification are manifested in the marginalisation and impoverishment of specific neighbourhoods. In both cases, the transformation is fuelled by immigration. The first kind of immigration is climate migration (either due to retirement or the adoption of a digital nomad lifestyle) and a change in the lifestyle of the social class of the Global North. The second type of migration refers to the population of the Global South, which is driven by the search for employment. Jesús M. González and Ismael Yrigoy examine this process in Chap. 18. Their analysis of two paradigmatic neighbourhoods of a touristified city reveals an ethnic-based clustering and a similar incidence of evictions, for a different reason, at both ends of the scale.

In relation to these transformations, the tourism functionalisation of cities also implies changes in land use, which, on the one hand, pave the way for the extension of the business, but, on the other hand, expose the marginalised social groups to risks and threats. Their vulnerability in the urban centres affected by touristification is the theme of Chap. 19, written by Alfonso Fernández, María José Piñeira and Carmen Mínguez. This study contributes to establishing the state of the question of the knowledge provided by the tourism geography studies in Spain.

In the same way, the generation of tourism employment in Spain is characterised by low remuneration, precariousness related to the strong seasonality, the enormous workload and feminisation due to the gender bias derived from the connection between women, care and hospitality. Ernest Cañada, the author of Chap. 20, analyses the studies related to the working conditions of room maids in Spain, establishing how they relate to the studies in the international context. His periodisation of the social and academic interest in the employment situation of this group reflects the structural changes determined by the economic cycles and the visibility of the group on social networks.

The spatial patterns of the tourism specialisation enable different geographical areas to be characterised, for example, based on the capacity of regulated accommodation. In Chap. 21, Maria Antònia Martínez and Ivan Murray analyse this pattern on a municipal level for the year 2018 and relate it to quantitative variables that cause socio-economic inequality. Their results provide evidence of a coincidence between the two patterns, the tourism specialisation and socio-economic inequality. The projection of the analysis, which goes beyond the start of the COVID-19 pandemic, enables the authors to establish the extent to which the tourism specialisation constituted the maximum socio-economic vulnerability which had to be mitigated by social protection provided by the State.

In addition, the resident population in the territories most specialised in tourism are also expressing their discontent due to the reduction in the quality of their environment arising from its monopolisation by the real estate growth to which it is associated. In Chap. 22 Enrique Navarro, Yolanda Romero and José María Romero not only analyse these complaints, but also the eco-social transformation proposals that the social movements have promoted, which transcend the growth paradigm (de-urbanisation, ecological regeneration, the protection of agricultural and natural spaces, innovative legal measures for environmental justice, among others) and which contribute to a higher quality and sustainability of the territory.

Mass tourism in Spain is related to air transport, with a predominance of the organised format of charter flights and tourist packages. The liberalisation of air traffic has favoured other types of supply such as the low-cost airlines. David Ramos scrutinises this transformation in Chap. 23, concluding that there has been an upturn in the charter model due to the financial crisis of 2008 as it contributes to spending control. On the other hand, there has also been an increase in public subsidies in this sector and the rescue of airlines after the COVID-19 pandemic.

Finally, relating accessibility and territorial smartness in the tourism sector is the approach adopted by Rosario Navalón-García, Ana Clara Rucci and Raquel Huete in Chap. 24. In order to analyse the behaviour of the so-called smart cities in relation to improvements in mobility, communication or information for disabled people, the authors have made a compilation of initiatives, good practices and awards with respect to accessible tourism. They associate the challenges of accessibility with their presence in the city plans. The implementation of the improvements and innovations detected may become smart destination plans and establish the bases of a solid relationship between smartness and accessible tourism.

1.4 Design of a Future Research Agenda

Although this work is only a sample of the broad range of current studies that address tourism geography in Spain, it does not prevent us from identifying challenges that we believe to be still pending and could constitute a future research agenda, from the perspective of the paradigm of an eco-social transformation. The first refers to the socio-territorial inequality and its structural connection with neoliberal capitalism. The second is concerned with the environmental and social risks derived from the current systemic vulnerability, aggravated in Spain by its tourism-real estate specialisation. The third challenge refers to the opportunities of democratic restructuring.

1.4.1 The Worsening of the Inequality Inherent in Tourism

As we can see in the different chapters, mass tourism in Spain is the result of industrial land use models, particularly concentrated in urban and coastal spaces (Chaps. 9 and 15). The urban development associated with mass tourism is based on a "growth machine" (Molotch, 1976), designed with the aim of satisfying the expansion of capital accumulation (Britton, 1991), through continuous processes of urban expansion and restructuring (Chap. 10) and supported by the increase in the capacity of the transport infrastructure, supply and waste management. Tourism growth has also taken place "inwards", incorporating aspects of daily life into the tourism business (Chaps. 9 and 17), as in the case of housing for tourist accommodation or the use of public streets for installing terraces, which has become known as global touristification (Cañada & Murray, 2019) (Chap. 7). Luxury tourism worsens this inequality through socio-spatial segregation (Chap. 18), which implies a greater consumption of energy, water (Chap. 13) and materials per capita.

However, this analysis does not dominate the studies of tourism geography. In fact, quite the reverse is the case. The academic contributions mostly address the tourism phenomenon as one derived from the demand side, fruit of social achievements, such as paid holidays and technological innovations, particularly transport. In this way, a trend has emerged analysing tourism from approaches focused on the supposed consumer sovereignty, sidelining the complexity of this industrial activity, with unequal social and power relations in the construction and future of tourism spaces (Murray, 2013). To this we should add the promotion of mass tourist destinations with stereotyped and idyllic images that evoke privileged spaces of the elite classes in tax havens and global cities (Méndez Gutiérrez del Valle, 2018). These semiotics of capitalism incorporate leisure time as a business, as "they also exploit through the production of signs and meanings that colonise the collective mind and conditions ways of being and feeling" (Moruno, 2013, p. 168).

The industrial intensification of the social metabolism of urban spaces that host mass tourism, often affected by saturation, requires the existence of a hinterland as a supplier of resources and labour, and a place to absorb the waste generated (Brenner & Katsikis, 2020). These operational spaces, located at the other end of the socio-spatial scale, are also characterised by undertourism, being neglected and socially abandoned, in the same way as "empty dying Spain". In cases where tourism is adopted as a development strategy of these disadvantaged spaces, its role for local stimulation is insufficient (Chap. 2). It remains to be seen whether the proximity tourism alternative will consolidate. It gained strength during the pandemic, constituting an opportunity and favouring the "local turn" (Chap. 3). In this unequal process of geographic development, the existence of operational spaces (Chap. 14) is fundamental for maintaining mass tourism on a metabolic level and as a supplier of labour (Chap. 20).

Overtourism and undertourism are two extremes of the polarisation intrinsic to capitalism. We can find uneven geographical development, on the one hand, empty and decaying interior areas while the coastal zones are densely populated with large investment flows. In the case of Spain, these latter spaces are characterised by a strongly industrialised model specialised in real estate, tourism and mega infrastructure projects. This is due to economies of agglomeration and scale, with a greater specialisation and division of labour, which impose compulsive and globalised mobility, stimulated by tourism (Urry, 2002). This uneven geographical development depends on accessibility and connectivity (Chap. 23). These two factors are the most threatened as a result of the health, energy and climate crises.

Tourism that is resilient to chronic emergencies should be equitable and fair, in terms of how it operates, its effects on people and the place and how us, as academics, teach and study the travel industry, particularly in the training of its current and future leaders. An ethical commitment to equality consists in making specific changes in the practices and decisions on different levels, together with the development of a broader framework. As a result, these actions oblige us to question consumerism and the capitalist point of view that has contributed to the massive growth of the tourism industry. Rethinking tourism from the perspective of de-growth implies a questioning of high speed, long distance and short stay trips, which do not consider the socio-economic environment (Blázquez-Salom et al., in press). The succession and overlapping of poli-crises (economic, climate, health, geopolitical, energy...) lead us to believe that it will be necessary to transition towards new forms of tourism, in accordance with a deeper transformation of livelihoods and lifestyles in line with post-capitalist principles (Fletcher et al., 2023; Higgins-Desbiolles, 2020) (Chap. 3). These topics constitute another vital focus of future research and praxis.

1.4.2 Capitalism Without Nature?

The COVID-19 health crisis revealed our dependency on natural cycles (Davis, 2020) and the fragility of the tourism model (Chap. 6). Anthropogenic climate change, the sixth extinction of living species or the exhaustion of hydrocarburates are symptoms of the socio-ecological crisis of the Anthropocene or Capitalocene (Blázquez-Salom, 2020). The intensification of the social metabolism of mass tourist destinations based on the development of the real estate environment, aggravates the eco-social crisis (Gössling & Peeters, 2015), particularly in terms of its contribution to the climate emergency (Lenzen et al., 2018) and territorial deterioration (Chaps. 12 and 13). The intention to disconnect the economy from the rest of nature shows a treatment of these contradictions as simple resolvable problems, as they are considered as externalities that can be relocated in space and deferred in time.

On the one hand, the socio-institutional responses to this situation developed in recent decades have essentially consisted in proposals of sustainable development and green capitalism. In the apt definition of Hall et al. (2021), the acronym BAU (normally, Business As Usual) corresponds to Bruntlandt As Usual. These solutions to the Anthropocene crisis take the form of mitigating and adapting measures in accordance with the growth dogma (Fletcher, 2023). Alternatively, the current political debate also contemplates the deployment of another economic, social and environmental project based on de-growth (Chap. 22), which questions capitalism based on diagnostics and radical proposals also in the field of tourism (Blázquez-Salom et al., in press). In the current scenario of a poli-crisis the proposal of alternatives to compulsory growth has become an intellectual challenge which is more pressing than ever (Chap. 4), in pursuit of post-growth, post-capitalism, post-development or de-growth (Fletcher et al., 2023).

1.4.3 Eco-Social Transformation Through Democratic Radicality

The structural crisis of tourism has been addressed through renovation, restructuring and reconversion operations (Vera-Rebollo & Rodríguez-Sánchez, 2012) (Chaps. 5, 11, and 24), aimed at bringing about a new return on investment (Blázquez-Salom et al., in press), but aggravating inequality through the segregation of cheap tourism and the resident population, in favour of tourists and immigrants from the Global North with a higher purchasing power and a luxury supply (Chap. 18). However, social inequality and vulnerability have only increased. They are also associated with tourism intensity and specialisation (Chap. 21) and have been worsened by the application of flexibility and austerity mechanisms after the 2008 crisis (Martínez-Caldentey & Murray, 2019). Thus, the recovery from the systemic crisis of 2008 was overtaken by a new tourism bubble after the COVID-19 pandemic. These crises have influenced the relaxation of the regulations, the collapse of the productive economy, the execution of speculative transactions, the deterioration of working conditions (Chap. 20) and the increase in social vulnerability (Chap. 19). Tourism sustainability implies finding a balanced solution not only in terms of economic profitability but also with respect to social well-being, inequality and ecological balance (Chap. 16).

Beyond the palliative measures to address the contradictions of capitalism, such as sustainable development or green capitalism, the de-growth-based alternatives provide practical solutions for a more sustainable tourism. For example, and in accordance with the proposals of Gössling and Higham (2020): prolonging the duration of the stay, obtaining supplies from nearby markets, rethinking food in order to use organic (and local) and seasonal produce, moving towards a high value-added model, seeking local benefit rather than that of foreign-owned global platforms, abandoning carbon-intensive trips, for example, cruises, etc. But the political project of tourism de-growth does not consist in distributing the burden of guilt equally among the population, appealing to their responsibility as consumers who should change their habits. Nor does it simply consist in planning or restructuring the built environment to reduce the number of beds in tourist accommodation. These dilemmas are not fairly resolved if, in order to combat saturation, we opt to concentrate wealth and power, excluding the disadvantages and making tourism only accessible to the richest segments. In order to de-grow in a fair way a more inclusive tourism should be promoted, which is accessible to the disadvantaged social classes, so that they can enjoy its benefits. This segment should also be given access to decision-making (Cañada, 2020). Only in this way can social and environmental justice be linked to grassroots democracy (Blanco-Romero, in press).

It would then be necessary to talk about democratic radicality also in the field of tourism management and planning. Community self-organisation in fields such as tourism promotes this democratic practice and governance, generating the best results through proximity (Chap. 3). This implies breaking away from the predominant formalist and liberal vision, opting for decision-making management, ensuring the redistribution of wealth and power (Aguiló Bonet, 2014), which an activity like tourism can generate. Unlearning and transforming would be possible with projects and campaigns that generate articulation and complementarity practices among the participative, radical and representative democracy (Aguiló Bonet, 2013), combating the monoculture created by the globalised liberal democracy based on a sole discourse. Within this context, new forms of managing and planning the tourism industry and the related territories should be conceived as broad and higher quality participation practices, in which the public sector constitutes an element of social change towards an economy based on the common good and transparency. In this way, it would be possible to establish alternative mechanisms of economic organisation, focusing on defending citizens' rights, reducing vulnerability and dependency on market mechanisms (García & Pradel, 2019) and shifting towards governance systems in the creation of appropriate tourism development models (Blanco-Romero, in press).

1.5 Conclusions

The Spanish group of specialists in tourism geography generates an abundance of studies, which are characterised by being based on a close and direct knowledge of this industry. This approach enriches these studies as it contributes to defining developmentalism, tourist saturation, the emptying of the interior of the peninsula, the current poli-crises (climate, energy, housing...) and the tourism monoculture of a good part of Spain's coastal regions. The two blocks into which this book is organised reveal the traditional approaches in Spanish tourism geography. The first block refers to the study of specific geographical areas affected on different scales by the expansion and intensification of the tourism industry or by its absence. The uneven geographical development forces the latter to act as operational spaces, supplying labour, food or energy. The second block of studies analyses the variables and processes that are particularly affected by touristification. As a whole, these contributions define a future research agenda oriented towards addressing the challenges arising from the environmental and social crisis, rooted in inequalities. We understand, therefore, that tourism geography in Spain could contribute to governance, palliating, in a consensual and planned way, these challenges by reducing inequality in the flow of materials, water and energy consumed by the tourism industry through its contraction and convergence.

Acknowledgements This work was supported by the Grant PID2020-114186RB- C21 funded by MCIN-AEI/ 10.13039/501100011033; and Grant RTI2018-094844-B-C31 funded by MCIN-AEI/ 10.13039/501100011033 & ERDF.

References

Abellán, N., Izcara, C., López, A., & Salvador, M. (2021). *Desigualdades de género en el mercado laboral turístico* (Informes en Contraste, 14). Alba Sud Editorial. https://www.albasud.org/publicacion/es/97/desigualdades-de-genero-en-el-mercado-laboral-turistico

Aguiló Bonet, A. J. (2013). Una cartografía abismal de la democracia: representación política y democracia de baja intensidad. In Instituto Internacional Casa de Mateus (org.), *Representação/Representation* (pp. 54–79). IICM.

Aguiló Bonet, A. J. (2014). Radicalidad democrática y nuevos movimientos populares: las luchas por otras democracias. *Ágora, 1*(2), 65–86. https://doi.org/10.6035/Kult-ur.2014.1.2.3

Alonso-Monasterio Fernandez, P. (2020). *Impacto del turismo en los espacios naturales y rurales.* Ecologistas en Acción. https://www.ecologistasenaccion.org/124062/informe-impacto-del-turismo-en-los-espacios-naturales-y-rurales/

Anton Clavé, S., López Palomeque, F., Marchena Gómez, M., & Vera Rebollo, F. (1996). La investigación turística en España: aportaciones de la Geografía (1960–1995). *Estudios Turísticos*, 129, 165–209. Instituto de estudios turísticos. http://hdl.handle.net/11441/44422

Barceló Pons, B. (1964). El turisme a les Illes Balears. *Serra d'Or*, VI(2–3), 37–39 VI(4), 23–25, VI (5)25–27. https://www.cervantesvirtual.com/obra-visor/serra-dor-2/html/b469a507-50cc-4972-8e10-0ba8d565d181_40.html

Blanco-Romero, A. (in press). The power of the neighborhoods. Governance and urbanism from below in a working-class district of Barcelona, In J. A. Rio Fernandes & R. C. Lois González (Eds.), *Urban Change in Iberian Peninsula.* Springer.

Blanco-Romero, A., & Mínguez, C. (2022). Research in Tourism Geography conducted by the Tourism Group of AGE (2012–2020). In Spanish Committee of the International Geographical Union (Ed.), *Spanish Geography at Contemporary Times. State of the question (1972–2022). Spanish contribution to 35th IGC. Paris 2022. Time for Geographers* (pp. 265–285). AGE. https://doi.org/10.21138/ugi.en.2022.13

Blázquez-Salom, M. (2020). Escaping the Capitalocene: sustainable development, green capitalism and degrowth. In Spanish Committee of the International Geographical Union (Ed.), *Spain, bridge between continents. Spanish contribution to 34th IGC, Istanbul 2020* (pp. 257–271). Centro Nacional de Información Geográfica. https://www.age-geografia.es/site/wp-content/uploads/2021/02/Spanish-Contribution-UGI-Istanbul-2020_ENG.pdf

Blázquez-Salom, M., & Cañada, E. (2011). *Turismo Placebo. Nueva colonización turística: del Mediterráneo a Mesoamérica y El Caribe. Lógicas espaciales del capital turístico*. Alba Sud. https://www.albasud.org/publicacion/ca/41/turismo-placebo

Blázquez-Salom, M., Murray, I., & Artigues-Bonet, A. A. (in press). Hegemonía hotelera. El rol del Estado en la reactivación de la tasa de beneficio hotelera tras la crisis de 2008. *Investigaciones Turísticas* (25), 220–249. https://doi.org/10.14198/INTURI.22122

Blázquez-Salom, M., Murray, I., Fletcher, R., Sekulova, F., Blanco-Romero, A., & Cañada, E. (in press). Tourism and degrowth: Beyond the capitalist growth imperative. In J. Wilson & D. Müller (Eds.), *New Routledge handbook of tourism geographies*. Routledge.

Brenner, N., & Katsikis, N. (2020). Operational landscapes: Hinterlands of the Capitalocene. *Architectural Design, 90*, 22–31. https://doi.org/10.1002/ad.2521

Britton, S. (1991). Tourism, capital, and place: Towards a critical geography of tourism. *Environment and Planning D: Society and Space, 9*(4), 451–478. https://doi.org/10.1068/d090451

Buades, J. (2014 [2006]). *Exportando paraísos. La colonización turística del planeta*. Alba Sud. https://www.albasud.org/publicacion/es/63/exportando-paraisos-la-colonizacion-turistica-del-planeta

Cañada, E. (2020). Desarrollo inclusivo del turismo, un debate sobre sus límites y posibilidades. In E. Cabanilla, E. Molina, & C. Garrido (Eds.), *El turismo repensado desde la pandemia* (p. 13). I Encuentro Internacional sobre Turismo y Territorio. Editorial Universitaria. https://repositorio.uce.edu.ec/archivos/ltorres/Publicaciones/Memorias_I_Encuentro_de_Turismo.pdf#page=13

Cañada, E., & Izcara, C. (Eds.). (2021). *Turismos de proximidad, un plural en disputa*. Icaria Editorial.

Cañada, E., & Murray, I. (Eds.). (2019). *Turistificación global. Perspectivas críticas en turismo*. Icaria Editorial.

Cañada, E., & Murray, I. (2021). *#TourismPostCOVID19. Turistificación confinada* (Colección Turismos, 7). Alba Sud Editorial. https://www.albasud.org/publicacion/es/103/tourismpostcovid19-turistificacion-confinada

Corral, J. A., & de San Eugenio, J. (2013). La investigación de la geografía del turismo en España a través del análisis bibliométrico de las actas del Coloquio de Geografía del Turismo, Ocio y Recreación (1990–2012). *Boletín de la Asociación de Geógrafos Españoles, 63*, 535–549. http://hdl.handle.net/10854/3366

Davis, M. (2020). *The monster enters: COVID 19, avian flu, and the plagues of capitalism*. OR Books.

Fernández-Mayoralas, G. (2021). *Informe sobre la investigación geográfica en España: 2013–2020*. Asociación Española de Geografía-AGE. https://www.age-geografia.es/site/wp-content/uploads/2021/12/INFORME_INVESTIGACION_2021.pdf

Fletcher, R. (2023). *Failing forward: The rise and fall of neoliberal conservation*. UC Press.

Fletcher, R., Blanco-Romero, A., Blázquez-Salom, M., Cañada, E., Murray, I., & Sekulova, F. (2023). Pathways to post-capitalist tourism. *Tourism Geographies, 25*(2–3), 707–728. https://doi.org/10.1080/14616688.2021.196520

García Hernández, M., & De la Calle Vaquero, M. (2004). La investigación geográfica española en materia de turismo (1997–2004). *Anales de Geografía de la Universidad Complutense, 24*, 257–277. https://www.ucm.es/data/cont/media/www/pag-36035/2004_Art%C3%ADculo%20ANALES.PDF

García Manrique, E. (1968). Las implicaciones del turismo en la economía de Mallorca. In *Aportación Española al XXI Congreso Geográfico Internacional (India)* (pp. 219–232). CSIC.

García Manrique, E. (1969). Sobre el turismo español. In *Homenaje al Dr. Cassellas* (pp. 503–530). Zaragoza.

García, M., & Pradel, M. (2019). Chapter 19: Bottom-linked approach to social innovation governance. In P. Van den Broeck, A. Mehmood, & C. Parra (Eds.), *Social innovation as political transformation* (pp. 97–98). Edward Elgar Publishing. https://doi-org.are.uab.cat/10.4337/9781788974288.00030

Gascón, J., & Cañada, E. (2005). *Viajar a todo tren: Turismo, desarrollo y sostenibilidad*. Icaria Editorial.

Gössling, S., & Higham, J. (2020). The low-carbon imperative: Destination management under urgent climate change. *Journal of Travel Research, 60*(6), 1167–1179. https://doi.org/10.1177/0047287520933679

Gössling, S., & Peeters, P. (2015). Assessing tourism's global environmental impact 1900–2050. *Journal of Sustainable Tourism, 23*(5), 639–659. https://doi.org/10.1080/09669582.2015.1008500

Hall, C. M., Lundmark, L., & Zhang, J. J. (2021). Conclusions—Degrowing tourism: Can tourism move beyond BAU (Brundtland-as-Usual)? In C. M. Hall, L. Lundmark, & J. J. Zhang (Eds.), *Tourism and degrowth: New perspectives on tourism entrepreneurship, destinations and policy* (pp. 239–248). Routledge. https://doi.org/10.4324/9780429320590-18

Higgins-Desbiolles, F. (2020). Socialising tourism for social and ecological justice after COVID-19. *Tourism Geographies, 22*(3), 610–623. https://doi.org/10.1080/14616688.2020.1757748

Lenzen, M., Sun, Y. Y., Faturay, F., Peng Ting, Y., Geschke, A., & Malik, A. (2018). The carbon footprint of global tourism. *Nature Climate Change, 8*, 522–528. https://doi.org/10.1038/s41558-018-0141-x

López Ontiveros, A., & Mulero Mendigorri, A. (1997). Síntesis de la investigación reciente sobre recreación rural en España (1960–1995). *Agricultura y Sociedad, 83*, 77–116. https://dialnet.unirioja.es/servlet/articulo?codigo=83041&orden=106141&info=link

López Palomeque, F., & Cánoves Valiente, G. (Ed.) Blanco-Romero, A., & Torres Delgado, A. (coord.). (2014). *Turismo y territorio. Innovación, renovación y desafíos*. Tirant lo Blanch. Valencia, 735 pp.

Luis Gómez, A. (1988). *Aproximación histórica al estudio de la geografía del ocio*. Antrophos.

Martínez-Caldentey, M. A., & Murray, I. (2019). Crisis y desarrollo geográfico desigual en la Unión Europea (2009–2015). *Revista de Geografía Norte Grande, 72*, 163–184. https://doi.org/10.4067/S0718-34022019000100163

Melendreras Gimeno, M. C. (1964). *El turismo en la provincia de Murcia*. Universidad de Murcia, Fac. de Filosofía y Letras, Sección de Historia, (PhD Thesis).

Méndez Gutiérrez del Valle, R. (2018). *La telaraña financiera. Una geografía de la financiarización y su crisis*. RIL Editores. https://estudiosurbanos.uc.cl/libro/la-telarana-financiera-una-geografia-de-la-financiarizacion-y-su-crisis/

Molotch, H. (1976). The city as a growth machine. Toward a political economy of place. *American Journal of Sociology, 82*(2), 309–332. https://www.jstor.org/stable/2777096

Moruno, J. (2013). 'American Beauty'. ¿Qué es el postfordismo? In P. Iglesias (Ed.), *Cuando las películas votan. Lecciones de ciencias sociales a través del cine* (pp. 159–175). Los libros de la Catarata.

Murray, I. (2013). Algunes notes sobre el turisme i la forma en què les ciències socials l'han abordat críticament. *Biblio 3W. Revista Bibliográfica de Geografía y Ciencias Sociales, XVIII*(1016). http://www.ub.es/geocrit/b3w-1016.htm

Murray, I. (2015). *Capitalismo y turismo en España. Del «milagro económico» a la «gran crisis»*. Alba Sud. https://www.albasud.org/publicacion/es/68/capitalismo-y-turismo-en-espana-del-quot-milagro-economico-quot-a-la-quot-gran-crisis-quot

Pons Granja, J. (1967). *Andorra y el turismo, Zaragoza*. Univ. de Zaragoza, Fac. de Filosofía y Letras, (Final degree Thesis).

Rexac Cirici, M. R. (1967). *Estudio del municipio de Sitges*. Universidad de Barcelona, Facultad de Filosofía y Letras, (Final degree Thesis).

Rodríguez De La Torre, F. (1965). *El turismo en la provincia de Alicante*. Universidad de Murcia, Facultad de Filosofía y Letras, Sección de Historia, (Final degree Thesis).

Rosa-Jiménez, C., & España, K. (Coord.). (2023). *Overtourist City. Estrategias para recuperar la habitabilidad del centro histórico de Málaga frente a la turistificación*. Junta de Andalucía. RU. Books.

Sarrión Martínez, V. (1962). *Benidorm, tema de actualidad.* Universidad de Murcia, Facultad de Filosofía y Letras, Sección de Historia, (Final degree Thesis).

Schenkel, E. (2019). *Turismo social en América Latina. Aprendizajes de las experiencias regionales.* Alba Sud. https://www.albasud.org/publicacion/es/91/turismo-social-en-am-rica-latina-aprendizajes-de-las-experiencias-regionales

Teberga, A. (2021). *Trabajo en cruceros. De la ampliación a la intensificación de las jornadas laborales.* Alba Sud Editorial, Informes en contraste, 15. https://www.albasud.org/publicacion/es/100/trabajo-en-cruceros-de-la-ampliacion-a-la-intensificacion-de-las-jornadas-laborales

Urry, J. (2002 [1990]). *The tourist gaze.* SAGE.

Vera-Rebollo, J. F., & Rodríguez-Sánchez, I. (2012). *Renovación y reestructuración de destinos turísticos en áreas costeras. Marco de análisis, procesos, instrumentos y realidades.* Publicacions de la Universitat de València.

Vilà Valentí, J. (1962). El valor económico del turismo en España. *Estudios Geográficos, 87*, 293–297.

Wilson, J., & Anton Clavé, S. (Eds.). (2013). *Geographies of tourism: European research perspectives.* Emerald Insight.

Asunción Blanco-Romero is a geographer and tenured professor at the Department of Geography at Universitat Autònoma de Barcelona (Spain). She is a member of the TUDISTAR research group (Tourism and socio-territorial dynamics) in which she has participated in several research projects. She has been president and vice-president of the "Tourism, leisure and recreation" working group of the Spanish Geography Association. She is also the editor or the scientific journal DAG (Documents d'Anàlisi Geogràfica). Her research and teaching interests include resilience tourism and local development, globalization and sustainability degrowth strategies, territorial planning, geographic and gender issues in cooperative development, and critical perspective in tourism studies.

Macià Blázquez-Salom is a professor in the Department of Geography at the University of the Balearic Islands. He teaches and researches on tourism, sustainability and land use planning. He has been visitor scholar in universities of Mexico (Toluca and La Paz), Nicaragua (UNAN), Dominican Republic (INTEC), Austria (Salzburg), Germany (Rurh-Bochum), Sweden (Mid-Sweden) and the Netherlands (Wageningen). As a way to link activism and research, within the framework of Radical Geography and Political Ecology, he collaborates with social movements in Spain, particularly in the Balearic Islands (https://www.gobmallorca.com/), but also in Latin America (https://www.albasud.org/).

Part I
Territorial Diversity

Chapter 2
Tourism in Emptied Spain

Xosé Somoza Medina and Marta Somoza Medina

2.1 Introduction

The importance of tourism in Spain has grown non-stop since the mid-twentieth century, when the development of post-war Europe and the openness of Franco's dictatorship offered the perfect combination for the arriving of more tourists each year at Spanish coasts. First advances were made by public impulse plus private initiative in Marbella, Benidorm, Mallorca and the Canary Islands. The economic benefit obtained by tourism in these areas was so important, that in the decades that followed, the entire national territory promoted its transformation into a tourist destination, including the rural regions that were far from urban systems. The impulse of the public administrations to rural tourism and tourism of nature was reflected in subsidies to private initiatives, the elaboration of planning instruments and the creation of new tourist infrastructures in sparsely populated areas, which formed the base for a further growth and diversification of the Spanish tourism model (Vera Rebollo, 1994; Cànoves Valiente et al., 2016). As a result, in January 2001 the National Institute of Statistics (INE) registered 4958 rural tourism establishments, irregular distributed throughout the territory, with a higher concentration of them-one of every three establishments- located in Asturias, Huesca, Navarra and Tenerife. In January 2022, according to the data published by the Statistics National Institute (INE), the total number of accommodations was tripled, 14,337 establishments, where Malaga, Asturias, Ávila and Navarra concentrated the 25% of the

X. Somoza Medina (✉)
Geography and Geology Department, University of León, León, Spain
e-mail: somoza@unileon.es

M. Somoza Medina
Architecture Department. CESUGA, University of San Jorge, Zaragoza, Spain

A. Blanco-Romero, M. Blázquez-Salom (eds.), *Spanish Tourism Geographies*, Geographies of Tourism and Global Change,
https://doi.org/10.1007/978-3-031-39780-6_2

offer. It is significant that, two of the provinces with the largest number of rural establishments in 2001 and 2022 are part of what it is known as the emptied Spain -Huesca and Ávila-, turned into destinations of inland tourism for the urban markets of near Zaragoza and Madrid. Together with these two provinces, there are two communities, that have promoted for decades this type of tourism (Asturias and Navarra) and there are two provinces of sun and beach tourism, that have spread their accommodation infrastructures inland (Tenerife and Malaga).

For the so rural Spanish environments, this process of becoming a tourist destination has been an objective encouraged by all public administrations: municipalities, provincial councils, autonomous communities and, above all, Local Action Groups (LOGs), the territorial governance of rural development set by the LEADER initiative. In some regions, the structural transformation of the territory has been well planned and has generated positive socioeconomic dynamics that are maintained over time, as in the examples of Taramundi in Asturias (Córdoba Azcárate, 2004) or the Alpujarras of Granada (Entrena Durán, 2006). However, in other cases the objective of transformation has remained only as a political will with few consequences on the territory, where the creation of tourist infrastructures supported by public funds have been abandoned shortly after. In this sense, tourism sustainability should be related to the reduction of possible impacts in areas of great demand, but also to the consideration of clear justification to avoid oversized supply in certain areas.

This dilemma raised in many regions of emptied Spain where, thanks to different public subsidies, the proliferation of tourist infrastructures have been generated without any prior or potential demand that could justify it. Thus, interpretive centers, auditoriums, museums, exhibition halls, hotels, spas, restaurants and other varied tourist facilities, which have been built or placed in buildings rehabilitated for this purpose, run into serious difficulties shortly after opening due to a lack of demand (González & Macías, 2014; Arcila & López, 2015; Navarro et al., 2016). Territorial resilience, the capacity to adapt and overcome a geographical space in the face of negative situations, should be promoted in these places of the emptied Spain, continuing to generate productive landscapes, instead of just contemplative ones to be consumed by urban tourists. Tourism can be an engine of development in certain areas close to demand markets and with privileged territorial resources, but not in all rural areas by definition or general rule.

This chapter aims to analyze the evolution of rural and nature tourism in those Spanish territories that have suffered the most from depopulation, contrasting generic rural accommodation projects with the so-called return tourism, the temporary or definitive return of former inhabitants or families from these places during their leisure time.

2.2 The Analysis of Rural Tourism

The vision of the European rural world changed since 1988, when the document of the European Commission "The future of rural society" explained the main disadvantages of these territories and the measures that should be implemented to improve the quality of life of their inhabitants. In this document, which would guide rural development policies throughout Europe, it was stated: "Among the services to be created in rural areas, rural tourism occupies a very special place and, a priori, the prospects seem very favorable" (European Commission, 1988, 11). From then on, rural development programmes in Europe were articulated through the LEADER Community initiative, later replicated in Spain with the PRODER Programme and other aid instruments of the autonomous communities, such as AGADER in Galicia. In this way, dozens of associated territories in rural areas were able to obtain aid to promote a type of development in which tourism was a fundamental element. In 1991, 52 LOGs promoted development actions in different territories of Spain, out of a total of 1288 municipalities. Twenty years later, there were 264 LOGs established in Spain, covering more than 7000 municipalities (Sancho Comíns, 2011). Nevertheless, at the beginning of the twenty-first century, tourism was no longer the star measure of the LOGs, and the actions have been focused more on the valorization of agricultural production, the development of new companies and the improvement of the quality of rural life. Nowadays, with regard to the community funding period 2021–2027, there are 241 registered LOGs, that follow the guidelines set by each autonomous community in its own Rural Development Programme (RDP). A clear result of rural development policy is the great increase in rural accommodation during the first few years (Fig. 2.1), much of which was co-financed with Europeans funds (Hortelano Mínguez, 2015).

Tourism has been evaluated in different investigations as an engine of rural development and a key line in public aid programs in these territories. Blanco Portillo and Benayas del Álamo (1994) analyzed the first experiences of tourism promotion of LEADER I, concluding that initiatives in this area were fundamental, both for the number of projects and for the investment made, although they criticized the lack of

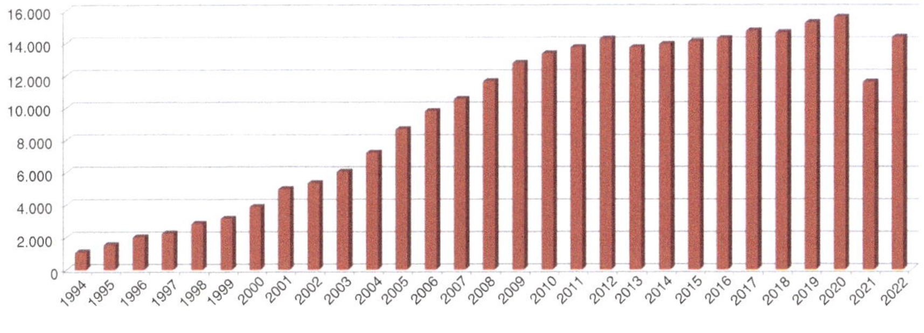

Fig. 2.1 Rural tourism establishments opened in January of each year in Spain (1994–2022). (Source: Ine.es)

studies on the real demand for these activities and the lack of tourism planning. This last element, the lack of planning, is one of the biggest conditions for the future sustainability of this type of tourism according to Pulido Fernández and Cárdenas García (2011). Ivars Baidal (2000) affirms that the supply of rural tourism grew in Spain as an unquestionable reality, although it highlights other realities in gestation, harsh realities and even fictions, such as considering that tourism is the key to rural development. A similar critical sense can be observed in the work of Cànoves et al. (2006), in which they affirm that rural tourism is not the solution to the problems of the Spanish rural environment, questioning the sustainability of many of the developed tourism initiatives. Vázquez Varela and Martín Gil (2011) also focus on the sustainability problems of Spanish rural tourism, pointing to the absence of planning, the structural characteristics of the sector, the low qualification, the undifferentiated offer, and the poor commercialization as the main elements to improve. For their part, Maroto and Pinos (2020), question whether rural tourism has been able to stop the processes of depopulation in southern Spain, analyzing the demographic evolution of rural Andalusian municipalities with more accommodation. These authors conclude that tourism alone is not capable of reversing the recessionary trends of rural territories, and that other productive economic processes are needed to achieve this objective (Maroto & Pinos, 2020). There are more analyses of specific cases, in which the development generated by tourism in other rural territories is studied, such as Castile and León, Catalonia, Extremadura, Aragon, Galicia or Murcia (Bachiller Martínez, 1994; Espejo Marín, 1997; Maya Frades, 1999; Andrés Sarasa, 2000; García Marín, 2011; Aparicio Guerrero, 2004; Rico González, 2005, Nieto Masot & Cárdenas Alonso, 2015; Cànoves et al., 2014; Millán Vázquez de la Torre et al., 2014; Jurado Almonte & Pazos-García, 2016; Font Garolera et al., 2018). In the latest research cited, the case of Teruel is analyzed, putting in relation the main public investments and the results in recent years. The authors wonder if tourism is the miracle sector that will manage to curb the negative dynamics of the province of Teruel and promote the development of this territory, reaching the conclusion that, although the data do not show a clear growth, the important investments made have allowed to mitigate the situation of socioeconomic crisis (Font Garolera et al., 2018). The most complete studies on the development of tourism generated through rural development programs in a specific territory are two doctoral thesis: Hortelano Mínguez (2015) and Engelmo Moriche (2022). The first is entitled "Rural development and tourism in Castile and León: successes and failures", highlighting from the title itself the ambivalence of this process. In the conclusions, Hortelano points out that rural tourism has not managed to stop the demographic bleeding in the most of the territories benefit from European aid (Hortelano Mínguez, 2015, 333). The second one is entitled "Analysis of the LEADER aids to rural tourism in Extremadura (1995–2015)". In the Final Remarks, Engelmo points out that "Despite the importance that the Leader Method has shown in rural areas, the main objective has not been met as rural areas continue to lose population... it should be avoided that rural areas end up being merely holiday areas, similar to the complexes snow resorts that during the high season have population and economic activity, but that once the season is over they are practically abandoned" (Engelmo Moriche, 2022, 247–248).

On the other hand, most of the researchers who focus their studies on rural/local development do recognize the importance of community initiatives in the mobilization of private investments in these territories, in the ability to transfer innovation and entrepreneurship processes and in the fact of placing the problems of less inhabited and dynamic spaces on the political agenda (Febles Ramírez et al., 2005; Esparcia Pérez et al., 2009). There is talk about a new rurality, in which the traditional functions of the countryside based on food production change towards the consideration of these territories as places where somebody can develop a vital project with higher quality than urban spaces and where improvements in communication infrastructures, physical and virtual, allow the development of innovative and productive projects (Trigueros et al., 2018). In this vision of the new rurality as a modern and innovative space of production and services, in which women have a fundamental role (Baylina Ferrè, 2019), tourism is just an activity within a wide range of socioeconomic and environmental development possibilities.

2.3 Theoretical Framework

The notion of sustainable development spread from the 1987 United Nations paper "Our Common Future", also known as the Brundtland Report due to the surname of the former Norwegian Prime Minister who chaired then the World Commission on Environment and Development (General Assembly, 1987). Later, at the meeting of this organization held in Rio de Janeiro in 1992, this type of development not focused on the economic, in contrast to the capitalist model, was extended by adding the social and environmental pillars and defining it as the ideal situation in which the needs of the present generation can be fulfil without compromising the ability of future generations to meet their own needs.

The OECD published in 2006 the study "The New Rural Paradigm", in which it laid the foundations of what has been defined as "new rurality" (OECD, 2006). According to this vision, the loss of employment in the traditional primary agricultural and livestock sector had generated in followed decades a deep crisis in the rural area of developed countries, which had been tried to alleviate with sectoral policies of agricultural subsidies, oriented from top to bottom, which had not yielded results, since the municipalities continued to lose population. The orientation that the OECD advised to give to rural policies was based on the generation of new economic engines, through investments and not subsidies, in which the territorial dimension (place-based policy) should be imposed on sectorial approaches and in which multilevel governance would have greater recognition. The five pillars that defended this new normal rural were: the local scale, diversity, the environment, cooperation and innovation.

The traditional functionality of the rural environment for producing food for the urban population, was transformed as countries were advancing in their development and the globalization of the economy were spreading, until a time when the lack of function became in abandonment. The rural exodus was not only the march of the youngest and potentially innovative population to the cities which benefited from this continuous contribution of new sap to be increasingly dynamic and transgressive, it was also the condemnation to a deferred immobility of a territory emptied of meaning, where people were getting older to finally disappear.

The only way to reverse this situation would be to bet on new functions for the rural environment, which could fix population and even attract new settlers. With the paradigm of the new rurality, the countryside has now come to be seen as a territory where anyone can put into practice the principles of sustainability, the "slow" movement or local development. A quiet territory, refuge from the current accelerated way of life, which offers better conditions for a postmodern society critical of the virtues and risks of living in the city.

By valuing the sustainability of rural tourism in the Spanish sparsely populated territories we try to question whether the efforts invested in the creation and pro-motion of tourism resources help to sustain the desired balance between economic growth, social development and environmental protection, which guarantee also the continuity of future generations, or is the new rurality, understood as the generalized development of services and the recovery of the fields' productivity, the key for the future of these territories.

In the current socio-economic context of permanent mutation sustainability is more fragile than ever. It must be associated with the flexibility and adaptability that is provided by being resilient. Actions must be directed towards the search for sustainable development as a constant improvement of society through the pro-motion of territorial resilience in a context of risks and constant unforeseeable changes. This statement, contemplated in the specific area to which this research refers, entails rethinking general maxims and theoretical models applied indiscriminately to any territory. We have to seek the differentiation and multiplicity of options of the rural spaces furthest from the hierarchical urban system, avoiding the pretension of returning to their original state and being adapted to the new and changing circumstances that define our present days.

2.4 Analysis of Data on Tourism in Rural Spain

Rural tourism has as a precedent in Spain, the public program Holidays in Farmhouses (Vacaciones en Casas de Labranza), initiated during the Franco dictatorship, in 1967, through an agreement between the General Directorate of Tourism Promotion and the General Directorate of Agricultural Training. This proposal offered the possibility for families of vacationers who lived in the cities to spend several days in the countryside. In that first year, 1152 tourist places were offered throughout Spain in 384 farmers' homes in 35 towns in two different provinces (Carazo García Olalla, 1982). In 1977, when political changes in Spain began to reduce institutional support for this Program, the places offered were 34,693, distributed in 448 locations in 41 provinces with 9410 rural houses (Olmedo & Carmona, 1997) (Fig. 2.2). Let us bear in mind that, in 1994, when the development of rural tourism promoted by the

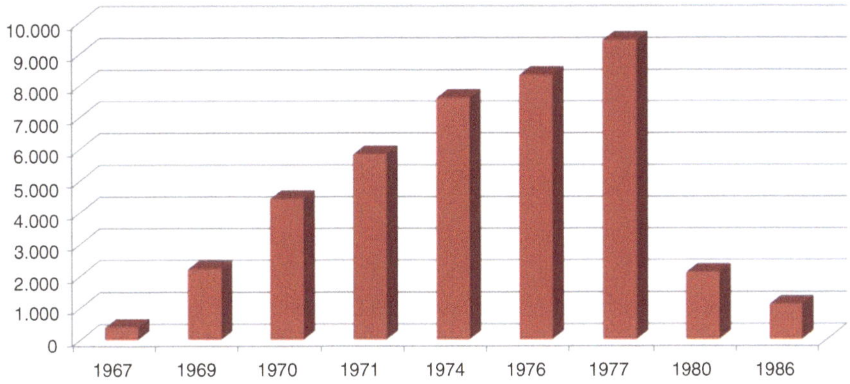

Fig. 2.2 Farmhouses offered in the program guides. (Sources: Carazo García Olalla, 1982; Olmedo & Carmona, 1997)

LOGs began, the number of establishments was 1074, and 9400 rural tourism accommodations will not be reached until 2006.

Does this mean that rural tourism was already a thriving and developed tourism modality in 1977 Spain? Obviously not. The reality is that in the Program "Holidays in Farmhouses", along with incipient examples of rural tourism in regions such as Las Alpujarras of Granada, the Jerte Valley in Cáceres or the Sierra de Gredos in Ávila, generic announcements made by the municipalities were included, which were propaganda exaggerations and not a structured and truthful offer, as in the guide of 1970, in which a place named Viver (2056 inhabitants in 1970) offered 500 farmhouses or Eslida (875 inhabitants) offered 300 houses (Olmedo & Carmona, 1997). On the other hand, the Agricultural Extension Services, as now the LOGs, promoted the diversification of the income of the rural population with institutional aid. With the Program, a series of aids were channelled, first subsidies and then interest-free loans, which the owners of the houses had to use to improve their houses, such as installing a bathroom (Canoves et al., 2005a, b), in order to complement agricultural income with tourism. The contract with the State Administration obliged the owners to offer the house in the guide a minimum of three years or until the loan was extinguished, which could have a maximum duration of 10 years. Once the obligation to offer the house was finished, if the business was not profitable, the tourist activity was closed without further contemplation.

The analysis of the tourist guides of farmhouses carried out by Olmedo and Carmona (1997) allows us to analyze the evolution of this accommodation model, whose most important indicators appear in Table 2.1. Observing this evolution, we can appreciate the irregularities manifested in these guides. For example, between 1974 and 1977, only four more locations were added to the catalogue, from 444 to 448, and yet this means the increase of almost 2000 farmhouses, from 7588 to 9410, because of the massive offers made by some mayors. The ratio of places/establishment is also revealing. In the first years, a room of each Farmhouse was offered to tourism, so the number of places were the 3 people who could occupy that room.

Table 2.1 Data from the guides of the vacations in farmhouses

Guide	Establishments	Locations	Places	Places/establishment
1967	384	35	1.152	3
1969	2.202	131	7.380	3,3
1971	5.818	319	18.763	3,2
1974	7.588	444	28.473	3,7
1977	9.410	448	34.693	3,7
1980	2.093	398	9.628	4,6
1985–86	1.096	458	8.852	8,1

Sources: Carazo García Olalla (1982) and Olmedo and Carmona (1997)

In 1980, when the vast majority of homes that had already paid the instalments of their loans were canceled and the generic offers of the municipalities disappeared, the ratio amounted to 4.6, and it was in the guide of 1985–86 when the number of establishments was closer to a professionalized offer and distributed throughout the country. Despite being the last guide published by the Program and having only 1096 houses, in this guide a greater number of locations are registered, 458, and the ratio of places per establishment is higher, 8.1.

The Vacations in Farmhouses Program was not transferred to the autonomous communities, when the administrative reorganization began to take its first steps and the different regions to project their tourism policy on the territory. Only Catalonia and the Valencian Community had some continuity in the offer of rural accommodation, while other regions took a few years to publish their own rural tourism guides, with a different categorization for each type of accommodation, depending on whether we were in Asturias, Andalusia, Galicia, Navarra or Castile and León (Olmedo & Carmona, 1997; Canoves et al., 2005a, b; del Río Posada, 2016). At this stage, the work of the LOGs and the regional policies tried to promote tourism as an economic and employment alternative to the agricultural sector, thinking that its rapid development would allow to fix population and reverse the recessive demographic dynamics.

With the arrival of the twenty-first century, policies to promote tourism in rural areas move away from investment in traditional accommodation and focus on the multiplication of recreational activities, enhancement of natural resources, cultural theming projects, promotion of active tourism companies, quality local productions, health tourism, gastronomic tourism or wine tourism (Cànoves et al., 2005a, b). The sector specializes and becomes professionalized, leaving aside the primitive model of income supplementation, especially in those interior areas with sufficient resources and proximity to demand centers, while, in other more peripheral rural territories, the few initiatives are abandoned and investments are redirected to other activities and services.

Tables 2.2 and 2.3 offer different indicators of the INE statistics of rural tourism accommodation and hotel tourism between 2001 and 2021. The numbers of establishments and places correspond to the data of the months of August of each year, when the offer is highest, arrivals and overnight stays are annual, and the average

Table 2.2 Supply and demand in rural tourism establishments

Rural Tourism	2001	2007	2014	2019	2021
Establishments	5.728	11.816	16.086	18.285	17.321
Places	46.354	107.270	152.206	179.171	168.714
Ratio Est/places	8,09	9,07	9,46	9,79	9,74
Passengers	1.210.890	2.645.237	2.822.946	4.421.397	3.350.225
Overnight stays	3.660.815	7.938.622	7.750.576	11.726.030	9.690.951
Average stay	3,02	3,00	2,74	2,65	2,89
Average occup.	21,6%	20,2%	13,9%	17,9%	15,7%

Source: ine.es

Table 2.3 Supply and demand in hotels

Hotel tourism	2001	2007	2014	2019	2021
Establishments	16.217	16.125	16.807	16.943	14.813
Places	1.292.417	1.566.782	1.765.755	1.856.265	1.641.955
Ratio Est/places	79,69	97,16	105,06	109,55	110,84
Passengers	59.904.301	84.423.433	87.814.529	108.716.047	60.437.585
Overnight stays	228.681.569	271.689.482	292.260.630	342.995.594	172.160.867
Average stay	3,81	3,21	3,32	3,15	2,84
Average occup.	48,4%	47,5%	45,3%	50,6%	28,7%

Source: ine.es

stay and occupation is obtained by crossing the capacity data of August with the annual totals. In the tables can be seen how the number of rural tourism establishments grows, while the number of hotels remains more stable. The data for 2021 correspond to the effects of the pandemic on tourism activity. Since 2019 there are more rural tourism establishments than hotels, although the number of places is much higher, around 10 hotel places for each rural tourism bed. The evolution of the ratio of establishments / places grows, both in rural tourism and in hotel tourism, from 8.09 to 9.74 and from 79.69 to 110.84, respectively, showing the process of adjustment and requalification of tourist establishments. Thus, the 14,813 hotels in 2021 offer more places than the 16,217 in 2001. In terms of demand, the number of arrivals and overnight stays grows at a faster rate in rural tourism, which practically multiplies these indicators by four between 2001 and 2019, compared to the more moderate growth in hotel tourism, which multiplies by two arrivals and by 1.5 overnight stays in the same period. The two indicators with the worst tourism results are those that appear below. The average stay has been reduced considerably, almost continuously between 2001 and 2021, both in rural and hotel tourism. Rural tourism accommodations had an average stay of 3.02 nights in 2001, registering 2.65 nights in 2019. On the other hand, the hotels had an average stay of 3.81 nights in 2001 and 2.84 in 2021. It is noteworthy how in 2021 rural tourism accommodations have for the first time an average stay slightly higher than the average hotel stay, 2.89 compared to 2.84. However, the last indicator reveals the chiaroscuros of rural tourism, since the average occupation is reduced in its year of greatest percentage,

2001, to 21.6%. In other words, during the entire period analyzed, out of every 10 rural tourism rooms at least 8 were empty, and in 2014 almost 9 out of 10. In the case of hotel tourism, if we exclude the 2021 data, the occupancy rate is more positive and stands at around 45–50%.

The combined analysis offered in Tables 2.2 and 2.3 shows the growth of the rural tourism offer in recent decades, compared to hotel infrastructure, but also the difficulties for economic sustainability that can be deduced from such low occupancy rates.

The concept of empty Spain, or emptied, as it is usually used with critical intention, tries to expose the contrast in the evolution of the distribution of the population compared to the full or filled Spain. On the one hand, it is a rural Spain with a sparsely inhabited interior with densities lower than 10 inhabitants/km^2 and on the other, urbanized and coastal Spain with high concentrations of population. The aforementioned density is usually used in Europe as a minimum threshold, defining as demographic desert the areas that have a lower density. That is why there is also mentioned in the media the Spanish Lapland (Cerdá, 2017), a large extension that runs through the Spanish Iberian system covering territories of ten provinces, from Zaragoza to Valencia, where the density does not reach 7 inhabitants/km^2 (Burillo Cuadrado et al., 2019).

Among the 50 Spanish provinces, in 2021 only 10 had a population lower than that one registered in 1900. In that year, just seven of them had a density of less than 20 inhabitants/km^2. These 10 provinces: Ávila, Cuenca, Huesca, Lugo, Ourense, Palencia, Segovia, Soria, Teruel and Zamora (Fig. 2.3) are according to statistic numbers the most emptied provinces in Spain. All are provinces in the North of Spain, interior except Lugo, and all have almost the same population in 1833 and 2021, except for Cuenca, Soria and Teruel, the heart of Spanish Lap-land, which today have many fewer inhabitants than 200 years ago. Table 2.4 shows the demographic evolution of these 10 provinces between 1833 and 2021. Most of them -Ávila, Lugo, Ourense, Palencia, Segovia, Soria and Zamora- show a positive evolution between 1833 and 1950. From this date, which roughly supposes the industrialization of Spain, they begin to decline. Cuenca, shows an irregular behavior, with two large declines, one in the first half of the nineteenth century and another in the second half of the twentieth century. Huesca reaches its maximum population in 1900 and shows in the last period a slight increase, being the only one of the 10 provinces that improves in the XXI century. Teruel, on the other hand, experienced a slight increase during the nineteenth century, reaching its peak in 1900. Then it initiated a moderate decline in the first half of the twentieth century that accelerated critically in recent decades.

The data provided by the Statistical National Institute (INE) allow a provincial disaggregation, so we can differentiate the statistics of rural tourist supply and demand in coastal provinces, inland provinces and "emptied" provinces, as showed in Figs. 2.4, 2.5, 2.6, 2.7, and 2.8.

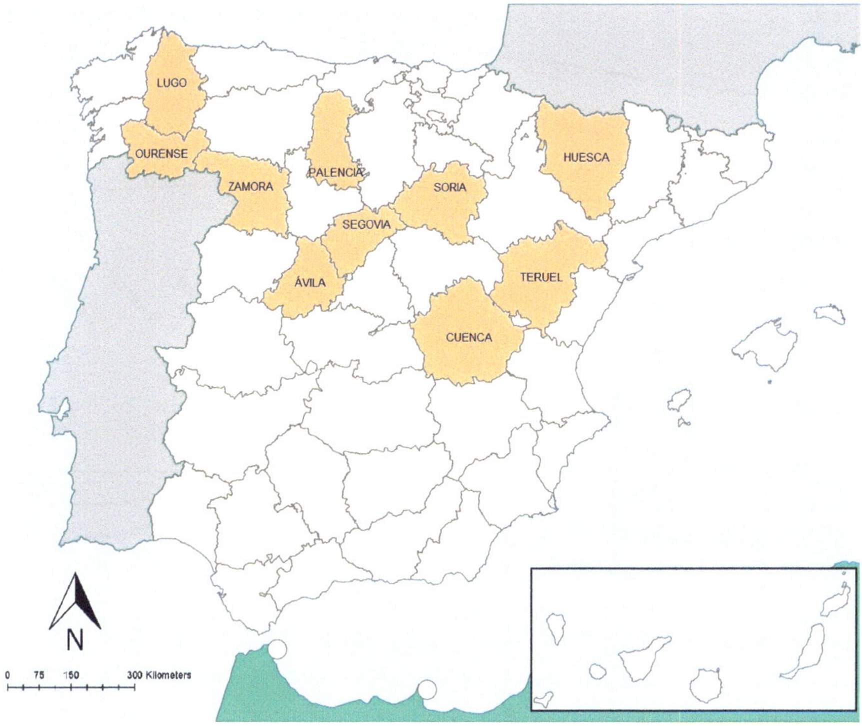

Fig. 2.3 Provinces of the emptied Spain. (Source: ine.es)

Table 2.4 Demographic evolution of the emptied provinces of Spain 1833–2021

	1833	1857	1900	1950	2000	2021
Ávila	137.903	164.039	200.457	251.030	165.615	158.854
Cuenca	334.582	229.959	249.696	335.719	201.141	199.004
Huesca	214.874	257.839	244.867	236.232	206.587	221.407
Lugo	357.272	424.186	465.386	508.916	362.565	326.537
Ourense	319.038	371.818	404.311	467.903	340.322	305.151
Palencia	148.491	185.970	192.473	233.290	176.371	158.549
Segovia	134.854	146.839	159.243	201.433	147.623	154.056
Soria	115.619	147.468	150.462	161.182	91.270	89.420
Teruel	218.403	238.628	246.001	236.002	137.086	133.253
Zamora	159.425	249.162	275.545	315.885	201.777	169.457
ESPAÑA	12.286.941	15.464.340	18.618.086	27.976.755	40.470.182	47.398.695

Source: ine.es

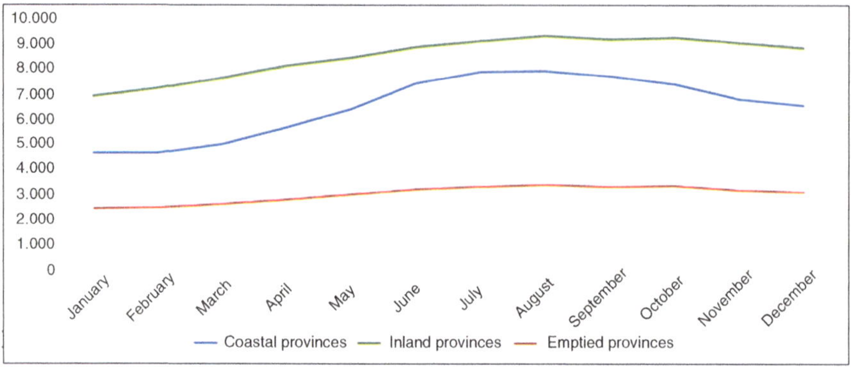

Fig. 2.4 Rural tourism establishments opened in Spain in 2021. (Source: ine.es)

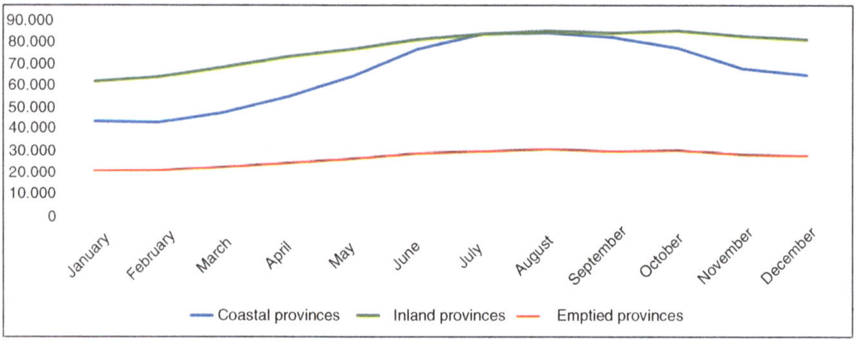

Fig. 2.5 Places in rural tourism establishments opened in Spain in 2021. (Source: ine.es)

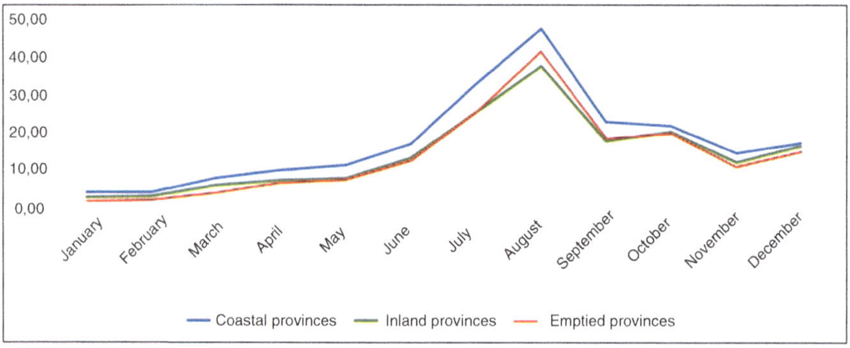

Fig. 2.6 Occupancy rate in rural tourism establishments in Spain in 2021. (Source: ine.es)

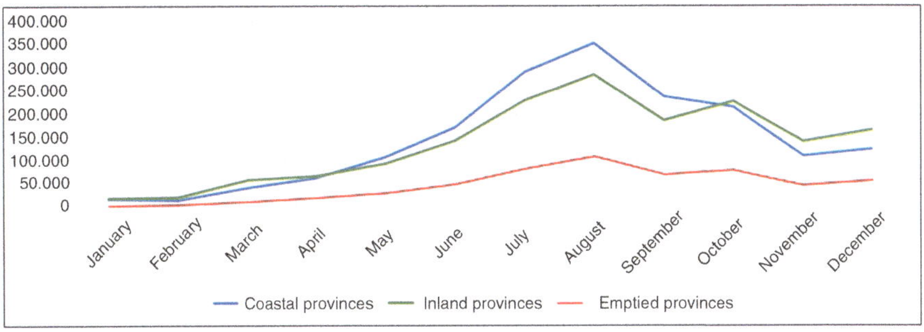

Fig. 2.7 Arrivals in rural tourism establishments in Spain in 2021. (Source: ine.es)

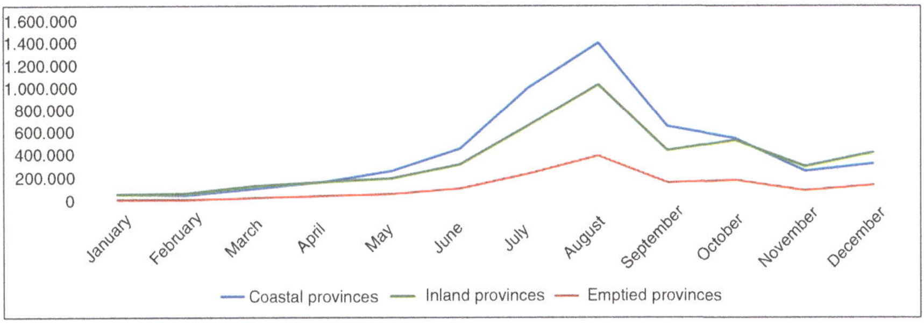

Fig. 2.8 Overnight stays in rural tourism establishments in Spain in 2021. (Source: ine.es)

If we return to the data with which we began this work, we find that two of the most depopulated provinces in Spain, Ávila and Huesca, are among the territories with the largest number of rural establishments. In August 2021, the INE statistics counted 731 rural tourism houses in Ávila and 663 in Huesca. It is related to the demand of the metropolitan areas of Madrid and Zaragoza, respectively. However, if we differentiate between rural tourism establishments in coastal, inland and emptied provinces as reflected in Fig. 2.4, it can be seen how the establishments of emptied Spain represent approximately 15% of the total, compared to 35% of those located in coastal provinces. Looking at the offered places, establishments in emptied provinces provide a greater number of places per unit, especially in August (Fig. 2.5). These numbers of rural tourism in coastal provinces cover a high percentage of accommodations, which depend more on their proximity to the coast in relation to sun and beach tourism than on the rural character of their location (Sánchez Sánchez & Sánchez Sánchez, 2021).

Figures 2.5 and 2.6 show how the supply varies according to the months, especially in the case of the coastal provinces. The number of establishments (Fig. 2.4) in the inland provinces was lowest in January, at 6993, compared with 9377 in August, an increase of 34%. Between June and December, the number of establishments in the inland provinces remained fairly stable. In the coastal provinces, January recorded 4661 establishments and August 7947, an increase of 70.5%, with a clear peak between June and September. The emptied provinces had 2442 establishments opened in January and 3409 in August, an increase of 39.6%. The seasonality of the offer and the different behavior of the coastal provinces can be seen more clearly in the supply of tourist places (Fig. 2.5). In the inland provinces, the difference in the supply of places between January and August represents an increase of 38.2%, in the empty provinces 48.8% and in the coastal provinces 93%.

According to the demand data, the occupation (Fig. 2.6) shows a very similar behavior between inland provinces and emptied provinces. There is a higher record in emptied provinces in August, thanks to the higher than 50% occupancy in Lugo, Huesca and Ourense, while Segovia or Cuenca do not exceed 27%. The number of arrivals and overnight stays (Figs. 2.7 and 2.8) show the low prominence of the 10 emptied provinces compared to the establishments located in the 22 coastal provinces.

2.5 Discussion

Rural tourism and nature tourism are tourist modalities that take place far from urban areas. Therefore, both can be a new activity for those Spanish territories placed outside urban-industrial development and weakened since the middle of the last century by a constant demographic bleeding. However, the poor accessibility of these places, which leads to other development deficits, is also a brake on their touristic possibilities. Moreover, due to the own characteristics of both tourist modalities, their economic development won't be never equivalent to that of other massive tourist modalities, such as the sun and beach model of the Spanish coasts. In consequence, it is inappropriate to believe that this tourism can become an engine of development for an extensive territory. The rural places that are near and well connected to the metropolitan areas will be raised as preferred destinations for tourists who wish to practice these modalities, reporting contributions to local economies. A high percentage of rural accommodation establishments, that are located in provinces with coastline, are actually oriented to sun and beach tourism. More remote regions depopulated and with fewer services and infrastructures will remain without tourists, in spite of the fact that there were many rural houses that were financed with European funds. This is confirmed in different investigations that focus their analysis on the effects of LEADER programmes in such areas as well as the ongoing existence of subsidized business based on tourism projects. (González & Macías, 2014; Arcila & López, 2015; Paül i Carril et al., 2016; Navarro Valverde et al., 2018; Moriche et al., 2021). The study of Sierra de Gata and the Jerte Valley shows that approximately the 50% of business projects related to tourist accommodation closed after few years of being open (Moriche et al., 2021).

However, these lights and shades of exogenous tourism in emptied Spain do not mean that this territory is a wasteland. Until few decades ago, these regions were the livelihood of thousands of people, so in a future they can return to be productive. If appropriate policies are promoted on these places according with its reality, they can be economically and socially active with a population lower than they had two centuries ago (Somoza Medina, 2004; Plaza Gutiérrez, 2006). Rural spaces have survived throughout their long history as productive spaces. They must remain as landscapes in production, otherwise we will continue to sorrow for their slow agony without giving any solution. Sine agriculture nihil, "without agriculture nothing" says the motto of the agronomists. It also should have been the motto of rural development from the beginning. New rurality must be in productive landscapes, not productivist, with sustainable and quality crops and cattle.

In relation of the tourist function in these emptied regions, it's more appropriate to encourage the so-called return tourism, the seasonal or definitive return of families with roots in these towns. Concrete measures should promote the implementation of productive activities and services for the resident population and the rehabilitation and habitability of second residences.

The resilience of sparsely populated regions lies in the identity bonds and in the attachment feeling of the emigrated population. It's worth to recover the cultivated space -the ager- with agro-livestock productions of good quality, designated with seals of origin and with high added value. In this context, the surplus value of being distributed in the chain of intermediaries has to be avoided thanks to short circuits of commercialization. Food security and valorization of the agro-livestock heritage of these regions must be promoted.

Women, men and their descendants, who return temporarily in the summer or permanently when they retire, should be the agents of change, if services and employment opportunities permit it.

Therefore many steps must be done, but always taking into account that the population to whom aid programs should be directed are the residents and the emigrated population that maintains links with their places of origin.

Delgado Urrecho (2018) highlights the importance of the population linked to these places. Emigrated people who maintain properties in the villages, and who periodically return to these villages, promote the creation of jobs and equipment. That was revealed in each province by the Survey of Infrastructures and Local Equipment. For Pérez Rubio and García García (2016), this population would be the protagonist of what they call "country tourism". In their study case placed in Extremadura, they study the periodic return of tens of thousands of people. The authors analyze through surveys the loyalty of these movements (i.e. 85% of the Extremaduran that emigrated to Madrid), their average expenditure (more than 1000 euros for the 41% of the interviewed persons) or the duration of stays (more than two weeks in 70% of the cases). This type of displacement in summer is typical of areas of emigration and it normally supposes the reactivation of economic and social activities. Every summer in the north of Portugal, thousands of emigrants with their families return from France to spend the holidays in remote villages of Tras Os Montes. Is easy to hear the French language in the streets and Portuguese authorities celebrate this return with dozens of parties to welcome them publicly (Santos, 2021).

In our research data show how the offer of rural and nature tourism has grown in Spain since the late twentieth century, driven by rural development programs financed with European funds. The increase of rural tourism establishments and public subsidies generated for some time a false conviction, according to which this activity could become an alternative development for rural areas located far from urban regions.

Data reveal the unsustainability of some rural touristic businesses due to the low occupancy and the high seasonality. Provinces near Madrid accumulate a relatively important demand for rural tourism, although this is insufficient to cover the thousands of accommodation places. It is true that in a pandemic context rural tourism is perceived as safer than others touristic modalities. This fact has increased recent occupancy rates and reawakened the idea of rural tourism as an alternative activity to agriculture (Dot Jutgla et al., 2022; Zaar, 2022). Probably the reduction of restrictions from 2022 and the increase in confidence in tourism demand will attenuate this slight growth of tourism in rural areas, especially in the emptied provinces of Spain. In these territories, resilience is based on affective links and patrimonial ties of the emigrated population.

This work has tried to show how the sustainability of rural and nature tourism in emptied Spain is threatened. The low socio-economic profitability of investments, that for two decades have tried to create new tourist destinations, must reorient the actions towards a tourism of returned people, which fill these villages during the summer. More studies of this type of tourism must be done to know its true impact in emptied Spain.

Acknowledgements Research carried out within the Research Challenges R + D + i Project. "Vulnerability, resilience and heritage reuse strategy in deindustrialized spaces". Ministry of Science, Innovation and Universities, Call 2018. Principal researcher Paz Benito del Pozo. Reference RTI2018-095014-B-100.

References

Andrés Sarasa, J. L. (2000). Aportaciones básicas del turismo al desarrollo rural. *Cuadernos de Turismo, 6*, 45–59.

Aparicio Guerrero, A. E. (2004). El turismo rural: una de las alternativas al desarrollo rural en la serranía de Cuenca. *Cuadernos de Turismo, 13*, 73–89.

Arcila, M., y López, J. (2015). Los centros de interpretación como motor de desarrollo turístico local, ¿un modelo fracasado? El caso de la provincia de Cádiz. *Boletín de la Asociación de Geógrafos Españoles, 67*, 143–165. https://doi.org/10.21138/bage.1821

Bachiller Martínez, J. M. (1994). El turismo rural como propuesta de revitalización económica de áreas rurales desfavorecidas: el suroeste soriano. *Ería. Revista Cuatrimestral de Geografía, 33*, 25–39.

Baylina Ferrè, M. (2019). La mujer como eje vertebrador de la nueva ruralidad. Un estado de la cuestión. *Perspectives on rural development, 2019*(3), 153–176.

Blanco Portillo, R., y Benayas del Álamo, J. (1994). El turismo como motor de desarrollo rural. Análisis de los proyectos subvencionados por LEADER 1. *Revista de Estudios Agrosociales, 169*, 119–147.

Burillo Cuadrado, M. P., Rubio Terrado, P., y Burillo Mozota, F. (2019). Estrategias frente a la despoblación de la Serranía celtibérica en el marco de la política de cohesión euro-pea 2021–2027. *Economía Agraria y Recursos Naturales, 19* (1), 83–97.

Cànoves Valiente, G., Prat Forga, J. M., & Blanco Romero, A. (2016). Turismo en España, más allá del sol y la playa. Evolución reciente y cambios en los destinos de litoral hacia un turismo cultural. *Boletín de la Asociación de Geógrafos Españoles*, 71. https://doi.org/10.21138/bage.2289

Cànoves, G., Herrera, L., & Villarino Pérez, M. (2005a). Turismo rural en España: paisajes y usuarios, nuevos usos y nuevas visiones. *Cuadernos de turismo., 15*, 63–76.

Cànoves, G., Herrera, L., & Blanco Romero, A. (2005b). Turismo rural en España: un análisis de la evolución en el contexto europeo. *Cuadernos de Geografía, 77*, 41–58.

Cànoves, G., Villarino, M., & Herrera, L. (2006). Políticas públicas, turismo rural y sostenibilidad: difícil equilibrio. *Boletín de la Asociación de Geógrafos Españoles, 41*, 199–217.

Cànoves, G., Villarino, M., Blanco Romero, A., de Uña, E., y Espejo, C. (Eds.). (2014). *Turismo de interior: renovarse o morir. Estrategias y productos en Catalunya, Galicia y Murcia.* Universidad de Valencia.

Carazo García Olalla, L. (1982). El turismo rural como recurso de la población agraria. *Re-vista de Estudios Agrosociales*, n° 120, 117–130.

Cerdá, P. (2017). *Los últimos. Voces de la Laponia española*. Pepitas.

Comisión Europea. (1988). El futuro del mundo rural.. Comisión Europea.

Córdoba Azcárate, M. (2004). A través de Taramundi: el ensayo de una nueva mirada teórica sobre el campo del desarrollo y el turismo rural. In *Turismo, ocio y deporte: VIII Congreso de Sociología, Alicante* (pp. 263–274).

del Río Posada, J. (2016). El turismo rural en Castilla y León: 1967–2017: Pasado, presente y futuro. *Gran Tour: Revista de Investigaciones Turísticas, 13*, 105–131.

Delgado Urrecho, J. M. (2018). Más allá del tópico de la España Vacía: Una Geografía de la Despoblación. In A. Blanco, J. A. López Ruiz, & S. Mora (coords.), *Informe España 2018* (pp. 232–295). Universidad Pontificia Comillas.

Dot Jutgla, E., Romagosa Casals, F., & Noguera Noguera, M. (2022). El incremento del turismo de proximidad en Cataluña en verano de 2020: una oportunidad para la consolidación del turismo rural como una forma de turismo sostenible y segura. *Investigaciones Turísticas, 2022*(23), 162–185. https://doi.org/10.14198/INTURI2022.23.8

Engelmo Moriche, A. M. (2022). *Análisis de las ayudas del método LEADER al turismo rural de Extremadura (1995–2015).* Tesis Doctoral, Accesible in https://dehesa.unex.es/handle/10662/14903

Entrena Durán, F. (2006). Turismo rural y desarrollo local: estudio de caso del Sur de España. *Revista Mexicana de Sociología, 68*(3), 511–549.

Esparcia Pérez, J., Noguera Tur, J., & Pitarch Garrido, D. (2009). *Gestión y promoción del desarrollo local.* Universidad de Valencia.

Espejo Marín, C. (1997). La oferta de turismo rural en el Noroeste de la Región de Murcia. *Geographical, 35*, 87–97.

Febles Ramírez, M. F., García Rodríguez, J. L., & Zapata Hernández, V. M. (2005). La iniciativa comunitaria LEADER en España. *Boletín de la Asociación de Geógrafos Españoles, 39*, 361–398.

Font Garolera, J., Sánchez Aguilera, D., & Coma Guitart, J. (2018). El turismo como milagro. Estrategias de dinamización de áreas en declive en Teruel. In X. Somoza (coord.), *De lugar geográfico a destino turístico* (pp. 369–380). AGE.

García Marín, R. (2011). Turismo y desarrollo rural en la comarca del Noroeste de la Región de Murcia: los programas europeos LEADER. *Cuadernos de Turismo, 27*, 419–435.

General Assembly. (1987). *Informe de la Comisión Mundial sobre el Medio Ambiente y el Desarrollo "Nuestro futuro común"*. Organización de las Naciones Unidas (ONU).

González, P., y Macías, A. (2014). Neoliberalismo corporativo y clientelismo en España: etnografía de la financiación europea del desarrollo rural a través de un proyecto fallido. *Revista de Antropología Iberoamericana, 9* (3), 223–250.

Hortelano Mínguez, L. A. (2015). *Desarrollo rural y turismo en Castilla y León: éxitos y fracasos*. Tesis Doctoral, Universidad de Salamanca.

Ivars Baidal, J. A. (2000). Turismo y espacios rurales: conceptos, filosofías y realidades. *Investigaciones Geográficas, 23*, 59–88.

Jurado Almonte, J. M., & Pazos-García, F. J. (2016). Población y turismo rural en territorios de baja densidad demográfica en España. *Boletín de la Asociación de Geógrafos Españoles, 71*, 247–272. https://doi.org/10.21138/bage.2282

Maroto, J. C., & Pinos, A. (2020). ¿El turismo rural freno de la despoblación? El caso del sur de España. In E. Cejudo & F. A. Navarro (Eds.), *Despoblación y Transformaciones Sociodemográficas de Los Territorios Rurales: Los Casos de España, Italia y Francia* (pp. 327–374). Universidad de Salento.

Maya Frades, A. (1999). Los programas de desarrollo rural en la provincia de Zamora: ¿iniciativas capaces de dinamizar el atraso socioeconómico de sus espacios rurales? *Polígonos, Revista de Geografía, 8*, 99–128. https://doi.org/10.18002/pol.v0i8.820

Millán Vázquez de la Torre, G., Morales Fernández, E., y Pérez Naranjo, L. M. (2014). Turismo gastronómico, denominaciones de origen y desarrollo rural en Andalucía: situación actual. *Boletín de la Asociación de Geógrafos Españoles, 65*, 113–137. https://doi.org/10.21138/bage.1746

Moriche, Á. M. E., Masot, A. N., & Aliseda, J. M. (2021). La sostenibilidad económica de las ayudas al turismo rural del Método Leader en áreas de montaña: dos casos de estudio españoles (Valle del Jerte y Sierra de Gata, Extremadura). *Boletín de la Asociación de Geógrafos Españoles, 88*. https://doi.org/10.21138/bage.3006

Navarro Valverde, F. A., Cejudo García, E., y Cañete Pérez, J. A. (2018). Análisis a largo plazo de las actuaciones en desarrollo rural neoendógeno. Continuidad de las empresas creadas con la ayuda de LEADER y PRODER en tres comarcas andaluzas en la dé-cada de 1990. *Ager: Revista de estudios sobre despoblación y desarrollo rural, 25*, 189–219.

Navarro, F., Woods, M., y Cejudo, E. (2016). The LEADER Initiative has been a victim of its own success. The decline of the bottom-up approach in Rural Development Programmes. The cases of Wales and Andalusia. *Sociologia Ruralis, 56* (2) 270–288.

Nieto Masot, A., y Cárdenas Alonso, G. (2015). El método LEADER como política de desarrollo rural en Extremadura en los últimos 20 años (1991–2013). *Boletín de la Asociación de Geógrafos Españoles, 69*, 139–162. https://doi.org/10.21138/bage.1893

OECD (Organization of Economic Cooperation and Development). (2006). *New Rural Paradigm: Policies and governance*. OECD.

Olmedo, V. G., & Carmona, A. V. (1997). "Vacaciones en casas de labranza". Un programa con muy buenas intenciones, pero carente de realidad. In M. Valenzuela (coord.), *Los turismos de interior: el retorno a la tradición viajera* (pp. 483–492). Universidad Autónoma de Madrid.

Paül i Carril, V., Lazovski, O., y Lois González, R. C. (2016). Más de veinte años de LEADER en Galicia. Un análisis de su desarrollo en las Comarcas de a Limia y a Baixa Limia. In A. R. Ruiz, et al. (coords.), *Treinta años de Política Agraria Común en España: Agricultura y multifuncionalidad en el contexto de la nueva ruralidad* (pp. 413–428). AGE.

Pérez Rubio, J. A., y García García, Y. (2016). El Turismo paisano. Un turismo rural sin atractivo turístico. In Fundación de Estudios Rurales, *Anuario 2016* (pp. 171–178). UPA.

Plaza Gutiérrez, J. I. (2006). Territorio, Geografía Rural y políticas públicas. Desarrollo y sustentabilidad en las áreas rurales. *Boletín de la Asociación de Geógrafos Españoles, 41*, 69–95.

Pulido Fernández, J. I., y Cárdenas García, P. J. (2011). El turismo rural en España. Orientaciones estratégicas para una tipología aún en desarrollo. *Boletín de la Asociación de Geógrafos Españoles, 56*, 155–176.

Rico González, M. (2005). El turismo como nueva fuente de ingresos para el medio rural de Castilla y León. *Cuadernos de Turismo, 16*, 175–195.

Sánchez Sánchez, F. J, y Sánchez Sánchez, A. M. (2021). Factores determinantes del turismo rural en espacios protegidos como impulso para el desarrollo rural en España. *Ager: Revista de estudios sobre despoblación y desarrollo rural, 31*, 139–176.

Sancho Comíns, J. (2011). Enfoque territorial. In Ministerio de Medio Ambiente y Medio Rural y Marino (MARM) (Ed.), *LEADER en España (1991–2011). Una contribución activa al desarrollo rural* (pp. 41–58). MARM. http://goo.gl/LmJXdX

Santos, R. (2021). Return migration and rural tourism development in Portugal. *Tourism Planning & Development*. https://doi.org/10.1080/21568316.2021.1953121

Somoza Medina, J. (2004). Implicaciones territoriales del desarrollo rural. In E. Pérez & R. Rodríguez (coords.), *Espacios y desarrollos rurales: una visión múltiple desde Europa y Latinoamérica* (pp. 67–80). Trea.

Trigueros, M. A., Hernando, F. M., & Prieto, E. M. (2018). La persistencia de la dualidad rural y el valor de la nueva ruralidad en Castilla y León (España). *Investigaciones Geográficas, 70*, 9–30.

Vázquez Varela, C., y Martín Gil, F. (2011). Problemas de sostenibilidad del turismo rural en España. *Anales de Geografía de la Universidad Complutense de Madrid, 31* (1), 171–194.

Vera Rebollo, J. F. (1994). El modelo turístico del Mediterráneo español: agota-miento y estrategias de reestructuración. *Papers de Turisme, 14–15*, 133–147.

Zaar, M. H. (2022). Del turismo de masas al turismo rural. La coyuntura española desde las políticas de desarrollo rural y la pandemia COVID-19. *Ar@cne. Revista Electrónica de Recursos en Internet sobre Geografía y Ciencias Sociales*, 26. https://doi.org/10.1344/ara2022.263.38364

Xosé Somoza Medina (1969). Geographer. Associated Professor of urban planning and tourism at the University of León since 1997. Doctor in Geography and History from the University of Santiago de Compostela. His main lines of research focus on rural and urban tourism, medium-sized cities, governance and rural development. He has taught at different universities in Spain, Europe and Latin America. Author of more than 100 publications, including books, magazines and book chapters. Outside the academic field, he has been director of the Center for Innovation and Services of Galicia in Ferrol and General Director of Urban Rehabilitation in the municipality of Ourense

Marta Somoza Medina (1971). Architect. PhD in Urbanism from the Polytechnic University of Catalonia, develops urban planning work at different scales. In 2019 he carried out landscape research at the Technische Universität München. Municipal architect in Allariz (Ourense. Spain), she has organized conferences, courses and exhibitions in the field of scientific dissemination and transfer, in addition to giving lectures at different universities, participating in congresses and being the author of various publications.

Chapter 3
Inland Spain: Challenges and Opportunities of Undertourism Contexts Within the Local Turn

Inmaculada Diaz-Soria, Asunción Blanco-Romero, and Gemma Cánoves Valiente

3.1 Introduction

In 2019, Spain was the destination of 83.7 million tourists (Instituto Nacional de Estadística, 2020), the second country in the world ranking after France before the pandemic. Despite these figures and the importance of the tourism sector for the country, historically, tourism has not been distributed uniformly across the territory but concentrated on the coasts, and, more recently, in the large cities. Meanwhile, the interior territories, far from the large cities, have remained on the margin, with a few exceptions. The scientific literature on tourism in Spain has become aware of this dichotomy between the coast and the interior, between destinations with *overtourism* problems (Goodwin, 2017; Papathanassis, 2017; Sheivachman, 2017; Erschbamer et al., 2018; Koens et al., 2018; Milano, 2018; Milano et al., 2018; Peeters et al., 2018; UNWTO, 2018; Dodds & Butler, 2019; Vargas-Sánchez, 2019; Cánoves Valiente et al., 2020; Pons et al., 2020; Picó Gutiérrez et al., 2022) and interior areas in *undertourism* that attempt to promote tourism as a strategy to invigorate the territory and slow down depopulation (Lampreu, 2020; Maroto Martos & Pinos Navarrete, 2019; Moriche et al., 2020; Sarrión Garrigós, 2020; Vayá & González, 2022). This interpretation in terms of inequality and territorial imbalance gives rise to a segmented conception of the urban and rural world (Molinero, 2020), which

I. Diaz-Soria (✉)
Department of Geography and Tourism School, Universitat Autònoma de Barcelona, Cerdanyola del Vallès, Spain
e-mail: inmaculada.diaz@uab.cat

A. Blanco-Romero · G. Cánoves Valiente
Department of Geography, Universitat Autònoma de Barcelona, Cerdanyola del Vallès, Spain
e-mail: asuncion.blanco@uab.cat; gemma.canoves@uab.cat

A. Blanco-Romero, M. Blázquez-Salom (eds.), *Spanish Tourism Geographies*, Geographies of Tourism and Global Change,
https://doi.org/10.1007/978-3-031-39780-6_3

contrasts with the objectives of the European policies in rural spaces, theoretically designed in keeping with the idea of territorial cohesion. That is, that the rural and urban environments share the same territory (Gómez-Pellón, 2018) and their problems should be understood from a holistic perspective. This territorial imbalance is manifested with an uneven distribution of the population across the Spanish territory, particularly after the rural exodus at the beginning of the twentieth century and during the 1960s (Esparcia, 2020).

The territorial dynamics occurring in the interior spaces are framed within a situation of growing depopulation (Cejudo & Navarro, 2019). The development of new information and communication technologies that facilitate remote working have enabled the arrival of different types of new rural inhabitants (Romita & Núñez Morales, 2014). With a limited influence on the depopulation curve, these residents modify the social and economic composition of the rural territories, even generating processes of elitisation and gentrification (Solana, 2006). The pandemic could have boosted the arrival of new inhabitants to rural spaces (Prados, 2020), although their importance and the effects on the idiosyncrasy of these populations have yet to be seen.

Within this context of growing depopulation, in many cases there has been a promotion of cultural and heritage tourism as a form of development (Pillet, 2012) to invigorate spaces that are today understood by their strategic value for food sovereignty (Iglesias, 2020) and in the framework of the ecological transition (Ortiz, 2020; Ibarra & Rabanaque, 2020). In this respect, it seems that tourism in inland areas does not alter the territory or the life of the community and that it is considered as the panacea for development. However, in its endeavour to develop tourism as an activity for the economic and social promotion of these territories, the public administration sometimes develops an excessive or inappropriate promotion that has negative effects (Cánoves et al., 2017). Over time, tourism and tourists have an effect which, depending on their management, can be multiplying or predatory.

This text seeks to analyse recent topics of the research on tourism in inland Spain over the last decade, including consolidated concepts such as inland tourism, depopulation, local and endogenous development and more recent concepts such as resilience, overtourism/undertourism, micro-clusters, social capital and proximity tourism. We understand that the dynamism that tourism can generate in inland areas is not comparable with that of the large coastal complexes, although it seems that tourism has positive repercussions for the population and the invigoration of the local economies, provided that it is managed in a way that adapts to each socio-territorial reality. The reinforcement of the heritage in the inland spaces seems to have increased their resilience in the face of the 2008 crisis, although the efficacy of this strategy in the current crisis caused by the COVID-19 pandemic has yet to be analysed (Prados, 2020). In relation to all of this, we consider that the interior territories are spaces of low tourism intensity, particularly international tourism, but which, however, are attractive spaces for proximity tourism which has gained prominence in this pandemic context (Vera-Rebollo & Ivars, 2020). At a time when displacements were restricted to areas close to the residence of the potential visitor, the proximity activities were revitalised (Araújo et al., 2020), together with the local culture and leisure in natural spaces (Pazos et al., 2020; Izcara & Cañada, 2021b), often related to interior territories. Over the last few decades, new clusters have emerged in these territories around innovative tourism products which can help local development and give rise to a change in the tourism model within a context of socio-ecological transition. Within this context, it is worth asking whether inland territories can consider proximity tourism as a means to improve their resilience, as one of the possible scenarios to contemplate within a context of tourism development and territorial reinvention.

Through a bibliographic review and based on the experience of the studies carried out by the research team (Cánoves et al., 2014, 2017), we seek to present the theoretical and practical evolution from inland tourism to the more recent concept of proximity tourism. The mix-methods studies conducted by the team over the last decade have focused on inland destinations in Spain. This data has provided us with the basic knowledge for analysing the resilience and sustainability of destinations, of inland territories, of the products developed in these contexts, and the clusters of actors related to certain products (Prat & Cánoves, 2017; Blanco-Romero & Cánoves, 2018; De Uña-Álvarez & Villarino, 2019; Cánoves Valiente et al., 2020). Based on all this, we will identify the social and territorial repercussions arising from the development of a tourism programme in inland territories as a development tool and propose possible lines of research so as to continue the critical analysis of the development of proximity tourism.

In the first part of this article, we differentiate between inland tourism and rural tourism and reflect on the concept of development in these contexts. We will also review other concepts which have been prominent in the Spanish scientific literature over the last decade, such as the concept of resilience and the adaptive cycle or the analysis of clusters and stakeholder networks. The first part ends with a reflection on roots tourism, the "village trips", which is a consolidated practice in Spain, although under-researched and on the evolution of these destinations in terms of *overtourism* and *undertourism*. The second part reviews the concept of proximity tourism within a post-pandemic context and the socio-ecological transition. This article seeks to identify spaces of opportunity for interior territories and destinations in Spain related to tourism. It also acknowledges the research carried out over the last decade by Spanish academics and proposes a possible research agenda in relation to these topics.

3.2 Inland Tourism: Key Concepts

Inland tourism is defined by certain authors as all tourism activity "in spaces far from coastal regions and cities that seeks the uniqueness, distinctiveness, the authenticity of the destination" (Blanco-Romero et al., 2021:303). It is, therefore, first defined from a geographic perspective, based on those places in which tourism is not developed yet which possess tourism features, with an emphasis on their authentic nature and the fact that it is based on endogenous resources (Jardón et al., 2022). This definition leads us to two essential points. On the one hand, the differentiation between the development of tourism in coastal regions and cities, which are also more usually associated with *overtourism* processes, and tourism development outside of these places, analysing the latter in a situation of *undertourism* (Blanco-Romero et al., 2020). On the other hand, the tourism binomial of the interior and endogenous development is inseparable. These territories seek to diversify their activities through the creation or reinforcement of their identity. Within this context, inland tourism would be the framework of strategies promoted by local actors, both public and private (Cánoves et al., 2017). Inland tourism and local development are, in this way, key elements for understanding the revitalisation of territories and the strategies of the local actors seem decisive for slowing down or driving the effects of this revitalisation.

3.2.1 Inland Tourism, Rural Tourism and Development

The concepts of inland tourism and rural tourism are often used indistinctly. Rural tourism, understood as "the series of activities undertaken by visitors in their tourist displacements to the rural environment", includes a diverse range of activities whose "common denominator is the rural space as an environment [...] of reference" (Blanco-Romero et al., 2021:335). In Spain, settlements with less than 10,000 inhabitants are considered as rural (Instituto Geográfico Nacional, 2019) which, historically, have been defined by a concentration of "several dozen houses [...], from where extensive agricultural or cattle farming spaces were managed" (Lois, 2019). These habitats have been evolving since the mid-twentieth century, conditioned by urbanisation, the modernisation of the agricultural activity and infrastructures (*ibid.*). European policies have played an important role in this evolution, including the LEADER and PRODER programmes, which have sought to drive endogenous development in rural areas understood in a broad sense and not necessarily limited to the traditional rural habitat. Similarly, cooperation operational programmes have contributed to the development of cross-border areas, with tourism as a tool (Márquez, 2014). The potential of the recreational use in these spaces is presented in these plans as a basis for the modernisation of the activity, the incorporation in the global economy, sustainable rural development (Bartual & Pareja, 2015) and the shaping of new destinations (rural, interior, heritage, nature) (Santos & Lopez, 2018). Within this context, tourism has emerged as a local development strategy in Spain and as a tool for compensating the disparities between territories (Cánoves et al., 2005; Del Valle, 2016).

From an excessively idealised perspective of the rural world, tourism is a complementary activity to those typical of the rural space, where the tourist seeks to come into contact with the autochthonous environment, which, in turn, implies the involvement of the local population in the tourism proposals. These characteristics partially coincide with the previous definition of inland tourism, particularly the idea of an experience in an authentic environment related, most of all, to natural and cultural values.

In spite of this, although inland tourism is associated with the promotion of this activity in those spaces suffering from an economic and demographic decline (Del Pino & Camarero, 2017), enhancing the resources and attractive images constitutes a strategy to favour smart endogenous development within a much broader approach, related to the capacities of innovation, entrepreneurship, digitalisation (Calvo et al., 2020; Martínez-Puche et al., 2020), resilience and sustainability (Alarcón & Mecha, 2020). These characteristics associated with inland tourism sometimes lead to confusion in its meaning with rural tourism. In fact, the studies seem to be divided when using one expression or another, although they are not completely equivalent (Cánoves et al., 2004).

In Spain, studies on inland tourism have been developed through the research and definition of the concept and highly varied analyses have been conducted on the activity in rural spaces located far from the principal nodes of development (Pillet, 1997; Priestley, 1997; Montiel Molina, 2003; Osorio Acosta, 2006; García et al., 2008; Pillet, 2011). At the end of the twentieth century, some geography research opted to vindicate inland tourism as the return to the travelling tradition (Valenzuela, 1997). However, more recently, different research projects and case studies have been framed more definitively in this denomination (Pillet, 2012; Cánoves et al., 2014). This name has become so popular that different public administrations and organisations refer to their destinations as such and have elaborated specific campaigns, such as the case of Turespaña, the Tourism Institute of Spain.

For these reasons, inland tourism is interpreted as a practice that has set the objective of territorial rebalance through the conditioning of different places for receiving visits, thereby constituting a perfect complement to the other economic sectors of the region. In the majority of cases, tourism has arisen as an intuitive project based on heritage structured by local agents. The first objective of the strategy is to construct a destination, based on a series of fundamental elements in the place. The formation of these tourist attractions implies the selection of components integrated into the elaboration of a discourse and the creation of a representative image to be consolidated (Hernández et al., 2013). Within this process, the objective is for tourism to end up constituting one of the functional alternatives with greatest weight in the interior territories and a transversal activity that generates change. In this way, we could say that the tourism activity accompanies innovation, combining quality, brand and projection (Cánoves et al., 2017).

Inland tourism covers many destinations and products, which vary depending on the resources existing in each territory. There are more traditional products, such as accommodation in rural houses, nature tourism, or the reconversion of the Parador Network (Bosch, 2019). There are also products characterised as hybrid (López Palomeque et al., 2014), products of a cultural nature, from music festivals (Diaz-Soria et al., 2014) to religious manifestations (Cebrián & García, 2014), historical memory activities (González, 2016), identity aspects or based on the landscape, industrial tourism (Prat & Diaz-Soria, 2014; Llurdés et al., 2016; Rodríguez-Zulaica, 2017), wine tourism, gastronomic tourism, adventure sports, archaeotourism,, senior tourism, astronomic tourism, or special types such as geocatching or hacks, in mountain ski areas (Cánoves et al., 2017). It includes the ever more popular film tourism, in which the interactions between local and global realities are manifested (Puche, 2015).

The many types of tourism practised in inland areas has favoured the use of the term in plural (inland tourisms), reflecting the variety of responses of the rural world subject to constant evolution. In any case, the different studies and research carried out show that, in order for the development of the different tourism activities to be truly successful in the interior territories, well-defined strategies must be defined. These strategies should favour the capacity of resilience of the territories thanks to the synergies between the actors and promoters involved, connect the reflection with the action and prioritise appropriate strategies to each case, creating territories with local leadership, social cohesion and identity.

3.2.2 *Resilience, Microclusters and Actors*

The Spanish interior territories have directed their strategies towards positioning the destinations based on their natural, heritage and cultural resources. This strategy favours the enhancement and protection of the landscape within the framework of social and environmental sustainability (Ivars, 2004). Tourism development theoretically involves the search for competitiveness, products and opportunities that form a global experience of the territory visited (Murphy, 2000), the capacity to offer added value in products and destinations, maintaining the position with respect to the competitors (Hassan, 2000) or the adaptation to changing situations of the environment (Pedreño & Ramón, 2009).

Continuing with this idea, territorial resilience is defined as this capacity of tourist destinations to absorb the impacts and crises and recover based on their resources, the adaptability of their structure and their flexibility. In this way, a resilient tourist destination is able to predict and anticipate crises, creating new strategies and conditions which enable it to recover from this situation both reinforced and reinvented (Sancho & Gutiérrez, 2010; Prados, 2020; Vélez, 2010) and favouring a new stage of development through the combination of inherited and new characteristics (Blanco-Romero & Blázquez-Salom, 2021a). Resilience management requires the collaboration between different actors of the territories, the local government, the residents and tourists in order to construct cohesive tourist destinations. Therefore, a low level of resilience makes destinations vulnerable and exposes them to all types of crises.

Of the three possible mechanisms of resilience identified by Lew (2013), we are talking about the opportunity to learn and prepare for future impacts, "an ecological approach", (Ranjan, 2012) and the opportunity to resist, transform and better adapt to the new circumstances, a "transformational approach" (Davidson, 2010; Martin, 2012; Davoudi, 2017). One way of analysing this resilience is through the adaptation of the territories through evolutionary cycles, which reflect the recovery and reorientation of destinations to survive adversities (Chien-yu & Chin-cheng, 2016; Espeso-Molinero & Pastor-Alfonso, 2020). The use of these adaptive cycles

(Holling & Gunderson, 2002; Lew, 2017) as an interpretative framework seems better suited for analysing the tourism reality of these territories than Butler's traditional life-cycle model (1980, revised and updated in 2004, 2009a, b), based on an economic perspective and infinite growth, in which reinvention serves for continued growth. In Butler's model, the alternatives contemplated when a disruptive moment is reached are decline or stabilisation interpreted as stagnation. This concept has negative connotations. Reality shows us, on the other hand, that not all territories experience constant growth and that not all destinations, products, tourism types or experiences should seek growth. This questioning is articulated in terms of economics and politics, through, for example, the Doughnut Economics (Raworth, 2018; Amat & Bargués, 2022). In this sense, it may be considered appropriate and, therefore, desirable, to reach a state of permanent stability. According to this logic, the processes of reinvention would be understood within an infinite cycle of permanent transformation, but unconnected to the objective of constant growth. This analysis combines different levels, approaches and dimensions to contemplate possible adaptation cycles of the tourism system and product within a succession of adaptive cycles linked together throughout the life of the destination.

At each moment within the adaptive cycle there are different levels of resilience, both territorial and of the destination and tourism product. These levels of resilience constitute territorial processes and very diverse dynamics, directly related to the agents involved in each of the socio-territorial metabolisms, some of which are not directly related to the system or tourism activity.

The concept of the tourism cluster, understood as a geographic space in which a group of agents interact around the product, enables the functional analysis of these metabolisms to be articulated (Blanco-Romero, 2017). The cluster is a dynamic unit, which, over time, emerges, transforms, evolves, grows or even disappears, following the logic of the previously described adaptive cycles.

The study of the dynamic evolution of the clusters and the phases of their life cycle has, on the whole, been addressed from an industrial perspective (Cruz et al., 2010). From geographic economics (Boschma & Fornahl, 2011; Crespo, 2011), the cluster is conceptualised as a whole, with an in-depth analysis of the economic and/or territorial aspects. Some authors affirm that the evolution of the cluster depends on the strategies developed, the relational dynamics of the agents, the training and transformation capacity of the companies and the territorial context (Belussi & Sedita, 2009; Hervas-Oliver et al., 2014), while others analyse it based on the cooperation between the actors (Weidenfeld et al., 2011) or in terms of innovation (Parra & Santana, 2011).

The cluster has also to be understood as the entrepreneurial capacity of the tourism agents (Power et al., 2017; Fu et al., 2019) and a perspective based on opportunities (Yachin, 2019). The life-cycle stages of a cluster vary depending on the authors (Van Klink & De Langen, 2001; Lorenzen & Maskell, 2005; Menzel & Fornahl, 2010), but they all coincide in that the fundamental phases are: introduction, growth, maturity and decline, following Butler's previously mentioned terminology of the tourism life cycle. Therefore, at the beginning of its development, the cluster is usually formed by small establishments that are barely related to one another. Subsequently, the greater cooperation between them, the increase in the relations with other agents in the territory and competitiveness facilitate its consolidation until it becomes a mature cluster.

In inland destinations the concept of tourism microcluster is used, maybe due to their low tourism relevance compared to other territories. The principal objective of this model is to economically sustain communities which have been affected by a socio-economic decline and which have few possibilities for development in the long term. The microcluster is attributed the role of contributing to sustaining the dynamics of cooperative development, even of a community nature. However, for this contribution to be effective, it is necessary to incorporate complementary activities and the decisive involvement of the local community in the tourism activities (Sáez, 2009).

Currently, within the functioning of the tourism microclusters, the social fabric, the capacity of social innovation and the empowerment of the local population play an essential role (Tárraga & Morales, 2021). In this sense, some authors indicate the importance of the networks made up of actors as a key element for creating supportive economic circuits and therefore achieve the sustainability of the tourism socio-economic metabolism developed (Romeiro & Costa, 2010; Contreras et al., 2018; Yubero, 2019) (Fig. 3.1).

In the same way, for the tourism economic circuits to be able to draw from the generation of endogenous knowledge, participation and promotion from the public administration is necessary for its consolidation and development. One of the most common manifestations of inland tourism in Spain are the village trips, conceptualised under the terms "residential tourism", "visiting home and familiar places", "return tourism" or "roots tourism" (Sánchez-Oro, 2020). This phenomenon widens the spectrum of actors and seems to have an influence on the social organisation of inland territories. Therefore, we will dedicate the next section to this phenomenon.

3.2.3 The "Village" and the Spaces in Overtourism and Undertourism

Trips to the village, to the region of origin of parents, grandparents or even older generations can be framed within the category of *Visiting Home and Familiar Places* or VHFP (Pearce, 2012; Griffin, 2017) and have specific characteristics: there is usually a second residence (Bachimon et al., 2015), the destinations are not necessarily very geographically close with respect to the usual place of residence (up to 1000 km), the trips are usually made by car (Barbier, 1965) (although certain age groups frequently use the train and bus) and the motivations are related to tradition, rest, the conservation of social and family ties (Blondy et al., 2016; García et al., 2008) and even regenerative experiences when the tourists actively contribute to the survival and dynamics of the village (Gascón, 2021).

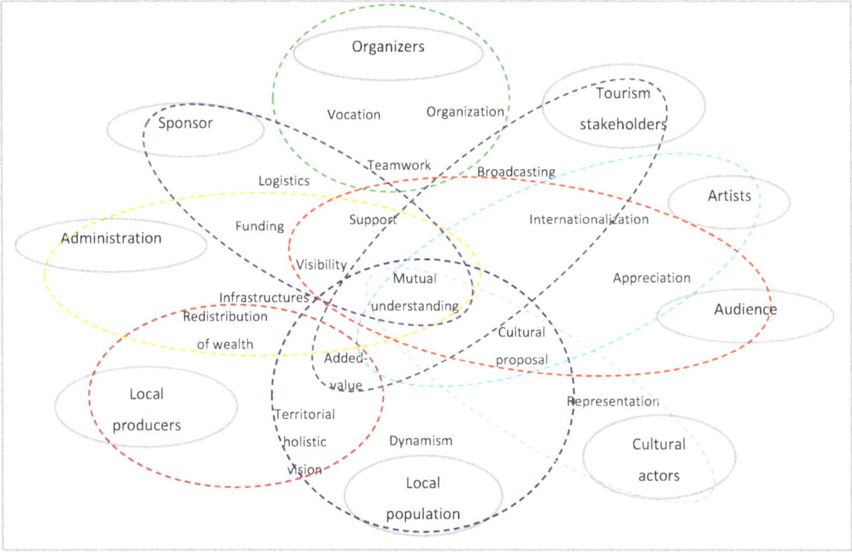

Fig. 3.1 Actors network analysed as a microcluster based on music festivals in a Spanish region. (Source: Own elaboration)

Holidays in the village are not covered by the principal indicators of the tourism market, as the accommodation is often not commercial and tourism activities, infrastructures or facilities are not used. For this reason, there are no data that rigorously measure the importance of this phenomenon. It consists in being at home, visiting families and friends and practising activities that do not differ much from those carried out by the local population: shopping, cleaning, walks and conversations, although in some cases spending is made on leisure (Sánchez-Oro, 2020).

This phenomenon has historically emerged in interior municipalities seasonally and periodically, traditionally with a degree of saturation at certain times during the summer and during the local annual festivals. In spite of this, during the years of the pandemic, with the restrictions to international mobility, some of these territories have become part of the proximity tourism markets in addition to their traditional role as return territories (Sánchez-Rivero et al., 2021). In other words, in these specific contexts, in which the effects of massification and the typical seasonality of *overtourism* could be glimpsed at certain moments, these effects have been intensified and prolonged by the attraction of a greater number of day trippers and domestic tourists in search of fresh air and escaping the cities and crowds during the pandemic (Benítez-Aurioles, 2021). Villages with charm and natural spaces have experienced processes and problems that are typically associated with coastal destinations and large cities, being overwhelmed at times by this proximity tourism (Medina-Chavarría et al., 2022).

Within this context, these spaces, often branded as vulnerable, are calling for help in order to control this success. They claim that they are not vulnerable and that they have their own idiosyncrasy and are simply not prepared to receive and manage intense flows of visitors who put the very nature of these spaces in danger. Therefore, they are calling for support to limit the development of tourism and adapt it (Font, 2021; Izcara & Cañada, 2021a). Suitably managed tourism, due to its contextualised nature, does not have a single definition or universally recognised data. However, we defend its usefulness when articulating this vindication and this understanding of tourism as a complementary economic activity. We believe that their value allows us to overcome the idea of *undertourism*, which, more than the absence of tourism activity, refers, most of all, to a specific model of socio-economic development and includes demographic stagnation.

These new dynamics could vary the reading that has been made, until now, of the tourist impacts, depending on the characteristics of the destinations. For the Spanish case, Gómez and Martín (2019) confirmed the hypothesis demonstrated in international studies (Doxey, 1975; Smith, 1977; Butler, 1980), according to which, the perception of the impacts varies depending on the characteristics of the destinations. In the years prior to the pandemic, they demonstrated that the residents of inland destinations had a more positive perception of the socio-cultural impacts of tourism, particularly in relation to the conservation of the heritage and the preservation of the cultural identity compared with the residents in coastal destinations, who "perceive the negative impacts generated by tourism more intensely" (Gómez & Martín, 2019:340). The context created by the pandemic and post-pandemic could have varied these perceptions.

3.3 From Inland Tourism to Proximity Tourism

The COVID-19 pandemic and its socioeconomic consequences have generated an unprecedented crisis for the tourism sector (Romagosa, 2020). Spain closed 2020 with less than 20 million international travellers and the tourism revenue fell by more than 75% with respect to the previous year. The paralysis of the sector limited the tourism practice to a perimeter which we can define as proximity, placing the notions of distance and proximity at the centre of our practices; distance between people and staying at home for all aspects of life (work, leisure). In this way, there has been a rediscovery of nearby places, sometimes unknown, reinforcing the emotional ties between people and their territories (Tuan, 1990) and generating a possible space of opportunity for inland territories (Benítez-Aurioles, 2022). Historically, proximity tourism has been practised, for example, under the umbrella of mountain day trips (Hagimont, 2014), trips to the home village (Gascón, 2021) or second residences (Blondy et al., 2016). "Domesticating" tourism (Blanco-Romero & Blázquez-Salom, 2021b) and the forms of consumption in a planned way are vindicating strategies which have been employed by ecologists and experts for decades (Brochard & Dumont, 2019). Within this crisis context, proximity tourism has become a concept that is considered from two perspectives: on the one hand, as a solution for the survival of the sector in the short term (Fernández, 2020; Cañada & Izcara, 2021) and, on the other hand, as an opportunity to re-think the development of tourism according to real sustainability criteria that enable its survival in the medium and long term (Palomo et al., 2020). This option is coherent with the phenomenon of localisation, a process of recovering the control of the resources and relations within the territories, a resistance movement against the dehumanisation of globalisation (Norberg-Hodge, 2020). It has yet to be seen whether it will finally consolidate as a preferred travel option or is simply a passing trend. From a tourism point of view, this movement is framed within a "local turn" in tourism (Higgins-Desbiolles & Bigby, 2022a), a redefinition of tourism from proximity which would be materialised, for example, through slow tourism in destinations close to the place of origin of the visitor. In principle, this interpretation benefits rural and interior territories which have accessible tourism markets (less than three hours away by land transport) which can be potentially revisited. This is the case of areas close to large cities promoted as quiet territories (Díaz Soria & Llurdés Coit, 2013).

On a conceptual level, the analysis is complex due to the ambiguity of the term "proximity". Therefore, the meanings are different, and it may be understood that proximity may be geographic when the displacement is short in terms of kilometres (Cañada & Izcara, 2021), when it represents a concentration or grouping of territorial actors (cluster, hub) or when it defines the physical distance with respect to spaces of centrality. Proximity tourism can also be used to understand the relations with familiar spaces (*Visiting Home and Familiar Places* or VHFP, *Visiting Friends and Relatives* or VFR) (Pearce, 2012; Griffin, 2017), although the values and impacts attributed to it differ from the first meaning. During the COVID-19 health crisis, while (geographic) proximity tourism has been reinforced within the context of restrictions on mobility, (familiar, emotional) proximity tourism has been limited or even paralysed.

This concept is not new (Davila García, 1988), neither is the fact that it is related to slowness (Germann Moltz, 2009), degrowth (Bourdeau & Berthelot, 2008) or post-tourism (Bourdeau, 2018). However, it is currently being contemplated as a short-term solution to slow down the losses of the sector while international mobility recovers (Exceltur, 2020). Within the context of the inland destinations, we consider it as a means of tourism development to be consolidated in the long term, with these territories benefiting from the movement of localisation which seeks to reduce hypermobility and contribute to the socio-ecological transition. Furthermore, this strategy would have the objective of developing resilient and appropriate forms of tourism, based on logics other than those of globalisation which have conditioned neo-liberal tourism (Higgins-Desbiolles, 2020; Fletcher et al., 2021a).

As a concept, it can be expressed in the singular form, replacing similar concepts such as local tourism (Amirou & Bachimon, 2000), intra-regional tourism (Jeuring & Diaz-Soria, 2017) or the staycation (Izcara & Cañada, 2020) and emphasising the idea of proximity as a positive value within a context of ecological transition. In practice, we can talk about proximity tourisms in plural, due to the many forms that may be adopted and the diversity of interests involved (Cañada, 2020).

Proximity tourism is not systematically ecological or equitable. Developed from a capitalist perspective, it generates negative impacts as a result of the commercialisation of common goods and spaces. These impacts can include the expulsion of certain groups from strategic areas, the extractivist use of resources or overcrowding in fragile spaces (such as natural spaces) which, to date, have received few tourists. In order for proximity tourism to be sustainable, beyond the behaviour of the visitors or the strategies deployed by the sector to manage specific impacts, it is necessary that it is developed within a geographic, institutional, political and economic environment that guarantees its adequacy and sustainability. It is necessary for proximity tourism to be developed within a new paradigm that disconnects travel and distance, in which tourism serves cultures and local economies, a 180° turn with respect to the current situation in which these are at the service of tourism development (Norberg-Hodge, 2022). This turnaround is framed within a context analysed as post-capitalist (Fletcher et al., 2021b), in which proximity tourism can constitute a transforming tool. The concept of *convivial conservation* (Büscher & Fletcher, 2019, 2020) drives social and environmental justice, it defends connection and respect as a basis for a new relationship with nature anchored in the local communities (Müller et al., 2022). Within this concept, proximity tourism has the potential to be mobilised as a way of accessing the nearby territory, learning about it and connecting with it, increasing respect for it and committing to its protection.

Therefore, we understand proximity tourism as an opportunity for inland areas with a situation of *undertourism* provided that it is developed under post-capitalist logics, inspired by existing initiatives, which have not been given the importance that they deserve due to their small economic volume compared to the dominant Fordist tourism (based on the sun and beach model or the mass cultural-urban attractions). These existing initiatives can serve as references and become emancipating experiences, inspiring new forms of tourism for rural and inland spaces. They include, for example, projects related to gastronomy and agriculture, such as those promoted by agricultural cooperatives (Cooperativa Viver in Valencia) and farms (Mas Casas Cruïlles in Girona), projects such as Gastronomía 360° at Fogar do Santiso in Galicia, or those initiatives connected with the *Slow Food* label. They also include nature tourism and historical memory routes (Izcara & Cañada, 2021a), alternative ways of organising accommodation and tourism services through cooperativism and social economy (Fernández, 2019; Vargas, 2019), responsible and sustainable tourism networks (Aethnic, Voyager autrement, The "Good Tourism" Blog), social tourism policies or governance and territorial management experiences based on social innovation (Iniciativa Legislativa Popular Mar Menor, "Travel to tomorrow", "Resilient destinations").

This perspective of proximity tourism does not seem to represent a definitive short-term alternative to the traditional tourism model based on unlimited growth. However, it could become a way of attracting visitors for interior destinations if the public administrations promote it through adequate planning. As we have seen, it is possible that the interior tourism clusters can derive potential positive impacts on the local economies and could contribute to slowing down depopulation, improving the territorial resilience. They can benefit from proximity tourism as an opportunity to evaluate and promote the existing local good practices in terms of respecting the environment, working conditions, the redistribution of profits and the social impact of economic activities.

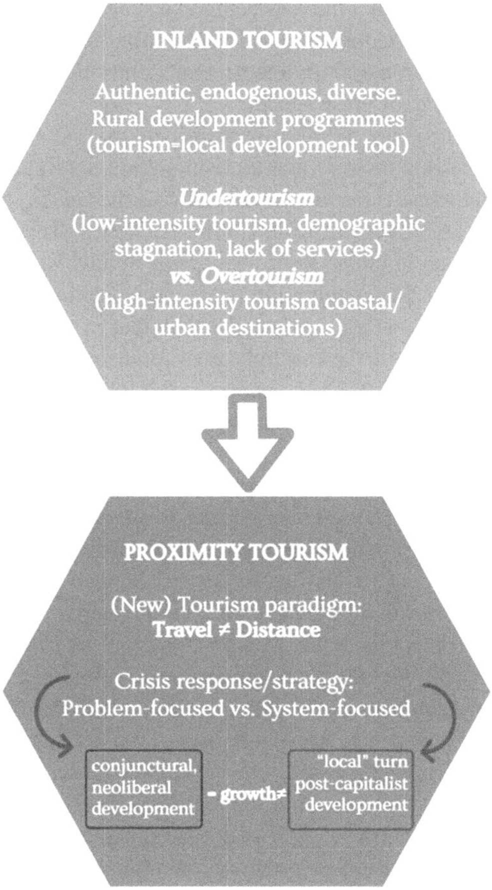

Fig. 3.2 From inland to proximity tourism theoretical framework. (Source: Own elaboration)

The existing and new proposals based on these principles represent a trend within the local development processes and sustainable and responsible tourism. Within this context, tourism could be considered as a tool for promoting territorial rebalance and the empowerment of the local communities in a context of socio-ecological transition. Figure 3.2 summarizes the evolution from inland to proximity tourism.

3.4 Agenda for a New Understanding of Tourism: Opportunities for Inland Tourism

Through an extensive bibliographic review, this article summarises some important elements for understanding the evolution and dynamics of tourism in inland spaces in Spain, particularly over the last decade. Unlike the mass tourism development concentrated in Spain on the coasts and, more recently, in urban centres, inland spaces are identified as being in a state of *undertourism*, which is not only indicative of a low frequentation of visitors, but denotes demographic, economic and institutional structural problems that contribute to the construction of a disenchanted narrative of these spaces.

There are infinite examples of the expectations of actors on different scales regarding the role that tourism could play in local development processes based on endogenous resources in interior territories, in the contribution that it could make to the improvement of territorial resilience. The many studies analysed indicate that tourism in inland territories seems to have positive repercussions for the population and for the invigoration of the local economy, although the dynamism that can be generated is not comparable with that of the large coastal complexes.

With a diverse supply that could be defined by its connection with open and natural spaces, the consumption of local products, the health benefits of tranquillity and the possibility of having significant experiences in a context of a health crisis and mobility restrictions, the interior destinations have recently become popular as destinations for proximity tourism, which confirms our principal conviction. This popularity has led some spaces to situations of overcrowding, obliging the adoption of severe contention measures and the demand for help by the local governments to manage an overwhelming sporadic success. Despite these episodes, proximity tourism cannot be understood as an alternative to mass tourism in terms of volume. It does not have the capacity to replace international tourism, nor should it aspire to do so. We advocate dimensioned proposals that are adapted to each territory, complementary to other activities together with which they may form a metabolism that flows and can be understood in a holistic way.

From this perspective, proximity tourism is presented as a lever for potential change when re-thinking human activities within a context of socio-ecological transition. Although it is not systematically sustainable, it could be if it is planned within localisation processes that place the population and the local ecosystems at the centre of the development processes. The dynamics generated by this tourism perspective (proximity activities, supply based on the local culture and on leisure in natural areas, synergies at the heart of the clusters of local actors) should be addressed from a political will with adaptation measures so as to achieve adapted tourism.

Re-sizing the incidence of the tourism activity without commodifying more spaces implies a proposal of de-growth in massified spaces. In this readjustment process, how should we plan or predict transformation processes of both the economic activities that are developed in these territories and the redirection of the workforce towards other activities? How do we make use of the infrastructure that is exclusively used for tourism?

Some studies call for the definition of a research agenda that redefines tourism from the local communities, their discourse, practices, governance and potential (Higgins-Desbiolles & Bigby, 2022b). After reviewing the principal topics in the case of inland tourism in Spain, we believe that these are essential themes that should be integrated into a research agenda that facilitate the re-thinking of tourism within new contexts and on a global level. It is, therefore, essential to adopt an approach based on the importance of the transformation towards an eco-social metabolism. Analysing how tourism is framed within more global proposals and actions, placing the local populations at the centre, in keeping with the principals of the *local turn*. Promoting the reactivation through diversification with mechanisms to avoid mono-culture and to enable the reorganisation of the tourism workforce within a context of socio-ecological transition. All of this through the use of infrastructures and opportunities within the framework of the circular economy.

In this new context, the rural and inland areas have a large potential, although an appropriate regional planning is required to induce the capacity of the tourism activity as a local, endogenous and small-scale catalyst. The lessons learned from the *overtourism* experiences in consolidated Spanish destinations emphasise the idea of the need to design an adapted tourism in these emerging tourism territories, according to the reality of each context, the existing resources and the needs and priorities of the local populations and ecosystems.

They can benefit from proximity tourism as an opportunity to evaluate and promote the existing local good practices in terms of respecting the environment, working conditions, the redistribution of profits and the social impact of economic activities. The existing and new initiatives, based on these principles, can represent a trend within the local development and sustainable and responsible tourism processes, appearing as spaces of hope in light of a change in the model in the long-term.

Acknowledgements This publication is part of the research project "Adaptation strategies and alternatives based on inland tourism destinations in times of change (ADAPTATUR_Interior)", Grant PID2020-114186RB-C21 funded by MCIN-AEI/ 10.13039/501100011033. And "Inland tourism in Spain: Challenges, valorization and strategies in the face of changing situations and crises for boosting tourism products and destinations. Case dynamics" Grant CSO2016-74861-R.

References

Alarcón, V., & Mecha, R. (2020). Desarrollo inteligente y sostenibilidad en la comarca despoblada de El Záncara (Cuenca) a partir de su SPL agroalimentario y turismo rural de calidad. In Asociación Española de Geografía & Grupo de Geografía Rural (Eds.), *Espacios rurales y retos demográficos: una mirada desde los territorios de la despoblación: ColoRural 2020*, III Coloquio Internacional de Geografía Rural, 418–433. https://geografiarural.age-geografia.es/wp-content/uploads/2021/03/ColoRural2020_Actas_Bajo-peso.pdf

Amat, D., & Bargués, P. (2022). De Glasgow a Sharm el-Sheij (COP27): ¿cómo podemos aumentar la ambición climática? *CIDOB Notes Internacionals, 267*, 1–8. https://doi.org/10.24241/NotesInt.2022/267/es

Amirou R., & Bachimon P. (dir.) (2000). *Le tourisme local. Une culture de l'exotisme*. L'Harmattan.

Araújo, N., Fraiz, J. A., & Toubes, D. R. (2020). El turismo poscoronavirus en Galicia. Plan de reactivación. In M. Simancas, R. Hernández, & N. Padrón (Eds.), *Turismo post-COVID-19. Reflexiones, retos, oportunidades* (pp. 201–209). Universidad de La Laguna.

Bachimon, P., Dérioz, P., & Vlès, V. (2015). Résidence secondaire ou dédoublement résidentiel? *Espaces, 326*. https://www.tourisme-espaces.com/doc/9480.residence-secondaire-dedoublement-residentiel.html

Barbier, B. (1965). Méthodes d'étude des résidences secondaires. L'exemple des Basses-Alpes. *Méditerranée, 6*(2), 89–111. https://www.persee.fr/doc/medit_0025-8296_1965_num_6_2_1155

Bartual, M. T., & Pareja, M. (2015). Procesos participativos de gestión para la sostenibilidad del desarrollo rural. El caso CETS (Carta Europea de Desarrollo Sostenible) en el Delta del Ebro, España. *Revista Interamericana de Ambiente y Turismo, 11*(1), 16–30. https://riat.utalca.cl/index.php/test/article/view/303

Belussi, F., & Sedita, S. (2009). Life cycle vs. multiple path dependency in industrial districts. *European Planning Studies, 17*, 505–528. https://doi.org/10.1080/09654310802682065

Benítez-Aurioles, B. B. (2021). El turismo de interior en España ante la crisis de la COVID-19. *Boletín económico del ICE*, 3139. https://doi.org/10.32796/bice.2021.3139.7285

Benítez-Aurioles, B. B. (2022). Impacto territorial de la COVID-19 sobre el turismo. Una oportunidad para el desarrollo rural en España. *Ager: Revista de estudios sobre despoblación y desarrollo rural=Journal of depopulation and rural development studies, 35*, 99–130. https://doi.org/10.4422/ager.2022.12

Blanco-Romero, A. (2017). Análisis de la sostenibilidad del Clúster y de la resiliencia territorial de la Comarca y del destino turístico: Ripollès. In G. Cánoves, A. Blanco-Romero, J. M. Prat, & M. Villarino (Eds.), *Turismo de interior en España: productos y dinámicas territoriales* (pp. 253–257). Universitat de València.

Blanco-Romero, A., & Blázquez-Salom, M. (2021a). Inland territorial and tourism resilience in a polarized world. In C. Bevilacqua, F. Calabrò, & L. Della Spina (Eds.), *New metropolitan perspectives. NMP 2020. Smart innovation, systems and technologies* (p. 178). Springer. https://doi.org/10.1007/978-3-030-48279-4_178

Blanco-Romero, A., & Blázquez-Salom, M. (2021b). Domesticar el turismo. La proximidad, la mejor vacuna. In E. Cañada & C. Izcara Conde (Eds.), *Turismos de proximidad, un plural en disputa* (pp. 59–67). Icaria Editorial.

Blanco-Romero, A., & Cánoves, G. (2018). Turismo en espacios interiores, capacidad de cambio y resiliencia. In X. Santos & L. Lopez (Eds.), *Turismo rural y de naturaleza. Una mirada al mundo* (pp. 75–101). Síntesis.

Blanco-Romero, A., Pons, G. X., Blázquez-Salom, M., Navalón-García, R., & Troitiño-Torralba, L. (2020). La sostenibilidad turística a debate. In G. X. Pons, A. Blanco-Romero, R. Navalón-García, L. Troitiño-Torralba, & M. Blázquez-Salom (Eds.), *Sostenibilidad Turística: overtourism vs undertourism*. Societat d'Història Natural de les Balears, 31: 11–20. Palma (Illes Balears). https://www.age-geografia-turismo.com/app/download/7752826511/Sostenibilidad+Tur%C3%ADstica%2C+Mon.+Soc.+Hist.+Nat+Balears%2C+31.pdf?t=1603352996

Blanco-Romero, A., Blázquez-Salom, M., De la Calle Vaquero, M., Fernández Tabales, A., García Hernández, M., Lois González, R. C., Mínguez García, C., Navalón García, R., Navarro Jurado, E., & Troitiño Torralba, L. (2021). *Diccionario de turismo*. Ediciones Cátedra.

Blondy, C., Vacher, L., & Vye, D. (2016). Les résidents secondaires, des acteurs essentiels des systèmes touristiques littoraux français? *Territoire en mouvement – Revue de géographie et aménagement, 30*. https://journals.openedition.org/tem/3344

Bosch, J. I. (2019). Paradores como eje dinamizador sociocultural de su territorio. *Estudios Turísticos, 217–218*, 165–170. https://turismo.janium.net/janium/Objetos/REVISTAS_ESTUDIOS_TURISTICOS/12%20eje%20dinamizador.pdf

Boschma, R., & Fornahl, D. (2011). Cluster evolution and a roadmap for future research. *Regional Studies, 45*, 1295–1298. https://doi.org/10.1080/00343404.2011.633253

Bourdeau, P. (2018). L'après-tourisme revisité. *Via, 13*. https://doi.org/10.4000/viatourism.1936

Bourdeau, P., & Berthelot, L. (2008). La décroissance pour repenser le tourisme? *L'autre voie, 5*. http://www.deroutes.com/AV5/bourdeaulib5.htm

Brochard, A., & Dumont, L. (2019). *Nous avons besoin de vivre toute l'année. Entretien croisé sur les luttes contre la touristification*. Jefklak. https://www.jefklak.org/nous-avons-besoin-de-vivre-toute-lannee/

Büscher, B., & Fletcher, R. (2019). Towards convivial conservation. *Conservation and Society, 17*(3), 283–296. https://doi.org/10.4103/cs.cs_19_75
Büscher, B., & Fletcher, R. (2020). *The conservation revolution: Radical ideas for saving nature beyond the Anthropocene*. Verso.
Butler, R. W. (1980). The concept of a tourist area cycle of evolution: Implications for management of resources. *The Canadian Geographer/Le Géographe canadien, 24*(1), 5–12. https://doi.org/10.1111/j.1541-0064.1980.tb00970.x
Butler, R. W. (2004). The tourism area life cycle in the twenty-first century. In A. A. Lew, C. M. Hall, & A. M. Williams (Eds.), *A companion to tourism* (pp. 159–170). Blackwell.
Butler, R. W. (2009a). Tourism in the future: Cycles, waves or wheels? *Futures, 41*(6), 346–352. https://doi.org/10.1016/j.futures.2008.11.002
Butler, R. W. (2009b). Tourism destination development: Cycles and forces, myths and realities. *Tourism Recreation Research, 34*(3), 247–254.
Calvo, J. L., Sánchez, C., & García, R. (2020). Emprendimiento Digital y Desarrollo Local. ¿Es posible el Emprendimiento Digital en la España Vaciada? *Emprendimiento local y desarrollo local*. https://www.mincotur.gob.es/Publicaciones/Publicacionesperiodicas/EconomiaIndustrial/RevistaEconomiaIndustrial/417/CALVO,%20SANCHEZ%20Y%20GARC%C3%8DA.pdf
Cañada, E. (2020, July 08). *Turismos de proximidad, un plural en disputa*. Alba Sud.
Cañada, E., & Izcara, C. (2021). Transformar el turismo desde la proximidad. In E. Cañada & C. Izcara (Eds.), *Turismos de proximidad: Un plural en disputa* (pp. 9–34). Icaria.
Cánoves Valiente, G., Blanco-Romero, A., & Díaz-Soria, I. (2020). Turismo de interior en España, del overtourism al undertourism In G. X. Pons, A. Blanco-Romero, R. Navalón-García, L. Troitiño-Torralba, & M. Blázquez-Salom (Eds.) *Sostenibilidad Turística: overtourism vs undertourism*. Societat d'Història Natural de les Balears, 31, 471–482. Palma (Illes Balears). https://www.age-geografia-turismo.com/app/download/7752826511/Sostenibilidad+Tur%C3%ADstica%2C+Mon.+Soc.+Hist.+Nat+Balears%2C+31.pdf?t=1603352996
Cánoves, G., Villarino, M., Priestley, G. K., & Blanco-Romero, A. (2004). Rural tourism in Spain: An analysis of recent evolution. *Geoforum, 35*(6), 755–769. https://doi.org/10.1016/j.geoforum.2004.03.005
Cánoves, G., Herrera, L., & Villarino, M. (2005). Turismo rural en España: paisajes y usuarios, nuevos usos y nuevas visiones. *Cuadernos de Turismo, 15*, 63–76.
Cánoves, G., Villarino, M., Blanco-Romero, A., De Uña, E., & Espejo, C. (Eds.). (2014). *Turismo de Interior: Renovarse o morir. Estrategias y productos en Catalunya, Galicia y Murcia*. Publicacions de la Universitat de Valencia PUV. https://roderic.uv.es/handle/10550/82595
Cánoves, G., Blanco-Romero, A., Prat, J. M., & Villarino, M. (Eds.). (2017). *Turismo de Interior en España. Productos y dinámicas territoriales* (Desarrollo Territorial, 19). Publicaciones de la Universidad de Valencia. https://roderic.uv.es/handle/10550/82702
Cebrián, A., & García, R. (2014). Del turismo religioso a las peregrinaciones permanentes. Diversificación turística en el sureste español. *CULTUR: Revista de Cultura e Turismo, 8*(2), 3–30. http://periodicos.uesc.br/index.php/cultur/issue/view/42
Cejudo, E., & Navarro, F. (2019). La despoblación rural como reto social. Algunos apuntes. In E. Cejudo & F. Navarro (Eds.), *Despoblación y transformaciones sociodemográficas de los territorios rurales: los casos de Italia, España y Francia. Perspectives on rural development, 3*, 17–40. Universidad del Salento. http://siba-ese.unisalento.it/index.php/prd/issue/view/1703
Chien-yu, T., & Chin-cheng, N. (2016). Vulnerability, resilience, and the adaptive cycle in a crisis-prone tourism community. *Tourism Geographies, 18*(1), 80–105. https://doi.org/10.1080/14616688.2015.1116600
Contreras, J., Mancheno Ponce, D., & Cifuentes Vega, S. (2018). Circuitos económicos solidarios turísticos un desafío de sostenibilidad. *RETOS. Revista de Ciencias de la Administración y Economía, 8*(15). https://doi.org/10.17163/ret.n15.2018.05.

Crespo, J. (2011). How emergence conditions of technological clusters affect their viability? Theoretical perspectives on cluster life cycles. *European Planning Studies, 19*, 2025–2046. https://doi.org/10.1080/09654313.2011.633824

Cruz, S., Santos, C., Teixeira, A., & Castro, A. (2010). The evolution of the cluster literature: Shedding light on the regional studies–Regional science debate. *Regional Studies, 44*, 1263–1288.

Davidson, D. J. (2010). The applicability of the concept of resilience to social systems: Some sources of optimism and nagging doubts. *Society and Natural Resources, 23*, 1135–1149. https://doi.org/10.1080/08941921003652940

Davila García, J. A. (1988). *Le tourisme de proximité dans la délégation de Xochimilco*. Tesis doctoral, Université d'Aix-Marseille 3. https://www.theses.fr/1988AIX32022

Davoudi, S. (2017). Resilience: A bridging concept of a dead end? *Planning Theory and Practice, 13*(2), 299–333. https://doi.org/10.1080/14649357.2012.677124

De Uña-Álvarez, E., & Villarino, M. (2019). Linking wine culture, identity, tourism and rural development in a denomination of origin territory (NW of Spain). *Cuadernos de Turismo, 44*, 93–110. https://doi.org/10.6018/turismo.44.404761

Del Pino, J. A., & Camarero, L. (2017). Despoblamiento rural: imaginarios y realidades. *Soberanía alimentaria, biodiversidad y culturas, 27*, 6–11. https://soberaniaalimentaria.info/numeros-publicados/58-numero-27/395-despoblamiento-rural

Del Valle, E. A. (2016). *El turismo rural en España: análisis de la evolución del modelo de desarrollo y perspectivas futuras*. Tesis Doctoral, Universidad de Vigo. http://www.investigo.biblioteca.uvigo.es/xmlui/bitstream/handle/11093/632/El_turismo_rural_en_Espa%C3%B1a.pdf?sequence=1

Díaz Soria, I., & Llurdés Coit, J. C. (2013). Reflexiones sobre el turismo de proximidad como una estrategia para el desarrollo local. *Cuadernos de Turismo, 32*, 65–88. https://revistas.um.es/turismo/article/view/177421

Diaz-Soria, I., Blanco-Romero, A., & Cánoves, G. (2014). Emporda music festivals as tourism diversification tools. *International Journal of Event and Festival Management, 5*(3), 297–310. https://doi.org/10.1108/IJEFM-11-2013-0036

Dodds, R., & Butler, R. (Eds.). (2019). *Overtourism: Issues, realities and solutions*. De Gruyter Oldenbourg. https://doi-org.are.uab.cat/10.1515/9783110607369-201

Doxey, G. (1975). A causation theory of visitor-resident irritants: Methodology and research inferences in the impact of tourism. In *Sixth annual conference proceedings of the travel research association* (pp. 195–198).

Erschbamer, G., Innerhofer, E., & Pechlaner, H. (2018). *Overtourism: How much tourism is too much*. Eurac Research, Bolzano, Italy.

Esparcia, J. (2020). La despoblación: emergencia y despliegue de politícas públicas en Europa y en España. In Asociación Española de Geografía & Grupo de Geografía Rural (Eds.), *Espacios rurales y retos demográficos: una mirada desde los territorios de la despoblación: ColoRural 2020*, III Coloquio Internacional de Geografía Rural, 75–149. https://geografiarural.age-geografia.es/wp-content/uploads/2021/03/ColoRural2020_Actas_Bajo-peso.pdf

Espeso-Molinero, P., & Pastor-Alfonso, M. J. (2020). Governance, community resilience, and indigenous tourism in Nahá, Mexico. *Sustainability, 12*(15), 5973. https://doi.org/10.3390/su12155973

Exceltur. (2020). *Revisión del impacto del Covid 19 sobre el sector turístico español*. https://www.exceltur.org/wp-content/uploads/2020/08/Revisi%C3%B3n-Impacto-COVID-y-cierre-2020-18-AGO-20.pdf

Fernández, R. (2019). *El largo viaje del patrón. Un análisis de los hoteles recuperados en Argentina*. Albasud. https://www.albasud.org/blog/es/1119/el-largo-viaje-del-patr-n-un-an-lisis-de-los-hoteles-recuperados-en-argentina

Fernández, M. T. (2020). Estrategias de marketing turístico destinadas al turista de proximidad tras el COVID-19. In M. Simancas, R. Hernández, & N. Padrón (Eds.), *Turismo pos-COVID-19. Reflexiones, retos y oportunidades* (pp. 499–505). Universidad de La Laguna.

Fletcher, R., Blanco-Romero, A., Blázquez-Salom, M., Cañada, E., Murray Mas, I., & Sekulova, F. (2021a). Pathways to post-capitalist tourism. *Tourism Geographies, 1*(22), 707–728. https://doi.org/10.1080/14616688.2021.1965202

Fletcher, R., Blanco-Romero, A., Blázquez-Salom, M., Murray Mas, I., & Sekulova, F. (2021b). Socialisation at scale: Post-capitalist tourism in a post-COVID-19 world. In F. Higgins-Desbiolles, A. Doering, & B. C. Bigby (Eds.), *Socialising tourism: Rethinking tourism for social and ecological justice* (pp. 229–243). Taylor and Francis AS. https://doi.org/10.4324/9781003164616-17

Font, M. (2021, May 15). De la massificació a la regulació de l'accés a la muntanya. *Público*. https://www.publico.es/public/medi-natural-massificacio-regulacio-l-acces-muntanya.html

Fu, H., Okumus, F., Wu, K., & Köseoglu, M. A. (2019). The entrepreneurship research in hospitality and tourism. *International Journal of Hospitality Management, 78*, 1–12. https://doi.org/10.1016/j.ijhm.2018.10.005

García, J. A., Cebrián, F., & Panadero, M. (2008). El turismo de segunda residencia en el interior peninsular. *Scripta Nova. Revista electrónica de Geografía y Ciencias Sociales, 270*(94). http://www.ub.es/geocrit/sn/sn-270/sn-270-94.htm

Gascón, J. (2021). *Turismo Doméstico de Diáspora: más allá del Turismo Rural*. Albasud. https://www.albasud.org/noticia/es/1329/turismo-domestico-de-diaspora-mas-alla-del-turismo-rural

Germann Moltz, J. (2009). Representing peace in tourism mobilities: Staycations. Slow travel and the amazing race. *Journal of Tourism and Cultural Change, 7*(4), 270–286. https://doi.org/10.1080/14766820903464242

Gómez, D., & Martín, C. (2019). Los impactos del turismo en España: diferencias entre destinos de sol y playa y destinos de interior. *Cuadernos de Turismo, 43*, 325–347. https://doi.org/10.6018/turismo.43.13

Gómez-Pellón, E. (2018). *Rural worlds, social sustainability and local landscapes in the globalisation era: Case studies in Southern Europe*. Aranzadi Thomson Reuters.

González, D. (2016). La patrimonialización de la memoria histórica: entre el deber social y la estrategia turística. Apuntes sobre el caso catalán. *Pasos, 14*(5), 1267–1280. https://doi.org/10.25145/j.pasos.2016.14.085

Goodwin, H. (2017). *The challenge of overtourism* (Working paper 4). Responsible Tourism Partnership. https://haroldgoodwin.info/wp-content/uploads/2020/08/rtpwp4overtourism012017.pdf

Griffin, T. (2017). Immigrant hosts and intra–regional travel. *Tourism Geographies, 19*(1), 44–62. https://doi-org.are.uab.cat/10.1080/14616688.2016.1169314

Hagimont, S. (2014). *De l'invention du tourisme à son institution: construction d'un territoire et enjeux de pouvoir (l'exemple des Pyrénées centrales, Second Empire-1914)*. Hypotheses. https://unhiv.hypotheses.org/55

Hassan, S. S. (2000). Determinants of market competitiveness in an environmentally sustainable tourism industry. *Journal of Travel Research, 38*(3), 239–245. https://doi.org/10.1177/004728750003800305

Hernández, M. L., Castelló, A., Alonso, M. P., & Pueyo, Á. (2013). Innovació i desenvolupament endogen en àrees rurals: el cas del Somontano de Barbastre (Osca, Espanya). *Documents d'Anàlisi Geogràfica, 59*(3), 501–522. https://doi.org/10.5565/rev/dag.58

Hervas-Oliver, J. L., Sempere-Ripoll, F., & Boronat-Moll, C. (2014). Process innovation strategy in SMEs, organizational innovation and performance: A misleading debate? *Small Business Economics, 43*(4), 873–886. https://doi.org/10.1007/s11187-014-9567-3

Higgins-Desbiolles, F. (2020). Socialising Tourism for social and ecological justice after COVID-19. *Tourism Geographies, 22*(3), 610–623. https://doi.org/10.1080/14616688.2020.1757748

Higgins-Desbiolles, F., & Bigby, B. C. (2022a). A local turn in tourism studies. *Annals of Tourism Research, 92*, 103291. https://doi.org/10.1016/j.annals.2021.103291

Higgins-Desbiolles, F., & Bigby, B. C. (2022b). Introduction: Embracing the local turn in tourism to empower communities. In F. Higgins-Desbiolles & B. C. Bigby (Eds.), *The local turn in tourism: Empowering communities* (pp. 1–27). Channel View Publications. https://doi.org/10.21832/9781845418809-008

Holling, C. S., & Gunderson, L. H. (2002). Resilience and adaptive cycles. In L. Gunderson & C. S. Holling (Eds.), *Panarchy: Understanding transformations in human and natural systems* (pp. 25–62). Island Press.

Ibarra, P., & Rabanaque, I. (2020). La marca de calidad territorial y el paisaje como estrategia para el turismo interior de la comarca del Matarraña (Teruel). In G.X. Pons, A. Blanco-Romero, R. Navalón-García, L. Troitiño-Torralba & M. Blázquez-Salom (Eds.), *Sostenibilidad Turística: overtourism vs undertourism*. Societat d'Història Natural de les Balears, 31, 69–81. Palma (Illes Balears). https://www.age-geografia-turismo.com/app/download/7752826511/Sostenibilidad+Tur%C3%ADstica%2C+Mon.+Soc.+Hist.+Nat+Balears%2C+31.pdf?t=1603352996

Iglesias, J. (2020). Prólogo: Localizando la felicidad. In H. Norberg-Hodge (Ed.), *El futuro es local*. Pol·len Edicions SCCL.

Instituto Geográfico Nacional. (2019). *España en mapas. Una síntesis geográfica*. Centro Nacional de Información Geográfica.

Instituto Nacional de Estadística. (2020). Un retrato de nuestros turistas. *Cuenta Satélite de Turismo de España*. https://www.ine.es/ss/Satellite?L=0&c=INECifrasINE_C&cid=1259952806229&p=1254735116567&pagename=ProductosYServicios%2FINECifrasINE_C%2FPYSDetalleCifrasINE

Ivars, J. A. (2004). Regional tourism planning in Spain: Evolution and perspectives. *Annals of Tourism Research, 31*(2), 313–333. https://doi.org/10.1016/j.annals.2003.12.001

Izcara, C., & Cañada, E. (2020). *Staycation: ¿una forma de entender el turismo de proximidad?* Albasud. https://www.albasud.org/noticia/es/1203/staycation-iquest-una-forma-de-entender-el-turismo-de-proximidad

Izcara, C., & Cañada, E. (2021a). Transformar el turismo desde la proximidad. ¿Cómo abordamos el debate en Cataluña? *Informes en contraste – Turismo Responsable*. Albasud. https://www.albasud.org/publ/docs/109.pdf

Izcara, C., & Cañada, E. (2021b). *Espacios naturales saturados: colisión de derechos*. Albasud. https://www.albasud.org/noticia/es/1369/espacios-naturales-saturados-colision-de-derechos

Jardón, E., del Campo, M., & Miguéns-Refojo, V. (2022). La sostenibilidad turística: el turismo de interior, Ourense. *ROTUR, 16*(1), 165–185. https://doi.org/10.17979/rotur.2022.16.1.7411

Jeuring, J., & Diaz-Soria, I. (2017). Introduction: Proximity and intraregional aspects of tourism. *Tourism Geographies, 19*(1), 4–8. https://doi.org/10.1080/14616688.2016.1233290

Koens, K., Postma, A., & Papp, B. (2018). Is overtourism overused? Understanding the impact of tourism in a city context. *Sustainability, 10*, 4384. https://doi.org/10.3390/su10124384

Lampreu, S. (2020). Strategies for territorial development and tourism against the depopulation of rural areas in Sardinia (Italy). *Revista Galega de Economía, 29*(2), 1–20. https://doi.org/10.15304/rge.29.2.6908

Lew, A. (2013). Scale, change and resilience in community tourism planning. *Tourism Geographies, 16*(1), 14–22. https://doi.org/10.1080/14616688.2013.864325

Lew, A. (2017). *Modeling the resilience adaptive cycle*. Collaborative for Sustainable Tourism and Resilient Communities Blog. https://www.academia.edu/31054408/Modeling_the_Resilience_Adaptive_Cycle

Llurdés, J. C., Diaz-Soria, I., & Romagosa, F. (2016). Patrimonio minero y turismo de proximidad: explorando sinergias. El caso de Cardona. *Documents d'Anàlisi Geogràfica, 62*(1), 55–77. https://doi.org/10.5565/rev/dag.231

Lois, R. C. (2019). Hábitat rural. In *Atlas Nacional de España*. Instituto Geográfico Nacional. http://atlasnacional.ign.es/wane/Discusi%C3%B3n:H%C3%A1bitat_rural

López Palomeque, F., Cánoves, G., Blanco-Romero, A., & Torres, A. (2014). *Turismo y territorio. Innovación, renovación y desafíos*. Tirant Humanidades.

Lorenzen, M., & Maskell, P. (2005). The cluster as a nexus of knowledge creation. In P. Cooke & A. Piccaluga (Eds.), *Regional economies as knowledge laboratories*. Edward Elgar Publishing. https://doi.org/10.4337/9781845423391.00012

Maroto Martos, J. C., & Pinos Navarrete, A. (2019). ¿El turismo rural freno de la despoblación? El caso del sur de España. In E. Cejudo & F. Navarro (Eds.), *Despoblación y transformaciones sociodemográficas de los territorios rurales: los casos de Italia, España y Francia. Perspectives on rural development, 3,* 327–374. Universidad del Salento. http://siba-ese.unisalento.it/index.php/prd/issue/view/1703

Márquez, J. A. (2014). Desarrollo difícil y turismo en la frontera Portugal-España. In A. J. Campesino & J. M. Jurado (Eds.), *Turismo de frontera (III): productos turísticos de la Raya Ibérica* (pp. 199–220). Universidad de Huelva.

Martin, R. (2012). Regional economic resilience, hysteresis, and recessionary shocks. *Journal of Economic Geography, 12*(1), 1–32. https://doi.org/10.1093/jeg/lbr019

Martínez-Puche, A., Amat, X., Cortés, C., Larrosa, J. L., Lorente, A., Ortiz, S., & Sancho, I. (2020). Turismo rural inteligente. Capacitación aplicada en el marco del proyecto europeo SmartRural. In G. X. Pons, A. Blanco-Romero, R. Navalón-García, L. Troitiño-Torralba, & M. Blázquez-Salom (Eds.), *Sostenibilidad Turística: overtourism vs undertourism*. Societat d'Història Natural de les Balears, 31, 359–368. Palma (Illes Balears). https://www.age-geografia-turismo.com/app/download/7752826511/Sostenibilidad+Tur%C3%ADstica%2C+Mon.+Soc.+Hist.+Nat+Balears%2C+31.pdf?t=1603352996

Medina-Chavarría, M. E., Gutiérre, A., & Saladié, O. (2022). The response to increasing numbers of visitors to Catalonia's protected natural areas during the COVID-19 pandemic: A review from publications in digital media. *Boletín de la Asociación de Geógrafos Españoles, 93.* https://doi.org/10.21138/bage.3183.

Menzel, M. P., & Fornahl, D. (2010). Cluster life cycles: Dimensions and rationales of cluster development. *Industrial and Corporate Change, 19*(1), 205–238. https://doi.org/10.1093/icc/dtp036

Milano, C. (2018). Overtourism, malestar social y turismofobia. Un debate controvertido. *PASOS: Revista de Turismo y Patrimonio Cultural, 16*(3), 551–564. https://doi.org/10.25145/j.pasos.2018.16.041

Milano, C., Cheer, J. M., & Novelli, M. (2018). Overtourism: A growing global problem. *The Conversation*. https://theconversation.com/overtourism-a-growing-global-problem-100029

Molinero, F. (2020). Dinámica, discursos, valores y representaciones: la diferenciación del espacio rural. In Asociación Española de Geografía & Grupo de Geografía Rural (Eds.), *Espacios rurales y retos demográficos: una mirada desde los territorios de la despoblación: ColoRural 2020,* III Coloquio Internacional de Geografía Rural, 7–37. https://geografiarural.age-geografia.es/wp-content/uploads/2021/03/ColoRural2020_Actas_Bajo-peso.pdf

Montiel Molina, C. (2003). El turismo de interior en el desarrollo socioeconómico de las comarcas forestales de la Comunidad Valenciana. *Investigaciones Geográficas, 31,* 15–36. https://www.redalyc.org/articulo.oa?id=17603102

Moriche, Á. M. E., Masot, A. N., & Rodríguez, N. R. (2020). Las ayudas al turismo rural y la despoblación rural: El caso del grupo de acción local la vera. In Asociación Española de Geografía & Grupo de Geografía Rural (Eds.), *Espacios rurales y retos demográficos: una mirada desde los territorios de la despoblación: ColoRural 2020,* III Coloquio Internacional de Geografía Rural, 376–390. https://geografiarural.age-geografia.es/wp-content/uploads/2021/03/ColoRural2020_Actas_Bajo-peso.pdf

Müller, N., Fletcher, R., & Blázquez-Salom, M. (2022). Convivial tourism in proximity. In F. Higgins-Desbiolles & B. C. Bigby (Eds.), *The local turn in tourism. Empowering communities* (pp. 215–231). Channel View Publications. https://doi.org/10.21832/HIGGIN8793

Murphy, P. (2000). The destination product and its impact on traveller perceptions. *Tourism Management, 21*(1), 43–52. https://doi.org/10.1016/S0261-5177(99)00080-1

Norberg-Hodge, H. (2020). *El futuro es local*. Pol·len Edicions SCCL.

Norberg-Hodge, H. (2022). Foreword. In F. Higgins-Desbiolles & B. C. Bigby (Eds.), *The local turn in tourism. Empowering communities* (pp. XXIII–XXV). Channel View Publications. https://doi.org/10.21832/HIGGIN8793

Ortiz, D. (2020). Los territorios rurales valencianos en un nuevo modelo económico de transición ecológica: retos para la gobernanza. *Revista de Treball, Economia i Societat, 98*, Monográfica Informe medio rural. Comité Econòmic i Social de la Comunitat Valenciana (CES-CV).

Osorio Acosta, E. (2006). *Gestión competitiva en los destinos turísticos de interior*. PhD, Departamento de Economía y Ciencias Sociales. Universitat Politècnica de València. https://doi.org/10.4995/Thesis/10251/1893.

Palomo, G., Navarro, E., Cerezo, A., & Torres, E. (2020). Turismo poscoronavirus, ¿una oportunidad para el poscrecimiento? In M. Simancas, R. Hernández, & N. Padrón (Eds.), *Turismo post-COVID-19. Reflexiones, retos, oportunidades* (pp. 161–173). Universidad de La Laguna.

Papathanassis, A. (2017). Over-tourism and anti-tourist sentiment: An exploratory analysis and discussion. *"Ovidius" University Annals, Economic Sciences Series, 17*(2), 288–293. http://stec.univ-ovidius.ro/html/anale/RO/2017-2/Section%20III/25.pdf

Parra, E., & Santana, A. (2011). Innovación en modelos organizativos: Un análisis a través de los Clúster de Turismo Españoles. *Papeles de Economía, 128*, 254–265. https://www.funcas.es/wp-content/uploads/Migracion/Articulos/FUNCAS_PEE/128art18.pdf

Pazos, M., Lopez, L., & Piñeiro, M. A. (2020). La oportunidad para los geoparques en la era poscoronavirus. Repensar a partir del caso de Montañas do Courel (Galicia). In M. Simancas, R. Hernández, & N. Padrón (Eds.), *Turismo post-COVID-19. Reflexiones, retos, oportunidades* (pp. 597–608). Universidad de La Laguna.

Pearce, P. (2012). The experience of visiting home and familiar places. *Annals of Tourism Research, 39*(2), 1024–1047. https://doi.org/10.1016/j.annals.2011.11.018

Pedreño, A., & Ramón, A. (2009). El turismo: globalización, competitividad y sostenibilidad. In J. Velarde (Coord.), *El futuro de la economía española*. Colección Mediterráneo económico, 16. Fundación Cajamar.

Peeters, P., Gössling, S., Klijs, J., Milano, C., Novelli, M., Dijkmans, C., Eijgelaar, E., Hartman, S., Heslinga, J., Isaac, R., Mitas, O., Moretti, S., Nawijn, J., Papp, B., & Postma, A. (2018). *Research for TRAN Committee – Overtourism: Impact and possible policy responses*. European Parliament, Policy Department for Structural and Cohesion Policies. http://www.europarl.europa.eu/thinktank/en/document.html?reference=IPOL_STU(2018)629184

Picó Gutiérrez, V., Sánchez Aguilera, D., & Coll Ramis, M. À. (2022). Políticas públicas y overtourism en destinos urbanos: un análisis comparado entre Barcelona y Palma. *Cuadernos de Turismo, 49*, 189–207. https://doi.org/10.6018/turismo.521871

Pillet, F. (1997). El turismo de interior como nueva fase de desarrollo loca, In M. Valenzuela (Coord.), *Los turismos de interior: el retorno a la tradición viajera* (pp. 249–260). Universidad Autónoma de Madrid Ediciones. https://www.age-geografia-turismo.com/app/download/5929678311/actas+41.zip?t=1379534168

Pillet, F. (2011). El turismo de interior y el patrimonio territorial en Castilla-La Mancha. *Cuadernos de Turismo, 27*, 725–741. https://revistas.um.es/turismo/article/view/140181

Pillet, F. (2012). El turismo de interior en la España peninsular: el patrimonio territorial como destino turístico. *Boletín de la Asociación de Geógrafos Españoles, 59*, 345–366. https://www.bage.age-geografia.es/ojs/index.php/bage/article/download/1461/1382

Pons, G. X., Blanco-Romero, A., Navalón-García, R., Troitiño-Torralba, L., & Blázquez-Salom, M. (Eds.). (2020). *Sostenibilidad Turística: overtourism vs undertourism*. Societat d'Història Natural de les Balears, 31. Palma (Illes Balears). https://www.age-geografia-turismo.com/app/download/7752826511/Sostenibilidad+Tur%C3%ADstica%2C+Mon.+Soc.+Hist.+Nat+Balears%2C+31.pdf?t=1603352996

Power, S., Di Domenico, M. L., & Miller, G. (2017). The nature of ethical entrepreneurship in tourism. *Annals of Tourism Research, 65*, 36–48. https://doi.org/10.1016/j.annals.2017.05.001

Prados, M. J. (2020). Estrategias y funciones de los espacios rurales: de territorios en crisis a territorios resilientes. In Asociación Española de Geografía & Grupo de Geografía Rural (Ed.), *Espacios rurales y retos demográficos: una mirada desde los territorios de la despoblación: ColoRural 2020* (pp. 151–178). III Coloquio Internacional de Geografía Rural. https://geografiarural.age-geografia.es/wp-content/uploads/2021/03/ColoRural2020_Actas_Bajo-peso.pdf

Prat, J. M., & Cánoves, G. (2017). Valoración de las motivaciones de los turistas para visitar los destinos de montaña. El caso de la Cerdanya. *Turydes. Revista de Investigación en Turismo y en Desarrollo Local Sostenible, 22*, 1–12. https://www.eumed.net/rev/turydes/22/turismo-cerdanya.html

Prat, J. M., & Diaz-Soria, I. (2014). La repetición de visitas en el turismo industrial. Aplicación de un modelo de ecuaciones estructurales en el Bages y el Berguedà. *Documents d'Anàlisi Geogràfica, 60*(2), 349–368. https://doi.org/10.5565/rev/dag.88

Priestley, G. K. (1997). El turismo de interior en el conjunto del turismo en España. In M. Valenzuela (Coord.), *Los turismos de interior: el retorno a la tradición viajera* (pp. 261–270). Universidad Autónoma de Madrid Ediciones. https://www.age-geografia-turismo.com/app/download/5929678311/actas+41.zip?t=1379534168

Puche, M. C. (2015). "Appelsinpiken" y el turismo experiencial en Andalucía: Geografía de la emoción. *Via*, 8. https://doi.org/10.4000/viatourism.464

Ranjan, R. (2012). Natural resource sustainability versus livelihood resilience: Mode of groundwater exploitation strategies in developing regions. *Journal of Water Resources Planning and Management, 138*(5), 512–522. https://doi.org/10.1061/(ASCE)WR.1943-5452.0000201

Raworth, K. (2018). *Doughnut economics: Seven ways to think like a 21st-century economist*. Cornerstone.

Rodríguez-Zulaica, A. (2017). Redefiniendo el concepto de Turismo Industrial. Comparativa de la terminología en la literatura castellana, francesa y anglosajona. *Pasos, 15*(2), 311–318. https://doi.org/10.25145/j.pasos.2017.15.019

Romagosa, F. (2020). The COVID-19 crisis: Opportunities for sustainable and proximity tourism. *Tourism Geographies, 22*(3), 690–694. https://doi.org/10.1080/14616688.2020.1763447

Romeiro, P., & Costa, C. (2010). The potential of management networks in the innovation and competitiveness of rural tourism: A case study on the Valle del Jerte (Spain). *Current Issues in Tourism, 13*(1), 75–91. https://doi.org/10.1080/13683500902730452

Romita, T., & Núñez Morales, S. (2014). Nuevas poblaciones rurales: "Rural users", trashumantes, nuevos habitantes. New rural population: Rural users, nomadic and new inhabitants. *Grand Tour, 10*, 4–22. https://dialnet.unirioja.es/descarga/articulo/5051980.pdf

Sáez, A. (2009). Los micro-clusters turísticos ¿una herramienta para el desarrollo turístico en zonas rurales deprimidas? *Semestre Económico, 12*(24), 57–68. https://repositorio.uam.es/bitstream/handle/10486/664287/Micro-clustersTuristicos_Saez_2009.pdf?sequence=1&isAllowed=y

Sánchez-Oro, M. (2020). Back-home tourism in Extremadura, Spain: Spending patterns and motivation. *Cuadernos de Turismo, 46*, 71–91. https://doi.org/10.6018/turismo.451501

Sánchez-Rivero, M., Rodríguez-Rangel, M. C., & Ricci-Risquete, A. (2021). Percepción empresarial de la pandemia por COVID-19 y su impacto en el turismo: un análisis cualitativo del destino Extremadura, España. *Estudios Gerenciales, 37*(159), 265–279. https://doi.org/10.18046/j.estger.2021.159.4427

Sancho, A., & Gutiérrez, C. (2010, September 22–24). ¿Crisis del sector turístico? Una actitud resiliente hacia la sostenibilidad. In *4th Latin American congress of tourism research*. Montevideo (Uruguay). https://www.uv.es/sancho/crisis%20del%20sector%20turisitco%20y%20resiliencia.pdf

Santos, X., & Lopez, L. (Eds.). (2018). *Turismo rural y de naturaleza. Una mirada al mundo*. Edit. Síntesis.

Sarrión Garrigós, A. I. (2020). *El turismo como herramienta contra la despoblación: el caso de La Canal de Navarrés*. PhD, Universitat Politècnica de València. http://hdl.handle.net/10251/151697

Sheivachman, A. (2017). *Proposing solutions to overtourism in popular destinations: A Skift Framework*. https://skift.com/2017/10/23/proposing-solutions-to-overtourism-in-popular-destinations-a-skift-framework/
Smith, V. L. (1977). *Host and Gest: An anthropology of tourism*. University of Pensylvania Press.
Solana, M. (2006). Nuevas dinámicas migratorias en los espacios rurales: vivienda, cambio social y procesos de elitización. El caso del Empordanet (Gerona). *Ager, 5*, 57–87. http://ruralager.org/wp-content/uploads/Ager-05_3.pdf
Tárraga, J. B., & Morales, S. (2021). Empoderamiento femenino y desarrollo local en la España vaciada. Caso de estudio: Teruel. *Rotur, 15*(2), 45–60. https://doi.org/10.17979/rotur.2021.15.2.7388
Tuan, Y. F. (1990). *Topophilia*. Columbia University Press.
UNWTO. (2018). *Overtourism? Understanding and Managing Urban Tourism Growth beyond Perceptions*. UNWTO. https://www.e-unwto.org/doi/book/10.18111/9789284419999
Valenzuela, M. (Coord.). (1997). *Los turismos de interior: el retorno a la tradición viajera*. Universidad Autónoma de Madrid Ediciones. https://www.age-geografia-turismo.com/app/download/5929678311/actas+41.zip?t=1379534168
Van Klink, A., & De Langen, P. (2001). Cycles in industrial clusters: The case of the shipbuilding industry in the Northern Netherlands. *Tijdschrift voor Economische en Sociale Geografie, 92*(4), 449–463. https://doi.org/10.1111/1467-9663.00171
Vargas, T. (2019). El papel del cooperativismo en el desarrollo local: realidades y perspectivas. *COODES, 7*(1), 1–4. http://coodes.upr.edu.cu/index.php/coodes/article/view/232
Vargas-Sánchez, A. (2019). Overtourism: A challenge for Smart Cities. In Universidad de Huelva (Eds.), *Actas de las II Jornadas ScienCity 2019. Fomento de la Cultura Científica, Tecnológica y de Innovación en Ciudades Inteligentes* (pp. 26–29). https://dialnet-unirioja-es.are.uab.cat/servlet/libro?codigo=785636&orden=0&info=open_link_libro
Vayá, F. S., & González, E. S. M. (2022). Efectos del turismo rural sobre la evolución demográfica en municipios rurales de España. *Ager: Revista de estudios sobre despoblación y desarrollo rural= Journal of depopulation and rural development studies*, (35), 131–163. https://recyt-fecyt-es.are.uab.cat/index.php/AGER/article/view/96121
Vélez, L. (2010). *Una apuesta metodológica para valorar la resiliencia turística en tiempos de crisis*. Master Thesis project, Economy Faculty, Universitat de València.
Vera-Rebollo, J. F., & Ivars, J. A. (2020). *El impacto del COVID-19 en la actividad turística de la Comunitat Valenciana*. Prospect 2030 Comunitat Valenciana. https://prospectcv2030.com/wp-content/uploads/2020/04/INFORME-TURISMO-Y-COVID19-CV.pdf
Weidenfeld, A., Butler, R., & Williams, A. W. (2011). The role of clustering, cooperation and complementarities in the visitor attraction sector. *Current Issues in Tourism, 14*(7), 595–629. https://doi.org/10.1080/13683500.2010.517312
Yachin, J. M. (2019). The entrepreneur–opportunity nexus: Discovering the forces that promote product innovations in rural micro-tourism firms. *Scandinavian Journal of Hospitality and Tourism, 19*(1), 47–65. https://doi.org/10.1080/15022250.2017.1383936
Yubero, C. (2019). *Periferias turísticas rurales: transferencias, redes y actores. Los casos de la Sierra de Albarracín (Teruel) y la Comarca de Las Vegas (Madrid)*. PhD, Universidad Complutense de Madrid. https://eprints.ucm.es/id/eprint/58193/

Inmaculada Diaz-Soria is a geographer and lecturer at the Department of Geography and at the Tourism School of Universitat Autònoma de Barcelona (Spain). She is also a lecturer at Ostelea Tourism School (UdL) and IQS (ULL). She is a member of the TUDISTAR research group (Tourism and socio-territorial dynamics) in which she has participated in several research projects. Her research and teaching interests include the proximity aspects of tourism, tourism in inland and natural areas, tourism policies, and tourism within the socioecological transition.

Asunción Blanco-Romero is a geographer and tenured professor at the Department of Geography at Universitat Autònoma de Barcelona (Spain). She is a member of the TUDISTAR research group (Tourism and socio-territorial dynamics) in which she has participated in several research projects. She has been president and vice-president of the "Tourism, leisure and recreation" working group of the Spanish Geography Association. She is also the editor or the scientific journal DAG (Documents d'Anàlisi Geogràfica). Her research and teaching interests include resilience tourism and local development, globalization and sustainability degrowth strategies, territorial planning, geographic and gender issues in cooperative development, and critical perspective in tourism studies.

Gemma Cànoves Valiente is a Professor at the Department of Geography at Universitat Autònoma de Barcelona (Spain). Her research focuses on cultural tourism, religious tourism and heritage, rural tourism, and how tourism affects inland areas. She is the head of the research group TUDISTAR and has published many books and articles in scientific academic journals. She is a member of the Editorial Board of the main journals specialised in tourism in Spain. Gemma has occupied responsibility positions in several academic bodies and has been head of research at the School of Tourism at Universitat Autònoma de Barcelona.

Chapter 4
Nature Conservation Policies and Tourism in Spain

Nora Müller and Macià Blázquez-Salom

4.1 Introduction

Neoliberal conservation has become the predominant approach on a global scale to protect nature and provide for leisure and recreation. It implies the withdrawal of the state and the broader implication of the private sector in conservation activities, like non-governmental organisations (NGOs), tourism enterprises or landowners (Brockington et al., 2010). The consequent privatisation of nature is critiqued to maintain unequal power relations along class, gender and ethnicity (Büscher & Fletcher, 2014; Apostolopoulou et al., 2021). It is based on the assumption that nature can only be saved through its subsumption to capital through the commodification of nature by making it exchangeable on the market, for example as a nature-based tourism product (McAfee, 1999; Castree, 2003, 2008; Büscher, 2009).

Ecotourism and nature-based tourism more generally as well as private protected areas are two instruments exemplifying this trend in conservation policies. In the Spanish state, ecotourism and land stewardship initiatives has been increasingly promoted in the last decades and receive growing attention from the academy (Basora-Roca et al., 2010; Vaccaro et al., 2013; López-Palomeque et al., 2016b; Romero-Torres, 2020; Capdepón-Frías, 2021). The objectives of this chapter is then to review the development of conservation policies linked to tourism in Spain, and how this reflects the broader conservation trends. Additionally, we discuss the critique to this approaches and aim to present alternatives for tourism and conservation based on the convivial conservation approach and the degrowth paradigm.

N. Müller (✉) · M. Blázquez-Salom
Universitat de les Illes Balears, Palma, Spain
e-mail: nora.muller@uib.eu; mblazquez@uib.cat

A. Blanco-Romero, M. Blázquez-Salom (eds.), *Spanish Tourism Geographies*, Geographies of Tourism and Global Change,
https://doi.org/10.1007/978-3-031-39780-6_4

We start by paralleling conservation development in the literature with the Spanish case. We then focus on nature-based tourism and private protected areas that exemplify the neoliberal conservation approach. We end with the presentation of postcapitalist alternatives for a transformative change in conservation and tourism and point out a future research agenda.

4.2 Historical Development of Conservation – Global Trends and Their Reflection in Spain

Conservationism has a long history. Hunting areas can be understood as the first protected areas with restricted access to a demarcated area for the aristocracy. The widespread application of conservation measures, however, can be dated back to the declaration of the Yellowstone National Park in 1872 in the USA. The approach of the so-called 'fortress conservation' is based on the idea to designate certain natural areas for the protection of nature from human interference but promoting nature contemplation and outdoor recreation, hence, there exist an established and intimate relationship between conservation and tourism. Rapidly this model has been exported to Europe and the rest of the world and has nowadays become established in conservation policies (Brockington et al., 2010; Vaccaro et al., 2013).

In Spain, the passing of the Law for the Creation of National Parks in 1916 marks the beginning of protectionist conservation initiatives. The first two national parks are established in 1918 – Montaña de Covadonga and Valle de Ordesa – both by initiative of the aristocracy. Santamarina-Campos (2019) refers to this beginnings as "elitist conservation" (see also: Vaccaro et al., 2013). The elitist character of conservation with the state (and the aristocracy) as guarantor of the natural heritage through the establishment of National Parks affects the local population, especially the vulnerable and marginalised collectives. This model is also related to the socio-economic transformation of rural areas (Cortés-Vázquez, 2017). It is founded on the concept of nature as a place to escape from the city, emphasizing the alienation between society and nature. From the beginning, protected areas are associated with tourism providing an "exciting spectacle" to be experienced (Múgica de la Guerra, 1993; see also: Igoe, 2010; Santamarina-Campos, 2019, p. 61). The natural heritage is detected as an economic potential to attract capital and considered as a source of income and educational practices even though the boom of nature-based tourism set out at the end of the twentieth century.

This initial phase of conservationism in Spain reflects the broader conservation trends that, in a Keynesian context, developed into 'fortress conservation'. The state is in charge to preserve nature through the enclosure of natural areas and the establishment of a sanctioning regime with the aim to prevent environmental deterioration (Büscher & Fletcher, 2020).

In the post-Fordist and neoliberal context of the 1970's conservation policies become more 'flexible'. Local governance, the externalisation of state functions and the application of subsidiarity principles reflect the neoliberal conservation approach coinciding with the withdrawal of the welfare state (Büscher & Fletcher, 2020). It arises the idea to protect nature through its commodification. Commodification is understood as the process through which different aspects of life are subsumed to capital and made exchangeable on the market as a commodity. It implies privatisation, displacement, valuation, abstraction, individuation and alienability (Castree, 2003). Commodification is related to the upsurge of private protected areas and exemplified by Payments for Ecosystem Services, or the promotion of ecotourism. Our interest here is particularly in its realisation through private protected areas and the promotion of (eco)tourism that promises economic local development and nature protection simultaneously.

The neoliberal turn can also be observed in the Spanish conservation policy. The establishment and promotion of protected areas and nature-based tourism provoke and increase in the frequentation of protected areas (González-Domingo et al., 2021). Simultaneously, the Spanish state encourages the entrance of private actors in the management of natural areas (Santamarina-Campos, 2019) and the instrument of land stewardship is presented to complement conservation actions of the public sector (García-Álvarez, 2016). In this context, Vaccaro et al. (2013) argue that the neoliberal turn in conservation can be understood as a step backwards to a 'quasi-fortress' model of conservation with the concentration of capital, science and political power in private hands and manifested in private protected areas or the relation between business ventures and the public conservation labor.

According to Vaccaro et al. (2013, p. 259) protected areas as fundamental conservation instruments are challenged with the questions: "Is the enjoyment of conservation spaces a basic right of all citizens and consequently, should access to protected areas be granted democratically? Or should access be limited, for reasons of environmental integrity, to a few privileged visitors?" In the next section we focus on this two questions in relation to the commodification of nature. First, revising the Spanish conservation policies, the establishment of private protected areas through instruments of land stewardship and the relation with ecotourism or nature-based tourism and recreation more generally. We undergo a critical revision of these initiatives in the framework of neoliberal conservation.

4.3 The Conservation-Tourism Nexus in a Neoliberal Context – Commodification and Privatisation of Nature

4.3.1 Protected Areas and the Supply of Public Outdoor Recreation for Social Wellbeing

> National parks, due to their public characteristic, are a resource whose enjoyment is accessible to all citizens. Additionally, they are related to natural and cultural heritage that attracts a large number of visitors specifically. (Múgica de la Guerra, 1993, p. 2)

The public use of natural areas is contemplated in the Spanish constitution as a fundamental right to enjoy nature. It is related to contributing to social well-being given the positive aspects of recreation in open spaces for physical and mental health (Múgica de la Guerra, 1993). The promotion and facilitation of public outdoor recreation constitutes one objective of protected areas, however, subordinated to the conservation of its natural values. In the context of the end of the lockdown due to the COVID-19 pandemics the fundamental need to connect with and be immersed in non-human environments has been evidenced in Spain (EUROPARC Federation, 2021) and in Europe (McGinlay et al., 2020).

The supply of natural spaces for recreation and contemplation implemented by the public administration has different modalities: the natural protected areas, including National Parks, Natural Parks, Nature Reserves, etc., are regulated in the Ley 42/2007. The central state administration manages the network of National Parks, while the autonomous communities have the competence to manage and organize National Parks inside their territories. Other protection figures, like Nature Reserves, Natural Monuments, Protected Landscapes, etc. are enacted and administered by the regional administration. The Natura 2000 network has been transposed into the Spanish legal framework without assigning management measures for its public use. Another figure acting as attraction for tourism and real estate development is related to international figures like the declaration as a World Heritage Site by the UNESCO, as in the cases of the Alhambra (Andalucía, Granada) or the Serra de Tramuntana (Balearics, Mallorca).

This different types of protected areas have the main goals to protect nature and to provide for public outdoor recreation and tourist attraction. Access to and use of Natural Reserves and Parks is regulated and organised in the Management Plan (Plan Sector de Uso y Gestión – PRUG) with the aim to prevent saturation and deterioration, however they are not always developed or often outdated (González-Domingo et al., 2021). Other protection figure, Natural Areas (Paraje Natural) for example, have to enact a Plan of Natural Resources (PORN). Protected areas generally are argued to provide for rural development by developing tourism activities (see also: Beltran-Costa & Santamarina-Campos, 2016; Simancas-Cruz, 2016). Even though, this can be critiqued by the fact, that the agricultural sector, basic for rural economies, is not supported but substituted by tourism in some cases (Vaccaro et al., 2013) coinciding with the commodification of nature in the aspect of displacement (Castree, 2003).

4.3.2 Neoliberal Conservation and Ecotourism – Spain Is Different

With the beginning of the twenty-first century and in the context of neoliberal conservation nature-based tourism has become increasingly important as an alternative to mass tourism and an instrument to finance protected areas, however, some argue that an adequate tourist management is still missing (Romero-Torres, 2020; González-Domingo et al., 2021). According to López-Palomeque et al. (2016b) tourism in natural spaces can be distinguished in five groups: (1) ecotourism; (2) cultural tourism; (3) sports and adventure tourism; (4) health tourism; and (5) rural tourism (for a detailed distinction between different types of nature-based tourism see: Romero-Torres, 2020, pp. 59–65). Bearing this variety in mind, in the following we use the term nature-based tourism to relate to this diversity, using specific terms in specific cases.

The ecotourism modality gained importance in the last decades in the Spanish context (López-Palomeque et al., 2016b). Ecotourism has been defined in the Declaration of Daimiel (2016) as "the travel to a natural area to know it, interpret it, enjoy it and explore it, while appreciating it and contributing to its conservation, without generating negative impacts on the environment, but having positive repercussions on the local population". Three aspects are argued to explain the increase of ecotourism: first the growing environmental consciousness in modern society. Second, the shift of the economy towards the service-based sector abandoning the primary sector. This contributes to the revalorisation of rural and natural areas for tourism purposes. Finally, the growth of urban population related to modernity is provoking an increased demand for spaces of outdoor recreation and natural (protected) areas are idealised to respond to this demand providing a mean to escape dense urban centres (Basora-Roca et al., 2010; Beltran-Costa & Santamarina-Campos, 2016; López-Palomeque et al., 2016b).

The development of nature-based tourism – and ecotourism specifically – in Spain have been accompanied and promoted by the private and public sector, for instance with LEADER or PRODER programs of the European Union. In the context of the financial crisis 2008 the promotion of ecotourism in Spain becomes particularly relevant and ecotourist businesses have proliferated since then (Beltran-Costa & Vaccaro, 2019). It is argued to provide economic development for rural areas and legitimised by the strategy of sustainable tourism (Vázquez-Varela & Martín-Gil, 2011; Beltran-Costa & Santamarina-Campos, 2016; López-Palomeque et al., 2016b). Additionally, the European Charter for Sustainable Tourism (ECST) first published in 1998 provides certifications for tourism enterprises that operate in protected areas to apply standards of sustainability and aims to give a guarantee of sustainability standards to the tourists (Simancas-Cruz, 2016). The Daimiel Declaration of Ecotourism (2016) in Spain encourages and promotes this standards and certificates and a guide for tourism enterprises to adhere to the ECST has been published in 2010. In 2013, the Club Ecotourism in Spain is established as an instrument to boost and evaluate the ecotourist product and ensure a commitment with sustainability of the enterprises and destinations that are part of it while it also serves as guaranty for the tourist (Blanco-Portillo, 2011). Ecotourism has also been included in political plans and programs such as the Tourism Plan of Nature and Biodiversity (RD 416/2014).

The promotion of nature-based tourism causes situations of overtourism in natural areas and related problems of deterioration of the environment, opposite to its aim to contribute to nature conservation. It can be related to the renovation of tourism with the unique aim to provide economic profit (Vázquez-Varela & Martín-Gil, 2011; López-Palomeque et al., 2016a). Especially in the post-lockdown scenario of the COVID-19 pandemics environmental degradation in natural (protected) areas has been observed as a consequence of overcrowding (González-Domingo et al., 2021). Additionally, (eco)tourism destinations around the globe suffered from the halt of tourism evidencing the dependency on its revenues embedded in volatile markets (Fletcher et al., 2020). Notwithstanding, ecotourism in proximity, and nature-based tourism more broadly is promoted as a solution to reactivate the Spanish tourism sector after the pandemics (Müller et al., 2021). The Balearic Islands, for example, design an Action Plan Ecotourism 2021 and the Spanish association of ecotourism understands the pandemics as an opportunity to further promote and strengthen the ecotourism market (Soy Ecoturista & Escapda Rural, 2021). Like this the accumulation frontier of tourism is expanded (Moore, 2015).

This development of ecotourism can be related to the neoliberal conservation approach by commodifying conserved nature as a tourist product. This is problematic as it simplifies nature by putting a price on it to be able to exchange it as an asset on the market (Armas-Díaz et al., 2020). Protected areas are crucial in this relationship. The valorisation of natural resources as tourist experience promises socioeconomic development and competitiveness in a global tourism market (López-Palomeque et al., 2016b; Patiño-Romarís & Lois-González, 2016). However, a critical review of ecotourism suggests that it works only "exceptionally" as a way to support conservation and the socioeconomic development (Krüger, 2005; Ojeda, 2019). This is explained by the incorporation of these spaces into the sphere of business, which has its spatial expression in the development of infrastructures and services that turns them into places of consumption despite the fact that they "appeal to an abstract and pristine nature" (Ojeda, 2019, p. 463). Likewise, Vázquez-Varela and Martín-Gil (2011) observe that the created imaginaries for tourism lead to a socioeconomic transformation in rural areas substituting agriculture with tourism. Beltran-Costa (2012) notes that ecotourism can work as a way to update older forms of outdoor recreation like aristocratic hunting areas or wildlife expeditions based on class relations. In the same direction Smith (2008, p. 80) argues that National Parks "are almost artificial cultural experiences of an environment in which substantial benefits are continually recorded". In the next section we delve into the analysis of private protected areas, particularly in relation to the approach to manage public access and nature-based tourism.

4.3.3 Private Protected Areas, (Eco)Tourism and the Revanchist Conservation

In the last decades private protected areas have proliferated. The definition of Dudley (2008) refers to them as areas under control or in ownership of individuals, cooperatives, NGO's or companies with for-profit or not-for-profit management objectives.

The "authority for managing the protected land and resources rests with the landowners, who determine the conservation objective, develop and enforce management plans and remain in charge of decisions, subject to applicable legislation" (Dudley, 2008, p. 26). In Spain, the first private initiatives of nature conservation appear during the 1970's and 1980's. They are related to the conservation instrument land stewardship – Custodia del Territorio – inspired by the British model. With the Declaration of Montesquieu in 2000 land stewardship becomes institutionalised in Spain and is consolidated in 2007 with its inclusion in the Spanish Law 42/2007 of Natural Heritage and Biodiversity (Villagrasa-Rozas, 2016).

Land stewardship is conceptualised as an instrument to involucrate landowners and land users in the conservation of the natural and cultural heritage and the preservation of the landscape. The conservation activities are supported by land trusts. The instrument consists of voluntary agreements of collaboration between landowner and land trust that aim to preserve the natural resources and the landscape (Basora-Roca & Sabaté i Rotés, 2006; Capdepón-Frías, 2016; Prada-Campaña & Fundación Biodiversidad, 2017). It is embedded in a broader debate to increase public participation in conservation politics (Cortés-Vázquez, 2017) through the implication of landowners, users, civil organisation and society in general (Villagrasa-Rozas, 2016). Land stewardship in this way promotes and facilitates taking on individual responsibility in managing and protecting the territory and natural resources (Capdepón-Frías, 2021).

The emergence of private protected areas and land stewardship particularly is related to the neoliberal turn in conservation and the transference of state responsibilities to the society and private sector (Capdepón-Frías, 2021). Furthermore, it corresponds to the claim of insufficiency of state activities in conservation (García-Álvarez, 2016). In this sense, Capdepón-Frías (2021, p. 73) describes land stewardship as a "valuable model to broaden the public networks of protected areas" and providing buffer zones for public protected areas or constituting ecological corridors which ensure a greater ecological and territorial integration (see also: Capdepón-Frías & Durá-Alemán, 2019). Likewise, Villagrasa-Rozas (2016) argues, that the inclusion of private land in conservation is essential as huge part of the Spanish biodiversity can be found outside public protected areas. Additionally, land trusts are argued to be more flexible, independent and efficient in comparison with the state apparatus benefitting biodiversity conservation's success in collaboration with landowners (Basora-Roca & Sabaté i Rotés, 2006).

There exist a variety of motivation for landowners to engage in land stewardship on their property. A whole research body is dedicated to investigate the willingness of landowners to participate in conservation (Morton et al., 2010; Farmer et al., 2011; e.g. Broch et al., 2013; Tyrväinen et al., 2014) as land stewardship – private protected areas more broadly – aims to combine productive uses and economic returns with conservation actions (Basora-Roca & Sabaté i Rotés, 2006). Reasons to engage in conservation on private land are: Altruism, personal convictions or the personal desire to conserve nature and culture values (Dudley, 2008); to maintain the property in family hand, to enjoy privacy or tax benefits (Rissman & Merenlender,

2008; Stolton et al., 2014); to preserve traditional uses (Brown & Mitchell, 1997); or to obtain social recognition and be part of personal and mutual learning (Basora-Roca & Sabaté i Rotés, 2006). Finally economic and commercial interests "to put in value the natural and cultural attraction, to foment tourist activities and provide education, leisure and recreation services" are further motives for the engagement in land stewardship (Capdepón-Frías, 2021, p. 73). Incentive schemes, like revenues from nature-based tourism, hunting concessions or tax reduction are argued to support the development of land stewardship initiatives on private properties (Basora-Roca & Sabaté i Rotés, 2006; Dudley, 2008).

The combination of tourism and land stewardship on private lands are understood as a potential and investment opportunity. Nature-based tourism is argued to provide for local development and the diversification of the tourism sector while it also promotes alternative mechanisms to finance conservation activities (Capdepón-Frías, 2016; García-Álvarez, 2016). In the Spanish case, the Tourism Plan of Nature and Biodiversity (RD 416/2014, p. 6) supports this idea: "land stewardship can contribute to alliances between the different stakeholders in order to advance positive synergies between tourism, biodiversity and conservation". Notwithstanding, only few land stewardship initiatives actually engage in nature-based tourism (Capdepón-Frías, 2021). One example can be found in the Natural Park Alt Pirineu in Catalonia. The project Pirineu Viu (Living Perrineaus) involves landowners, forest managers and the civil society. It aims to improve the management and the conservation of the forest and the implementation of tourist itineraries and infrastructure to support public use, amongst other things (Basora-Roca et al., 2010).

From a critical perspective, the private engagement in conservation, like land stewardship initiatives and the promotion of nature-based tourism, facilitates the privatisation of nature and the creation of exclusive and elitist spaces only accessible to certain sectors of society, contrary to the general objectives of conservation areas to contribute to social well-being. Additionally, the economic dependence on subsidies, donations and tourism revenues may work against the durability of private protected areas when they lack economic viability (Apostolopoulou et al., 2021; Capdepón-Frías, 2021).

In the case of the modification of the Asturian Law of Protected Area's Management analysed by Cortés-Vázquez (2017) the implication of private property holders in the conservation management is legitimised through the discourse of public participation. Accordingly, it implies the maintenance of existing power relations by reducing the local population to the collective of landowners without recognizing the heterogeneity of the local community and the complexity of problems that rural and natural areas are facing. The maintenance of unequal power relation in processes of private involvement in conservation policies is also critiqued by Apostolopoulou et al. (2021, p. 248) as it is "often leading to the exclusion of low-income, working class and marginalised communities aggravating inequality along lines of class, gender, ethnicity and race".

The displacement of low-income residents can be related to environmental gentrification (Checker, 2011). It is built on the material and discursive success of the urban environmental movement and legitimised adopting the sustainability discourse. The scarcity and exclusivity of natural areas assign them a high value attracting affluent urban populations related to environmental gentrification (Vaccaro et al., 2013). This is even accentuated on insular territories, where the production of real estate tourism is based on accumulation by dispossession and sociospatial segregation (Clark et al., 2007; Blázquez-Salom, 2013; Valdivielso & Moranta, 2019).

From this perspective, Artigues-Bonet and Blázquez-Salom (2016) define "revanchist conservationism" as the privatisation of the enjoyment of natural spaces. They argue, that elites use their power position as landowners and the support of conservation to add these commons to their accumulation by dispossession. Likewise Apostolopoulou and Adams (2019) argue that the environmental degradation is used as an investment opportunity and for the accumulation of capital. In this line, the instrument of land stewardship can serve to formalize alliances unfavourable for the public use of private natural areas. Artigues-Bonet and Blázquez-Salom (2016) extend then the theory of revenge from the context of urban regeneration (Smith, 1996) to the privatisation and enclosure of natural spaces. They argue that global elites take, figuratively, the revenge of the environmental movements that, in previous times, protected natural spaces from claims of urban or tourist development or forest and mining exploitation. Although conservation practices put a stop to the deterioration of nature through legal protection, the revenge consists in the appropriation of real estate for elitist enjoyment of open and natural spaces (Artigues-Bonet & Blázquez-Salom, 2016).

In summary, the development of nature conservation has been materialised – mostly, if not exclusively – in the form of protected areas. The enclosure of land for conservation purposes is presented as a response to solve the environmental problems caused by capitalist development around the globe without accepting that environmental impacts do not stop at park limits. Additionally, protected areas are understood as a resource for nature-based tourism and recreation through its commodification. The ongoing subsumption of nature to capital is widely critiqued for causing displacement and dispossession, reinforcing social inequality and accentuating nature-humanity alienation (McAfee, 1999; Cortés-Vázquez & Apostolopoulou, 2019; Büscher & Fletcher, 2020). The approach of neoliberal conservation is thus grounded on the idea that "nature can only be 'saved' through its submission to capital and its subsequent revaluation in capitalist terms" (Büscher, 2009, p. 91). In this way, the resolution of the internal contradictions of capitalism is pursued without transforming capitalism and the causes for crises of overaccumulation, the deterioration of the environment or social inequality (Fletcher & Neves, 2012). In the next section, we thus present alternative approaches to conservation and tourism based on conviviality and fair degrowth. This is important in the context of the described critique of mainstream conservation practices and in light of the chronical socioecological crisis and related threats of social injustices while trying to attack environmental problems. Finally, we hint to the future research agenda.

4.3.4 Post-Capitalist Proposals for Nature and Tourism and the Future Research Agenda

The convivial conservation approach (Büscher & Fletcher, 2020) offers an alternative to the predominant neoliberal conservation that is characterised, as described, by the subsumption of nature to capital, making conservation necessary for capital accumulation. The predominant neoliberal conservation ignores the interconnectedness and dependence between all natures – human and non-human – while maintaining and exacerbating social inequalities (Büscher & Fletcher, 2014; Apostolopoulou et al., 2021). In the same line, Vaccaro et al. (2013, 264) conclude that "the conservationist world must be more receptive to the obvious fact that ecology and society cannot be understood or managed independently", related directly to the critique to the human-nature divide intrinsic in conservation practices. While Cortés-Vázquez (2017) states that a just distribution of the benefits of conservation activities has to be realised with a transparent and accessible management that recognizes the diversity and heterogeneity of the local community.

To attack this critique – established human-nature alienation and the commodification of nature aggravating social inequality – convivial conservation is inspired by a diversity of ideas, like degrowth, "buen vivir" or the commons. It proposes a post-capitalist approach aiming at social and environmental justice and grounded in the commons and the embedded value of nature in the specific context (Büscher & Fletcher, 2020). In this sense, conviviality is contrary to the commodification of nature that aspires at making nature exchangeable on the market by putting a price on it, through tourism inter alia. As a consequence of commodification, nature is abstracted and alienated from its human and non-human environment (Castree, 2003). Additionally, market logics tend to socioeconomic segregation and the displacement of people, opposite to the expressed aim of social justice.

In the realm of tourism, a convivial approach proposes the promotion of long-lasting, socially and environmentally just visitations in proximity and the appreciation of mundane environments, going beyond the "spectacle" of nature (Igoe, 2010). In this way, it proposes an alternative to commodify nature. Proximity tourism have been the respond to the end of the COVID-19 lockdown and the impossibility of long-distance travelling (Diaz-Soria & Llurdés-Coit, 2013; Müller et al., 2021). It is also related to the claim of "A local turn in tourism studies" (Higgins-Desbiolles & Bigby, 2021) and has obtained attention, especially in the pandemics context (Cañada & Izcara, 2021).

The degrowth discourse in tourism relates directly to the proposal to going beyond the "spectacle" of nature with the claim to decommodify (nature) tourism (Blanco-Romero, 2019; Blanco-Romero et al., 2019; Higgins-Desbiolles et al., 2019). It appears as a reaction to situations of overtourism and the negative environmental and social impacts of tourism on a global scale (Fletcher et al., 2019). Notwithstanding, the conceptualisation of decommodification and its translation into practices is just in the beginning (Fletcher et al., 2021). A review of this proposal would contribute a prepositive perspective to the conservation-tourism nexus, complementary to the convivial conservation approach.

Degrowth proposes decolonizing leisure time from its appropriation by capitalism to turn it into "time for living": loneliness, boredom and thinking; democracy; encounters and relationships; the game, the party and all the other autotelic activities; feel alive enjoying nature and poetry; crafts and DIY; beauty and wisdom; or rebellion and dissidence (Puche, 2010). Individual moderation allows one to focus on care, social reproduction, and relationships, participating in the democratically determined social expenditure (D'Alisa et al., 2015). From the degrowth perspective leisure in the public sphere is part of the improvement of social well-being and tourism can be considered one of tis expressions. Sharing the commons is one of the defining principles of degrowth. In relation to leisure the commons can be expressed, for example, in the enjoyment of the coastline or the beach, that is guaranteed – at least in the Spanish case – to everyone regardless of their purchasing power (Kallis, 2018).

Life is not limited to biological, functional or utilitarian sustenance. It is the expenditure of energy for non-productive activities that defines us as humans (Romano, 2015). But when the excess energy is spent for private and individualistic accumulation, it is opposed to the collective and political surplus energy. In the vision of degrowth the latter one gives sense to live, because it is based on a culture of sharing, care, experimentation, and co-determination. On the other hand, the overaccumulation of capital ends in crisis, devaluing and destroying the surplus in order to continue growing and demobilizing protest (Kallis, 2018). At the individual level, overaccumulation is destroyed with conspicuous consumerism to feign status (Veblen, 1934), as in luxury tourism. The ecofeminist approach to degrowth also raises this contrast of models, extending its critique of capitalism, linking it to patriarchy in the Androcene. Its division differentiates what unites us: symbiosis, solidarity, cooperation, collaboration, redistribution, affections, emotional connection and care; from what separates us: domination, competition, exploitation, status, appropriation, privatisation, ambition, hierarchical order, individualism or violence (Herrero, 2021; Aragón, 2022).

Residents of famous tourist destinations claim for degrowing tourism as a contestation of the tourist commodification causing displacement in different spheres of life: housing, public roads, ports, squares, local retail shops, etc. In addition to this complaints, the previous diagnosis of inequality and sociospatial segregation is associated with international and extractivist tourism of the consumer class of the global North. For instance, between the two shores of the Mediterranean basin or between the inner land of Spain, its archipelagos and the east coast (Blanco-Romero, 2019). Tourism degrowth should begin by promoting proximity and everyday mundane environments (Cañada & Izcara, 2021) through contraction and convergence by means of public regulation (Fletcher et al., 2019; Fletcher et al., 2021).

To prevent social exclusion in tourism and conservation activities, the collectivisation and democratic decision-making in conservation and tourism politics is suggested (Blanco-Romero et al., 2019; Büscher & Fletcher, 2020).

A transformation of conservation and tourism towards a commons regime imply to engage with the definition of just exclusion criteria, as commons does not equal open access (de Angelis, 2017). The implied commoning processes in conservation and tourism need further research.

The future research agenda can then be summarised in three issues:

1. Decommodification of nature and tourism: Conceptualisation and practical implications.
2. Commoning processes in conservation and tourism: Search for existing structures for transformative change.
3. Definition of exclusion criterion in natural areas: Based on both environmental and social justice criteria.

A transformation based on the degrowth paradigm and conviviality as a response to the negative social and environmental impacts of the predominant neoliberal conservation model, has to consider social justice criterion. The implication of the civil society into conservation effort through models of land stewardship and the establishment of private protected areas runs the risk to exclude marginalised groups of the society that do not have access to land and property, inter alia. The application of criteria of social justice when limiting access to concrete areas defending biodiversity conservation and preventing negative impacts of overcrowding has to be strengthened. We argue, that all kind of limiting access to natural areas based on arguments of biodiversity conservation have to consider social justice criteria, too. This is important to prevent the instrumentalisation of participation (Cortés-Vázquez, 2017) or conservation (Artigues-Bonet & Blázquez-Salom, 2016) to enclose and privatize nature contributing to social injustices and the maintenance of unequal power relations of domination in conservation and tourism practices (Apostolopoulou et al., 2021).

Acknowledgements This publication is part of the research project "Overtourism in Spanish Coastal Destinations. Tourism Degrowth Strategies", grant number RTI2018-094844-B-C31 funded by MCIN/AEI/ 10.13039/501100011033 and by "ERDF A way of making Europe".

References

Apostolopoulou, E., & Adams, W. M. (2019). El capitalismo neoliberal y la conservación en la era postcrisis: La dialéctica del acaparamiento "verde" y "antiverde" en Grecia y Reino Unido. In J. A. Cortés-Vázquez & O. Beltran-Costa (Eds.), *Estudis d'antropologia social i cultural: Vol. 21. Repensar la conservación: Naturaleza, mercado y sociedad civil* (pp. 27–56). Edicions de la Universitat de Barcelona.

Apostolopoulou, E., Chatzimentor, A., Maestre-Andrés, S., Requena-i-Mora, M., Pizarro, A., & Bormpoudakis, D. (2021). Reviewing 15 years of research on neoliberal conservation: Towards a decolonial, interdisciplinary, intersectional and community-engaged research agenda. *Geoforum, 124*, 236–256. https://doi.org/10.1016/j.geoforum.2021.05.006

Aragón, V. (2022). *Ecofeminismo y decrecimiento*. Los Libros de la Catarata.
Armas-Díaz, A., Sabaté-Bel, F., Murray, I., & Blázquez-Salom, M. (2020). Beyond the right to the Island: Exploring protests against the neoliberalization of nature in Tenerife (Canary Islands, Spain). *Erdkunde, 74*(4), 249–262. https://doi.org/10.3112/erdkunde.2020.04.02
Artigues-Bonet, A.-A., & Blázquez-Salom, M. (2016). Huídas al Paraíso y la Realización Mercantil del Sueño. In N. Benach, M. Zaar, & M. Vasconcelos (Eds.), *Actas del XIV Coloquio Internacional de Geocrítica: Las utopías y la construcción de la sociedad del futuro*.
Basora-Roca, X., & Sabaté i Rotés, X. (2006). *Custodia del Territorio en la práctica: Manual de introducción a una nueva estrategia participativa de conservación de la naturaleza y el paisaje*. Fundació Territori i Paisatge – Obra Social Caixa Catalunya; Xarxa de Custòdia del Territori.
Basora-Roca, X., Romero-Lengua, J., Sabaté i Rotés, X., & Sogues-Marco, M. (2010). La valorización del patrimonio forestal como recurso ecoturístico en zonas de montaña: El caso del Mig Pallars y el Parque Natural del Alt Pirineu (Pirineo de Lérida, Cataluña). *Ager: Revista De Estudios Sobre Despoblación Y Desarrollo Rural, 9*, 31–62.
Beltran-Costa, O. (2012). La naturaleza en venta: Imágenes culturales y mercado en el turismo amazónico del Sur del Perú. In J. M. V. del Río (Ed.), *Colección PASOS edita: Vol. 6. Amazonia: Viajeros, turistas y poblaciones indígenas* (pp. 123–154). Asociación Canaria de Antropología; PASOS, Revista de Turismo y Patrimonio Cultural.
Beltran-Costa, O., & Santamarina-Campos, B. (2016). Antropología de la Conservación en España. Balance y perspectivas. *Revista De Antropología Social, 25*(1), 85–109. https://doi.org/10.5209/rev_RASO.2016.v25.n1.52626
Beltran-Costa, O., & Vaccaro, I. (2019). La conservación de la naturaleza en Cataluña: La tensión entre lo público y lo privado. In J. A. Cortés-Vázquez & O. Beltran-Costa (Eds.), *Estudis d'antropologia social i cultural: Vol. 21. Repensar la conservación: Naturaleza, mercado y sociedad civil* (pp. 57–72). Edicions de la Universitat de Barcelona.
Blanco-Portillo, R. (2011). El club de producto Ecoturismo en España: Un método para su configuración y evaluación. *Estudios Turísticos, 187*, 59–106.
Blanco-Romero, A. (2019). Tourism degrowth. In E. Cañada (Ed.), *Contrast reports Serie: Vol. 9. Tourism in the geopolitics of the Mediterranean* (pp. 66–69). Albasud.
Blanco-Romero, A., Blàzquez-Salom, M., Morell, M., & Fletcher, R. (2019). Not tourism-phobia but urban-philia: Understanding stakeholders' perceptions of urban touristification. *Boletín De La Asociación De Geógrafos Españoles*. Advance online publication. https://doi.org/10.21138/bage.2834
Blázquez-Salom, M. (2013). More villas and more barriers: Gentrification and the enclosure of rural land on Majorca. *Méditerranée, 120*, 25–36. https://doi.org/10.4000/mediterranee.6638
Broch, S. W., Strange, N., Jacobsen, J. B., & Wilson, K. A. (2013). Farmers' willingness to provide ecosystem services and effects of their spatial distribution. *Ecological Economics, 92*, 78–86. https://doi.org/10.1016/j.ecolecon.2011.12.017
Brockington, D., Duffy, R., & Igoe, J. (2010). *Nature unbound: Conservation, capitalism and the future of protected areas*. Earthscan.
Brown, J., & Mitchell, B. (1997). Extending the reach of national parks and protected areas: Local stewardship initiatives. In J. G. Nelson & R. Serafin (Eds.), *NATO ASI series: Vol. 40. National Parks and protected areas: Keystones to conservation and sustainable development* (pp. 103–116). Springer.
Büscher, B. (2009). Letters of gold: Enabling primitive accumulation through neoliberal conservation. *Human Geography, 2*(3), 91–94. https://doi.org/10.1177/194277860900200309
Büscher, B., & Fletcher, R. (2014). Accumulation by conservation. *New Political Economy*. Advance online publication. https://doi.org/10.1080/13563467.2014.923824
Büscher, B., & Fletcher, R. (2020). *The conservation revolution: Radical ideas for saving nature beyond the Anthropocene*. Verso.
Cañada, E., & Izcara, C. (Eds.). (2021). *Turismos de proximidad: Un plural en disputa*. Icaria Antrazyt.

Capdepón-Frías, M. (2016). La custodia del territorio como instrumento de gestión medioambiental y turística: Una revisión desde el análisis de experiencias. In M. Blázquez-Salom, M. Mir-Gual, I. Murray, & G. X. Pons (Eds.), *Monografies de la Societat d'Història Natural de les Balears: Vol. 23, Turismo y crisis, turismo colaborativo y ecoturismo* (pp. 321–333). Societat d'Història Natural de les Balears.

Capdepón-Frías, M. (2021). Las áreas protegidas privadas como escenarios para el turismo: Implicaciones y cuestiones clave. *Cuadernos Geográficos*, 72–90. https://doi.org/10.30827/cuadgeo.v60i2.15336

Capdepón-Frías, M., & Durá-Alemán, C. J. (2019). Introducción al concepto de la conservación privada: "Nuevas" herramientas para la protección de la biodiversidad. *Ciudad Y Territorio Estudios Territoriales, 51*(199), 27–42.

Castree, N. (2003). Commodifying what nature? *Progress in Human Geography, 27*(3), 273–297. https://doi.org/10.1191/0309132503ph428oa

Castree, N. (2008). Neoliberalising nature: The logics of deregulation and reregulation. *Environment and Planning A: Economy and Space, 40*(1), 131–152. https://doi.org/10.1068/a3999

Checker, M. (2011). Wiped out by the "Greenwave": Environmental gentrification and the paradoxical politics of uUrban sustainability. *City & Society, 23*(2), 210–229. https://doi.org/10.1111/J.1548-744X.2011.01063.X

Clark, E., Johnson, K., Lundholm, E., & Malmberg, G. (2007). Island gentrification and space wars. In G. Baldacchino (Ed.), *A world of islands: An islands studies reader* (pp. 481–510). Institute of Island Studies University of Prince Edward Island.

Cortés-Vázquez, J. A. (2017). Conservación ambiental, participación pública y crisis económica: Entre la amenaza de privatización encubierta y la búsqueda de políticas más equitativas para los espacios naturales protegidos. *Quaderns De L'institut Catala D'antropologia, 22*(2), 232–241.

Cortés-Vázquez, J. A., & Apostolopoulou, E. (2019). Against neoliberal natures: Environmental movements, radical practice and "the right to nature". *Geoforum, 98*, 202–205. https://doi.org/10.1016/j.geoforum.2019.01.007

D'Alisa, G., Demaria, F., & Kallis, G. (Eds.). (2015). *Degrowh: A vocabulary for a new era*. Routledge.

de Angelis, M. (2017). *Omnia sunt communia: On the commons and the transformation to postcapitalism*. Zed Books.

Diaz-Soria, I., & Llurdés-Coit, J. C. (2013). Reflexiones sobre el turismo de proximidad como una estrategia para el desarrollo local. *Cuadernos De Turismo, 32*, 65–88.

Dudley, N. (Ed.). (2008). *Guidelines for applying protected area management categories*. IUCN.

EUROPARC Federation (Ed.). (2021). *Healthy parks and healthy people Europe: health & well-being benefits from parks & protected areas*. https://www.europarc.org/nature/healthy-parks-healthy-people-europe/

Farmer, J. R., Knapp, D., Meretsky, V. J., Chancellor, C., & Fischer, B. C. (2011). Motivations influencing the adoption of conservation easements. *Conservation Biology, 25*(4), 827–834. https://doi.org/10.1111/j.1523-1739.2011.01686.x

Fletcher, R., & Neves, K. (2012). Contradictions in tourism: The promise and pitfalls of ecotourism as a manifold capitalist fix. *Environment and Society, 3*(1). https://doi.org/10.3167/ares.2012.030105

Fletcher, R., Murray, I., Blanco-Romero, A., & Blázquez-Salom, M. (2019). Tourism and degrowth: An emerging agenda for research and praxis. *Journal of Sustainable Tourism, 27*(12), 1–19.

Fletcher, R., Büscher, B., Massarella, K., & Koot, S. (2020). 'Close the tap!': COVID-19 and the need for convivial conservation. *Journal of Australian Political Economy, 85*, 200–211.

Fletcher, R., Blanco-Romero, A., Blázquez-Salom, M., Cañada, E., Murray, I., & Sekulova, F. (2021). Pathways to post-capitalist tourism. *Tourism Geographies, 25*, 1–22.

García-Álvarez, G. (2016). Instrumentos territoriales y protección de la biodiversidad: una perspectiva jurídica. *Revista Aragonesa De Administración Pública, 16*, 11–31.

González-Domingo, A., Fosse, J., & Costa-Salavedra, C. (2021). *Evitar la masificación del turismo en los parques nacionales de España: Diagnóstico y directrices*. Proyecto NATUR: Conservar y Evitar la Masificación. eco-union.

Herrero, Y. (2021). *Los cinco elementos: Una cartilla de alfabetización ecológica*. Arcadia.

Higgins-Desbiolles, F., & Bigby, B. C. (2021). A local turn in tourism studies. *Annals of Tourism Research*. Advance online publication. https://doi.org/10.1016/j.annals.2021.103291

Higgins-Desbiolles, F., Carnicelli, S., Krolikowski, C., Wijesinghe, G., & Boluk, K. (2019). Degrowing tourism: Rethinking tourism. *Journal of Sustainable Tourism, 24*(4), 1–19. https://doi.org/10.1080/09669582.2019.1601732

Igoe, J. (2010). The spectacle of nature in the global economy of appearances: Anthropological engagements with the spectacular mediations of transnational conservation. *Critique of Anthropology, 30*(4), 375–397. https://doi.org/10.1177/0308275X10372468

Kallis, G. (2018). *Degrowth (1st). The economy/key ideas*. Agenda Publishing.

Krüger, O. (2005). The role of ecotourism in conservation: Panacea or Pandora's box? *Biodiversity and Conservation, 14*(3), 579–600. https://doi.org/10.1007/s10531-004-3917-4

López-Palomeque, F., Font-Garolera, J., Sanchez-Aguilera, D., Torres-Delgado, A., Elorrieta-Sanz, B., Coma-Gutart, J., Font-Urgell, X., Simeonova, B., & Romero-Torres, J. (2016a). Turismo y crisis: el comportamiento del inmobiliario turístico en la montaña española durante la última década: Estudio de casos: la Val d'Aran y Sierra Nevada. In M. Blázquez-Salom, M. Mir-Gual, I. Murray, & G. X. Pons (Eds.), *Monografies de la Societat d'Història Natural de les Balears: Vol. 23, Turismo y crisis, turismo colaborativo y ecoturismo* (pp. 43–53). Societat d'Història Natural de les Balears.

López-Palomeque, F., Font-Urgell, X., & Romero-Torres, J. (2016b). El papel de la administración pública en el fomento y planificación del ecoturismo en la montaña española. In M. Blázquez-Salom, M. Mir-Gual, I. Murray, & G. X. Pons (Eds.), *Monografies de la Societat d'Història Natural de les Balears: Vol. 23, Turismo y crisis, turismo colaborativo y ecoturismo* (pp. 293–303). Societat d'Història Natural de les Balears.

MAPAMA, Ministerio de Agricultura y Pesca, ALimentación y Medio Ambiente, METAD, Ministerio de Energía, Turismo y Agenda Digital, & Red de Parques Nacionales (Eds.) (2016). *Declaración de Ecoturismo de Daimiel*.

McAfee, K. (1999). Selling nature to save it? Biodiversity and green developmentalism. *Environment and Planning D: Society and Space, 17*(2), 133–154. https://doi.org/10.1068/d170133

McGinlay, J., Gkoumas, V., Holtvoeth, J., Fuertes, R. F. A., Bazhenova, E., Benzoni, A., Botsch, K., Martel, C. C., Sánchez, C. C., Cervera, I., Chaminade, G., Doerstel, J., García, C. J. F., Jones, A., Lammertz, M., Lotman, K., Odar, M., Pastor, T., Ritchie, C., et al. (2020). The impact of COVID-19 on the management of European protected areas and policy implications. *Forests, 11*(11), 1214. https://doi.org/10.3390/f11111214

Moore, J. W. (2015). *Capitalism in the web of life: Ecology and the accumulation of capital*. Verso.

Morton, L. W., Regen, E., Engle, D. M., Miller, J. R., & Harr, R. N. (2010). Perceptions of landowners concerning conservation, grazing, fire, and eastern redcedar management in tallgrass prairie. *Rangeland Ecology & Management, 63*(6), 645–654. https://doi.org/10.2111/REM-D-09-00041.1

Múgica de la Guerra, M. (1993). *Modelos de demanda paisajística y uso recreativo de los espacios naturales* [PhD, Universidad Autónoma de Madrid, Madrid]. dialnet.unirioja.es

Müller, N., Blázquez-Salom, M., & Fletcher, R. (2021). Ecoturismo de proximidad: El giro convivencial. In E. Cañada & C. Izcara (Eds.), *Turismos de proximidad: Un plural en disputa* (pp. 127–138). Icaria Antrazyt.

Ojeda, D. (2019). La playa vacía, el bosque exuberante y el otro exótico: Herramientas para el análisis crítico del turismo de naturaleza. In I. Murray & E. Cañada (Eds.), *Icaria Antrazyt. Análisis contemporáneo: Vol. 493. Turistificación global: Perspectivas críticas en turismo* (1st ed., pp. 463–473). Icaria.

Patiño-Romarís, C. A., & Lois-González, R. C. (2016). El producto turístico de naturaleza en el litoral gallego como instrumento de sostenibilidad ambiental. In M. Blázquez-Salom, M. Mir-Gual, I. Murray, & G. X. Pons (Eds.), *Monografies de la Societat d'Història Natural de les Balears: Vol. 23, Turismo y crisis, turismo colaborativo y ecoturismo* (pp. 345–355). Societat d'Història Natural de les Balears.

Prada-Campaña, Ó., & Fundación Biodiversidad. (2017). *Informe del 5° Inventario de Iniciativas de Custodia del Territorio en España*. Plataforma de Custodia del Territorio de la Fundación Biodiversidad, Ministerio para la Transición Ecológica.

Puche, P. (2010). Decrecimiento y ocio: Decrecimiento y tiempo para la vida. In C. Taibo (Ed.), *Decrecimientos: Sobre lo que hay que cambiar en la vida coti-diana* (pp. 183–197). Los Libros de la Catarata.

Rissman, A. R., & Merenlender, A. M. (2008). The conservation contributions of conservation easements: Analysis of the San Francisco bay area protected lands spatial database. *Ecology and Society, 13*(1). https://doi.org/10.5751/ES-02329-130140

Romano, O. (2015). Dépense. In G. D'Alisa, F. Demaria, & G. Kallis (Eds.), *Degrowh: A vocabulary for a new era* (pp. 136–143). Routledge.

Romero-Torres, J. (2020). *El desarrollo del ecoturismo en los destinos de montaña en España (2000–2019): Análisis, evaluación y perspectiva*. Universitat de Barcelona.

Santamarina-Campos, B. (2019). El inicio de la protección de la naturaleza en España: Orígenes y balance de la conservación. *Revista Española De Investigaciones Sociológicas, 168*, 55–72.

Simancas-Cruz, M. (2016). Caso 4. La planificación del turismo en áreas protegidas. In M. Simancas-Cruz (Ed.), *La planificación y gestión territorial del turismo* (pp. 259–300). Sintesis.

Smith, N. (1996). *The new urban frontier: Gentrification and the revanchist city*. Routledge.

Smith, N. (2008). *Uneven development: Nature, capital, and the production of space* (3rd ed.). Georgia Press.

Soy Ecoturista, & Escapda Rural. (2021). *Tendencias Turísticas 2021: Ecoturismo y turismo rural.*

Stolton, S., Redford, K. H., & Dudley, N. (2014). *The futures of privately protected areas*. IUCN.

Tyrväinen, L., Mäntymaa, E., & Ovaskainen, V. (2014). Demand for enhanced forest amenities in private lands: The case of the Ruka-Kuusamo tourism area, Finland. *Forest Policy and Economics, 47*, 4–13. https://doi.org/10.1016/j.forpol.2013.05.007

Vaccaro, I., Beltran-Costa, O., & Paquet, P. (2013). Political ecology and conservation policies: Some theoretical genealogies. *Journal of Political Ecology, 20*(1), 255–272. https://doi.org/10.2458/v20i1.21748

Valdivielso, J., & Moranta, J. (2019). The social construction of the tourism degrowth discourse in the Balearic Islands. *Journal of Sustainable Tourism, 27*(12), 1–18. https://doi.org/10.1080/09669582.2019.1660670

Vázquez-Varela, C., & Martín-Gil, F. (2011). Problemas de sostenibilidad del turismo rural en España. *Anales De Geografía De La Universidad Complutense, 31*(1), 171–194. https://doi.org/10.5209/rev_AGUC.2011.v31.n1.8

Veblen, T. (1934). *The theory of the leisure class*. Modern Library.

Villagrasa-Rozas, M. D. M. (2016). Custodia del territorio y Política Agraria Común: Oportunidades de conservación y desarrollo económico y social en el medio rural. *Revista Aragonesa de Administración Pública, 16*, 115–150.

Nora Müller is PhD-student in Tourism Geographies at the University of the Balearic Islands (Spain) titled "Private protected areas and the touristification and commodification of nature through neoliberal conservationism". Previously, she obtained the B.Sc. in Environmental Sciences from the Leuphana University (Germany) and the M.A. in Human Geography from the Eberhard Karls University of Tübingen (Germany). From the perspective of critical geography, she is interested in the relations between human and nonhuman natures, especially in the realm of tourism.

Macià Blázquez-Salom is a professor in the Department of Geography at the University of the Balearic Islands. He teaches and researches on tourism, sustainability and land use planning. He has been visitor scholar in universities of Mexico (Toluca and La Paz), Nicaragua (UNAN), Dominican Republic (INTEC), Austria (Salzburg), Germany (Ruhr-Bochum), Sweden (Mid-Sweden) and the Netherlands (Wageningen). As a way to link activism and research, within the framework of Radical Geography and Political Ecology, he collaborates with social movements in Spain, particularly in the Balearic Islands (https://www.gobmallorca.com/), but also in Latin America (https://www.albasud.org/).

Chapter 5
Learnings for Spain from the Experience of the Good Practices of the European Project SmartRural

Antonio Martínez-Puche, Luis Alfonso Hortelano Mínguez, and Xavier Amat-Montesinos

5.1 Introduction

The European project SmartRural (2019–2021),[1] dedicated to smart rural tourism, aimed to develop a MOOC with applied content, and to gather good practices and case studies that would facilitate its understanding and dissemination. To this end, the improvement of training in the field of smart rural tourism was envisaged, as well as creating a common training program between 7 European universities. Hence the need to undertake, from training, the possibility of training and sensitizing local populations and rural entrepreneurs, of the possibility of applying innovations in rural areas. It is not easy, and we have verified it in the current project. The digital gap, but also the cultural one between the countryside and the city, is notable, and the approaches that prevail in this matter (from the partner Universities of the SmartRural project), are excessively technological and theoretical (IoT), and not so much territorial and applied (differentiated rural contexts and with the need for Participatory Strategies).

The concept of smart tourism involves technological integration in all the elements that make up the tourism system, with the aim of exploiting synergies between technology and users to enrich the tourism experience. This perspective aims to

[1] 2019–1- TRO1-KA203–073781. "Towards an intelligent development of rural tourism in Europe (K203-Erasmus +).

A. Martínez-Puche (✉) · X. Amat-Montesinos
Universidad de Alicante, Alicante, Spain
e-mail: antonio.martinez@ua.es; xavier.amat@ua.es

L. A. Hortelano Mínguez
Universidad de Salamanca, Salamanca, Spain
e-mail: sito@usal.es

A. Blanco-Romero, M. Blázquez-Salom (eds.), *Spanish Tourism Geographies*, Geographies of Tourism and Global Change,
https://doi.org/10.1007/978-3-031-39780-6_5

ensure maximum connectivity in destinations through technological applications and is a determining factor for the competitiveness of companies in the tourism sector. Moreover, the emerging notion of intelligence is often linked to concepts such as innovation and sustainability, which implies a real conceptual renewal in the field of tourism planning. In rural environments, some practices and experiences have shown that the application of new technologies is often associated with the creation of new tourism products, favours the dissemination of in-formation and ultimately contributes to the promotion of rural destinations. Thus, in the case of Spain, there are proposals for models which take advantage of the possibilities offered by new technologies in the design of products through the incorporation of GPS, QR codes, mobile applications and augmented reality. The book chapter presented here proposes a critical analysis of the concepts of intelligence, innovation and sustainability, as well as their practical application in the field of rural tourism. Some results of the project are also presented SmartRural (where the University of Alicante has participated as a partner along with six other European universities). It has been a project in which the need to include the local population, to make rural tourist destinations unique and where technologies must improve the living conditions of the inhabitants (more services), beyond improving the tourist experience, has been noted.

It is undoubtedly necessary to establish guidelines and incorporate improvements in tourist destinations, in aspects such as mobility, connectivity, access to facilities and services, or the offer of local resources, without detriment to the overall experience of tourists and the coverage of the needs of the local population. Furthermore, in the new tourism scenario resulting from the crisis caused by COVID-19, there is a need to accelerate the incorporation of technologies in the management of rural destinations in the short term and to explore new products in the field of sustainable rural tourism.

5.2 Theoretical Framework

The rise of rural tourism is a complex social construction (Figs. 5.1 and 5.2). A territory is a space where the economic, the social, the cultural, the environmental and the political are present. In these cases, a destination image can spread even further through various channels, such as residents, tourists, advertising brochures, tourist guides, novels, travel writing, literary works, the media, and image campaigns (Kim & Oh, 2009). Technology (IT) is also heavily involved in imaging tourist destinations (Choi et al., 2017). Smart tourism is recognized as a term that describes the convergence of IT with the tourism experience. It includes forms of information exchange in tourism, based on mobile technology and Internet connectivity. As a result, the tourism public culture is no longer totally dependent on information provided in printed or even official form and commercial websites. Visitors now exchange user-created content via social media in the form of storytelling, tips, and shared comments. Nowadays, concepts and terms such as "e-

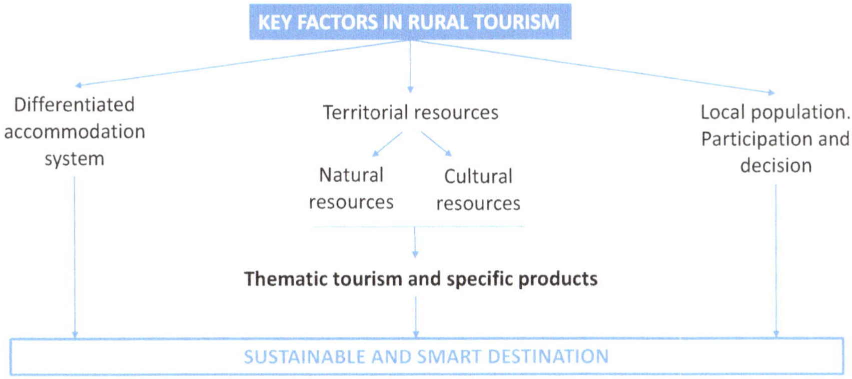

Fig. 5.1 New trends in rural tourism. (Own elaboration)

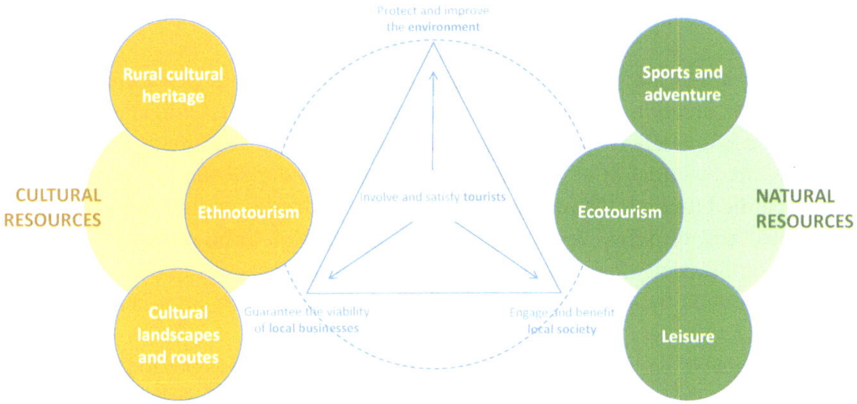

Fig. 5.2 SmartRural project elements. (Own elaboration)

tourism", "connected" tourist, "social" tourist, "prosumer", "ewom", "electronic commerce", "tourist applications", "geographical location" or all those related concepts with intelligence or "smart": "smart cities", "smart destinations", "wired cities", "smart destinations" have become benchmarks in the study, planning and management of destinations, as well as in the business tourism (Beltrán, 2018).

The concept of smart tourism implies the technological integration in all the organizations and entities that make up a tourist destination, with the aim of exploiting the synergies between technologies and users to enrich tourist experiences (Buhalis & Amaranggana, 2013). This perspective is based on guaranteeing maximum connectivity in destinations through new technological applications (Ballina et al., 2019) and is crucial for the competitiveness of companies in the tourism sector (Polo & Frías, 2010). In this sense, the academic literature agrees that smart tourism must inevitably address the needs of travelers before, during and after their trip, with the tourist experience being the central nucleus in the development of smart tourism destinations (Buhalis & Amaranggana, 2013; Ballina, 2019), which only in this way can be competitive (Luque et al., 2015; Ballina et al., 2019).

The conceptualization of a destination as smart has undergone an accelerated transformation in recent years, maturing from a limited approach to tourism promotion through web pages and on-line management of reservations (San Martín & Herrero, 2012), to another much more complex and diverse approach, which refers to the integral management of tourist destinations and incorporates multiple and disparate categories. Thus, for example, important elements are revealed such as political and governance conditions, available infrastructures, the way of analyzing the data generated by tourists or the way in which sustainability variables are incorporated, taking into account that the implementation of the smart paradigm in tourist destinations implies an inescapable commitment to their habitat and the quality of life of its inhabitants (Sigalat-Signes et al., 2020). Thus, the new approach to smart destinations requires a global project that ideally incorporates five areas of action: governance, sustainability, connectivity, information systems and innovation, its adaptation to each territorial environment being unavoidable (Ivars-Baidal et al., 2016).

In rural settings, experience has shown that the application of new technologies is related to the creation of new tourism products, favors the dissemination of information and ultimately contributes to the promotion of rural destinations. Thus, for example, the potential offered by new technologies in the creation of products has been identified through the incorporation of GPS, QR codes, mobile applications or augmented reality. However, in most cases the use of technology is limited to information functions for visitors, while the rest of the stages of tourism consumption still show an incipient development in the digital tourism environment (Coma et al., 2016).

In this context, development strategies through smart tools aspire for the moment to complement the existing tourism offer, by adding new products and services to the generic offer of rural tourism (Romão & Neuts, 2017). However, smart tourism specialization seems to show more and more relevant impacts in rural areas where tourism already plays a prominent role, incorporating improvements in aspects such as mobility, access to facilities and services, or the supply of local resources. In addition, new technologies could also be improving relations between rural tourism and other productive sectors, by exploring the synergies that can be established in the tourist appreciation of goods and services produced in the same region.

In short, the notion of intelligence is still diffuse and incorporates disparate characteristics, variables and dimensions. The conditions to affirm a territory or a productive sector as intelligent are not exempt from academic and institutional controversy (Desdemoustier et al., 2019) and, in the absence of an adequate conceptualization, many villages, tourist destinations or rural regions are pushed to define themselves as intelligent without a basis to justify such consideration. Consequently, the academic literature and applied research through projects such as SmartRural face the challenge of conceptualizing through rigorous analysis the notion of intelligence and its application to tourist destinations in rural areas, but starting from education and training. There is an excess of technological components in the conceptualisation of smart tourism in Spain (Flores et al., 2018). There-fore, we have three research questions: What are the attributes of the concept of smart rural tourism in relation to the experience of the SmartRural project? On the other hand, is the local population in rural environments taken into account in these renewed tourism typologies, and is the role of the different public and private actors (stakeholders) considered from a systemic viewpoint? Is it necessary to identify Good practices to learn lessons and establish basic criteria for a conceptualization of intelligent rural tourism base don experiencies and case studies (bottom up)?

5.3 Results and Proposals: Good Practices in the Framework of the SmartRural Project

Tourism in rural areas should contemplate those natural and cultural values (architecture, local festivals, landscapes, forests, etc.) that the territory has, as resources that can be used for tourism in order to establish a differentiation in the destination with the aim of facilitating market positioning (Fig. 5.3). In addition, smart rural tourism has to have the necessary technological coverage to ensure the real-time connection of tourists and their interaction with the rest of the components of the tourism system. Collaboration between public and private actors, citizens and tourists is very important; for example, tourism businesses must actively participate.

They must be competitive companies with a capacity for innovation and with the ability to assimilate technology, provide content to it and transfer knowledge, to take advantage of the generated information to make decisions and to continue to drive advances. Companies must incorporate the technology that allows them to interconnect and implement a "dashboard" (destination intelligence platform), which will receive inputs from various interconnected sources and transform the data into information and services.

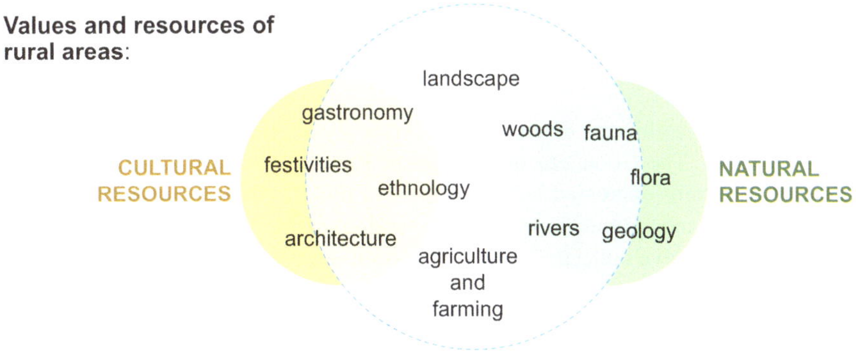

Fig. 5.3 Cultural and natural resources of rural areas. (Own elaboration)

Some of the reflections shared here, in relation to smart rural tourism, are based on an ERASMUS+ project. The project started in 2019 (October) and ended in 2021. It consisted of seven partners, led by Eskisher Osmangazi University, Faculty of Tourism (Turkey) and also by:

1. VDU Faculty of Bioeconomics Development (Lithuania)
2. University of Warmia and Mazury in Olsztyn (Poland)
3. University of Ljubljana (Slovenia)
4. Departament of Human Geography of Alicante University (Spain)
5. Unipegaso University (Italy)
6. University of Economics (Czech Republic)

The contributions of each of the partners have been important in the context of the pandemic (COVID-19). The work was organized online because the final Meeting was held in person in Naples (November 2021).

The development of rural communities depends on the enhancement of their resources and their adaptation to new market conditions and consumer habits. The emergence of the Internet and ICTs has brought about a change in tourist destinations, including in rural areas, which have become a digital tourism scenario (Jacobsen & Munar, 2012). For their part, tourists have also changed, as they are now hyper-connected and multi-channel people closely linked to smartphones and the use of information and activity management apps (Xiang et al., 2015; Buhalis & Foerste, 2014). Tourists make use of technology to obtain personalised services according to their needs and interests. The new generation of tourists highly values being able to digitally integrate with the destination before, during and after the visit (Fernández et al., 2017). This leads to the empowerment of tourists thanks to the democratisation of ICT, which allows them to make decisions throughout the tourist trip, thus being the protagonist of the creation of their own experiences (Celdrán-Bernabeu et al., 2018). Hence the importance, in the SmartRural project, of identifying, making visible and transferring good practices. The demonstration effect of the actions to be very important to continue advancing the concept of SmartRural.

5.3.1 Why a Manual of Good Practices? What For?

One of the tasks within the project of the University of Alicante was the Manual of Good Practices. The Good Practices in Smart Rural Tourism is a response to the needs and concerns expressed by the partners of this ERAMUS+ project regarding the promotion and articulation of smart rural tourism. Understood as a useful instrument to achieve collective benefits for rural communities and renewed entrepreneurs in European rural areas. All this in order to collect and capture those experiences or initiatives that respond to the needs of the territory such as:

- Promoting economic activity and diversification, through small enterprises.
- Promoting entrepreneurship and new business models through the association of tourism activity with ICTs.

- Making environmental conservation compatible with tourism development (sustainability).
- Creating an offer of accommodation and complementary activities, not concentrated and small-scale, but coordinated with the local environment.
- Favouring cooperation and co-governance networks between stakeholders in the territory.
- Developing more advanced tourism management based on knowledge and a new framework of relations between tourism agents.
- Facilitating connectivity and accessibility to rural destinations, both physically and digitally.
- Improving the proposals in the offer of activities based on creativity through Design Thinking.
- Differentiating the destination and positioning it in the market.
- Favouring the retention of socio-demographic staff and attracting new settlers.
- Improving the efficiency of destination management in all areas (marketing, environmental management, mobility, etc.).

The experiences collected in the manual that we have developed in the project are examples of good practices that stand out for their innovation, rein-venting new actions to promote smart rural tourism. Not only for applying new technologies (connectivity), but also for being more sustainable, cultural, environmental, creative and making destinations more accessible (digitalisation). It also covers how to undertake and design new tourism pro-jects (activities). Hence, this manual of good practices seeks transferability, i.e. the transferability to other geographical areas of the rural environment, by having a demonstrable impact (demonstrative example). They are experiences that have been evaluated and have a positive impact on the community and the territory. For this, good practices must be well planned, have an integrated approach, and be efficient in the use of local resources. Furthermore, they must be transformative and integrated actions, beyond the project proposed, and with a long-term perspective (durability and permanence over time):

- Phase 1. Definition of Smart Rural Tourism Best Practice. In order to establish a methodology for the selection of Best Practices, the first step to follow is to define the concept. A Smart Rural Tourism Best Practice is understood as an experience or initiative that responds in an innovative and satisfactory way to the objectives set out above.
- Phase 2. Drawing up a summary sheet. Brief sheet for the collection of in-formation by the partners participating in the project. The purpose of this sheet is to collect the name of the initiative, as well as to summarise what the activity consists of, its location, objectives, the attributes of being a go-od practice, the difficulties and problems detected and the contact details of the person/s or bodies directly related to it.
- Phase 3. Compilation and validation of experiences, which have been collected by theme, two for each of the partners (Table 5.1).

Table 5.1 Attributes of good practices in smart rural tourism, according to the European project SmartRural

Responsible university	Themes of the good practice	Description of attributes for good practice
University of Alicante, Spain	Experience, design and co-creation of smart rural tourism	Definition of tourism products and services Strategy adapted to the characteristics and singularities of the territory and its resources (cultural and natural) Institutional support and multilevel governance Networking within and outside rural areas, with universities, financial and cultural institutions, etc.
University of Ljubliana, Slovenia	Use of ICT and the internet of things for rural tourism (Digitalisation)	Digital tools are used in a bidirectional way and are an essential component of the tourism experience. In addition to consuming, tourists in-form themselves and create information about the destination through online resources (web 3.0 and social networks) The digital experience is functional, interactive and fully satisfying. Abundance of information and interaction in real time (incorporation of "advanced" tools: WIFI networks, QR, beacons, augmented reality...)
University of Economy, Czech Republic	Sustainability	Local communities create a coordinating body. Tourism stakeholders cooperate with the aim of offering tourists an integrated (tourism) pro-duct, to join the regional label and quality (certification) system The tourism product is marketed and promoted through a brand recognised in the market for its certified quality guarantee. Place branding
Eskisher Osmangazi University, Turkey	Accessibility	The agents collaborate in a network. Initially, some tourist resources and services have been adapted in order to improve territorial and universal accessibility. Adapted activities and tourist information for people with other disabilities have already been implemented In rural areas there is a tourist offer fully adapted to universal accessibility, and jobs have been created for people with other disabilities Accessibility to the rural environment stands out for its complete connectivity with other territories that can be the origin of visitors and tourists

(continued)

Table 5.1 (continued)

Responsible university	Themes of the good practice	Description of attributes for good practice
University of Warmia and Mazury in Olsztyn, Poland	Creativity and smart ideas for rural tourism destinations	There are business associations for creative tourism. They have the distinctive seal, or some kind of thematic tourism certificate. Recognition has been obtained through an official body They are part of a formally established strategy with partners and actors in the territory (national, regional, local) The network offers creative activities, experiential products and a clear programme of activities (festivals and events) in the territory. There is an international network with other smart destinations
Unipegaso, Italy	Strategies and business models	Consolidation and expansion of a business model that achieves an attractive and shared territorial image, go-od living conditions, with resources and job creation

Own elaboration

5.3.2 *Methodological Design of the Guide*

In each of the good practices that have been considered, and which are structured by thematic modules: Introduction, co-creation, digitalisation, sustainability, accessibility, creativity and entrepreneurship. The key issues to identify are:

– General data about the project: name, location, etc.

For information purposes, the general data of the project considered as a good practice are given below

– Introduction and context.

In the introduction you will find some brief basic introductory information about the project

– Why is it a good practice?

This section specifies the reasons why the project has been considered a go-od practice. In short, it explains why it has been classified as such.

– Participants in the project.

In the following section, the participants in the project must be listed. This is essential to consider the economic and social agents that have participated directly or indirectly in the project developed, in order to assess the degree of impact of the project from a socio-economic point of view.

– Lessons learnt.

This section of the fiche indicates the learnt lessons after the development of the project. Here you can find the key issues of each project and that can surely be useful to be considered in other projects.

– Developed activities.

The developed activities section specifies the actions that have been carried out in each of the partners' projects. The list of these activities can help other territories to identify some of them that can also be considered for the development of similar proposals and that work (benchmarking).

– Results

In this section of the sheet, we have considered the effects and achievements of the development projects that have been considered as a good practice. You will be able to see the results obtained after the implementation of the project. A brief explanation of why the good practice has had a social and territorial impact has been included here. In addition, the impact has been measured and each partner has rated from 1 to 5, according to the achievements obtained.

However, the concept of good practice in smart rural tourism has evolved in recent years, from a limited approach to the partial use of new technologies, towards a much more complex and diverse approach. According to our project experience, smart rural tourism incorporates elements such as: political and governance conditions, infrastructures, analysis of data generated by visitors, the way in which sustainability variables are included, in addition to the unavoidable commitment to quality of life of the local inhabitants.

Within this Erasmus+ project, six areas have been worked on: (1) design and co-creation of the tourism experience, (2) digitization, (3) sustainability, (4) accessibility, (5) creativity and (6) entrepreneurship.

Firstly, the use of technology can facilitate community participation in the design and co-creation processes of new economic activities in rural areas, such as tourism. The smart rural tourism experience is based on the links between public and private entities of a territory, capable of generating networks and rich and diversified governance structures. Collaboration, communication and active participation of the different entities (city councils, companies, associations, cooperatives, etc.) in the process of design and co-creation of the smart rural strategy will determine its economic, social and territorial impact.

Secondly, the use of ICT and internet of things for rural tourism. Digitalisation. In rural areas, digitization facilitates the creation of tourist products, helps to disseminate information and favors the promotion of destinations. A good practice in smart rural tourism must be properly integrated in a digital environment. On the one hand, new technologies provide dynamic and personalized information for visitors, who are capable of being involved in the tourist experience. On the other hand, the ability to collect and analyze data generated by tourists, allows destinations to learn about behavior patterns and adapt the offer in real time to the needs of visitors. The abundance of information and the capacity for interaction make the digital tourism experience more functional, interactive and satisfying.

Third place, sustainable rural tourism. The definition of sustainability for a rural tourist destination implies the adequate use of natural resources (natural heritage and biodiversity), respect for the social and cultural identity of the host communities and the consolidation of a viable and lasting economic model (stable employment, poverty eradication, etc.). The smart tourism strategy is based on the reasonable and weighted use of a territory's own re-sources, with decisive support for goods and services produced on a local or regional scale according to the parameters of sustainability. Digital tools can become the support of tourism marketing, helping to spread the values of sustainability and raise awareness among local inhabitants and visitors.

Fourth, accessible rural tourism. The smart strategy must have relevant impacts in rural areas. It implies incorporating improvements in aspects such as mobility, access to facilities and services, or the inclusion of people with specific needs. Accessibility is a multidimensional concept that has an impact on people's daily lives in many areas (building, transport, social activities, communication, etc.) and on the tourist experience. Accessibility refers, therefore, to the design of products and environments accessible to as many people as possible. Technology can improve territorial and universal access to tourism resources and services, so good practice requires the existence of plans to ensure accessibility (physical and virtual) to rural destinations.

Fifth, creativity and smart ideas for rural tourism destinations. Smart ideas in the field of tourism allow the creation of innovative products and services based on an understanding of the problems and needs of visitors. Therefore, a good practice must be adapted to the needs of the demand, be technologically feasible and be economically justified. The entities that make up the tourist destination carry out a continuous search for innovation and develop original solutions based on demand trends. In this way, the offer consists of creative activities and experiential products based on its own territorial resources, supported and promoted from the institutional and business spheres.

Sixth place, strategies and business models. Rural spaces are no longer exclusively associated with agricultural production but are seen as multifunctional places for the stimulation of new socio-economic activities that often incorporate tourism. In this context, development strategies through smart tools allow to expand and complement the tourist offer, by adding new products and services based on technology. But in addition, digitization also improves relations between rural tourism and other productive sectors, by generating synergies in the tourist valorization of goods and services produced within the same territory (online marketing, marketing platforms, etc.).

Below are the two Spanish good practices, with their objectives, which form part of the rest of the 14 good practices in smart rural tourism (Table 5.2).

Table 5.2 Summary of good practices of the SmartRural project

Good Practice	Main Objectives
Camperola Tours-Rural Community Tourism, Cooperativa Agroecològica La Camperola (Region of Alicante, Spain)	Enable the viability and sustainability of agro-ecological agriculture in the territory Revitalization of a diversified activity based on new guidelines for the production, distribution and consumption of agroecological, seasonal, healthy and quality food Build and articulate short marketing circuits, and promote a cooperative culture of work and decent life in rural areas Promote rural community tourism Valuation of local products and local commerce
Cluster Alicante Wine Route (Region of Alicante, Spain)	It promotes local development processes through the eco-nomic diversification of the territories It works in networks (public-private management) It improves the tourism experience in the territory. Furthermore, the wine routes revalue the natural and cultural heritage and can help to provide complementary economic in-come for the wineries and other companies in the area It improves the place branding. Deseasonalization of the tourist demand that is not exclusively focused on sun and be-ach

Main objectives. Own elaboration. Source: Martínez-Puche, A. et ali. IO3 Report and Good Practices country Intellectual Output n°3, carried out by the University of Alicante. http://smartrural.uwm.edu.pl/

Some of the solutions to implement smart rural tourism initiatives include the definition of a Strategic Plan, together with the development of protocols to help rural tourism entrepreneurs to develop smart projects. This should be complemented with financial support from public administrations to support tourism entrepreneurs. More training and information (advice) is needed to address tourism needs and activities in relation to smart tourism in rural areas (good practices and case studies with demonstrative effects). In addition, a healthy and active lifestyle should be promoted, encouraging outdoor and leisure activities in nature, discovering the different local re-sources. Encourage the creation of micro-enterprises related to ecotourism. Disseminate actions through social networks, creating thematic events with public-private initiatives.

The use of technological tools is undoubtedly important for making rural and inland tourism destinations accessible. The Valencian Community has been working on this for some years, establishing models and technological tools to measure, quantify, make accessible, safer and more sustainable, potential sustainable tourist destinations (Table 5.3).

Table 5.3 Model on the attributes of a smart inland tourist destination according to Invat.tur (Generalitat Valenciana)

Aspects	Accessibility, security and sustainability
Audio guides	From the point of view of accessibility, adapted information is provided. Information on routes and explanations of traditional sites
Electric Mobility	From sustainability, set up recharging points Rental and supply of bicycles App to control routes
Weareables	Within security, localisation of persons Real-time monitoring Allergy, eHealth and humidity monitoring
Ibeacons (Blinds)	Within accessibility. Conventional audio guidance Step counter Direction indicator and Guidance system
APPs Cyclists	Accessibility, emergency warnings Guided routes Gamification of routes
Road sensors	Accessibility and cognitive information Personalised route data Geopositioning
Augmented reality	Accessibility and interpretation of the landscape Vision of the historical heritage Historical recreation
Ultraviolet sensors	Sustainability and measuring sun exposure time Risk of skin exposure Real time
APPs	Accessibility, Emergency warnings Presence of insects Guided tours Gastronomic routes
Panels	Accessibility, Information Location of points of interest Microbiological test results Water, showers
Air Quality Sensors	Sustainability, presence of ozone pollutant gases Potentially settled dust or particles Pollen index Environmental comfort index Temperature
Drones Surveillance	Security Occupation of protected areas Early detection of fires Monitoring and surveillance of vegetation cover
Wifi	Accessibility, Connectivity Downloading of information PUSH information system Alert system

Source: Own elaboration. Adapted from Giner et al. (2021)

5.4 Discussion of Results: Brief Reference to the Spanish Case

Traditionally, innovations have been generated in places where resources and knowledge were concentrated in physical proximity. Large cities or business clusters ensured sufficient physical density to generate economic growth and territorial development processes. However, the digital trans-formation has prioritised virtual density over physical density (Gallardo, 2016), which implies greater independence with respect to the location of the various groups of people or companies, providing an extraordinary opportunity for rural areas. It is in this context that the notion of territorial intelligence associated with research, innovation and knowledge has emerged as a means to achieve economic growth supported by sustainable development (Naldi et al., 2015), with increasingly noticeable implications for rural regions.

Smart rural development is seen as part of the strategies introduced in response to low growth and productivity rates in rural areas (Naldi et al., 2015), but also in response to the dichotomy caused by the digital divide between urban and rural (Gallardo, 2016) and even as a stimulus to meet the aspirations of rural youth in marginalised territories (Fennell et al., 2018). But, in addition to all this, the smart paradigm incorporates technology as a component for the transformation of economic activities in rural areas, and some productive sectors such as tourism are proving to be particularly prone to the rapid expansion of the smart phenomenon. Smart tourism or smart destinations are thus a further step in the evolution of smart cities (Buhalis & Amaranggana, 2013). Can the same be true for smart tourism activities in rural areas? Answering this question requires specific and concrete references to smart rural tourism products beyond the description, the shop window and the catalogue. In this sense, and after the consulted references, we can state that there is a lack of clear systematisation in Spain, based on common criteria, to determine the rural area of action, the characteristics of the attributes established in rural smart tourism destinations and the scale of their management beyond SEGITTUR and other regional, local and supra-local bodies.

The creation of the Smart Tourism Destinations Network (DTI Network) on tenth October 2018, by the *Sociedad Mercantil Estatal para la Gestión de la Innovación y las Tecnologías Turísticas*, S.A. (SEGITTUR), ai-med to create a working network in order to promote collaboration and joint action among the 68 destinations that initially signed up. Initially, the network sought to generate synergies, coordinate actions, establish a framework for promotion, exchange best practices and achieve Spain's leadership in the field of tourism intelligence. Over the years, following the launch and development of various pilot projects for the transformation of Smart Tourism Destinations, the network has expanded its membership and expectations. The number of tourism areas participating in the network, both inside and outside Spain, has exceeded 325 destinations (the total number of network members is close to 500 members with the sum of organisations, associations, foundations, companies and observers). Currently, the objectives are as follows: to promote the conversion of Spanish tourist destinations into smart destinations; to promote public-public and public-private collaboration in the development of products, services and actions in the DTIs (Smart Tourism Destinations); to contribute to Spain's leadership in the field of tourism intelligence and to ensure the quality and evolution of the DTI project. The achievement of these goals is supported, from an internal perspective, through advice on the application of the methodology and the exchange of experiences and best practices among its members and, externally, through the visibility of the DTI network.

In this respect we will mention to several tourist destinations in Castilla y León that are members of the network aspire to transform their traditional tourism model, based on a diagnosis and the definition of an action plan, through a commitment to innovation, technology, sustainability and universal accessibility. The participation of the regional tourism areas is aligned with Axis 3 on Innovation and Tourism Intelligence of the Strategic Tourism Plan of Castilla y León 2019–2023. This line of action makes it compulsory to apply innovative measures and Smart criteria to territories, products and entrepreneurs in order to ensure the competitiveness of tourist destinations in the future. The objectives to be achieved with the actions are to position Castilla y León as an innovative destination; to have a methodology and working ecosystems that favour the development of innovation, and to promote the Centre for Tourism Analysis towards smart destinations and Smart criteria applied to the product. In this sense, the number of destinations in Castilla y León that are part of the DTI model is increasing every year; specifically, in the first half of 2022, there are 29 destinations that are part of the network (19 of a rural nature). From Castilla y León, the transformation of tourist destinations with the incorporation of Smart criteria is also seen as an opportunity to settle and attract population and help meet the demographic challenges. The Castilla y León tourism sec-tor has a wealth of endogenous resources and infrastructures that cannot be relocated but need innovative management.

The COVID-19 pandemic has generated new opportunities in terms of the offer of tourist activities and services with added value due to their location and the possibility of taking place in the open air. The hope for the provinces and regions ravaged by demographic emptying is that smart tourism destinations will become a "tractor" measure for the local economy in order to recover population numbers.

Already in July 2017, the regional administration of Castilla y León under-took the implementation of the Smart Rural Territory of Castilla y León (Smart Region) project, co-financed by the European Regional Development Fund (ERDF), following the implementation of a platform that develops intelligent software capable of managing public services 4.0 through sensors in an intelligent, efficient, comprehensive and coordinated manner (Smart Rural platform).

The project is part of the Castilla y León Digital Municipalities Network initiative, with the adhesion of the nine Provincial Councils and the main town councils of the Autonomous Community, with the general objective of promoting online public services in the local environment for its citizens, companies and organisations using new information and communication technologies. Telefónica España's wireless communication technology sup-port, with LPWA (Low Power Wide Area) cellular technologies of the Nar-row Band IoT (Internet of Things) network, allows all kinds of devices to be connected with minimum battery consumption and rapid implementation in the operators' network. A practical application in the tourism sector, wi-thin the framework of the Smart Rural platform of Castilla y León, has been the study of the profile, origin, behaviour and economic level of tourists on the Camino de Santiago, in the Arribes del Duero Natural Park (Zamora and Salamanca), in the Ribera del Duero (Soria, Burgos and Valladolid), in the Valle de Iruelas Nature Reserve (Ávila) and Ruta de la Plata (Salamanca), with the application of Telefónica's AURA Big Data unit. The generation of this data allows tourism branches in rural areas to promote promotional campaigns for specific segments and adapt services to visitor demand. Undoubtedly, in order to see the final results in relation to the expectations generated by the concept of Intelligent Rural Tourism, we will have to wait to see the acceptance of tourism demand and the synergies produced in the territories where these initiatives are being developed, as we have pointed out in Castilla y León.

5.5 Conclusions

The concept of smart rural tourism implies the technological integration of all the entities that make up a tourist destination. The objective is to exploit the synergies between the territory and the visitors to enrich the tourist experience. This perspective is required to achieve maximum connectivity in destinations through new technological applications. In addition, it is decisive to achieve the competitiveness of the tourism sector (Table 5.4).

The maturity reached by tourism in Spain explains why the desire to con-vert tourist destinations into smart destinations has been important. In addition, we have been among the first to develop legal instruments, as well as a White Paper on smart destinations, which makes us an international benchmark and gives us added value. Institutions such as SEGITTUR are driving the theoretical development of the

Table 5.4 Traditional rural tourism *versus* Smartrural

Traditional destination	Smart destination
It is all about commercial relations	What matters are the interactions between all components of the tourism system (supply and demand)
Creation of standardised products and unidirectional promotion	Co-creation (based on conversation) and product customisation
Stable destination configuration. Defined products and predictable tourist flows	Dynamic configuration of the destination. Continuous adaptation and self-production of products and flows
Relatively simple behaviour of tourist flows with a tendency to overcrowding	More complex behaviour and very limited predictability
Limited (and delayed) access to information	Abundance of information (in real time)

Source: Own elaboration from (Espeso-Molinero, 2019)

concept and promoting its practical application, emphasising issues of sustainability, connectivity and tourist mobility. A smart destination must meet a series of requirements such as technology, com-munity participation, sustainability, measuring and personalising the tastes of the demand, improving the living conditions of tourists and inhabitants, as well as the projection of a place branding.

However, the notion of intelligence applied to rural tourism remains diffuse and incorporates disparate characteristics, variables, territorial contexts and dimensions. The conditions for affirming a territory or a productive sector as smart are not exempt from academic and institutional controversy (Desdemoustier et al., 2019) and, in the absence of an adequate conceptualisation, many villages, tourist destinations or rural regions are pushed to de-fine themselves as smart without a basis that justifies such consideration. Consequently, academic literature and applied research through projects such as SmartRural face the challenge of conceptualising through rigorous analysis the notion of intelligence and its application to tourism destinations in rural areas, but starting with training and capacity building.

The elaboration of a manual of good practices, derived from the Erasmus + SmatRural project, can help to catalogue experiences, which, if properly measured and justified, can be a good reference of the characteristics and attributes of smart rural tourism in some parts of Europe and can serve as references for the development of future projects. Therefore, the experiences together with training (MOOC), reinforced by research, the participation of local stakeholders and various bodies and administrations (multilevel governance), can favour the development of smart tourism projects in rural areas, adjusted to the needs of the local population, and not only of tourists.

References

Ballina, F. J. (2019). Smart tourism destination, experiencia phygital y turismo rural. *International Journal of Information Systems and Tourism, 4*(1), 41–52.

Ballina, F. J., Valdés, L., & Del Valle, E. (2019). Tourism Destination: Urban versus Rural Technological Behaviours. *Revista Internacional de Turismo, Empresa y Territorio, 3*(2), 16–37.

Beltrán, G. (2018). *Turismo rural online. Páginas web y redes sociales*. UOC.

Buhalis, D., & Amaranggana, A. (2013). Smart tourism destinations. In *Information and communication technologies in tourism* (pp. 553–564). Springer International Publishing.

Buhalis, D., & Foerste, M. (2014). SoCoMo marketing for travel and tourism: Em powering co-creation of value. *Journal of Destination Marketing and Management, 4*(3), 151–161.

Celdrán-Bernabeu, M. A., Mazón, J. N., Ivars-Baidal, J. A., & Vera-Rebollo, J. F. (2018). Smart Tourism. Un estudio de mapeo sistemático. *Cuadernos de Turismo, 41*, 107–138.

Choi, S., Lehto, X. Y., & Morrison, A. M. (2017). Destination image representation on the web: Content analysis of Macau travel-related websites. *Tourism Management, 28*, 118–129.

Coma, J., Elorrieta, B., & Torres, A. (2016). La incidencia de las TIC en destinos turísticos de la montaña española. Un análisis de casos. *ARA, 6*(2), 75–86.

Desdemoustier, J., Crutzen, N., & Giffinger, R. (2019). Municipalities understand ing of the Smart City concept: An exploratory analysis in Belgium. *Technological Forecasting and Social Change, 142*, 129–141.

Espeso-Molinero, P. (2019). Tendencias del turismo cultural. *PASOS Revista de Turismo y Patrimonio Cultural, 17*(6), 1101–1112.
Fennell, S., Prabhjot, K., Jhunjhunwala, A., et al. (2018). Examining linkages be tween Smart Villages and Smart Cities: Learning from rural youth accessing the internet in India. *Telecommunications Policy, 42*, 810–823.
Fernández, A., López, J. M., Moreno, L., et al. (2017). Innovación y destinos inteligentes: oportunidad para el Know How turístico español. *ICE, Revista de Economía, 894*, 137–150.
Flores, D., Perogil, J., & Miedes, B. (2018). Smart Destinations or intelligent territories? *Study of cases in Spain, Revista de Estudios Regionales, 113*, 193–219.
Gallardo, R. (2016). *Responsive countryside: The digital age and rural communities*. University Extension Center for Technology Outreach.
Giner, D., Fuster M., & Pastor, A. (2021). Hacia la conceptualización de los espacios naturales inteligentes en el actual escenario truistico. In *Una aproximación al caso DTI de la Comunidad Valenciana, Estudios Turísticos* (n° 222 (2° S2021), pp. 113–126).
Ivars-Baidal, J. A., Solsona, F. J., & Giner, D. (2016). Gestión turística y tecnologías de la información y la comunicación (TIC): El nuevo enfoque de los destinos inteligentes. *Documents d'Anàlisi Geogràfica, 62*(2), 327–346.
Jacobsen, J. K. S., & Munar, A. M. (2012). Tourist information search and destination choice in a digital age. *Tourism Management Perspectives, 1*(1), 39–47.
Kim, H. K., & Oh, S. J. (2009). A Q methodological approach to tourism motivation and preference typology. *Journal of Tourism and Leisure Research, 21*(2), 69–88.
Luque, A. M., Zayas, B., & Caro, J. L. (2015). Los Destinos Turísticos Inteligentes en el marco de la Inteligencia Territorial: conflictos y oportunidades. *Investigaciones Turísticas, 10*, 1–25.
Naldi, L., Nilsson, P., Westlund, H., et al. (2015). What is smart rural development? *Journal of Rural Studies, 40*, 90–101.
Polo, A. I., & Frías, D. M. (2010). The relationship between business characteristics and ICT deployment in the rural tourism sector. The case of Spain. *The International Journal of Tourism Research, 12*(1), 34–48.
Romão, J., & Neuts, B. (2017). Territorial capital, smart tourism specialization and sustainable regional development: Experiences from Europe. *Habitat International, 68*, 64–74.
San Martín, H., & Herrero, A. (2012). Influence of the user's psychological factors on the online purchase intention in rural tourism: Integrating innovativeness to the UTAUT framework. *Tourism Management, 33*, 341–350.
Sigalat-Signes, E., Calvo-Palomares, R., Roig-Merino, B., et al. (2020). Transition towards a tourist innovation model: The smart tourism destination, reality or territorial marketing? *Journal of Innovation & Knowledge, 5*, 96–104.
Xiang, Z., Magnini, V. P., & Fesenmaier, D. R. (2015). Information technology and consumer behavior in travel and tourism: Insights from travel planning using the internet. *Journal of Retailing and Consumer Services, 22*, 244–249.

Antonio Martínez-Puche Graduate in Geography and History (specialising in Geography). Lecturer in the Department of Human Geography at the University of Alicante (UA). PhD in Geography. Department of Human Geography. Academic Co-Director of the Official Master's Degree in Local Development and Territorial Innovation (DELEITE-UA). Director of the AVANT-GVA Inter-University Chair (Anti-depopulation) at the University of Alicante. President of the Local Development Working Group of the Association of Spanish Geographers (GTDL-AGE). Researcher at the Inter-University Institute of Geography and the Institute of Social Studies of Latin America. Vice-president of the Re-search Council of REIMA (Ibero-American Environmental Network). Director of the Human Geography Research Group-UA. He has been a research collaborator at the MMSH – Mediterranean House of Human Sciences (Aix-en-Provence), at the Institute of Cultural Heritage of the University of Laval (Quebec), the Department of Political Science of the Aldo Moro University of Bari, and University of Nariño (Udenar-Colombia).

Luis Alfonso Hortelano Mínguez PhD from the University of Salamanca. Lecturer in the Department of Geography. Director of the Official Master's Degree in Cultural Heritage Evaluation and Management at the University of Salamanca (USAL). Representative of the USAL, in the anti-population chair which also includes the univer-sities of León, Burgos, Valladolid. Co-author of the strategic report on depo-pulation in Castilla y León. Member of the Territory, Innovation and Development Research Group of the USAL. Member of the Permanent Commissi-on of the Local Development Working Group (GTDL-AGE). Researcher of the research groups, Academic Culture, Heritage and Social Memory, and of the Territory, Innovation and Development group.

Xavier Amat-Montesinos is an assistant professor at the Department of Human Geography, University of Alicante, Spain, and a geographer with a post-graduate in local development. He has contributed to various publications on territorial planning, local development, sustainability and tourism. His current areas of research interest include environmental problems and territorial development. He has participated as a researcher in the project "SMARTRURAL, Towards an intelligent development of rural tourism in Europe". He has been a collaborator researcher at the School of Geography (University of Leeds, UK) and the Postgraduate Institute (Technical University of Manabí, Ecuador).

Chapter 6
The Fragility of Successful Tourism Destinations Linked to Cultural Itineraries: The Example of the Way of St. James

Lucrezia Lopez, Rubén Camilo Lois-González, and Rossella Moscarelli

6.1 Introduction

Cultural itineraries have unique features within tourist destinations, since they are not destinations created in an attractive place or stretch of coast (Gusmán Correia et al., 2017; Lois González & Lopez, 2012). The destination is a linear axis of several tens or hundreds of kilometers, which travelers visit calmly, contemplating the landscape, enjoying the monuments or rural spaces which have not been transformed in recent times (Moscarelli et al., 2020; Sánchez Carretero, 2015). Cultural itineraries respond to a political practice that has been promoted since the late 1980s, first by the Council of Europe and then by government bodies. These itineraries emerged after the declaration of The Way of St. James as the first European Cultural Itinerary, and one which has always been set as an example of promotion and management (Lois González, 2013). Also, because of its importance within the Spanish territory and its linear arrangement, it is considered a route that should be travelled. Consequently, overtourism is a relatively limited phenomenon in these itineraries (a destination so widespread between localities and regions is less conducive to crowds) but does still exist. In fact, the city of Santiago de Compostela, like other equally promoted towns or sanctuaries on the route, suffers from massification, the trivialisation of an in principle unique cultural experience, and the impact of an excess of visitors (Lopez et al., 2019). In any case, overtourism is a seasonal

L. Lopez (✉) · R. C. Lois-González
Department of Geography, University of Santiago de Compostela, Santiago de Compostela, Spain
e-mail: lucrezia.lopez@usc.es; rubencamilo.lois@usc.es

R. Moscarelli
Department of Architecture and Urban Studies, Politecnico di Milano, Milan, Italy
e-mail: rossella.moscarelli@polimi.it

A. Blanco-Romero, M. Blázquez-Salom (eds.), *Spanish Tourism Geographies*, Geographies of Tourism and Global Change,
https://doi.org/10.1007/978-3-031-39780-6_6

phenomenon, particularly seen in the final localities of the itineraries or in very specific sections of the routes.

The UNWTO (2018) defines overtourism as the impact a tourist has on a destination, which influences the quality of life perceived by its citizens and, at the same time, negatively affects the quality of tourist experiences. This concern about the sustainability of a tourist destination also reinforces the concept of degrowth, which has evolved due to its use in multi-disciplinary academic theoretical and empirical research and is meant as regulation and control of the rate of growth within a destination (Canavan, 2013; Weiss & Cattaneo, 2017). Based on these premises, the main objective of this work is to contribute to the debate regarding the fragility of tourist destinations and products, especially after COVID-19. In general, the pandemic has been a threat to the balance of tourism activity (undertourism) and cultural itineraries have suffered the impact of these new dynamics. Thus, this proposal focuses on an international tourism reality, that is the cultural itineraries and routes of pilgrimage, but especially on The Way. From a methodological point of view, we adopt a qualitative-interpretative approach, presenting a review of previous studies on cultural itineraries, routes of pilgrimage and especially The Way. We adopt a qualitative comparative netnographic approach to compare the perception associated with The Way in 2020 and today, in 2022, as collected in the forums concerning The Way. Such research methodology is a pilot case study that allows us to investigate extremely recent and still ongoing phenomena. We advance some reflections on the future and trends associated with The Way, that can be applied to other pilgrimage and cultural routes.

6.2 The Success of the Way of St. James Pre-COVID-19, as a Model of the Cultural Itineraries in the World: The Risk of Overtourism

The Way of St. James is the First European Cultural Itinerary established by a European Council Declaration, and one of the most important medieval pilgrimage routes in the world. The expression The Way is used to refer to a network of routes that large numbers of pilgrims walk or ride to get to the city of Santiago de Compostela, the final destination. The urban historical centre of Santiago and some of The Ways have been appointed UNESCO World Heritage Sites (WHS); firstly, in 1993 the French Way was declared a WHS, then in 2015 the Northern *Ways* received the same international accolade. Since 1993, The Way has been strongly promoted by the regional government, thus in recent years there has been continuous growth in the number of pilgrims arriving to Santiago: from 99,436 in 1993 to 347,578 in 2019 (Pilgrims' Reception Office).

The current Way is a reinterpretation of a medieval religious route that linked the Kingdom of the Franks, and all of Europe, with the sanctuary of Santiago de Compostela (Santos & Lois González, 2011). For this reason, the French Way is the first and most important of all, as established in Book V of the *Codex Calixtinus* that the Pope ordered to be written, to support a Christian route conceived to reach the relics of the Apostle (Lois González, 2013). Consequently, the memory of the French Way has been recovered throughout the twentieth century, and alongside it more routes have appeared such as maritime-fluvial routes, the North Way (along the coast of the Cantabrian Sea), the Portuguese Way, the Silver Route and others. The Ways have been delimited in recent decades, restored and embellished, as well as the towns and monuments of the places they cross (Lois González, 2000). Thus, new walking routes are created, but always with a renowned historical-religious reference as a background.

On numerous occasions, we have insisted on the character of The Way as a model or example to imitate (Gusmán Correia et al., 2017). Indeed, other Spanish cultural itineraries have followed a series of guidelines taken directly from this great route of sacred origin. On the one hand, all routes seek an important historical, historical-cultural or religious-cultural justification. The existence of relics, buildings that keep first editions of a codified tradition, documented evidence of historical pilgrimages and rites or celebrations at the place of destination, are common examples (Lois González et al., 2016). On the other hand, it is necessary to follow a path similar to what it was in the past (Lopez, 2014). Walking, by bicycle or by any other slow means, is an indisputable element of the itinerary (Santos & Lois González, 2011). Finally, the originality and apparent non-massification of the product is combined with festive days where religious or secular mass celebrations take place and which can be accompanied by situations of overtourism which, as long as they do not exceed a few very specific days, are undervalued by the popular perception (where the cultural, historical and ritual values stand out) (Lois González & Santos Solla, 2015).

As stated in the questionnaire prior to receiving the *Compostela*, the certificate that proves that the traditional pilgrimage has been completed, pilgrims can choose between eight route options (French Way, Portuguese Way, English Way, Silver Way, Northern Way, Primitive Way and Portuguese Way of the Coast). Until the beginning of the pandemic, the French Way was the busiest route and problems related to its carrying capacity and the saturation of public shelters began to manifest (Fraternidad Internacional del Camino de Santiago, 2014; Martín Duque, 2017; Porcal et al., 2012). The second most travelled route was one of the three variants of the Portuguese Way, an option introduced in 2016. We could thus assume that one consequence of the massification of the main route is the progressive diversification of the itineraries. The *Plan Director y Estratégico del Camino de Santiago en Galicia 2015–2021* (Xunta de Galicia, 2015) includes the promotion of all the pilgrimage routes to Santiago de Compostela as a line of action through their revitalisation and the dissemination of their values, thus showing the uniqueness of the less popular ones (Martínez Roget et al., 2018). This decision by a regional government, which to date had only been interested in increasing the number of pilgrims, seems to show a serious concern about *overtourism*, which has already appeared in reports by the "Observatorio Turístico de Santiago" (CETUR and SA Xacobeo, 2007–2010) and in previous work that we have carried out on the issue (Lopez & Lois González, 2021).

The success of The Way may be due to its inclusive nature (Moscarelli et al., 2020). This has resulted in a remarkable international visibility, which has transformed it into a multi-faith and multi-religious route, along which pilgrims and tourists with different beliefs and motivations share a single space (Lois González, 2013; Santos & Lois González, 2011). In fact, pilgrims' motivations seem to diversify according to the historical and cultural moment, following new religious, spiritual and secular trends (Farias et al., 2019; Possamai, 2003). The meaning of faith in The Way has become diversified in recent times; although spirituality remains one of the main motivations to complete The Way and visit Santiago, the Pilgrimage is an opportunity to rediscover oneself, to live a balanced relationship alongside nature, and a different way of experiencing one's social life, time and the landscape (CETUR and SA Xacobeo, 2007–2010; Lois González et al., 2016). The leisurely way of travel along the route, and the slow pace, make it possible to calmly contemplate the landscapes, connect with them and enjoy the hundreds of rural localities it passes through. Slow mobility allows internal introspection, favours thinking about oneself and gives a marked liminal character to the walking experience (Turner, 1979). The availability of time to reflect on oneself, liminality and the creation of a space of tolerance have favoured the *Caminonization.*[1] All the aforementioned attributes are related and have contributed to the internationalisation of The Way, as confirmed by the diversification of the countries of origin (Lois González & Santos Solla, 2015).

At a national and international level, cultural itineraries have set similar objectives for their promotion (Cerruti & Dioli, 2013; Jørgensen et al., 2020a, b; Lemmi and Siena Tangheroni 2015). In fact, the diversity of visitors is essential for a society that claims to be modern, democratic and tolerant. Therefore, the routes of the Jewish Quarters or the cities of Al-Andalus seek a majority of Christian or secular tourists, eager to recover the legacy of cultures violently eradicated from the country. The sanctuaries of La Liébana or Caravaca de la Cruz, are promoted as identity signs of regions such as Cantabria or Murcia. All these routes and landmarks seek closeness with the landscape and monuments, while enjoying the free time to reflect on oneself and celebrate collectiveness. In addition, there is an interest by the authorities to diversify the origin of tourists, to place the itinerary, the monument or the sanctuary on the global map of memorable places or holy cities, often seen as additional features (Lois González, 2013; Possamai, 2003).

For all these reasons, The Way continues to be a central tourism product in Spain, whose market is stimulated from time to time by the celebration of Holy Years, accompanied by the promotional campaigns of the *Xacobeo*, and is heavily internationalised, generating strong markets from non-European countries (such as Korea, United States or Brazil, among others). This convergence of factors has

[1] This expression recalls the term *Caminonization*, through which the anthropologist Peter Margry (2015) refers to the internationalisation of models and values associated with The Way. Although the concept defined by the anthropologist refers in the first instance to the territorial and social implications of The Way, this expression is also used to refer to the "Way effect" in different spheres of cultural life.

Table 6.1 Number of pilgrims collecting their *Compostela* (2012–2022)

Year	Number of pilgrims
2012	192,488
2013	215,880
2014	237,886
2015	262,516
2016	277,854
2017	301,036
2018	327,378
2019	347,578
2020	54,144
2021	178,912
2022	438,823

Source: Authors' elaboration based on data published by the Pilgrim's Reception Office in Santiago de Compostela

placed the number of pilgrims arriving in Santiago at increasingly high levels (more than 200,000 in 2015, 300,000 in 2017 and almost 350,000 in 2019), which has contributed to the issue of *overtourism* (Table 6.1).

It should be noted that overcrowding is incompatible with the enjoyment of the authentic walking experience that involves slow mobility, the contemplation experience, visiting an unsaturated heritage listed site or getting to know the local population. Before the COVID-19 pandemic, *overtourism* was only located in the streets of the historic centre of the city of Santiago, in particular in the streets surrounding the Cathedral, as well as in some localities along the French Way, and was beginning to be publicly condemned (Lopez et al., 2019).

Our intention to link *overtourism* with The Way focuses on four main aspects. Firstly, the massification of some places along The Way and the city of Santiago puts into question the experience of solitude, of re-encountering oneself, which contemporary pilgrims consider to be the main characteristic of the experience. Secondly, the difficulty of finding sufficient shops or personal services in the localities of The Way, make some overbooking situations more typical of mass tourism than of a type of tourism that is mainly cultural and in touch with nature (Blanco Romero et al., 2019; Lopez et al., 2019). Thirdly, the increase in the number of walkers translates into greater amounts of garbage and waste, unplanned surges in access to public facilities and an increase in expenditure on customer service systems for pilgrims, to avoid overflows. Finally, the massive arrival of walkers completely alters the daily rhythms of life for the local population who begin to show dissatisfaction with a tourism product that had been promoted as experienced and respectful with the environment (Milano et al., 2019; Mínguez García et al., 2019). Ultimately, *overtourism* carries the danger of "The Way dying of success" and that all its original essence is distorted. Therefore, the analysis of this hugely successful European cultural route should draw on the theories and characterisations of holiday destination saturation studies to learn about its problems and propose sustainable solutions.

In other Cultural Itineraries, the characteristics and problems associated with overtourism are similar to the example of The Way. These are conversion processes of products and tourism destinations originally defined by culture, spiritual retreat, and the calm contemplation of quality landscapes and monuments, into mass tourism (Jørgensen et al., 2020a, b; Mariotti, 2012; Trono, 2012; Trono & Oliva, 2013). Overtourism causes the saturation of places, which in turn becomes overcrowding. Its origin lies in promotional campaigns of sanctuaries or culture itineraries, which from the beginning have been obsessed with attracting the largest number of visitors, either in general or on specific dates, and where the control of the carrying capacity has been absent. The idealised vision of these destinations always mitigates the feeling of saturation, but the persistence of the problem ends up with the rejection of tourism by residents and in more or less organised protests by local groups (Milano et al., 2019; Mínguez García et al., 2019).

Another side of overtourism is the fragility of places crowded with tourists when exceptional crisis situations occur (Milano et al., 2019; UNWTO, 2018). In fact, COVID-19 has highlighted the vulnerability of regions with an economy focussed on tourism and, for the first time, has raised the issue of the weaknesses of cultural itineraries, routes of pilgrimage and especially The Way as a successful product that can have its function interrupted. Therefore, if over- and undertourism combine to explain a great contemporary Cultural Route, the planning that both public institutions and private agents carry out in the promotion of the historical route must be established on three scenarios: (a) the normal situation of growth, diversification of routes and the search for the improvement of services, a balance that is increasingly difficult to achieve; (b) overtourism, which requires drastic actions to contain flows and establish measures to regulate the carrying capacity of the entire historic centre of Santiago de Compostela and some sections of the French Way, and (c) undertourism, which will take place in periods of global crisis, in a century defined by uncertainty and a variety of environmental, health and financial issues, which spread in a context of globalisation.

The COVID-19 pandemic caused a significant decline in visits to cultural itineraries and pilgrimage routes throughout Europe, and in Spain (Mróz, 2021). The phenomenon was more evident in 2020, although in 2021 the conjunctural strength of local destinations alleviated the situation. In general, undertourism is not seen as a problem, largely because in relatively recent destinations promoted with strong public support, the noticeable and punctual decrease in visitors is not considered a threat to products that still need to be consolidated. This happens especially in places that are still not consolidated tourist destinations, where the economic expectations generated by the growth of tourism are not so ingrained in localities.

In the current, not entirely post-pandemic scenario, The Way is being confirmed as an opportunity for reflection, leisure and fun. It allows one to enjoy some freedom of movement and the continuous diversification of its offer seems to respond to the demands of tourists and pilgrims that are nourishing a certain holistic motivation (Lopez & Lois González, 2021). From interviews to online forums, and throughout travel journals, The Way represents an answer to postmodern concerns. A fact that seems to remind us that The Way has the features of a "post-secular therapy route", and perhaps "post-Covid". According to Mikaelsson (2012), the therapeutic route has two elements: the understanding of a personal issue (such as divorce, death, loss of work, etc.) and the experience of renewal or transformation. Both elements converge in this historical moment, conceiving The Way as a gradual therapy that brings benefits for physical, mental and spiritual healing. And in this exercise, the relevance of the landscapes should be noted, as they become *therapeutic landscapes* in which physical and constructed environment, social conditions and human perceptions produce an atmosphere conducive to healing (Gesler, 1996).

6.3 Methodology

In social sciences, qualitative methods aim to explain, predict, describe or explore the "why" of events and, in this way, advance possible responses to the problems arising from the research work (Álvarez Gayou, 1999; Denzin & Lincoln, 1994). Therefore, we have opted for a qualitative methodology with interpretive value through which to describe, analyse and synthesise facts and events related to the case study (Creswell, 1998; Denzin & Lincoln, 2018; Van Maanen, 1983; Vasilachis, 2006). We reviewed previous studies related to The Way before the pandemic in order to understand the extent of pre-pandemic Jacobean pilgrimages. This interpretive exercise follows explanatory narrative patterns that respond to the objectives of our research, since they allow us to approach outstanding issues and phenomena, that act as keys for understanding (Creswell, 2014).

Regarding the qualitative netnographic methodology (Aitamurto, 2013), we have chosen a comparative approach between two different historical moments. First, we introduce the results of previous research in which the impressions and emotions of the pilgrims are analysed in their first days after the confinement of 2020 (Lopez & Lois González, 2020) and then we analyse these perceptions today, when The Way seems to have regained its relevance. Following Kozinets (2002), we set six methodological phases:

- *Research planning*: we assume the ethnography exercise as a way of observing context (Costello et al., 2017; Kozinets, 2002, 2015; Trainor, 2018).
- *Entrée*: we distinguish two phases in which we set the research questions and select those ones which present more detailed or descriptive data (Tables 6.2 and 6.3).
- *Data collection*: two phases are distinguished. First, in 2020, ten pilgrims' online communities were selected, whose entries date back to the period 24 February and 3 June 2020 (Table 6.2). The second and more recent set of textual data refers to ten other pilgrims' online communities, whose participants published their entries between 6 December 2021 and 19 June 2022 (Table 6.3). Tables 6.2 and 6.3 include the title of each discussion, the language, the number of participants and the date. Considering that there are numerous communities, we chose the most active ones, whose entries are more recent. As far as the conversations are

Table 6.2 Sources details of the Online Communities 2020

Forum titles & sources	Language	Participants	Dates of analysed entries
(1) COVID-19: Coronavirus (Round 6)	English	34	26/4/2020–25/5/2020
(2) COVID-19: Coronavirus (Round 7)	English	26	27/5/2020–03/06/2020
(3) Thinking about tents	English	58	25/5/2020–03/06/2020
(4) When would you consider going back on the Camino?	English	97	5/6/2020–03/06/2020
(5) Coronavirus and The Way of St. James	Spanish	7	24/2/2020–10/5/2020
(6) Reflections from the lockdown	Spanish	4	4/5/2020–10/5/2020
(7) In the current pandemic situation an autocaravan is an option to take into account	Spanish	3	23/4/2020–25/4/2020
(8) Is it the end of Hospitality?	Spanish	4	8/4/2020–11/4/2020
(9) When will The Way be accessible again?	Spanish	40	8/5/2020–20/5/2020
(10) How the post-coronavirus Way is going to affect us.	Spanish	38	31/03/2020–14/05/2020

Authors' elaboration.

Table 6.3 Sources details of the Online Communities 2022

Forum titles & sources	Language	Participants	Dates of analysed entries
(1) Self Isolating on the Camino	English	39	3/4/2022–19/6/2022
(2) Camino Mask Strategies?	English	27	6/8/2022–6/12/2022
(3) Walking the Camino when immuno-compromised or suppressed, so extremely vulnerable to C19	English	34	4/20/2022–5/29/2022
(4) Vac Requirements for Spain	English	30	4/22/2022–5/15/2022
(5) Current Covid impacts on the Camino?? Input from current walkers appreciated!	English	61	4/25/2022–5/4/2022
(6) Vaccine Passport Requirements in Spain travelling from the US	English	79	2/2/2022 -1/3/2022
(7) France and the COVID pass an autocaravan is an option to take into account	English	48	2/15/2022–3/3/2022
(8) Getting Covid somewhere along The Way – where to go?	English	19	14/1/2022–12/2/2022
(9) Anyone Caught COVID on the Camino	English	56	29/12/2021–4/1/2022
(10) The Omicron variant and its implications for the Camino	English	37	6/12/2021–6/1/2022

Authors' elaboration.

concerned, after reading the most recent ones for both periods, our selection satisfied the following criteria: (1) Discussion topic about COVID-19; (2) Value and interest of the content; (3) International profile of their users.

- *Data analysis*: we fix a set of interpretative categories to point out the main relevant issues emerging on pilgrims' online communities. These categories are shown in the following section, where we include a cross-sectional presentation of these topics in the conversations indicated and referenced in the bibliography.
- *Ethical standards*: we consider the web content as public published content, protecting individuals' names (Kozinets, 2015; Langer & Beckman, 2005).
- *Research representation*: the results are presented in two periods (2020 and 2022). In each period we highlight the issues that pilgrims are most concerned about in online communities. These changes reflect a different image of The Way.

Applying the netnographic methodology to the study of a cultural itinerary is a pilot experiment of particularly interest since it allows us to analyse extremely recent and still ongoing phenomena. Furthermore, considering the complex context that the COVID-19 epidemic has unexpectedly created, such methodology allows us in part to overcome the difficulties associated with field investigations which, especially during the various lockdown periods, were impossible to carry out.

Nevertheless, it should also be remembered that this analysis, precisely because it is experimental and innovative, has limits in the results that are not intended to be exhaustive. Rather, it is a first reconstruction of how much COVID-19 has changed the life of a territory strongly influenced by the presence of a tourist and cultural itinerary. It should also be remembered that we choose to focus on the analysis of one of the main cultural routes (the first and also the most popular before the pandemic period), but the results highlighted are not fully extendable to all other contexts and similar cases. Focusing on a case study was however a necessity taking into consideration the specific nature of this survey methodology. Indeed, case study analysis constitutes a research strategy whose distinctive features "consist of the fact that it [the case study] seeks to examine a) a contemporary phenomenon in its real context, especially when b) the boundaries between the phenomenon and its context are not clearly distinguishable" (Yin, 1981, p. 59). And as Bruno Dente (2011, p. 190) states "it [the case study] differs deeply both from the experimental approach, as the latter assumes the full separability of the phenomenon from the context, and from historical research, because it refers to episodes of the past". The choice of defining case studies through which to develop the arguments of the thesis responds exactly to the need to analyse phenomena still in progress, closely linked to the context in which they occurred.

6.4 Results and Discussion

Tourism has been the economic sector most affected by the pandemic since it is based on the movement of people (Febrero Paños & Bermejo Patón, 2021). The territories dependent on tourism activity have been more fragile and vulnerable, especially in the case of international and mass destinations (UNWTO, 2020).

Likewise, The Way has also experienced this tourism crisis differently as perceived by pilgrims. In this section, we present the results of comparative netnographic exploration in order to assess the aspects of undertourism and the current tourism recovery through the perception of pilgrims in virtual communities.

6.4.1 Signals of Undertourism Along the Way: The Effects of the Pandemic

One of the main problems caused by the pandemic has been the progressive loss of pilgrims and the fall of pilgrimages towards the main European Catholic destinations during and after the first wave of the pandemic (Mróz, 2021). At a European level, the number of pilgrims has reduced along cross-border routes (The Way, Via Francigena, etc.), since pilgrims have chosen more local and national routes (Mróz, 2021). During the lockdown in the months of April and May 2020, no pilgrims were registered collecting their *Compostela* and the historical series of the Office of the Pilgrim was interrupted. According to the report *Previsións de impacto do COVID-19 no sector turístico gallego* published in April 2020 (Clúster de Turismo, 2020), in scenario number 1 post-COVID-192020, that is, with the lifting of the lockdown from the end of April or beginning of May, and with the recovery of the domestic market, arrivals would reach around 131,062 pilgrims, but the reality was harsher and 53,905 pilgrims finally collected their *Compostela* in 2020 from the Pilgrims' Reception Office. In addition, the closure of The Way and the progressive loss of pilgrims had an impact on the "microeconomics of The Way"; that is, on that set of economic activities that had flourished thanks to the Jacobean route, taking advantage of a consumption behaviour based on local goods and services and generating lucrative processes for the local economy (Martínez Roget et al., 2018). Among the symbolic activities of The Way are the shelters (public and private), thus the *Plan Director y Estratégico del Camino de Santiago en Galicia 2015–2021* (Xunta de Galicia, 2015) focussed on promoting the creation of new shelters or the rehabilitation of others.

We could summarise all these negative effects related to the closure of The Way with the expression *undertourism*, to refer to its fragility and to the set of problems generated by the crisis of international mobility caused by the pandemic. It is interesting to note that the signs of this *undertourism* were already being noticed by pilgrims and commented on in online forums. Since the beginning of the pandemic, pilgrims understood the negative consequences that the health crisis could have for The Way, so that the topics of their conversations ceased to be about advice and logistical issues linked to route, to become reflections on the consequences of COVID-19. Indeed, we could affirm that the following three aspects allude to undertourism signs along The Way: (1) The economic impacts; (2) Hostels; (3) The new profile of The Way.

As far as the economic consequences at this historic moment are concerned, there has also been an international shift from *overtourism* to *undertourism*. Also, pilgrims had been worried about the negative economic impacts of the pandemic on villages and activities dependent on the pilgrimage, especially considering that many of them had invested in expensive improvements with an eye on the Holy Year Xacobeo 2021.[2] In a general perspective in the case of The Way, as well as other cultural itineraries, economic concerns explicitly refer to issues faced by: hostels, hotels, bars, restaurants, shops, backpack carriers, taxi drivers and other businesses. For this reason, they were willing to contribute to the reactivation of "The Way economy" with a "tiny grain of sand", thus helping to sustain the economic recovery of these territories, spending their money on accommodation and the small supermarkets that mark route.

Of all the economic activities of The Way, the greatest concern of the pilgrims was the shelters. Indeed, their viability has been much debated in recent months. Not only the pilgrims, but also the shelters' volunteers believe that in the short-and-medium-term, they might undergo a dangerous metamorphosis that could completely change their essence and charm. Shelters and hospitals have always been symbols of moments of encounter for pilgrims and walkers, a common feature to all cultural itineraries. Therefore, all the health restrictions were considered a threat to their economic and symbolic survival, alluding to what these historical structures mean.

In the pilgrims' opinion, social-distancing and the obligation to reduce capacity will hinder their survival due to the costs of adaption. It seemed to be totally unfeasible to those private hostels who had undergone significant renovation works to reduce capacity to 30%.

Besides the numerous problems, pilgrims thought that the restrictions along The Way would have been a sort of "filter" to determine real and convinced pilgrims, and those who were not. In addition, peoples' intentions and willingness to walk The Way in summer 2020 depended on fear (contagion, availability of a vaccine, lack of proper control measures, trust in the system, etc.), motivations and, obviously, on nationalities. As a matter of fact, this last aspect temporarily changed the profile of The Way, and other cultural itineraries due to mobility restrictions. In this first phase of the pandemic, pilgrims foresaw a sort of "new phase" of The Way, mainly characterised by signals of undertourism that, while reducing massification, ensured the recovery of its essence and charm. In fact, pilgrims tried to look for a positive aspect in the effects of the pandemic in the short term. In line with what was stated by Galvani et al. (2020), according to which tourism degrowth would have implied more meaningful and less hedonistic trips, pilgrims hoped that health and hygienic restrictions would contribute to a "stricter selection" of pilgrims. Also, the President of the FEAACS (2020) considered that in the short term The Way would have

[2]The Holy Jacobean Year is essentially a Jubilee year which only covers the city of Santiago de Compostela. It is also called Jacobean Years and is celebrated every 6, 5, 6 and 11 years when the feast of Saint James (25th July) falls on a Sunday.

become a more personal, more spiritual and more cultural experience. Their forecasts were fulfilled and, as a clear expression of undertourism, the mass phenomenon of recent years has been reduced and the few pilgrims on The Way had firm convictions. Likewise, the inevitable loss of internationalisation became an opportunity to enjoy the paths again, especially along the French Way (Lopez & Lois González, 2020).

6.4.2 Back to a New Normal: The Way at Present

On June 21, 2020, the end of the state of alarm in Spain was declared and on July 1, 2020, The Way was officially reopened. From this moment, a slow and progressive flow of pilgrims began. As we have already seen, until the present historical break due to COVID-19, a growing number of *Compostelas* have been presented over the last 10 years (Pilgrim's Reception Office 2009–2019). The data confirms that this trend, which also influenced processes of *overtourism*, was suddenly interrupted in the period 2019–2020. Nevertheless, it is possible to note an increasing trend in the previous months: from 53,905 pilgrims in 2020 to 178,912 pilgrims in 2021 (data from Pilgrim's Office). It is necessary to remember that the year 2021 was an important and long-awaited year for many because, after 11 years, it was the *Año Santo Compostelano (*or *Año Xacobeo*, since July 25th, the day of the Apostle, fell on a Sunday). Assuming that the 2021 Holy Year would not meet expectations, exceptionally, the Holy Year was extended to 2022, to regain the benefits lost due to the pandemic.

This resumption of pilgrims seems to suggest a sort of return to normal of The Way, or rather, of a *new normal* in which the pilgrims and the places of The Way are adapting to the current pandemic conditions. The Way is a "world", insofar as it is characterised by its humanity (that is, by different groups that live it). For this reason, it requires special attention, and, for its reopening, a series of measures were designed which could impact its dynamics, rites and practices.

From the analysis of the online forums, pilgrims have shown a strong will to experience The Way once more, even if they need to understand how to manage the risks which are still present today. Compared to the other period of analysis, pilgrims seem less worried about the possibility of contracting COVID-19 along The Way. However, at the beginning of 2022, it was possible to detect some concern regarding the Omicron variant and the spread of the infection. Despite the fear linked to the contraction of the virus, it is possible to detect a growing interest in undertaking the journey anyway: some discussions focus on how to deal with COVID-19 in case you contract it while travelling (e.g. how to manage self-isolation, where to go and who to contact). It therefore seems that the main concern of pilgrims has become not only how to protect themselves from the virus, but also how to manage the disease and its course. Moreover, the desire of pilgrims to return to travel along The Way is also found through the analysis of the discussions about the official requirements (masks, vaccine, etc.) needed to access hostels and other places along The Way. Such typology of discussion is very recurrent in 2022 since it follows the evolution of the rules with respect to the spread of the virus that each country has promoted. Indeed, the different vaccination campaigns have contributed to the reduction in the climate of uncertainty that characterised the whole international mobility in general terms. The geo-political struggle characterised by the reciprocal interaction between vaccines and economic developments has been a determining factor for the worldwide recovery of international travel. This aspect also affects tourism along The Way: in 2022, it has been possible to note a sort of resumption of the extra-European international flow from the online discussions. In fact, there are many conversations in which pilgrims from the United States or Australia worry about the documentation (especially vaccines) necessary to enter Spain, France or Portugal. This interest seems to suggest a return to the period before 2020, when the data referring to the origin of the pilgrims revealed a significant contraction from the intercontinental markets. For example, the total number of American pilgrims arriving during 2020 was 438, and most of them (343) were from Brazil. Mobility restrictions created a reality made up mainly of Spanish pilgrims (36,965), followed by Portuguese pilgrims (2949), and the Italian market in third place (2694), as was the case before the pandemic (Pilgrim's Office, 2021).

In summary, the analysis shows a strong desire for pilgrims to once more experience The Way, even with the new rules and practices that COVID-19 has imposed. Thus, although in the medium term there will be more European than Asian or American pilgrims, the present trends show a slow recovery of international mobility, and this indicates a slow recovery of The Way as physical and metaphysical mobility benchmark.

6.5 Conclusions

The COVID-19 pandemic has forced the need to rethink priorities and directions aimed at ensuring sustainable future development, actively review previous studies and reports on unsustainable tourism practices to avoid the usual mistakes. In short, it should teach us how equity, ethics, inclusion, sustainability, responsibility, social and environmental justice, and especially resilience are the new sustainable values (Benjamin et al., 2020; Edelheim, 2020; Fletcher et al., 2020; Gössling et al., 2020; Higgins Desbiolles, 2020; UNWTO, 2020).

This article contributes to the debate on the effects of COVID-19 on tourism and in particular on that linked to cultural itineraries. The Way is chosen as a case study of this phenomenon. It appears as a particularly relevant case both for its international importance and for the critical conditions that it seemed to have reached before the pandemic in some of its busiest sections at risk of overtourism. The case is studied through a netnographic methodology with which the main issues that have emerged from the online pilgrim forums are compared in two key periods: in 2020, when the pandemic was in full swing and due to a long lockdown it was not possible to travel along The Way; and in 2022, attempting to observe the simultaneous resumption of flows of travellers and the definition of a new normal of The Way. The results show how the Jacobean route can be considered a resilient resource for the territory, as it adapts to the new forms in which pilgrims face the journey and the pilgrimage.

Currently, The Way is regaining its position and popularity, recovering the international character it had lost over the last 2 years. In fact, it has once again become an alternative form of sustainable travel, far from the crowded centres and in full contact with the countryside. The figures recorded in recent months of 2022 confirm that it has become fashionable again and that the positive trends reinforce its international dimension, in addition to responding to plural and transversal consumption schemes. As a matter of fact, the pilgrim's motivations seem to be adapted (and updated) according to the historical and cultural moment, thus interpreting the paradigms of new religious, spiritual and secular trends (Farias et al., 2019; Moscarelli et al., 2020; Possamai, 2003). As a result of what has been experienced in the last 2 years, motivations may be increasingly diversified, perhaps some motivations specifically related to a "post-pandemic therapy" will appear, based on contact with nature and the outdoors. The Way has once again become a chance to rediscover nature thanks to the enjoyment of different resources along the route. Thus, expanding its already varied offer and increasing the difficulties in recognising the limits between pilgrims or tourists, and with it the characterisation of an increasingly complex profile. This will require route planning and management that sets priorities to ensure sustainable development.

In conclusion, COVID-19, despite having severely undermined the complex system of The Way, has at the same time provided a unique opportunity to rethink some design and territorial planning approaches with the possibility of avoiding or limiting the risky phenomena of over-tourism that occurred before the pandemic. This analysis and this case can become a useful pilot study for other cultural itineraries, even in a less advanced stage of development than The Way.

Authorship Attribution All the authors have shared the concept of the paper. L.L. has written the paragraphs 2, 3, 4.1 and 5. R.C.L.G. has written the paragraphs 1, 2 and has reviewed the paper. R.M. has written the paragraphs 3, 4.2 and 5.

References

Aitamurto, T. (2013). Balancing between open and closed. *Digital Journalism, 1*, 229–251. https://doi.org/10.1080/21670811.2012.750150

Álvarez Gayou, J. L. (1999). Investigación cualitativa. *Archivos Hispanoamericanos de sexología, 5*, 117–123.

Benjamin, S., Dilette, A., & Alderman, D. H. (2020). We can't return to normal Committing to tourism equity in the post-pandemic age. *Tourism Geographies, 22*(3), 476–483. https://doi.org/10.1080/14616688.2020.1759130

Blanco Romero, A., Blázquez Salom, M., Morell, M., & Fletcher, R. (2019). Not tourism-phobia, but urban-philia: Understanding stakeholders' perception of urban touristification. *Boletín de la Asociación de Geógrafos Españoles, 83*, 3. https://doi.org/10.21138/bage.2834

Canavan, B. J. (2013). Sustainable tourism: Development, decline and de-growth. Management issues from the Isle of Man. *Journal of Sustainable Tourism, 22*(1), 127–147.
Cerruti, S., & Dioli, I. (2013). Via francigena mountain itineraries: The case of Piacenza valleys. *International Journal of Tourism and Pilgrimage, 1*, 83–92.
CETUR, & Xacobeo, S. A. (2007–2010). Observatorio Estadístico do Camiño de Santiago 2007, 2008, 2009 e 2010. Universidade de Santiago de Compostela: Xunta de Galicia y Centro de Estudios Turísticos.
Clúster de Turismo de Galicia. (2020). *Revisións de impacto do Covid-19 no sector turístico gallego*. https://clusterturismogalicia.com/wp-content/uploads/2020/04/2020_04_14_Previsión_COVID_v4_FINAL.pdf
Costello, L., McDermott, M., & Wallace, R. (2017). Netnography: Range of practices, misperceptions, and missed opportunities. *International Journal of Qualitative Methods, 16*, 1–12. https://doi.org/10.1177/1609406917700647
Creswell, J. W. (1998). *Qualitative inquiry and research design. Choosing among five traditions*. Sage Publications.
Creswell, J. (2014). *Research design: Qualitative, quantitative and mixed methods approaches*. Sage Publications.
Dente, B. (2011). *Le decisioni di policy*. Il Mulino.
Denzin, N. K., & Lincoln, Y. S. (Eds.). (1994). *Handbook of qualitative research*. Sage Publications.
Denzin, N.K., Lincoln, Y.S. (2018). The sage handbook of qualitative research. Sage Publications.
Edelheim, J. (2020). How should tourism education values be transformed after 2020? *Tourism Geographies, 22*(3), 547–554. https://doi.org/10.1080/14616688.2020.1760927
Farias, M., Coleman, T. J., Bartlett, J. E., Oviedo, L., Soares, P., Santos, T., & Bas, M. (2019). Atheists on the Santiago way: Examining motivations to go on pilgrimage. *Sociology of Religion, 80*(1), 28–44. https://doi.org/10.1093/socrel/sry019
FEAACS (Federación Española de Asociaciones de Amigos del Camino de Santiago). (2020). Toca repensar el Camino de Santiago, 31/05/2020. https://www.caminosantiago.org/cpperegrino/prensa/verprensa.asp?PrensaID=15157&fbclid=IwAR0qFQ4iYteR6yN3Q6PFwi0w5bT4p6HGVki2e2gKFSpYtdruxWuVCvJsl8c
Febrero Paños, E., & Bermejo Patón, F. (2021). Estimating the impact of COVID-19 on the Spanish economy with input-output analysis. *Revista Galega de Economía, 30*(1), 1–19. https://doi.org/10.15304/rge.30.1.7105
Fletcher, R., Murray, I. M., Blázquez-Salom, M., & Blanco Romero, A. (2020). Tourism, degrowth, and the COVID-19 crisis. *POLLEN Ecology Network*, 15/05/2020. https://politicalecologynetwork.org/2020/03/24/tourism-degrowth-and-the-covid-19-crisis
Fraternidad Internacional del Camino de Santiago. (2014). Manifiesto de Villafranca del Bierzo. *I Foro Internacional "El Legado de Elías Valiña". Villafranca del Bierzo. 14 de diciembre de 2014*. Santiago de Compostela: Fraternidad Internacional del Camino de Santiago.
Galvani, A., Lew, A. A., & Sotelo Pérez, M. (2020). COVID-19 is expanding global consciousness and the sustainability of travel and tourism. *Tourism Geographies, 22*(3), 567–576. https://doi.org/10.1080/14616688.2020.1760924
Gesler, W. (1996). Lourdes: Healing in a place of pilgrimage. *Health and Place, 2*(2), 95–105.
Gössling, S., Scott, D., & Hall, C. M. (2020). Pandemics, tourism and global change: A rapid assessment of COVID-19. *Journal of Sustainable Tourism, 29*(1), 1–20. https://doi.org/10.1080/09669582.2020.1758708
Gusmán Correia, M. I., Lopez, L., Lois González, R. C., & Santos, X. M. (2017). The challenges of the first European cultural itinerary: The way to St. James. *Almatourism – Journal of Tourism, Culture and Territorial Development, 8*(6), 1–19. https://doi.org/10.6092/issn.2036-5195/6034
Higgins Desbiolles, F. (2020). Socialising tourism for social and ecological justice after COVID-19. *Tourism Geographies, 22*(3), 610–623. https://doi.org/10.1080/14616688.2020.1757748

Jørgensen, N. N., Eade, J., Ekeland, T.-J., & Lorentzen, C. A. N. (2020a). The motivations of pilgrimage walking the St. Olav Way in Norway. *International Journal of Religious Tourism and Pilgrimage, 8*(8), 110–126.

Jørgensen, N. N., Eade, J., Ekeland, T. J., & Lorentzen, C. A. N. (2020b). The processes, effects and therapeutics of pilgrimage walking the St. Olav Way. *International Journal of Religious Tourism and Pilgrimage, 8*(1), 33–50.

Kozinets, R. (2002). The field behind the screen: Using netnography for marketing research in online communities. *Journal of Marketing Research, 39*, 61–72. https://doi.org/10.1509/jmkr.39.1.61.18935

Kozinets, R. (2015). *Netnography: Redefined.* Sage.

Langer, R., & Beckman, S. (2005). Sensitive research topics: Netnography revisited. *Qualitative Market Research, 8*, 189–203. https://doi.org/10.1108/13522750510592454

Lemmi, E., & Siena Tangheroni, M. (2015). The via Francigena as a tourist product for local development: The case of Lucca and its province. In G. Bambi & M. Barbari (Eds.), *The European pilgrimage routes for promoting sustainable and quality tourism in rural area* (pp. 437–453). Firenze University Press.

Lois González, R. C. (2000). Dotaciones e infraestructuras del Camino de Santiago. Una aproximación geográfica. In L. López Trigal (Ed.), *Ciudades y villas camineras jacobeas: III Jornadas de Estudio y Debate Urbanos* (pp. 225–245). Universidad de León, Secretariado de Publicaciones.

Lois González, R. C. (2013). The Camino de Santiago and its contemporary renewal: Pilgrims, tourists and territorial identities. *Culture and Religion, 14*, 8–22. https://doi.org/10.1080/14755610.2012.756406

Lois González, R. C., & Lopez, L. (2012). El Camino de Santiago: una aproximación a su carácter polisémico desde la geografía cultural y el turismo. *Documents d'Anàlisi Geogràfica, 58*(3), 459–479.

Lois González, R. C., & Santos Solla, X. (2015). Tourists and pilgrims on their way to Santiago. Motives, Caminos and final destinations. *Journal of Tourism and Cultural Change, 13*(2), 149–164. https://doi.org/10.1080/14766825.2014.918985

Lois González, R. C., Castro Fernández, B. M., & Lopez, L. (2016). From sacred place to monumental space: The mobility along the way to St. James. *Mobilities, 11*(5), 770–788. https://doi.org/10.1080/17450101.2015.1080528

Lopez, L. (2014). Riflessioni sullo Spazio Sacro: il Cammino di San Giacomo di Compostella. *Rivista Geografica Italiana, CXXI*(3), 289–309.

Lopez, L., & Lois González, R. C. (2020). New tourism dynamics along the Way of St. James. From undertourism and overtourismto the Post-COVID-19 Era. In G. X. Pons, A. Blanco-Romero, C. Navalon, L. Troitiño, & M. Blázquez-Salom (Eds.), *SostenibilidadTurística: overtourism vs undertourism. XVII ColoquioInternacional de Turismo AGE-UGI. Mon. Soc. Hist. Nat. Balears* (Vol. 31, pp. 541–552). http://ibdigital.uib.es/greenstone/sites/localsite/collect/monografiesHistoriaNatural/index/assoc/Monograf/iesSHNB_/2020vol0/31p541.dir/MonografiesSHNB_2020vol031p541.pdf

Lopez, L., & Lois González, R. C. (2021). La nueva normalidad en el Camino de Santiago: reflexiones para el futuro. *Revista Galega de Economía, 30*(3), 7568–7516. https://doi.org/10.15304/rge.30.3.7568

Lopez, L., Pazos Otón, M., & Piñeiro Antelo, M. A. (2019). ¿Existe Overtourism en Santiago de Compostela? Contribuciones para un debate ya iniciado. *Boletín de la Asociación de Geógrafos Españoles, 83*. https://doi.org/10.21138/bage.2825

Margry, P. J. (2015). To be or not to be. . .a pilgrim. Spiritual pluralism along the Camino Finisterre and the urge for the end. In C. Sánchez Carretero (Ed.), *Heritage, pilgrimage and the Camino to Finisterre* (pp. 175–211). Springer.

Mariotti, A. (2012). Local system, networks and international competitiveness: From cultural heritage to cultural routes. *Almatourism, Journal of Tourism, Culture and Territorial Development, 5*, 81–95.

Martín Duque, C. (2017). Los impactos del turismo en el Camino de Santiago Francés: una aproximación cualitativa. *Methaodos. Revista de Ciencias Sociales, 5*(1), 62–73. https://doi.org/10.17502/m.rcs.v5i1.155

Martínez Roget, F., Castro Domínguez, N., & Fraiz Brea, J. A. (2018). O turismo como fonte de riqueza en Galicia: análise dalgunhas experiencias. *Revista Galega de Economía, 2*(3), 65–80. https://doi.org/10.15304/rge.27.3.5437

Mikaelsson, L. (2012). Pilgrimage as post-secular therapy. *Scripta Instituti Donneriani Aboensis, 2*, 10.30674/scripta.67418–10.30674/scripta.67273.

Milano, C., Cheer, J., & Novelli, M. (2019). Overtourism and tourismphobia: A journey through four decades of tourism development, planning and local concerns. *Tourism, Planning & Development, 16*(4), 353–357. https://doi.org/10.1080/21568316.2019.1599604

Mínguez García, C., Piñeira Mantiñán, M. J., & Fernández Tabales, A. (2019). Social vulnerability and touristification of historic centers. *Sustainability, 11*(16), 4478. https://doi.org/10.3390/su11164478

Moscarelli, R., Lopez, L., & Lois González, R. C. (2020). Who is interested in developing the way of Saint James? The pilgrimage from faith to tourism. *Religions, 11*(1), 24. https://doi.org/10.3390/rel11010024

Mróz, F. (2021). The impact of COVID-19 on pilgrimages and religious. Tourism in Europe during the first six months of the pandemic. *Journal of Religious Health, 60*, 625–645. https://doi.org/10.1007/s10943-021-01201-0

Pilgrims' Reception Office. (1982–2021). *Informes estadísticos 2009–2020/statistical reports 2009–2020*. Oficina del Peregrino. www.archicompostela.org

Porcal, M. C., Díez, A., & Junguitu, J. (2012). Dimensión territorial y turística de la ruta norte del Camino de Santiago en el País Vasco: distintas concepciones, valoraciones y propuestas de intervención sobre un fenómeno multifacético. *Boletín de la Asociación de Geógrafos Españoles, 58*, 177–204. https://doi.org/10.21138/bage.2064

Possamai, A. (2003). Alternative spiritualties and the cultural logic of late capitalism. *Culture and Religion, 4*(1), 31–45. https://doi.org/10.1080/01438300302807

Sánchez Carretero, C. (Ed.). (2015). *Heritage, pilgrimage and the Camino to Finisterre. Walking to the end of the world*. Springer.

Santos, X. M., & Lois González, R. C. (2011). El Camino de Santiago en el contexto de los nuevos turismos. *Estudios Turísticos, 189*, 87–110.

Trainor, A. A. (2018). Community conversation as a method of gathering and analyzing qualitative data. *Journal of Disability Policy Studies, 29*(1), 2–6. https://doi.org/10.1177/1044207317739403

Trono, A. (Ed.). (2012). *Via Francigena. Cammini di Fede e Turismo Culturale*. Congedo.

Trono, A., & Oliva, L. (2013). Percorsi religiosi tra turismo culturale e strategie di pianificazione sostenibile: ricerca e innovazione. *Annali del turismo, II*, 9–34.

Turner, V. (1979). *Process, Performance and pilgrimage: A study in comparative symbology*. Concept Publishing Company.

UNWTO. (2018). *Overtourism? Understanding and managing urban tourism growth beyond perceptions*. UNWTO.

UNWTO. (2020). *Supporting jobs and economies through travel and tourism*, 10/05/2020. https://webunwto.s3.eu-west-1.amazonaws.com/s3fs-public/2020-04/COVID19_Recommendations_English_1.pdf

Van Maanen, J. (1983). *Qualitative methodology*. Sage.

Vasilachis, I. (2006). *Estrategias de investigación cualitativa*. Gedisa.

Weiss, M., & Cattaneo, C. (2017). Degrowth – Taking stock and reviewing an emerging academic paradigm. *Ecological Economics, 137*, 220–230. https://doi.org/10.1016/j.ecolecon.2017.01.014

Xunta de Galicia. (2015). *Plan Director y Estratégico del Camino de Santiago en Galicia 2015–2021*. Recuperado de: https://www.turismo.gal/canle-profesional/plans-e-proxectos/plan-director-camino-desantiago?langId=es_ES#:~:text=El%20Plan%20Director%20del%20Camino,privados%2C%20implicados%20en%20el%20mismo

Yin, R. K. (1981). The case study crisis: Some answers. *Administrative Science Quarterly, XXVI*(1), 58–65.

References of the Online Communities

Camino de Santiago. *SelfIsolating on the Camino*. 20/06/22, https://www.caminodesantiago.me/community/threads/self-isolating-on-the-camino.74049/

Camino de Santiago. *Camino Mask Strategies?*. 20/06/22, https://www.caminodesantiago.me/community/threads/camino-mask-strategies.75379/

Camino de Santiago. *Walking the Camino when immuno-compromised or suppressed, so extremely vulnerable to C19*. 20/06/22, https://www.caminodesantiago.me/community/threads/walking-the-camino-when-immuno-compromised-or-suppressed-so-extremely-vulnerable-to-c19.74393/

Camino de Santiago. *Vac Requirements for Spain*. 20/06/22, https://www.caminodesantiago.me/community/threads/vac-requirements-for-spain.74437/

Camino de Santiago. *Current Covid impacts on the Camino? Input from current walkers appreciated!* 20/06/22, https://www.caminodesantiago.me/community/threads/current-covid-impacts-on-the-camino-input-from-current-walkers-appreciated.74506/

Camino de Santiago. *Vaccine Passport Requirements in Spain traveling from the US*. 20/06/22, https://www.caminodesantiago.me/community/threads/vaccine-passport-requirements-in-spain-traveling-from-the-us.72919/

Camino de Santiago. *France and the COVID pass*. 20/06/22, https://www.caminodesantiago.me/community/threads/france-and-the-covid-pass.73147/

Camino de Santiago. *Getting Covid somewhere along the way - where to go?* 20/06/22, https://www.caminodesantiago.me/community/threads/getting-covid-somewhere-along-the-way-where-to-go.72695/

Camino de Santiago. *Anyone Caught COVID on the Camino*. 20/06/22, https://www.caminodesantiago.me/community/threads/anyone-caught-covid-on-the-camino.72537/

Camino de Santiago. *The Omicron variant and its implications for the camino*. 20/06/22, https://www.caminodesantiago.me/community/threads/the-omicron-variant-and-its-implications-for-the-camino.72393/

Camino de Santiago. *COVID-19: Coronavirus (Round 6)*. 3/06/2020, https://www.caminodesantiago.me/community/threads/covid-19-corona-virus-discussion-round-6.67460/

Camino de Santiago. *COVID-19: Coronavirus (Round 7)*. 3/06/2020, https://www.caminodesantiago.me/community/threads/covid-19-corona-virus-discussion-round-7.67716/#post-853779

Camino de Santiago. *Thinking about tents*. 3/06/2020, https://www.caminodesantiago.me/community/threads/thinking-about-tents.67703/page-2#post-853755

Camino de Santiago. *When would you consider going back on the Camino?* 3/06/2020, https://www.caminodesantiago.me/community/threads/when-would-you-consider-going-back-to-the-camino.67555/

Foro Camino. *El Coronavirus y el Camino de Santiago*. 3/06/2020, https://forocamino.com/index.php?threads/el-coronavirus-y-el-camino-de-santiago.396/

Foro Camino. *¿El fin de la hospitalidad?*. 3/06/2020, https://forocamino.com/index.php?threads/%C2%BFel-fin-de-la-hospitalidad.416/

Foro Camino. *Pues oye, con este tema del virus la caravana es una opción a tener en cuenta.* 3/06/2020, https://forocamino.com/index.php?threads/pues-oye-con-este-tema-del-virus-la-caravana-es-una-opci%C3%B3n-a-tener-en-cuenta.422/

Foro Camino. *Reflexiones desde el confinamiento.* 3/06/2020, https://forocamino.com/index.php?threads/reflexiones-desde-el-confinamiento.425/#post-1945

Gronze. *Para cuando estará viable el Camino?* 3/06/2020, https://www.gronze.com/foros/general/para-cuando-estara-viable-camino

Gronze. *Como nos va a afectar el Camino post-coronavirus.* 3/06/2020, https://www.gronze.com/foros/general/como-nos-va-afectar-camino-post-corona-virus

Lucrezia Lopez is Lecturer of Geography at the Department of Geography of the University of Santiago de Compostela (USC) and secretary of the CETUR (Centro de Estudos Turísticos). She is academic coordinator in the USC of the International Master's Degree in *City Challenges*, and secretary of its international academic commission. She has worked and is a member of various research projects at international, national and regional level. Her research lines concern the Camino de Santiago, studied by a plurality of geographic, cultural and touristic approaches. Additionally, she deals with cultural and Mediterranean studies, tourism geography, heritage and heritage management, and geohumanities.

Rubén Camilo Lois González is Professor of Geography at the University of Santiago de Compostela (Spain) and Vice President of the International Geographical Union (UGI). He has been visiting professor at universities in France, Norway, Italy and the Federal University of Bahia-UFBA (Brazil). He is Principal Investigator of various European Union and international cross-border cooperation projects. He is the author of nearly 200 books and book chapters, and more than 150 articles in academic journals. Notable among them are Scripta Nova, Finisterra, Annales de Géographie, Sud-Ouest Européen, Die Erde, European Planning Studies, Land, Use & Policy, Revista Geográfica del Norte Grande, Mobilities, Territory, Politics and Governance, Sociologia Ruralis and Gender, Place & Culture. He has been Dean of the Faculty of Geography and History of Santiago de Compostela. He is a member of the Board of Directors of the Spanish Association of Geography (1997–2011 and from 2017 to today) and founder of the College of Geographers of Spain (1999).

Rossella Moscarelli is PhD in Urban Planning, Design and Policies, and Assistant Professor at Politecnico di Milano, Department of Architecture and Urban Studies. Her research focuses on the walking paths, analysed as linear tools able to activate innovative planning and design approaches. In this context, the Camino de Santiago was studied during a research period at the Universidad de Santiago de Compostela, Spain, with a focus on the main public policies and actors involved in the development of the project since the early Nineties till now, and on the effects generated on the places crossed by the line.

Chapter 7
Turistification and Urban Heritage in Spanish Historic Centres: Permanence and Changes in a Long Process

Manuel de la Calle-Vaquero and María García-Hernández

7.1 Introduction: Context, Approach to the Topic and Methodology

Two different meanings can be given to the term touristification (Blanco Romero et al., 2020). The first is broad in scope and neutral in nature and alludes to the transformation process of a place in a tourist space and the associated effects. This transformation, which takes place on an attraction and destination level, implies an adaptation of the place to the visitor, whether tourism is the dominant activity of this space or not. Its repercussions, which can be positive or negative, affect the environment, the landscape, the economy and social and cultural aspects. The second meaning is more limited and has negative connotations: touristification is a form of tertiarization which threatens the urban and heritage safeguarding of historical centres. Although it has an impact on the shaping of the urban space, its most palpable effects can be seen in the socio-economic, symbolic and functional fabric. It is a process of territorial specialisation in tourism activities, the negative effects of which far outweigh the positive impacts, at least from the perspective of part of the local population. This text addresses the touristification process of the historical centres of Spanish cities. The process has been studied considering the double meaning of the term, depending on the authors.

Historical centres contain most of the urban heritage of cities (García & Calle Vaquero, 2019). This heritage includes archaeological sites, monumental buildings, parks and gardens, squares and other public spaces, as well as different manifesta-

M. de la Calle-Vaquero (✉) · M. García-Hernández
Complutense University of Madrid, Madrid, Spain
e-mail: mcalleva@ucm.es; mgarciah@ucm.es

A. Blanco-Romero, M. Blázquez-Salom (eds.), *Spanish Tourism Geographies*, Geographies of Tourism and Global Change,
https://doi.org/10.1007/978-3-031-39780-6_7

tions of intangible heritage and historical memory. At the same time, these centres are heritage sites in themselves (Rodwell, 2022). They are declared *Conjuntos Históricos* according to Spanish heritage legislation and, in some cases, are part of the World Heritage List promoted by UNESCO (Alcalá de Henares, Ávila, Cuenca, La Laguna, Segovia, Tarragona, Toledo... and, from 2021, the Paseo del Prado-Retiro in Madrid). The consideration of these places as heritage sites is not only due to historical-artistic aspects, or to their physical or morphological characteristics. It implies an evaluation as singular urban spaces that also takes into account social, functional and identity dimensions. They are living urban pieces that continue to constitute the symbolic heart of the city. This comprehensive approach is present in the notion of the "historic urban landscape", a notion not without controversy (Lalana Soto, 2022) that permeates all of UNESCO's recent approaches to city and urban heritage.

The touristification of historic centres is a very present topic in contemporary academic literature. It is the subject of a debate in which different branches of geography (urban geography, economic geography, social geography, geography of tourism...), but also different approaches from related disciplines (urban sociology, tourism policy, anthropology of heritage, political economy, etc.) intersect, overlap and complement each other. In this sense, the study of tourism and the city, and therefore the processes of touristification, has become a completely transdisciplinary field of work. Aware of this fact, this paper, however, has chosen to focus on the analysis of this phenomenon from the point of view of Geography: geography by Spanish geographers or by geographers who have worked on Spanish cities. This choice is justified by the need to limit, for reasons of space, the scope of the review of a subject that has been studied from multiple approaches and, to a certain extent, is so fashionable.

At present, the debate on touristification is closely intertwined with reflections on gentrification, tourist gentrification and overtourism. It is presented as a recent phenomenon associated with the current phase of neoliberal capitalism. Our perspective, which is based on the study of tourism development in Mediterranean heritage cities, is somewhat different. We consider that touristification is an ancient process, closely linked to the heritagization of these places. Cities such as Cordoba, Granada, Seville and Toledo became tourist destinations in the nineteenth century and from the beginning of the twentieth century, they undertook heritage recovery operations to promote tourism development (Calle Vaquero & García, 2016; Cocola-Gant, 2011; Fernández Tabales, 2018; García Álvarez, 2007). In the mid-twentieth century, with the emergence of sun and beach tourism, these cities were somewhat eclipsed by tourism. This was the time of excursions and tours by private bus, which represented the maximum manifestation of Fordist tourism in its urban and cultural version. Although tourism was a significant phenomenon in the dynamics of many historical centres, it hardly aroused interest at the political, social or academic level. Only a few works in urban geography paid attention to this activity (e.g. Bosque

Maurel, 1962) and the geography of tourism was still a very incipient field (Luis Gómez, 1988). Interest in these issues increased in the seventies of the twentieth century, when some studies on the hotel function began to be considered (Gutiérrez Ronco, 1984), which were parallel to the first approaches to historic centres as heritage sites. This situation changed in the 1990s when culture and heritage gained popularity as tourist assets and attention turned to urban spaces with a greater concentration of cultural attractions. Since then, and in a general context of growing tourist flow, the previously existing processes of touristification have been accentuated.

This text focuses on recent processes of touristification. In terms of time, two major cycles are identified: the first, from the 1990s to the 2008 crisis, and the second, from the crisis to the present day. The first is related to the revival of urban tourism at the end of the twentieth century and involves the visibility of a phenomenon with almost 200 years of history. The second involves an intensification of this phenomenon, but also its expansion towards neighbourhoods with a lesser tourist tradition. It also presents as a singular feature the proliferation of tourist homes (Calle Vaquero & García Hernández, 2020), an accommodation formula associated with sun and beach destinations before the birth of platform companies such as Airbnb. The text is organised into five sections. The first corresponds to this presentation, which includes the background, purpose and focus of the work. The second section deals with the first of the recent cycles of touristification and the third with the second cycle. Each of these two sections is organised in two parts. Firstly, the main features of the cycle are identified with allusions to the current tourism situation based on data and information extracted from secondary sources (official statistics, reports, previous work by the authors and also some other academic publications). Secondly, an account is given of the main contributions in Geography made during this cycle, in the form of a state of the art at the time of the period. Due to its orientation, the contributions published in scientific journals and, to a much lesser extent, monographs and book chapters have been selected. In contrast to a more conventional model, which separates the state of the art from the analysis of the phenomenon, we have preferred to present the dynamics of tourism and its approaches to study in an intertwined manner, thus highlighting the features of each cycle in terms of tourism and the city, but also its modes of study as far as Geography is concerned. In this sense, he reflects on how the phenomenon of the touristification of Spanish urban centres has been identified, analysed and studied in the field of Geography. Finally, the section of conclusions that closes the chapter ends by pointing out permanences and changes in the process of touristification and its interpretation in disciplinary terms and alludes to the situation of uncertainty generated after the outbreak of the COVID-19 pandemic.

7.2 From the 1990s to the Crisis of 2008: The First Recent Cycle of Touristification – The Rebirth of Urban Tourism and the Consolidation of Tourism as a Key Function of Historic Neighbourhoods

Echoing the evolution of the industry on a global level, in the 1990s there was a "rediscovery" of urban tourism in Spain, with significant growth in tourism demand and a growing interest in tourism among local governments and academic researchers, including those from the field of geography. According to data from official sources (FRONTUR and FAMILITUR) a total of 35 million foreign tourists visited Spain. This figure is quite a lot lower than the number of domestic tourism trips (115.5 million in 1999). By 2007, these figures had increased considerably, particularly those of international tourists (58.6 million foreign tourists). A significant part of these tourists visited urban destinations. In terms of the number of visitors, the most popular cities were the metropolitan tourism destinations, such as Barcelona and Madrid. According to data from the Hotel Occupancy Survey of the National Institute of Statistics (hereinafter EOH), the number of tourists staying in hotels in Barcelona grew from 3.3 million in 1999 to 5.5 million in 2007 (increase of 66.9%) with a rising share of international tourists. In Madrid, which is more associated with national capital tourism, the increase was 57.26% (7.3 million tourists in 2007). The growth also affected large heritage cities with a long tourism tradition, such as Santiago de Compostela (11.7%) and Seville (11.5%) but was particularly noteworthy in cities with barely any tourism tradition that used this activity as the main thrust of their development. For example, Malaga doubled its tourist figures between 1999 and 2007 when it accommodated 720,000 tourists in hotels (INSTITUTO NACIONAL DE ESTADÍSTICA, 1990a, b). In any case, the figure for international visitors would have been much higher, as those tourists staying in other types of accommodation must be added, together with those staying in nearby towns and also day trippers, who were particularly numerous in destinations close to the large metropolitan cities and the principal tourist areas on the coast. This positive tourist behaviour was favoured by the expansion of low-cost flights and affected all types of city-related tourism.

In parallel with this growth in tourism demand, there was a reorientation of the national tourism policy. On the one hand, the objective was to diversify the Spanish tourism product so as to break its excessive dependence on sun and beach tourism. Therefore, it particularly focused on other types of tourism, such as cultural tourism and urban tourism, which are repeatedly referred to in the national tourism plans of these years (FUTURES I and II, *Plan Integral de Calidad Turística Española*, and *Plan de Turismo Español Horizonte 2020*). On the other hand, the Ministry of Tourism promoted the creation of local tourism plans which were particularly numerous and important in the urban tourism destinations (Calle Vaquero & García Hernández, 2016).

In addition to this reorientation of the national tourism policy, tourism was incorporated into the urban political agenda (Calle Vaquero, 2008). As in the case of other western European cities, the urban authorities began to consider leisure, culture and tourism as potential drivers of development: the physical recovery of degraded spaces, socio-economic invigoration due to the generation of income and jobs and the external projection of highly positive images in an increasingly competitive city destination market. Consequently, the tourism argument was used to justify many operations, from the opening of large museums to the development of mega-events. The cities began to develop explicit tourism policies based on strategically oriented tourism plans and increasingly sophisticated tourism management entities. Barcelona led the way: it approved the first strategic tourism plan and established a Tourism Consortium which constituted a reference imitated by all local governments (Calle Vaquero & García Hernández, 2016).

Although the cities acquired the status of top tourist destinations, the affluence of visitors and the tourism activities continued to be concentrated in the historic centre. After protection systems had been installed in the preceding decades, in the 1980s, a new cycle of heritagization was carried out which was highly influenced by Italian urban planning (Troitiño Vinuesa, 1992). Special plans were approved and a whole host of urban regeneration operations were implemented: the opening of museums, the pedestrianisation of streets, the improvement in urban furniture, etc. Moreover, with respect to the social and cultural dimension, the idea of the productive use of urban heritage was promoted; a renovated asset that can contribute to an improved competitiveness of the city as a whole. Therefore, the declaration of a historic centre as a UNESCO World Heritage Site was no longer a mere cultural issue but an urban marketing strategy. And the cities which had been awarded this label, organised into the Spanish Group of World Heritage Cities, acted together and lobbied to capture state funds, mainly for international tourism promotion.

As a result of the growing tourist flows, many of the historical quarters underwent a considerable touristification process which was manifested in different ways. It gave rise to a substantial increase in the presence of tourists at monuments, museums and parts of the city corresponding to the "historic-tourist city". The accommodation supply increased considerably, particularly hotels and similar. According to the data from EOH, between 1999 and 2007, the number of beds increased by 50%, with the construction of new hotels and the transformation of heritage buildings into boutique hotels. In. parallel, the number of different types of tourism businesses also grew: souvenir shops, tourist guide services, tourist transport services, etc. And a good part of the hotel, restaurant and retail trades in the historic districts adapted to incorporate tourists into their clientele.

This touristification process affected all types of historic neighbourhoods of urban destinations. It was highly pronounced in the centres of large metropolitan destinations, where strictly cultural tourism coexisted with other types of tourism that incorporated visits to heritage sites. The historic districts of coastal cities, such as Barcelona, Malaga, Palma de Mallorca and Valencia were filled with day trippers from the sun and beach destinations and cruise ship passengers. However, the process was particularly prominent in the centres of the heritage cities, specialised in a highly traditional heritage tourism with a high incidence of visitors in groups. It is these cities which recorded the highest levels of tourism intensity. The number of overnight stays per resident in 2007 was 12.31 in Santiago de Compostela, 11.48 in Granada, 9.44 in Toledo and 9.06 in Segovia. These figures are higher than those recorded for Barcelona (7.32), Madrid (4.53) and other large cities such as Valencia (4.02) (Troitiño Vinuesa et al., 2007). In many cases, these destinations were medium-sized cities in the interior of the country, experiencing a socio-economic sluggishness. This functional weakness determined their particular vulnerability to touristification. Similarly, the socio-economic fragility of certain historic neighbourhoods made the touristification process more evident.

In these years, tourism became a specific field of study within Spanish Geography, with one of its focuses being urban spaces. The Working Group on Geography of Tourism, Leisure and Recreation of the Spanish Geographical Association (AGE) was created in 1989 with a significant participation of urban geographers. Since 1990, workshops and congresses have been held approximately every two years, bringing together Spanish specialists in the geography of tourism, who also play a prominent role in the Spanish Association of Scientific Experts in Tourism (AECIT). The incorporation of tourism studies in universities, on a graduate and post-graduate level, promoted the touristification of geography, as tourism was becoming an academic field of opportunity. The expansion of the teaching of material related to the geography of tourism was accompanied by an increase in basic research, but also with more applied and even strictly consultancy research related to different aspects of tourism development.

With respect to previous phases, the geography of tourism during these years addressed the complex relationships between tourism and the city. Based on urban geography approaches, the effects of this new urban function on the pre-existing structures were analysed, particularly in terms of their impact on historic districts. Furthermore, there was a progressive shift towards studies more specific to the geography of tourism, which addressed the possibilities of tourism development based on urban heritage. These issues appear in some of the first manuals of geography of tourism (López Palomeque et al., 1997; Calabuig & Barrado, 2001), which include specific chapters on cities as one of the principal tourism spaces. These studies repeatedly cite the contributions of English-speaking geographers, such as Stephen Page, Colin Michael Hall, Christopher M. Law and Douglas G. Pearce. However, references to French geographers were given equal weight (Georges Cazes, Philippe Duhamel, Rémy Knafou, Jean-Pierre Lozato-Giotart, etc.) as were those of other countries (Myriam Jansen-Verbeke, Daniel Hiernaux, etc.). On the other hand, the external projection of the geography of tourism carried out in Spain was scarce and limited to very specific authors and mainly oriented towards Latin America.

In addition to general reflections, more specific studies began to emerge on different types of urban destinations. Marchena Gómez (1995) analyses the significance of metropolitan tourism, an advanced state of urban tourism with two indispensable references: (a) the identifying traits of its historical and cultural heritage, manifested in monuments, urban ambience, exhibitions and festivals;

(b) the holding of cyclical or exceptional events, with a calendar of fairs, congresses, cultural, social and sporting events or, for example, winning the bid for holding the Olympic Games. Vera Rebollo and Dávila Linares (1995) indicate that heritage concentrated in the historic centres made these types of destination much more unique than more standardised attractions (congress centres, exhibition halls, shopping centres, etc.). However, there are few conceptual studies and a predominance of case studies, with general presentations of the city as a tourist destination or more focused on specific topics. López Palomeque (1995) reflect on Barcelona, a city that underwent a profound urban and tourism transformation from when it hosted the Olympics in 1992. The studies on Madrid are more specific and focus on its heritage and culture as the principal tourist attractions of the capital. Valenzuela Rubio (1999) highlights the positive implications of the urban regeneration programmes of the historic centre for the growth of cultural tourism, even though this growth was not among the explicit objectives of these programmes. Vacas Guerrero (2000, 2005) analyses the tourism dimension of museums, particularly the Paseo del Arte, an area concentrating the principal painting collections of the city.

In general terms, these studies portray a highly positive perspective. The contribution of urban heritage was extolled as it increased the attractiveness of cities to tourists. The local governments recognised culture, leisure and tourism as new drivers of urban development. This, in turn, justified the regeneration of historic neighbourhoods, which included the refurbishment of monumental buildings, the opening of museums and other cultural centres and the improvement of the public space. The most critical views came from the cultural field, which denounced the superficiality of the tourist gaze on the principal manifestations of cultural heritage. These approaches combine elitist intellectual traditions, which differentiate between traveler and tourist, with the positions that Jafari (2005) included in the so-called "precautionary platform" that emerged in tourism studies in the 70s of the twentieth century.

This positive perspective was particularly pronounced in the historic cities or art cities. They were medium-sized cities whose historic centres constituted the nucleus of their urban identity and the principal tourist attraction. From an urban geography perspective, with a wealth of applied research projects, Troitiño Vinuesa (1995, 1998a, b) indicate the importance of tourism as a driver in transforming these spaces and advocate for a sustainable development that takes into account urban and cultural considerations, not just tourism growth objectives. From a more critical point of view, Brandis and Del Río (1995) identify the impacts of tourism on the urban environment, from a landscape point of view. Barke et al. (1996) use the Ashworth and Tundbridge model in their analysis of the Spanish historical-tourist city. This model also serves as a reference for Calle Vaquero (2002), which provides a global perspective of the historic cities as tourist destinations. Among other aspects, it reflects on the effects of the insertion of the tourism function in economic and socio-cultural terms analyzing the touristification processes affecting certain areas of the historical centres of Granada and Toledo. García Hernández (2000) reflects on the concept of the tourism carrying capacity and its application to the management of heritage sites. This management implies the control of the internal flow of visitors (García Hernández, 2003) by entities with competencies going beyond tourism (García Hernández, 2007).

In addition to general reflections, during these years many case studies were also published. Although they provide a general perspective of each city, they usually focus on certain aspects which are particularly relevant. Almeida-García (2007) analyses the structure and tourism dynamics of the Spanish Heritage Cities. Troitiño Vinuesa (2002) addresses the case of Aranjuez, whose historic centre forms part of the cultural landscape declared a World Heritage Site. Galí Espelt and Donaire Benito (2006) analyse images and the tourist use of Girona. Santos Solla (1999) reflects on tourism in Santiago de Compostela, referring to the Holy Years and the Xacobeo cultural programme. Along similar lines, Marchena Gómez (1998) examines the contribution of Seville Expo '92 to the transformation of the historic city of Seville into a metropolitan destination.

7.3 From the Crisis of 2008 to COVID-19: The Most Recent Cycle of Touristification and Touristification 2.0

The period of crisis associated with the great recession had short-lived effects on the indicators of tourism activity. In 2010, the visitor figures of 2007 had been surpassed, particularly in the cities that were better positioned in international markets (Calle Vaquero & García Hernández, 2014). Moreover, the Great Recession reinforced the commitment of the local governments to tourism as a factor of urban well-being, driving a new cycle of touristification which put the safeguarding of the cultural heritage in the historic neighbourhoods at risk and generated different forms of local rejection. The field of Spanish Geography registered these changes and the traditional lines of study were enriched with contributions from critical studies and those based on the use of big data.

The new cycle of touristification was driven by strong growth in the affluence of visitors. During the period 1999–2019, according to the EOH data, the number of tourists staying in hotels multiplied by 1.81 in Malaga, 1.62 in Seville, 1.51 in Barcelona, 1.45 in Santiago de Compostela and 1.31 in Madrid. These were years of strong growth in international demand, particularly of distant markets; of the cheapening of travel and accommodation; of marketing-oriented urban policies and particularly tourism promotion; and also strong hotel investment. The growth in demand led to a significant increase in tourism pressure. The number of overnight stays per resident in 2019 reached 15.06 in Santiago de Compostela, 15.02 in Granada, 10.75 in Toledo and 10.34 in Segovia. However, the greatest relative increases were recorded in the large cities, such as Valencia (5.48), Madrid (6.38) and particularly Barcelona (13.05).

The emergence of the concept of tourist short rental housing (tourist dwellings) constituted a differential feature of this new cycle of touristification. The action of companies such as Airbnb, whose digital platforms facilitate contact between supply and demand on a global level, has favoured the conversion of housing into a tourism good. The studies conducted by the tourism organisation EXCELTUR (2015) estimate an accommodation capacity of P2P platforms equivalent to 137,196 beds in Barcelona (64% of total capacity), 40,319 in Madrid (30%), 18,907 in Seville (45%), 28,763 in Malaga (73%) and 14,254 in Donostia-San Sebastián (70%). Given the liquidity of this supply, its quantification is uncertain. However, there is a general belief that its growth has been breath-taking. This new accommodation formula has provided an alternative for tourists at a cheaper price: for the hotel sector, a new competitor; for property owners and investors, a source of income in terms of the sharing economy or as a business option through the economic use of residential assets; for the local governments, a source of conflict; and for academics, a new field of research that has generated a wealth of studies that address cases such as Barcelona (Arias Sans & Quaglieri Domínguez, 2016; Gutiérrez et al., 2017), Bilbao (Aguado-Moralejo et al., 2021), Cáceres (Rengifo Gallego et al., 2020), Madrid (Martínez Caldentey et al., 2020) or Seville (Camarillo-Naranjo et al., 2021). This new "Airbnb geography" basically examines the spatial distribution of the tourist short rental housing in each city, highlighting its concentration in historic centres. And this concentration, expressed through density and intensity indicators, creates tension in the residential markets, particularly in the rental property market.

During these years the tourist pressure on historic centres increased substantially and the term "overtourism" began to be used. The new cycle of touristification was manifested in very different ways (Calle Vaquero, 2019): (1) A greater presence of visitors in the central spaces of the city; (2) the increase in activities directly related to tourist consumption, such as accommodation and souvenir shops; (3) The reorientation of an increasingly broad range of businesses towards foreign clients; (4) The conversion of properties into a new tourism good; and (5) The creation of a landscape or urban scene dominated by tourism elements. In some cases, this process only implied an intensification of long-term phenomena; in others, it responded to more recent mutations in the tourism activity, such as the emergence of tourist short rental housing. Its effects are also known. On a morphological level, there has been a deterioration in the landscape. The higher presence of visitors has led to an increase in congestion and pollution, particularly in port cities with cruise terminals. Touristification has also led to functional simplification, breaking the mix of uses characteristic of the historic city. It also favours residential decline and the substitution of the permanent population with a temporary population with a higher purchasing power.

These effects were particularly intense in the cities with the most dynamic tourism activity, well-connected with the international markets and sometimes with large cruise ports. In these cities the touristification of "off the beaten track" spaces intensified but also affected the surrounding neighbourhoods which, until recently, had not received visitors. As in the case of other European cities, this extension of the tourist footprint has been fostered by a new tourists who seeks a different kind of contact with the city: one that is supposedly more authentic with the daily life and removed from the pre-existing tourism bubbles. The "more authentic" neighbourhoods are precisely those that are undergoing more intense processes of urban regeneration. Largely residential neighbourhoods are seeing their housing stock being transformed into tourist rental properties, the proliferation of businesses oriented to temporary visitors (laundrettes, locker rooms, etc.) and, in general, a reorientation of the retail and restaurant supply towards the tastes of foreign consumers.

As a result, it is in these neighbourhoods where there has been a greater degree of local rejection of tourism. These signs of rejection are labelled with the term "tourismphobia". As on other occasions, Barcelona paved the way. For some time, the more negative effects associated with the tourism success of the city have been questioned. Gradually, certain social movements began to incorporate tourism into their agendas. Taking a critical position of the current tourism development models, they promoted actions such as public debates, citizen protest rallies, the painting of slogans on walls and even symbolic attacks on tourist facilities. Inserted in a globalisation mindset, manifestations of this kind have formed the basis of the activity of platforms such as the *Assemblea de Barris per un Turisme Sostenible* (Barcelona), *Málaga no se Vende* (Málaga), *Lavapiés dónde Vas* (Madrid) or *Ciutat per a qui l'habita, no per a qui la visita* (Palma).

The rise in the number of urban problems related to an increase in tourism pressure led to a significant turnaround in the local tourism policies, particularly in cities with left-wing governments. This change was influenced by the action of social movements but also by the circulation of ideas between local governments, influenced by the example of Barcelona (Russo & Scarnato, 2018). The main idea was to promote a more sustainable tourism development in harmony with the city, containing the tourism activity in the central areas with greatest pressure and expanding the activity to the peripheral areas. These approaches have been included in some of the strategic plans and have taken a more specific form in measures related to marketing, the management of transport infrastructures, smart tourism destination programmes, etc. Measures have also been adopted in urban planning, a wholly local area of competence which has been used as a principal instrument to contain the expansion of accommodation in the central areas of the city (Calle Vaquero et al., 2021). Instruments of this type have been approved in Barcelona, Madrid, San Sebastián, Santiago de Compostela or Valencia. Among other aspects, attempts have been made to regulate the implementation of properties for tourism use through the adoption of a zonation model in accordance with the levels of tourism pressure.

During these years there was an increase in the number of studies on city tourism, with significant contributions from the field of geography and a greater orientation towards publishing in English and transnational approaches. The geography of city tourism became a fully consolidated field, incorporating new research lines and new data sources. López Palomeque (2015) provides an excellent outline of the evolution of tourism in Barcelona. Rengifo Gallego et al. (2015) offer a similar study on Cáceres, Escudero Gómez (2018) on Toledo and Troitiño Vinuesa and Troitiño Torralba (2009) on the Heritage cities of Castilla y León. More specifically Benach Rovira (2016) examines the impact that the promotion and sale of urban spaces to international tourism can have on the complex meanings of these spaces, García-

Hernández et al. (2017) analyse the problems of tourism pressure of the Old Quarter of San Sebastián, Santos & Lopez investigate the tourism management of Santiago de Compostela (2019) and Mínguez et al. (2019) examine the relationship between touristification and social vulnerability in Seville. Almeida-García et al. (2019) address the perception of "overtourism" of the population of the historical quarter of Malaga based on surveys conducted among the residents. Lopez et al. (2019) examine this same perception in Santiago de Compostela, although based on interviews with stakeholders. The perspective of the actors is also adopted by Valenzuela and Hidalgo (2010) in a study on the historic centre of Madrid and by Barrado-Timón and Hidalgo-Giralt (2019) in an analysis of Lavapiés, one of the neighbourhoods experiencing touristification and the associated conflict most recently.

These studies are enriched by the use of new data sources and the inclusion of new perspectives emerging in the critical studies which are habitual in urban geography research. With respect to the first issue, extracting mass data from digital platforms and analysing them through geographical information systems is common practice. In addition to studies on properties used for tourism, these sources are also used to record the tourism footprint (Salas-Olmedo et al., 2018) in order to determine visit repetition levels in monumental sites (Galí Espelt & Donaire Benito, 2018), to identify situations of congestion at heritage sites (Donaire et al., 2020) and to record the narratives with respect to Airbnb (Wilson et al., 2021), etc.

With regard to the incorporation of new perspectives based on critical studies, the first contributions interpreted touristification in terms of tourism gentrification in accordance with the approaches developed by Kevin F. Gotham in his study on the French Quarter of New Orleans. This approach has generated a considerable academic production, with fuzzy boundaries between the different social sciences and a clear orientation towards activism. From the field of geography, for the first time, García Herrera et al. (2007) examine the relationship between gentrification, displacement and tourism in a study on Santa Cruz de Tenerife. Based on this initial reference, the studies multiplied, although many addressed the situation in Barcelona. Paying particular attention to the effects of tourist short rental housing, Cocola-Gant reflects on the displacement processes and the local resistance to gentrification (Cocola-Gant & Pardo, 2017; Cocola-Gant et al., 2021). Together with López-Gay, this author addresses the socio-demographic effects of these processes, for which the Gothic Quarter of Barcelona is used as a reference (Cocola-Gant & Lopez-Gay, 2020; López-Gay et al., 2021). Blanco-Romero and Blázquez-Salom examine the relationship between the property bubble and the transformation of housing into a tourism good, also in Barcelona (Blanco-Romero & Blázquez-Salom, 2018; Blanco-Romero et al., 2018). The specific mechanisms of this relationship are studied by Arias Sans (2018) and Yrigoy (2019), in the case of the Old Quarter of Palma in Mallorca. Norfe analyses the relationship between nightlife and tourism in La Barceloneta (Nofre et al., 2018), but also the misalignments between touristification and gentrification (Sequera & Nofre, 2018). On Madrid, the work of Cabrerizo et al. (2016) focuses on the processes of touristification and social conflict in the historic centre. And also, in their studies on Seville, Jover and Díaz-Parra (2019, 2020) address these differences, concluding that transnational gentrification and touristification are new urban strategies and practices to enhance the value of real estate and appropriate urban surplus in unique urban areas.

7.4 General Conclusions and an Unfinished Epilogue: COVID-19, Undertourism and an Uncertain Future

The aim of this text is to trace the recent dynamics of the touristification of the historic centres of Spanish cities, pointing out the general and specific features of the two recent cycles of touristification and the analyses and interpretations made of this phenomenon from the perspective of geography. It is therefore a chronicle of a process that affects urban centres but also of the dominant research approaches at each moment, pointing out aspects of permanence and change.

As pointed out at the beginning of the chapter, tourism is a long-standing phenomenon at the urban level and is linked in a complex way to the process of heritagization (Gravari Barbas, 2020). As soon as tourists arrive in cities, there are processes of touristification, which are part of the general framework of tertiarisation. In Spain, these processes have been accentuated since the resurgence of city tourism in the 1990s. This resurgence and the incorporation of tourism into the urban political agenda led to the development of tourism-city relations as a specific field of analysis, which was nourished by important contributions from urban geography that later evolved into the geography of city tourism. The research methodologies were built on the foundations of urban analysis, specifically of intra-urban space, increasingly enriched by the analytical proposals that were being developed at the international level. With a strong case study orientation and direct links to planning work commissioned by administrations, they presented the virtues and shortcomings that Hiernaux (2006) pointed out for a geography of tourism based on traditional spatial analysis and a clear professional ('applied') orientation.

This orientation is in line with how tourism is incorporated into the urban agenda as a vector of opportunity and is aligned with the dominant approaches in public policies and international academic production in the field of tourism and the city. The combination of culture, leisure and tourism is conceived as a potential vector of urban development: of physical regeneration of degraded spaces and of social and functional revitalisation of areas in decline. This mix can be used for the recovery of industrial cities in crisis. But also heritage cities that are beginning to see heritage not so much as a burden but as a resource that needs to be mobilised to generate urban well-being. This is the case in most of Spain's historic cities. Tourism was looked upon in a friendly way, although there was an awareness of the more negative effects of touristification. At the environmental level, a greater presence of visitors and their vehicles favoured an increase in congestion and pollution. A growing weight of tourism-related activities directly (accommodation...) and indirectly (hotels, commerce...) encouraged functional simplification, breaking the mix of uses characteristic of urban centres. In addition, the residential decline was encouraged, with the displacement of residents to more peripheral neighbourhoods and other municipalities. The very safeguarding of these centres as heritage assets of urban nature could be threatened by the process of touristification, promoting processes such as the musealisation of urban space and/or its conversion into theme parks. However, within the dominant political and academic discourse, the risks of this process were seen as outweighed by the potential benefits for urban regeneration. For a long time, this friendly vision of urban tourism prevailed, confronting its benefits with the negative impacts generated by mass tourism on the Spanish coastline.

The growth of the tourism activity after the Great Recession intensified the process of touristification that was in progress. However, a series of new phenomena also emerged, which made these years unique. On the one hand, housing were being transformed into tourism goods. On the other hand, the tourism footprint was being expanded towards residential neighbourhoods "off the beaten track". These two aspects were closely related and gave rise to a new cycle of touristification with its own features. The action of digital platforms such as Airbnb enabled the emergence of a new accommodation supply with a completely disruptive scope. It enabled the entry of a stock of properties in the urban centres into the market, giving rise to a liquid supply. From a manifestation of the sharing economy it has become a new business niche in which property owners, property promoters, investment funds, property package managers and complementary services companies operate. The conversion of homes into tourism goods constitutes an added pressure on the residential nature of the centres, as it reduces the stock of housing available for the resident population of the city, both as owners and tenants. The risk that touristification represents to maintaining this residential condition (noise, retail transformation, massification of public space...) has become a direct threat, with a displacement of residents, flat by flat, block by block and neighbourhood by neighbourhood.

The tourism rental houses tend to be concentrated in the urban centre, even more so than the rest of the accommodation supply. However, this particularly affects a series of neighbourhoods in these centres. These areas, until now, have been removed from the tourist centre of the city and have a strong residential component, in many cases of a popular nature and sometimes with an ethnic mix. These neighbourhoods have been the object of powerful urban regeneration programmes which have also increased their attractiveness to recent settlers who initiated the incipient gentrification processes. Cosmopolitan, trendy and cool places have also become attractive to a type of visitor seeking a more unique experience. They stay in tourist short rental housing in their search for a supposedly more authentic contact with the place. The gentrification process has become tourist gentrification, fostering residential and retail displacement. To a large extent, the social resistance movements against tourism growth emerged in these neighbourhoods, where the defending the right to housing and the city against tourism speculation takes on its full meaning.

Institutional positions on these processes are plural. Competition, free market and digital economy authorities welcome new tourism operators, a position also defended by tourism administrations. Local governments present more restrictive positions, especially left-wing governments and the areas of urban planning and culture. Spurred on by social movements, they argue that tourism growth associated with VUTs cannot be carried out against the residents' right to the city. In academia, too, there is a plurality of positions and approaches. The geography of tourism in the traditional format and applied orientation continues to be important, but the critical turn, linked to social activism, with broad global connections and continuous references to the work of David Harvey and Henri Lefebvre, among others, is also widely present. The critique of tourism is a form of critique of the system in its most contemporary formulation, and the idea of gentrification is one of its most recurrent concepts. This approach allows us to link up with the tradition of the most critical urban studies, which draws on different disciplinary approaches and has recently incorporated tourism into its research agenda. On the other hand, approaches in terms of cultural, performative and new mobilities are still limited, although they permeate the work of emerging researchers. Many of these new turns involve thematic openness and significant methodological innovations. In some cases, they relate to approaches centred on the practices and relationships that lead to the production of tourist space and move from a positivist geography to a more critical, relational and post-structural geography of tourism in line with the general evolution of the discipline. The use of big data is common to all approaches.

As has been pointed out throughout the text, the processes of heritagization and touristification have been closely related. The conservation and recovery of urban cultural assets reinforce the tourist attractiveness of historic centres. And tourism development has been the argument repeatedly used to justify the protection of and investment in urban heritage. However, the relationship between the two processes is not free of contradictions (Martínez Yáñez, 2022). The growth of tourism has led to an improvement in the physical dimension of historic centres. On the other hand, in those cities where this growth has been very large and rapid, tourism has broken the pre-existing functional balances, favouring processes of simplification and displacement. In this sense, the safeguarding of historic centres as urban heritage is threatened by growing touristification. As a consequence, entities such as UNESCO, the World Heritage Centre or ICOMOS place tourism as one of the potential vectors of pressure on heritage assets and defend the need to adjust the growth of tourist activity to the carrying capacity of the sites.

The touristification process, which was peaking in 2019, completely stopped with the outbreak of the COVID-19 pandemic. In addition to the fear of infection of the potential travellers, the public health measures imposed to control the virus affected the nucleus of the tourism activity in the cities: limitations on movement and social distancing. The lockdowns in the spring of 2020 gave rise to an unprecedented situation of "zero tourism". Since then, the tourism activity has registered fluctuations associated with the successive waves of the pandemic and the public health measures adopted. In any case, the levels of activity are still much lower than pre-pandemic times. With respect to 2019, the overnight stays in hotels in 2021 were 60% lower in Barcelona and 50% lower in Madrid and Seville (according to the EOH data). Given that the flow of foreign visitors and those of business trips have particularly decreased, the more negative dynamics have corresponded to the destinations with a greater international projection, while the historic cities of the interior have recorded a less negative evolution.

The conversion of an "overtourism" situation into one of "undertourism" is having diverse effects, with a scope that is still uncertain. The abrupt fall in demand has particularly affected conventional accommodation, with temporary closures of establishments and low levels of occupation. The difficulties experienced by many companies are giving rise to the sale of property assets to international capital, which is taking advantage of the depreciation in the value of the properties in the central areas of the cities. The tourist short rental housing are faring better as they are perceived by tourists as a safer environment within a pandemic context. Furthermore, the liquid nature of this accommodation supply fosters its rapid conversion from tourist rental to residential rental and vice-versa. Therefore, it is still an attractive sector for property investors. In this respect, the approach of many of the operations in progress should be one focused on property production and the financialization of the built environment (Casañas Vilamala & Blázquez-Salom, 2021), which means adopting perspectives that are more in line with economic geography and urban geography than tourism geography, at least as it has been developed until now. On another level, souvenir sales, transport and tourist guide businesses, among others, have also been affected. And the reduction in the activity of retail and restaurant establishments, with the resulting closures, is giving rise to a substitution of traditional businesses with international brands, with companies that work in the medium term with a much higher financial capacity.

The reduction in the tourism activity has removed touristification and its effects from social debate. Similarly, these issues have also disappeared from the government agenda in the principal destinations. The measures adopted to reduce tourism pressure in the historic centres has given rise to actions to sustain the business fabric of tourism and leisure, even providing advantages sometimes questioned by the neighbourhood groups. The speed of the changes hinders their monitoring in terms of research. As a result, there are still very few studies that address this new stage with an empirical analysis. On the other hand, there are many general reflections alluding to the future relationship between tourism and territory and, more specifically, between tourism and the city and historic districts (Cañada & Murray, 2021; Escudero Gómez, 2021). In the debate on this future, there are those who consider the need to recover levels of activity similar to those before the pandemic, which would lead to a resumption of the path of touristification; there are also those who consider the current situation as an opportunity to establish a harmonious and sustainable model for inserting tourism into the city and its heritage spaces; and those who defend tourism de-growth within the framework of a more global change in economic and social organisation. The future is uncertain, but our desire is that from geography we can work together to achieve the best version.

Acknowledgements Grant PID2020-114186RB-C22 funded by MCIN-AEI/ 10.13039/501100011033.

References

Aguado-Moralejo, I., Echebarria, C., & Barrutia, J. M. (2021). Airbnb: ¿economía colaborativa o nuevo modelo de negocio? Evidencia empírica reciente desde una ciudad de tamaño medio como Bilbao. *Investigaciones Geográficas, 76*, 97–118. https://doi.org/10.14198/INGEO.18910

Almeida-García, F. (2007). Análisis de la estructura turística de las Ciudades Patrimonio de la Humanidad de España. *Baetica, 29*, 57–72.

Almeida-García, F., Cortés-Macías, R., & Balbuena-Vázquez, A. (2019). Tourism-phobia in historic centres: The case of Malaga. *Boletín de la Asociación de Geógrafos Españoles, 83*(2823), 1–32. https://doi.org/10.21138/bage.2823

Arias Sans, A. (2018). Turismo y gentrificación. Apuntes desde Barcelona. *Papers. Regió Metropolitana de Barcelona. Territori, estratègies, planejament, 60*, 130–139.

Arias Sans, A., & Quaglieri Domínguez, A. (2016). Unravelling airbnb: Urban perspectives from Barcelona. In A. P. Russo & G. Richards (Eds.), *Reinventing the local in tourism: Producing, consuming and negotiating place* (pp. 209–228). Channel View Publications. https://doi.org/10.21832/9781845415709-015

Barke, M., Towner, J., & Newton, M. Thomas. (1996). *Tourism in Spain: Critical issues*. CAB International.

Barrado-Timón, D. A., & Hidalgo-Giralt, C. (2019). Golden hordes or mere barbarians? Discourses on tourism, touristification, and tourismophobia in Madrid's Lavapiés neighborhood. *Boletín de la Asociación de Geógrafos Españoles, 83*(2824), 1–36. https://doi.org/10.21138/bage.2824

Benach Rovira, N. (2016). ¿Ciudades en el mapa o en la guía turística? Venta de la ciudad y sentido del lugar. *Revista CIDOB d'Afers Internacionals, 113*, 89–105.

Blanco Romero, A., Blázquez Salom, M., de la Calle Vaquero, M., Fernández Tabales, A., García Hernández, M., Lois González, R. C., Mínguez García, M. C., Navalón García, R., Navarro Jurado, E., & Troitiño Torralba, L. (2020). *Diccionario de turismo*. Cátedra.

Blanco-Romero, A., & Blázquez-Salom, M. (2018). The commodification of housing and urban planning in Barcelona. *Sud-Ouest Europeen, 46*, 9–22. https://doi.org/10.4000/soe.4568

Blanco-Romero, A., Blázquez-Salom, M., & Cànoves, G. (2018). Barcelona, housing rent bubble in a tourist city. Social responses and local policies. *Sustainability, 10*(6), 1–18. https://doi.org/10.3390/su10062043

Bosque Maurel, J. (1962). *Geografía urbana de Granada*. CSIC – Departamento de Geografía Aplicada del Instituto Juan Sebastián Elcano, Zaragoza. Reimpresión de 1988.

Brandis, D., & Del Río, I. (1995). Turismo y medio ambiente de las ciudades históricas. *Anales de Geografía de la Universidad Complutense de Madrid, 15*, 157–167.

Cabrerizo, C., Sequera, J., & Bachiller, P. G. (2016). Entre la turistificación y los espacios de resistencia en el centro de Madrid: Algunas claves para (re) pensar la ciudad turística. *Ecología Política, 52*, 78–82. Retrieved from https://www.ecologiapolitica.info/?p=6775

Calabuig, J., & Barrado, D. A. (Eds.). (2001). *Geografía mundial del turismo*. Síntesis.

Camarillo-Naranjo, J.-M., Vallejo-Villalta, I., Fernández-Tabales, A., & Santos-Pavón, E. (2021). Where is tourist housing actually located? New approaches and sources for detailed scale analysis. *European Planning Studies*, 1–25. https://doi.org/10.1080/09654313.2021.2002825

Cañada, E., & Murray, I. (Eds.). (2021). *TourismPostCOVID19. Turistificación confinada*. Alba Sud Editorial.

Casañas Vilamala, F., & Blázquez-Salom, M. (2021). De casal a hotel boutique al Centre Històric de Palma (2012–2020). *Kult-Ur, 8*(15), 99–122. https://doi.org/10.6035/kult-ur.2021.8.15.3

Cocola-Gant, A. (2011). El Barrio Gótico de Barcelona. De símbolo nacional a parque temático. *Scripta Nova. Revista Electrónica de Geografía y Ciencias Sociales, XV* (371).

Cocola-Gant, A., & Lopez-Gay, A. (2020). Transnational gentrification, tourism and the formation of 'foreign only' enclaves in Barcelona. *Urban Studies, 57*(15), 3025–3043. https://doi.org/10.1177/0042098020916111

Cocola-Gant, A., & Pardo, D. (2017). Resisting tourism gentrification: The experience of grassroots movements in Barcelona. *Urbanistica Tre, Giornale Online di Urbanistica, 5*(13), 39–47.

Cocola-Gant, A., Hof, A., Smigiel, C., & Yrigoy, I. (2021). Short-term rentals as a new urban frontier –evidence from European cities. *Environment and Planning A: Economy and Space, 53*(7), 1601–1608.

Calle Vaquero, M. de la (2002). *La ciudad histórica como destino turístico*. Ariel.

Calle Vaquero, M. de la (2008). El turismo en las políticas urbanas. Aproximación a la situación de las ciudades españolas. En J. A. Ivars y J. F. Vera (coord.) *Espacios Turísticos. Mercantilización, Paisaje e Identidad* (pp. 507–529). Universidad de Alicante.

Calle Vaquero, M. de la (2019). Turistificación de centros urbanos: clarificando el debate. *Boletín de la Asociación de Geógrafos Españoles, 83*(2829), 1–40. https://doi.org/10.21138/bage.2829

Calle Vaquero, M. de la & García Hernández, M. (2014). Las ciudades históricas españolas ante la crisis: dinámica de la afluencia de visitantes y acciones de política turística. En E. Navarro Jurado et al. (coord. y ed.), *Espacios turísticos e inteligencia territorial. Respuestas ante la crisis* (pp. 339–355). Universidad de Sevilla y Universidad de Málaga.

Calle Vaquero, M. de la & García Hernández, M. (2016). Políticas locales de turismo en ciudades históricas españolas. Génesis, evolución y situación actual. *PASOS Revista de Turismo y Patrimonio Cultural, 14*(3), 691–704. https://doi.org/10.25145/j.pasos.2016.14.045

Calle Vaquero, M. de la & García Hernández, M. (2020). Turismo, patrimonio y centros urbanos: ¿hacia un nuevo ciclo de turistificación?. En Pons, G.X., Blanco-Romero, A., Navalón-García, R., Troitiño-Torralba, L. & Blázquez-Salom, M. (eds.). *Sostenibilidad turística: overtourism vs undertourism* (pp. 135–146). Mon. Soc. Hist. Nat. Balears, 31 ,Palma (Illes Balears).

Calle Vaquero, M. de la, García Hernández, M., & de Miguel, S. M. (2021). Urban planning regulations for tourism in the context of overtourism. Applications in historic centres. *Sustainability, 13*(1), 1–21. https://doi.org/10.3390/su13010070

Donaire, J. A., Galí, N., & Gulisova, B. (2020). Tracking visitors in crowded spaces using zenith images: Drones and time-lapse. *Tourism Management Perspectives, 35*, 100680. https://doi.org/10.1016/j.tmp.2020.100680

Escudero Gómez, L. A. (2018). Realities and problems of a major cultural tourist destination in Spain, Toledo. *PASOS. Revista de Turismo y Patrimonio Cultural, 18*(3), 617–636. https://doi.org/10.25145/j.pasos.2018.16.045

Escudero Gómez, L. A. (2021). Cultural tourism in cities post-COVID-19: A perspective and proposals for an alternative model. *Boletín de la Asociación de Geógrafos Españoles, 91*, 1–51. https://doi.org/10.21138/bage.3138

EXCELTUR. (2015). *Alojamiento turístico en viviendas de alquiler: impacto y retos asociados*. Retrieved from: https://www.exceltur.org/uncategorized/alojamiento-turistico-2/

Fernández Tabales, A. (2018). Turismofobia "Avant la lettre" en la Sevilla de 1929: el viejo debate entre autenticidad y mercantilización del patrimonio. *Revista Otras Arqueologías, 3*, 131–148.

Galí Espelt, N., & Donaire Benito, J. A. (2006). La imagen percibida por los turistas de la ciudad de Girona. *Estudios Turísticos, 168*, 123–140.

Galí Espelt, N., & Donaire Benito, J. A. (2018). First-time versus repeat visitors' behavior patterns: a GPS analysis. *Boletín de la Asociación de Geógrafos Españoles, 78*, 49–65. https://doi.org/10.21138/bage.2711

García Álvarez, J. (2007). Paisajes nacionales, turismo y políticas de memoria: Toledo (1900–1950). *Ería, 73–74*, 193–212.

García Hernández, M. (2000). Turismo y medio ambiente en ciudades históricas: de la capacidad de acogida turística a la gestión de los flujos de visitantes. *Anales de Geografía de la Universidad Complutense, 20*, 131–148.

García Hernández, M. (2003). *Turismo y conjuntos monumentales: capacidad de acogida turística y gestión de flujos de visitantes*. Tirant lo Blanch.

García Hernández, M. (2007). Entidades de planificación y gestión turística a escala local. El caso de las Ciudades Patrimonio de la Humanidad de España. *Cuadernos de Turismo, 20*, 79–102.

García Hernández, M., & de la Calle Vaquero, M. (2019). Urban heritage. In B. Warf (Ed.), *Oxford bibliographies in geography*. Oxford University Press. https://doi.org/10.1093/OBO/9780199984002-0208

García Herrera, L. M., Smith, N., & Mejías Vera, M. Á. (2007). Gentrification, displacement, and tourism in Santa Cruz de Tenerife. *Urban Geography, 28*(3), 276–298. https://doi.org/10.2747/0272-3638.28.3.276

García-Hernández, M.,, Calle Vaquero, M. de la, & Yubero, C. (2017). Cultural heritage and urban tourism: Historic city centres under pressure. *Sustainability, 9*(1346). https://doi.org/10.3390/su9081346

Gravari Barbas, M. (2020). Heritage and tourism: From opposition to coproduction. In M. Gravari Barbas (coord.), *A research agenda for heritage tourism*. Edward Elgar Publishing Limited.

Gutiérrez Ronco, S. (1984). *La función hotelera de Madrid*. CSIC, Instituto de Geografía Aplicada.

Gutiérrez, J., García-Palomares, J. C., Romanillos, G., & Salas-Olmedo, M. H. (2017). The eruption of Airbnb in tourist cities: Comparing spatial patterns of hotels and peer-to-peer accommodation in Barcelona. *Tourism Management, 62*, 278–291. https://doi.org/10.1016/j.tourman.2017.05.003

Hiernaux, D. (2006). Geografia del Turismo. In D. Hiernaux & A. Lindón (Eds.), *Tratado de geografía humana* (pp. 401–432). Anthropos.

INSTITUTO NACIONAL DE ESTADÍSTICA (INE). (1990a-nowadays). *FAMILITUR – Encuesta de Turismo de Residentes*. https://www.ine.es/dyngs/INEbase/es/operacion.htm?c=Estadistica_C&cid=1254736176990&menu=ultiDatos&idp=1254735576863

INSTITUTO NACIONAL DE ESTADÍSTICA (INE). (1990b-nowadays). *FRONTUR – Movimientos turísticos en frontera*. https://www.ine.es/dyngs/INEbase/es/operacion.htm?c=Estadistica_C&cid=1254736176996&menu=ultiDatos&idp=1254735576863

INSTITUTO NACIONAL DE ESTADÍSTICA (INE). (several years). *EOH – Encuesta de Ocupación Hotelera*. https://www.ine.es/dyngs/INEbase/es/operacion.htm?c=Estadistica_C&cid=1254736177015&menu=ultiDatos&idp=1254735576863

Jafari, J. (2005). El turismo como disciplina científica. *Política y Sociedad, 42*(1), 39–56.

Jover, J., & Díaz-Parra, I. (2019). Gentrification, transnational gentrification and touristification in Seville, Spain. *Urban Studies, 57*(15), 3044–3059. https://doi.org/10.1177/0042098019857585

Jover, J., & Díaz-Parra, I. (2020). Who is the city for? Overtourism, lifestyle migration and social sustainability. *Tourism Geographies, 24*(1), 9–32. https://doi.org/10.1080/14616688.2020.1713878

Lalana Soto, L. (2022). Paisajes urbanos históricos. De la ciudad como patrimonio al patrimonio en la ciudad. *Revista PH, 107*, 106–125. https://doi.org/10.33349/2022.107.5201

López Palomeque, F. (1995). La estrategia del turismo metropolitano: El caso de Barcelona. *Estudios Turísticos, 126*, 119–141.

López Palomeque, F. (2015). Barcelona, de ciudad con turismo a ciudad turística. Notas sobre un proceso complejo e inacabado. *Documents d'anàlisi geográfica, 61*(3), 483–506. https://doi.org/10.5565/rev/dag.296

López Palomeque, F., Antón, S., Marchena Gómez, M., & Vera Rebollo, J. F. (1997). *Análisis territorial del turismo: una nueva geografía del turismo* (1a. ed.). Ariel.

Lopez, L., Pazos Otón, M., & Piñeriro Antelo, M. de los Á. (2019). ¿Existe overtourism en Santiago de Compostela? Contribuciones para un debate ya iniciado. *Boletín de la Asociación de Geógrafos Españoles, 83*(2825), 1–48. https://doi.org/10.21138/bage.2825

López-Gay, A., Cocola-Gant, A., & Russo, A. P. (2021). Urban tourism and population change: Gentrification in the age of mobilities. *Population, Space and Place, 27*(1), e2380. https://doi.org/10.1002/psp.2380

Luis Gómez, A. (1988). Geografía española: paisaje y análisis de las actividades de ocio. *Baética, 11*, 71–100.

Marchena Gómez, M. (1995). El turismo metropolitano: una aproximación conceptual. *Estudios Turísticos, 126*, 7–21.

Marchena Gómez, M. (1998). El impacto de las grandes exposiciones en la actividad turística: las consecuencias de Sevilla 1992. *Estudios y perspectivas en turismo, 7* (1–2).
Martínez Caldentey, M. A., Murray, I., & Blázquez Salom, M. (2020). En la ciudad de Madrid todos los caminos conducen a Airbnb. *Investigaciones Turísticas, 19*, 1. https://doi.org/10.14198/inturi2020.19.01
Martínez Yáñez, C. (2022). Los efectos del turismo en el patrimonio cultural: paradojas, contradicciones y desequilibrios en un momento de cambio. En R. Martínez, F. Cabrales, M. de la Calle Vaquero, M. García, C. Mínguez, & L. Troitiño (coords.), *Leyendo el territorio. Homenaje a Miguel Ángel Troitiño* (pp. 730–742). Universidad de Guadalajara.
Martínez-Caldentey, M. A., Blázquez-Salom, M., & Murray, I. (2020). Habitabilidad y Airbnb: el alquiler de la vivienda en el distrito Centro de Madrid. *Cuadernos de Turismo, 46*, 367–393. https://doi.org/10.6018/turismo.451881
Mínguez, C., Piñeira, M. J., & Fernández-Tabales, A. (2019). Social vulnerability and touristification of historic centers. *Sustainability, 11*(16), 1–24. https://doi.org/10.3390/su11164478
Nofre, J., Giordano, E., Eldridge, A., Martins, J. C., & Sequera, J. (2018). Tourism, nightlife and planning: challenges and opportunities for community liveability in La Barceloneta. *Tourism Geographies, 20*(3), 377–396. https://doi.org/10.1080/14616688.2017.1375972
Rengifo Gallego, J. I., Campesino Fernández, A. J., & Sánchez Martín, J. M. (2015). El turismo en la ciudad de Cáceres (1986–2010): un cuarto de siglo emblemático. *Boletín de la Asociación de Geógrafos Españoles, 2015*(67), 375–401. https://doi.org/10.21138/bage.1830
Rengifo Gallego, J. I., Campesino Fernández, A. J., Sánchez Martín, J. M., Salcedo Hernández, J. C., & Martín Delgado, L. M. (2020). Los apartamentos turísticos de la ciudad de Cáceres: Rehabilitación y refuncionalización del centro histórico. *Cuadernos Geográficos, 59*(3), 238–263. https://doi.org/10.30827/cuadgeo.v59i3.11136
Rodwell, D. (2022). Sustainable urban heritage vs heritage orthodoxy. In M. T. Albert, R. Bernecker, C. Cave, A. C. Prodan, & M. Ripp (Eds.), *50 years world heritage convention: Shared responsibility – Confict & reconciliation* (pp. 125–138). Springer. https://doi.org/10.1007/978-3-031-05660-4_10
Russo, A. P., & Scarnato, A. (2018). "Barcelona in common": A new urban regime for the 21st-century tourist city? *Journal of Urban Affairs, 40*(4), 455–474. https://doi.org/10.1080/07352166.2017.1373023
Salas-Olmedo, M. H., Moya-Gómez, B., García-Palomares, J. C., & Gutiérrez, J. (2018). Tourists' digital footprint in cities: Comparing Big Data sources. *Tourism Management, 66*, 13 25. https://doi.org/10.1016/j.tourman.2017.11.001
Santos Solla, X. M. (1999). Mitos y realidades del Xacobeo. *Boletín de la Asociación de Geógrafos Españoles, 28*, 103–117.
Santos Solla, X. M., & Lopez, L. (2019). Gestión turística en ciudades históricas. Santiago de Compostela. *Estudios Turísticos, 218*, 195–204.
Sequera, J., & Nofre, J. (2018). Shaken, not stirred: New debates on touristification and the limits of gentrification. *City, 22*(5–6), 843–855. https://doi.org/10.1080/13604813.2018.1548819
Troitiño Vinuesa, M. A. (1992). *Cascos antiguos y centros históricos: problemas, políticas y dinámicas urbanas*. Ministerio de Obras Públicas y Transportes.
Troitiño Vinuesa, M. A. (1995). El turismo en las ciudades históricas. *Polígonos, 5*, 49–65. https://doi.org/10.18002/pol.v0i5.1131
Troitiño Vinuesa, M. A. (1998a). Turismo y desarrollo sostenible en las ciudades históricas con patrimonio arquitectónico-monumental. *Estudios Turísticos, 137*, 5–53.
Troitiño Vinuesa, M. A. (1998b). Turismo y desarrollo sostenible en ciudades históricas. *Ería: Revista Cuatrimestral de Geografía, 47*, 211–228.
Troitiño Vinuesa, M. A. (2002). Aranjuez: Patrimonio Cultural, recuperación urbana y turismo. *Anales de Geografía de la Universidad Complutense*, 495–518.

Troitiño Vinuesa, M. A., & Troitiño Torralba, L. (2009). Turismo y patrimonio en Castilla y León: las ciudades patrimonio de la humanidad (Ávila, Salamanca y Segovia) como destinos turísticos de referencia. *Polígonos, 19*, 145–178.

Troitiño Vinuesa, M. A., de la Calle Vaquero, M., García Hernández, M., & Troitiño Torralba, L. (2007). *La situación turística del Grupo Español de Ciudades Patrimonio de la Humanidad: bases para la puesta en marcha del Observatorio Turístico*. Ed. Grupo Español de Ciudades Patrimonio de la Humanidad / Universidad Complutense de Madrid, León.

Vacas Guerrero, T. (2000). Los museos madrileños como oferta turístico-cultural. *Cuadernos de Turismo, 5*, 105–111.

Vacas Guerrero, T. (2005). Los museos del "Paseo del Arte": espacio turístico-cultural de ámbito internacional. *Estudios Turísticos, 165*, 97–129.

Valenzuela, M., & Hidalgo, C. (2010). Tourism in historic cores, conflict or opportunity? The stakeholders point of view in Madrid's case. *Rivista di Scienze del Turismo – Ambiente Cultura Diritto Economia, 1*(2), 281–300.

Valenzuela Rubio, M. (1999). Turismo cultural y políticas urbanas sostenibles en el casco histórico de Madrid. *Revista de Arte, Geografía e Historia, 2*, 387–419.

Vera Rebollo, J. F., & Dávila Linares, J. M. (1995). Turismo y patrimonio histórico y cultural. *Estudios Turísticos, 126*, 161–177.

Wilson, J., Garay-Tamajon, L., & Morales-Perez, S. (2021). Politicising platform-mediated tourism rentals in the digital sphere: Airbnb in Madrid and Barcelona. *Journal of Sustainable Tourism*, 1–22. https://doi.org/10.1080/09669582.2020.1866585

Yrigoy, I. (2019). Rent gap reloaded: Airbnb and the shift from residential to touristic rental housing in the Palma Old Quarter in Mallorca, Spain. *Urban Studies, 56*(13), 2709–2726. https://doi.org/10.1177/0042098018803261

Manuel de la Calle-Vaquero is associate professor in the Department of Geography (Complutense University of Madrid). He has a PhD in Geography and a Bachelor in Sociology. He is the Coordinator of the PhD Programm in Tourism of the Complutense University (tourism and heritage research line). He is part of the research team "Tourism, Heritage and Development" (www.ucm.es/geoturis). He is the author of numerous publications on these subjects and he has participated in several research projects of competitive calls (EU; National Plan of I + D + i, Regional Governement Madrid, AECID).

María García-Hernández PhD is associate professor of Human Geography at Complutense University of Madrid. She has a PhD in Geography, a bachelor in Geography and a bachelor in Documentation (UC3M). She is the director of the research group "Tourism, Heritage and Development" (www.ucm.es/geoturis) focused on cultural tourism and tourism planning of heritage sites. Her researches are focused on: cultural tourism, tourism carrying capacity indicators, management of visitor flows in cultural attractions and destinations, urban tourism, and heritage destinations typologies.

Chapter 8
Risks of the Current State of Overtourism in Medium and Small Cities of Spain's Interior: An Analysis Through Indicators

Carmen Hidalgo-Giralt, Antonio Palacios-García, Diego A. Barrado-Timón, and Francisca Cea-D'Ancona

8.1 Introduction

Over the past two decades, Spain consistently broke all records in terms of international tourist arrivals, reaching a peak in 2019 (just prior to the pandemic) of over 83.5 million travelers, according to the Border Survey of Tourist Movements (FRONTUR) from Spain's National Institute of Statistics (INE). According to the Panorama of International Tourism report (2020) from the World Tourism Organization (UNWTO), Spain was in that same year the second most-visited country in the world, surpassed by only France with 90.2 million international tourists. Adding to this 173 million trips by Spanish tourists within the country's borders (INE Survey of Resident Tourists 2020), total figures compute to an astounding 263.2 million in 2019.

Although 'sun and beach' tourism continues to predominate in Spain as the main tourist modality,[1] urban tourism has also undergone significant development in

[1] According to the Border Survey of Tourist Movement from the INE (2019), the Autonomous Communities that received a greater number of international tourists were those with high specialization in coastal tourism: Catalonia received 23.2% of total international tourists, followed by the Balearic Islands (16.4%), the Canary Islands (15.7%), and Andalusia (14.4%).

C. Hidalgo-Giralt (✉) · A. Palacios-García · D. A. Barrado-Timón
Department of Geography, Faculty of Philosophy and Letters, Universidad Autónoma de Madrid, Madrid, Spain
e-mail: carmen.hidalgog@uam.es

F. Cea-D'Ancona
Department of Applied Economics, Faculty of Economics and Business Administration, Universidad Autónoma de Madrid, Madrid, Spain

A. Blanco-Romero, M. Blázquez-Salom (eds.), *Spanish Tourism Geographies*, Geographies of Tourism and Global Change,
https://doi.org/10.1007/978-3-031-39780-6_8

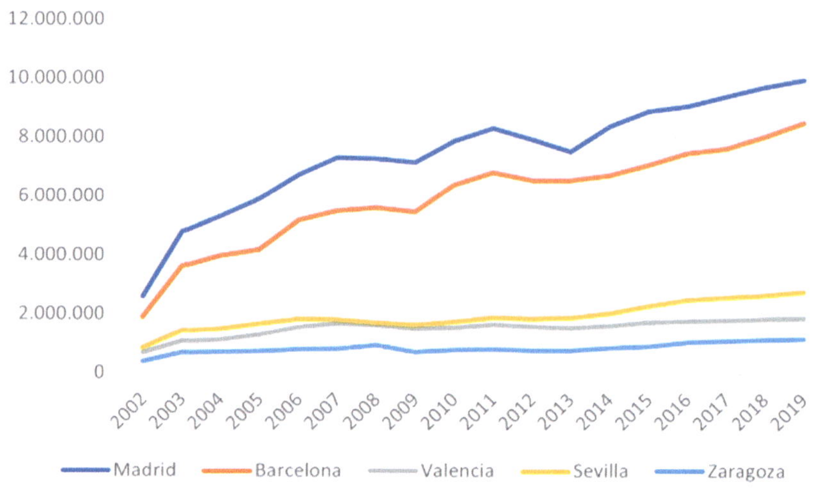

Fig. 8.1 Travelers entering through tourist points (Madrid, Barcelona, Seville, Valencia, and Zaragoza). (Source: Authors' elaboration, taking as reference the INE Survey on Hotel Occupancy)

recent decades, as shown in Fig. 8.1, which depicts the increase in travelers during 2002–2019 in five of the most populous cities: Madrid, Barcelona, Seville, Valencia, and Zaragoza. Although this rise has generated countless positive impacts on the territory (increases in employment and GDP, recovery of cultural heritage and local traditions, etc.), when managed inefficiently, such expansion can affect destinations in negative ways, as through overtourism, touristification, and tourismophobia – three dynamics addressed in this study from a theoretical point of view.

On an international scale, cities where the problem of overcrowding has become more visible include Barcelona in Spain, as well as Venice, Amsterdam, and Lisbon; but the effects of increased tourism are likewise evident in other Spanish destinations such as Madrid, Valencia, Palma de Mallorca, and Malaga (Fig. 8.2). Considering this context, it must be assumed that a process of intensification of tourist flows is also taking place in Spain's most touristic small and medium cities, and thus the parallel phenomenon of tourist massification or overtourism. The objective of this research is to explore the pressure exerted by these tourist flows in smaller Spanish cities, in order to determine via relevant indicators the risks of overtourism to which they are exposed.

8.2 Conceptual Aspects of Overtourism

The term 'overtourism', defined as "the impact of tourism on a destination or part of it, which is perceived to have an excessive and negative influence on the quality of life of citizens and/or on the quality of the experience of visitors" (UNWTO, 2018), is both relatively new and widely recognized by society, due to extensive mention by

Fig. 8.2 Examples of cities with high concentrations of visitor flows. (Source: Author (Carmen Hidalgo-Giralt))

the media (Ali, 2016; Gretzel, 2019; Koens et al., 2018). However, its conceptual configuration dates back to the 1960s, when reflection – both academic (Forster, 1964; Butler, 1974; Cohen, 1988; Swarbrooke, 1999; Clawson & Knetsch, 2011) and from international organizations (WCED, 1987; Cape Town Declaration on Responsible Tourism, 2002; Quebec Declaration on Eco-tourism, 2002) – commenced on the negative impacts of mass tourism and the need to align it with the new paradigms of sustainability. Thus overtourism presently refers to a rapid and unprecedented increase in the number of tourists concentrated into city centers, where the majority of tourist resources and services are located (Goodwin, 2017; Seraphin et al., 2018), transforming the destination city from various perspectives (social, cultural, economic, landscape, etc.). The term also alludes to elements such as the ability to correctly manage tourist congestion, the reasons for increases in flows, or how this phenomenon transcends tourist activity itself (Milano, 2018; Navarro & Capote, 2021).

An interesting debate has emerged around the limits of overtourism. At what point can an historic center be considered to suffer from overtourism? This is extremely difficult to measure from a general perspective, since every tourist destination comprises a multitude of complex factors. However, forms of measurement must be established that can reveal the risks of overtourism in order to apply containment measures; hence, in the empirical portion of this research, indicators are proposed that focus on that precise task. On the other hand, overtourism is a phenomenon that is partly subjective (García Hernández et al., 2019; Navarro & Capote, 2021), depending on a perception of saturation by the local population, by tourists, and by the agents both public and private involved in its management, as well as the interests of all those actors, along with any evident negative effects (garbage, noise, loss of local identity, etc.).

Reasons for overtourism in cities respond to multiple and complex factors (de la Calle-Vaquero, 2019; Judd & Fainstein, 1999). Undeniable, for instance, is the advancing influence of urban societies through globalization, or the role that the urban way of life has come to play (Judd, 2003). Cities have also seen their accessibility strengthened by the rise of low-cost airlines and the proliferation of new models of accommodation based on platform economies, such as Airbnb (Wilson et al., 2021). Similarly, international tourist arrivals have risen notably thanks to the incorporation into the tourism market of emerging countries, especially from the Asia-Pacific region. Nor can one overlook the fact that since the global recession of 2008, increased deregulation through neoliberal policies (Colomb & Novy, 2016) has favored the commodification of cities and their transformation into entertainment centers (Barrado-Timón & Hidalgo-Giralt, 2019), causing what Ardura and Sorando (2016) term the "creative destruction of cities."

Regarding the impacts of overtourism, it must be borne in mind that two distinct functions overlap in urban centers. On the one hand, these are spaces inhabited by a resident population; on the other, they are destinations consumed by tourists. An uncontrolled increase in tourism threatens the balance between living space (that is, "the right to the city" expressed by Levebfre) and tourist space, thereby compromising the pillars of sustainability and accelerating touristification. In the particular concern of sustainability, overtourism generates profound changes in traditional economic activities, and especially commerce, which ultimately targets the tourist and not the resident. Overtourism also contributes to speculative increases in property prices and rents. Socially, together with other urban dynamics like gentrification, overtourism favors expulsion of the local population and its replacement by a very volatile tourist population that establishes minimal roots in a neighborhood. In terms of cultural aspects, overtourism transforms the identity of neighborhoods and their customs, traditions, landscapes, etc., as well as the types of cultural activities they develop (Palacios-García et al., 2020a, b, 2021). From an environmental perspective, noise pollution, environmental degradation, and increases in solid waste are only a few of the most immediate effects.

A large proportion of these impacts contribute to transforming the space to welcome tourists without concern for the resident population, generating 'touristification' of destinations. This concept refers to changes made in historic city centers to satisfy the requirements of tourists, leaving citizen needs aside (Clancy, 2019; López-Villanueva & Crespi-Vallbona, 2021; Miriam-Hermi, 2019). For all these reasons, it is not surprising when a population begins to perceive tourism as a negative activity and protest against it (Garay et al., 2020). Moreover, not only can an increase in tourism negatively affect the living conditions and space of a local population, but it can drive a deterioration of the tourist experience as the visitor encounters trampled destinations, saturated attractions, and absent cultures.

If overtourism and its impacts are the problems considered in this research, then tools for management of tourism are understood as parts of the solution. Before delving into these, it must be noted that the poorly managed intensification of tourism is a rather difficult problem to solve, since each destination develops its own casuistry that can only be resolved individually (Goodwin, 2017). There is no single model of overtourism, nor is there a single exit from it. However, many authors point out that success in mitigating the problem will lie, to a great extent, in the public and private actors involved in the touristic development of destinations. The implementation of effective management measures will therefore address the impacts of tourism from a sectoral as well as a strategic logic (García Hernández et al., 2019; Palacios-García et al., 2020a, b).

The contribution of this research to understanding the phenomenon of mass tourism includes the analysis of diagnostic tools relating to overtourism, and specifically indicators of risk. It should be clarified that measurement of the negative impacts of tourist activity is nothing new, with antecedents stretching back to the application of 'carrying capacities' for tourist destinations (*i.e*., quantification of the optimal number of tourists that a given area can contain). Furthermore, numerous studies have been made of the impacts of tourism development on destinations, most notably Doxey's 'Irritation Index' that determines changes in the attitudes of residents toward tourism based on the number of visitors; also pertinent is Butler's life-cycle of tourist destinations.

As a result of the visibility of the phenomenon, numerous tools have been developed to measure tourist pressure. In addition to traditional tools addressing the impact of tourism on the GDP of destinations, or the evolution in percentages of tourist flows over a given period, newer tools consider the threats to cultural heritage, the overloading of tourist infrastructures, or the number of tourist apartments (World Travel & Tourism Council, 2017). As specified in the Methodology section, the fundamental objective of this research is to develop an indicator (articulated from other indicators) that allows determination of the risks to a destination posed by a rise in poorly planned tourism that exceeds the limits of capacity and load.

8.3 Overtourism in the Case of Spain: A Bibliographic Approach

Studies of overtourism or tourist overload in Spain date back to the final decades of the twentieth century, when increased tourist flows to cities had become evident and tourism policies began to be integrated into urban agendas (de la Calle-Vaquero, 2019). It was then that urban tourism caught the interest of academia, and many specialists began to analyze from diverse perspectives the urban transformations created by tourism (Valenzuela Rubio, 1992; Troitiño Vinuesa, 1995; Zárate Martín, 1995; López Palomeque, 1995). At the dawn of the new century, studies on urban tourism multiplied, at various scales and using many approaches (Molina Collado et al., 2007; López & Pazos Otón, 2009; Troitiño Vinuesa & Troitiño Torralba, 2009; Barrado-Timón, 2010; Manero Miguel, 2011), with special emphasis on the drawbacks (de la Calle-Vaquero, 2019). Scholars turned their attention to tourist overload of destinations – a direct reference to overtourism – and the implementation of management, measurement, and mitigation, for example through the determination of carrying capacity (García Hernández, 2000; Bonilla & Bonilla, 2008; García Hernández & de la Calle-Vaquero, 2012).

As mentioned in the previous section, since the economic recession that began in 2008, numerous Spanish tourist destinations have been compromised by increases in visitor flows. Parallel to this phenomenon, a tremendously fruitful academic debate has opened around tourist overload, and the term 'overtourism' is now accepted by most researchers. Milano (2018) was among the first specialists in Spain to focus the debate on overtourism and its origins and implications. Other forums for discussion have since acquired urgency. Two of note include Issue 83 of the Bulletin of the Spanish Association of Geography (BAGE), 'Transformations of urban centers induced by tourism: contractions and controversies' (2019) – with the theme of 'tourist overload' as the backbone (de la Calle-Vaquero et al., 2019) – and publication of the Proceedings of the VII International Tourism Colloquium AGE-IGU held in Maó, Menorca, in October of 2020, entitled 'Tourist sustainability: overtourism vs undertourism' and now a point of reference on this subject in Spain (Pons et al., 2020).

The published bibliography related to this critical issue for urban tourism can be classified into five large groups. The first refers to conceptual aspects of overtourism, reflections on aspects related to tourist overload, and characterization of the territories where this phenomenon persists (Milano, 2018; de la Calle-Vaquero, 2019; Piñeira et al., 2020; de la Calle-Vaquero & García-Hernández, 2020; Valero-Escandell & García-Tortosa, 2020). The second group of studies, among the most prolific, is related to tools that aim to measure or analyze possible solutions to overtourism from different points of view: urban instruments (de la Calle-Vaquero et al., 2020; Velasco-González et al., 2019), measurement indicators (Pareto Boada et al., 2020; Guitart & Serrat, 2020; Hidalgo et al., 2020) or technological solutions (Gutiérrez Puebla et al., 2016; García Hernández et al., 2019). The third case groups both the perceptions that local agents have of overtourism (Blanco-Romero et al., 2019; Barrado-Timón & Hidalgo-Giralt, 2019) and the emergence of neighborhood resistance movements against the uncontrolled rise of tourism in cities (Milano, 2017; Romero-Padilla et al., 2019). A fourth branch takes a very critical view of the commodification of tourist housing in historic centers (Gil & Sequera, 2018; Garay et al., 2020; Martínez-Caldentey et al., 2020; Garay-Tamajón et al., 2022; Gil, 2021), and a fifth group focuses on specific case studies that demonstrate the transformations that overtourism compels in certain city neighborhoods (Palacios-García et al., 2020a, b; Palacios-García et al., 2021).

The bibliography centered on tourist overload in Spain is not only recent – the phenomenon has attracted the attention of researchers since the 1990s, when urban tourism began to develop in current terms. However, it was not until the 2008 crisis that the term 'overtourism' appeared regularly in scientific publications, addressing the idea of tourist overload but also focusing on its impacts on destinations (social, cultural, economic, etc.) and on ways to measure the phenomenon and seek possible solutions through analysis. As noted, the topics have been quite varied and take diverse methodological approaches, but all tend to develop highly critical views.

8.4 Overtourism in Small and Medium Spanish Cities: An Empirical Case

Taking into account the problems generated by increased tourist pressure in the urban centers of large cities, the central hypothesis of this research proposes that this may likewise be the case with small and medium-sized in Spain's interior; the phenomenon is not exclusive to large cities or 'sun and beach' destinations but has been a characteristic feature of Spanish tourism for a decade or more. The precise main objective of this work consists in analyzing and characterizing the evolution of tourist flows in medium and small non-coastal Spanish cities from the perspective of overtourism, with special attention paid to the risks that this generates. The years 2007, 2014, and 2021 have been taken as references for study, representing a range sufficiently wide to visualize specific instances such as the behavior of tourism following the recession of 2008, or the impact of COVID-19, yet near enough in time to determine current trends such as increased tourist pressure.

To verify this working hypothesis and meet the stated objectives, from a methodological point of view, the report 'Coping with success. Managing overcrowding in tourist destinations' (2017) prepared by the World Travel & Tourism Council has been consulted. This report diagnoses the risks of overtourism in 68 cities worldwide, through a set of five indicators: (1) Alienated residents; (2) Degraded tourist experience; (3) Overloaded infrastructure; (4) Damage to nature; and (5) Threats to culture and heritage. For the present research, five additional indicators – structure, tourist intensity, tourist seasonality (I), tourist seasonality (II), and cultural heritage – have been designed, structured into three dimensions – alienation of the local population, overloading of tourist infrastructures, and threats to cultural heritage – in order to reveal potential risks of overtourism in medium and small Spanish cities (Table 8.1).

8.5 Analysis of Results

Once the indicators were established and results obtained, two types of analysis were carried out:

Analysis 1. Tourism Index (TI). This index, based on the dimensions and the risk indicators of overtourism, offers a general view of the phenomenon. Statistical justification of the TI can be found in Annex 1.

Table 8.1 Indicators of risks of overtourism

Table of Contents	
Dimension 1. Alienation of local population	
Indicator 1. Structure	Ratio 1. Month with more personnel/less personnel employed
Indicator 2. Tourist intensity	Ratio 2. Travelers/local population (sum of non-resident travelers and residents in Spain)
Dimension 2. Overloading of tourist infrastructures	
Indicator 3. Tourist seasonality (I)	Ratio 3. Degree of occupancy by place
Indicator 4. Tourist seasonality (II)	Ratio 4. Month with more overnight stays/fewer overnight stays
Dimension 3. Threats to cultural heritage	
Indicator 5. Cultural heritage	Ratio 5. No. of Goods of Cultural Interest (BICs)[a]/Travelers

Source: Authors' elaboration with reference to the World Travel and Tourism Council (2017)
[a]BIC signifies the highest degree of protection of cultural heritage in Spain

Analysis 2. Diagnosis Matrix (DM). This matrix of the risks of overtourism helps in discovering the state of vulnerability in which this type of tourist destination finds itself. Unlike the TI, the DM is based on structuring of the indicator results into quintiles, rather than factorial grouping.

Through these two approaches of analysis, the issue of overtourism in small and medium-sized inland Spanish cities can be examined. The dual-approach makes possible the combination of relatively micro (urban scale) and relatively macro (national scale) perspectives.

8.5.1 Tourism Index (TI)

The results of the TI analysis for the year 2007 reveal that eight of the 34 cities analyzed (23% of total) are at high risk of overtourism (Jaca, Antequera, Ronda, Soria, Teruel, Toledo, and Zamora), with group mean of 53.79 (Annex 2). As a characteristic feature of this year, the cities show notably high and medium TI values (Índice), above the global average, in the five indicators. However, the structural factors ('Structure'-Estructural- and 'Tourist seasonality II'-Ocupación-) influence the TI less, presenting their lowest values with respect to tourist occupancy (the 'Intensity' and 'Tourist seasonality I' indicators) and to the 'Cultural heritage'-factor (Cultural) (Fig. 8.3). We must highlight the case of Granada, presented as atypical and closer to the group of cities showing a high TI, given that its tourist occupancy is very high, despite its positioning in the group of cities with a medium TI result.

In 2014, 13.3% of the cities analyzed (six out of 45 – Jaca, Antequera, Ronda, Segovia, Toledo, and Zamora) are at high risk of overtourism, with a group TI average of 61.09 (Annex 3). In this case, the cities with a very low TI have a higher

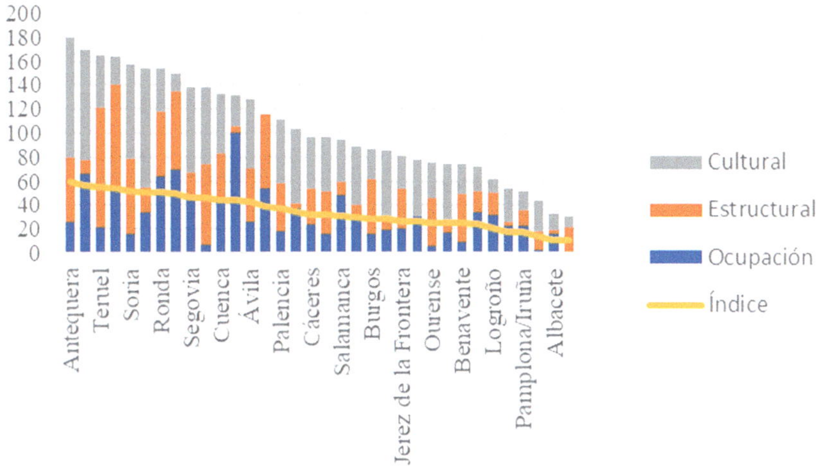

Fig. 8.3 TI (2007). (Source: Authors' elaboration)

tourist occupancy, while in those destinations with a medium and high TI result, the components with the greatest impact on higher scores are 'Cultural heritage' and, to a lesser extent, the tourist occupancy factors ('Intensity' and 'Tourist seasonality I') followed by the structural factors ('Structure' and 'Tourist seasonality II') (Fig. 8.4).

For the year 2021, the number of cities at risk of overtourism drops to 9.3%, or three out of 32 total (Segovia, Toledo, and Zamora); these show a group TI average risk of overtourism of 74.64 (Annex 4). Generally speaking, cities in a low or medium situation present less weight in the 'Cultural heritage', 'Structure', and 'Tourist intensity' indicators (Fig. 8.5).

8.5.2 Diagnostic Matrix (DM)

Taking into account those medium and small Spanish cities that scored highest in the TI during the years 2007, 2014, and 2021, in-depth analysis was made of the risks of overtourism in six specific cities: Zamora, Segovia, Toledo, Jaca, Antequera, and Ronda. The first three were selected for scoring high on the TI during all three years, while the other scored high in two. The results of the indicators, structured in this case in quintiles to facilitate study and graphic representation, were first collected into a Diagnostic Matrix, with the most representative cases then illustrated in radial charts.

The 2007 DM indicates that the cities with a high risk of overtourism together scored 115 points out of a possible 150. The selected destinations obtained very similar results, with Jaca scoring highest (at 22). Zamora, Segovia, Toledo, and

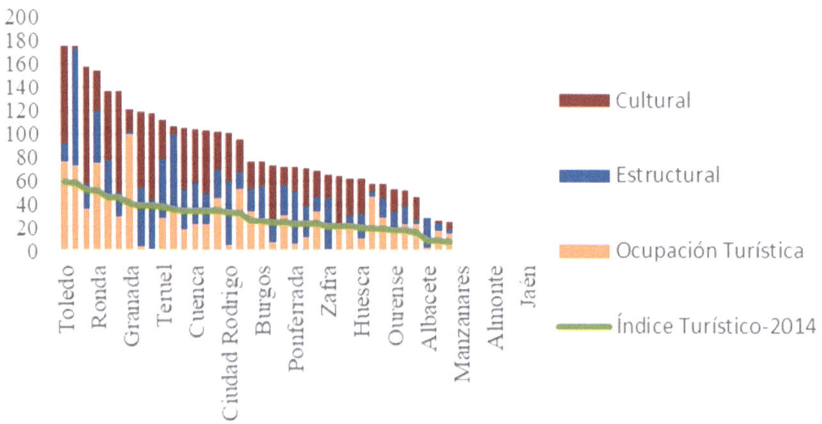

Fig. 8.4 TI (2014). (Source: Authors' elaboration)

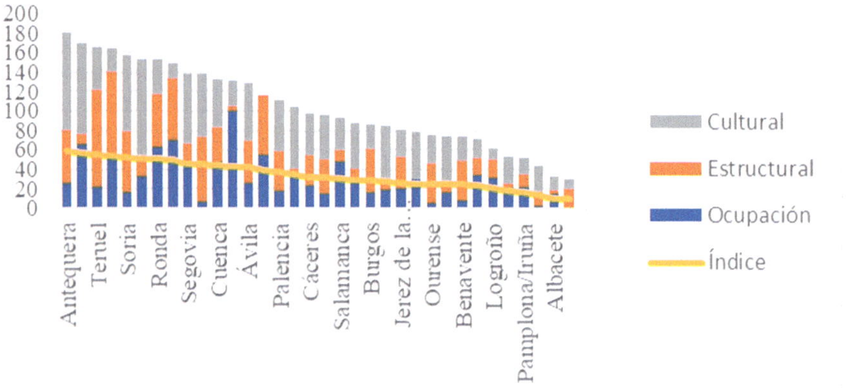

Fig. 8.5 TI (2021). (Source: Authors' elaboration)

Antequera scored 18 of a possible 25 points each, the lowest position in this group (Table 8.2). For 2014, the results are very similar to those of 2007, with practically all the cities obtaining the same score. Ronda scored highest (21), Jaca and Antequera lowest (17). The overall group score is 113 points out of 150 (Table 8.3). Finally, that same situation repeats in 2021. The results are very close to one another, although in this year data from only three cities appear: Zamora, Segovia, and Toledo (Table 8.4). This was doubtless due to the fact that in 2021, the effects of the pandemic were still being felt, so that cities including Soria, Ronda, and Teruel left the high-risk group. Other cities such as Jaca and Antequera provided no data.

Table 8.2 Diagnosis matrix of medium and small inland cities in Spain at high risk of overtourism (2007)

Small and medium Spanish cities	Dimension 1. Alienation of the local population		Dimension 2. Overloading of tourist infrastructures		Dimension 3. Threats to cultural heritage	Total
	Indicator 1. Structure	Indicator 2. Tourist intensity	Indicator 3. Tourist seasonality (I)	Indicator 4. Tourist seasonality (II)	Indicator 5. Cultural heritage	
Zamora	3	2	4	4	5	18/25
Segovia	3	4	3	3	5	18/25
Toledo	1	5	5	2	5	18/25
Jaca	5	5	5	5	2	22/25
Antequera	5	3	1	4	5	18/25
Ronda	5	5	4	4	3	21/25
TOTAL	22/30	24/30	22/30	22/30	25/30	115/150

Source: Authors' elaboration

Table 8.3 Diagnosis matrix of medium and small inland cities in Spain at high risk of overtourism (2014)

Small and medium Spanish cities	Dimension 1. Alienation of the local population		Dimension 2. Overloading of tourist infrastructures		Dimension 3. Threats to cultural heritage	Total
	Indicator 1. Structure	Indicator 2. Tourist intensity	Indicator 3. Tourist seasonality (I)	Indicator 4. Tourist seasonality (II)	Indicator 5. Cultural heritage	
Zamora	4	3	4	3	5	19/25
Segovia	2	4	5	5	4	20/25
Toledo	2	5	5	2	5	19/25
Jaca	5	5	4	2	1	17/25
Antequera	3	3	3	3	5	17/25
Ronda	5	5	5	3	3	21/25
TOTAL	21/30	25/30	26/30	18/30	23/30	113/150

Source: Authors' elaboration

Examining the most recent year, 2021, Segovia and Toledo score highest in the risk of overtourism, with 20 points out of 25. Zamora is in third place, with 18 points. The radial graphs indicate that 'Cultural heritage' and 'Tourist seasonality II' show the highest risk of overtourism for these cities; 'Structure' shows the least risk (Figs. 8.6, 8.7, and 8.8).

Table 8.4 Diagnosis matrix of medium and small inland cities in Spain at high risk of overtourism (2021)

Small and medium Spanish cities	Dimension 1. Alienation of the local population		Dimension 2. Overloading of tourist infrastructures		Dimension 3. Threats to cultural heritage	Total
	Indicator 1. Structure	Indicator 2. Tourist intensity	Indicator 3. Tourist seasonality (I)	Indicator 4. Tourist seasonality (II)	Indicator 5. Cultural heritage	
Zamora	4	2	4	3	5	18/25
Segovia	2	4	5	5	4	20/25
Toledo	3	5	5	2	5	20/25
TOTAL	9/15	11/15	14/15	10/15	14/15	58/75

Source: Authors' elaboration

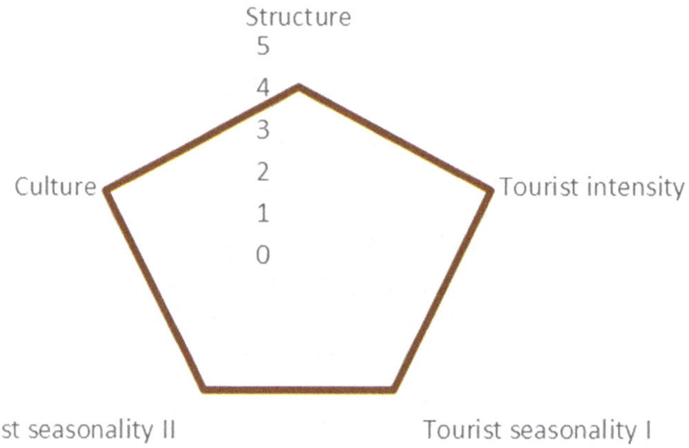

Fig. 8.6 Radial graph of the risk-of-overtourism indicators in the city of Toledo (2021). (Source: Authors' elaboration)

8.6 Main Conclusions

Before drawing specific conclusions from this research, it should be again noted that in 2021, there was a lower contribution of data by certain tourist destinations (Jaca and Antequera) to the INE survey, likely due to the effects of the pandemic. Although this has not conditioned the research in a significant way, it must be taken into account. At the same time, notable variability was detected in the case studies selected. For 2007, the DM extends the risk of overtourism to Segovia but excludes Soria and Teruel. Variability from a chronological perspective is also observed. The TI indicates that Zamora and Toledo remained at high risk in all three selected years, while the risk for Segovia, Soria, Teruel, and Ronda dropped in subsequent years. Although, as before, this does not significantly affect the research, this fact should not be neglected in the event of implementing policies for planning and managing overtourism in the destinations.

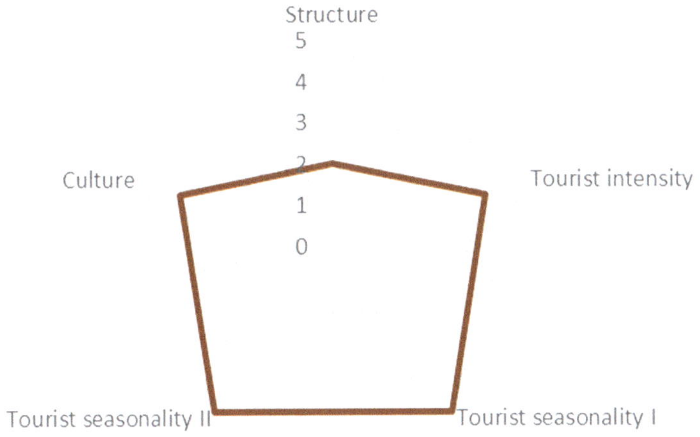

Fig. 8.7 Radial graph of the risk-of-overtourism indicators in the city of Segovia (2021). (Source: Authors' elaboration)

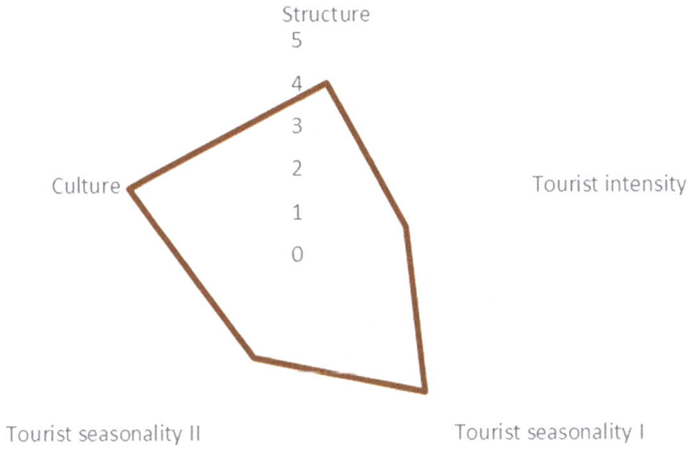

Fig. 8.8 Radial graph of the risk-of-overtourism indicators in the city of Zamora (2021). (Source: Authors' elaboration)

The time perspective reveals that there were no large fluctuations in the risks of overtourism across the period. Risks were reflected in the 2007 data, as well as in the data for 2014 and 2021, with few differences. The same cities tended to score high in the risks of overtourism throughout the analyzed period, suggesting two conclusions: that the risks of overtourism have been present in these Spanish cities since at least since 2007, and possibly much earlier; and that, were any policies indeed implemented to mitigate tourist saturation, they appear not to have been very effective, at least from a macro perspective.

Concerning the indicators, both the TI and DM reveal the vulnerability of cultural heritage. In fact, the 'Cultural heritage' indicator proves to be a decisive factor in city groupings in the TI as well as in the DM, where the cities always score quite high. On the other hand, 'Tourist seasonality' does not seem to greatly influence the TI, and it is among the indicators where cities score lowest in the DM. Nevertheless, as far as indicator results are concerned, observations are largely general, with no specific patterns detected. The radial diagrams for 2021 (Figs. 8.6, 8.7, and 8.8) show the three cases to be very different.

Beyond these analyses of the risks of overtourism experienced by medium and small cities in inland Spain, certain reflection has been given to the location factors that can influence tourist intensity. From a geographical point of view, three different patterns of behavior have been established, the first of which is related to cultural heritage. All the cities examined possess a significant number of Assets of Cultural Interest (BICs), and some (including Toledo and Segovia) have been declared World Heritage cities. The second pattern detected relates to the proximity to population centers capable of emitting significant flows of visitors, as in the cases of both Toledo and Segovia, not far from Madrid, or else proximity to neighboring countries, as in the cases of Jaca, Teruel, and Zamora. The third pattern detected in the spatial distribution of tourist spots with high intensity relates to proximity to some area of strong tourist concentration, such as the Costa del Sol, as in the cases of Ronda or Antequera.

8.7 Discussion

As mentioned in connection with conceptual aspects, the main contribution of this research into the phenomenon of mass tourism lies in the development of a series of indicators that allow evaluation of the risks of overtourism in medium and small non-coastal cities in Spain. These can be added to the existing set of management tools at the service of both tourism planners and researchers assessing this phenomenon. These indicators have here been employed in two types of analysis, the TI and the DM, and this permits a combined interpretation of results at the national as well as urban scales, further enriching the research.

Thanks to this approach, it becomes possible to verify the working hypothesis that overtourism affects not only large cities or 'sun and beach' destinations but is also present in small and medium-sized cities in Spain's interior. Indeed, certain destinations have experienced overtourism as a prolonged and almost structural phenomenon that will demand profound changes to tourism models in order to be addressed successfully. The results here obtained corroborate the conclusions of other studies related to increased tourist flows in Spain (Hidalgo & Maene, 2017), to the risks of overtourism (Barrado-Timón et al., 2020), and to the peculiar case of medium and small Spanish cities (Barrado-Timón et al., 2022).

Nevertheless, this study suffers from a few limitations, most notably those related to data collection. Statistics from Spain's National Institute of Statistics (INE) are not available for all cities for the entire period analyzed, compromising a global assessment of the risks of overtourism. In order to continue exploring this phenomenon, future lines of research are encouraged, along with reinforcement of analyses already made, to develop case studies for specific cities that can verify whether a risk of overtourism is truly present, and how this manifests on an urban scale. Likewise, it should be recalled that overtourism usually occurs in the historic centers of cities, and working with city-wide municipal data can ostensibly distort the studies; thus, to obtain more realistic results, it would be necessary to adjust the indicators to data concentrated on historic centers.

Annexes

Annex 1

The Tourist Index model considers three dimensions: 'alienation of the local population', 'overloading of tourist infrastructures', and 'threats to cultural heritage'. Each dimensions consists of a group of components: Structure indicator and an Intensity indicator (Dimension I), a Seasonality indicators (I and II) (Dimension 2), and a Cultural indicator (Dimension 3) (Table 8.1), from which the following ratios are integrated (Fig. 8.9): (R1) month with more employed personnel / month with

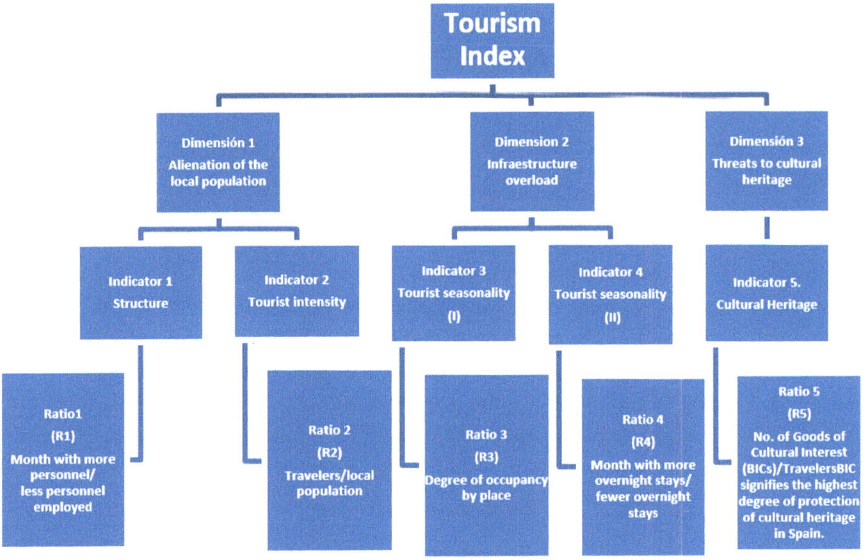

Fig. 8.9 Dimensions, indicators, and ratios used for TI. (Source: Authors' elaboration)

fewer employed personnel; (R2) travelers (sum of non-resident travelers and residents in Spain) / population; (R3) degree of occupancy per place; (R4) month with most overnight stays / month with least overnight stays; and (R5) BICs / travelers.

The methodology of the index allows measurement of each dimension while offering a general score and an aggregate classification by cities in a certain aspect of tourism; it also raises the possibility of undertaking comparative evaluations. As will be shown, the established separation by dimensions does not provide admissible results, so only the ratios have been used. A significant relationship is observed between the information of the 'Seasonality II' index and the 'Structure' index, along with other relationships such as the 'Intensity' ratios with 'Seasonality I'.

The TI produces values of 0–100, the highest values corresponding to higher scores for the indicators of each city. In general terms, the index is calculated as a simple average of the values achieved in each of the three considered dimensions (cultural vitality, creative economy, and socio-spatial context), which in turn take values between 0 and 100, the highest values corresponding to the best performance. The dimension values are estimated as a simple average of the components that comprise them, also assuming values of between 0 and 100, where 100 represents the maximum value obtained in the cities analyzed.

In turn, each dimension is calculated by means of various indicators. The score for each component is estimated through a procedure known as 'principal component analysis' or 'factor analysis,' by which an aggregate score can be calculated from multiple indicators related to a common concept. For optimal construction of the Tourism Index we sought an equal weighting of the indicators in each component, using factor analysis to calculate these weights. This process of factor analysis weighted many indicators nearly equally within their components, thus indicating a satisfactory selection of indicators for measuring each component. Some components, however, proved more difficult to measure due to insufficient or inherently diverse data across cities. This justified the decision to use factor analysis weighting. It is important to compensate for differences in available data and distinct indicators between components, all throughout the index, allowing the factor analysis to weight indicators appropriately, in order to arrive at the best composite measure of each component.

Before implementing factor analysis, the 'fit' between the indicators within a component is assessed by calculating Cronbach's alpha for the indicators in each component. Cronbach's alpha provides a measure of the internal consistency between indicators. A rule of thumb applied by professionals is that this alpha value must be above 0.7 for grouping of variables to be considered valid (Bland & Altman, 1997). In this case, the value of the alpha with the ratios considered is 0.726. Although Cronbach's alpha is a good way to achieve a conceptual fit, it does not provide a direct measure of the goodness of fit of factor analysis (Manly & Navarro, 2016).

Reliability statistics		
Cronbach's alpha	Cronbach's alpha based on standardized elements	Number of elements
0.726	0.728	7

Factor analysis (FA) uses the shared covariance between all the indicators of each component to calculate a set of weights that permits creation of a single added value from different indicators (Manly & Navarro, 2016). This combined value is called a 'factor'. If the indicators of a component are well chosen, the factor will extract a score that can be used as a valid synthetic measure of the component between cities. Factor analysis provides a set of weights for the underlying variables within each component to control for the fact that these variables sometimes correlate with one another. In this case, factor analysis is used to weight the indicators within similar groups of signals.

After performing factor analysis on each component, the goodness of fit was assessed using the Kaiser-Meyer-Olkin measure of sampling sufficiency. Results from this analysis are provided in the following table:

KMO and Bartlett tests		
Kaiser-Meyer-Olkin measure of sampling adequacy		0.658
Bartlett's sphericity test	Approximate Chi-squared	40.884
	gl	15
	Sig.	0–000

In general, KMO scores should be above 0.5. In our data, the mean KMO score is greater than 0.5, so that the grouping of indicators selected as constituents for the index appears to provide a good measure of the underlying concept, especially for exploratory (rather than confirmatory) factor analysis.

From the normalized data, the value of each component is calculated through the sum of the scores of each of the indicators, weighted by the specific weight obtained from factorial analysis previously carried out within each component:

$$\text{Gross Component} = \Sigma(w_i * indicator_i)$$

The final step in calculating each component is to provide transparency and comparability between the components. The goal is to transform the values so that each result can be easily interpreted, both relative to other components and across cities. To this end, best- and worst-case scores are computed based on the data, along with data for individual cities. Best- and worst-case scores are defined at the indicator level according to the definition of each data point. This process allows cities to be rated on a scale of 0–100, where 100 is the best-case estimate and 0 is the worst-case estimate at the component level. The formula used is expressed as follows:

$$Component = \frac{X_j - X_{worst}}{X_{best} - X_{worst}}$$

where *Xj* is the gross component for city *j*.

From this analysis, as shown in the following table, we find three factors:

Factor 1, comprised of the ratios defined by the 'Intensity' and 'Seasonality I' indicators, related to tourist occupancy.

Factor 2, defined by the 'Structure' and 'Seasonality II' ratios, which can be interpreted as structural.
Factor 3, formed by the 'Cultural' indicator.

Factor weight matrix			
	Factors		
Ratios	1	2	3
R5	0.890		
R3	0.736		
R2	0.630	0.567	
R6		0.872	
R1		0.807	
R7			0.984

Extraction method: principal component analysis
Rotation method: Varimax with Kaiser normalization

From this analysis in which the factors are established, the index for each city is obtained as a simple average of the components and of each dimension:

$$Tourism\ Index = \frac{1}{3}\sum_{Dim=1}^{3}\left(\frac{1}{h}\sum_{h} Factor_h\right)_{Dim}$$

Annex 2

Grouping	City	2007 index
Very low	Vitoria-Gastéiz	15.19
	Albacete	
	Badajoz	
	Lleida	
	Logroño	
	Pamplona/Iruña	
Low	Burgos	28.33
	Cáceres	
	Plasencia	
	Jerez de la Frontera	
	Ciudad Real	
	Guadalajara	
	Huesca	
	Úbeda	
	León	
	Ourense	
	Oviedo	
	Salamanca	
	Benavente	

(continued)

Grouping	City	2007 index
Medium	Ávila	42.66
	Arcos de la Frontera	
	Santiago de Compostela	
	Cuenca	
	Granada	
	Palencia	
	Segovia	
High	Jaca	53.79
	Antequera	
	Ronda	
	Soria	
	Teruel	
	Toledo	
	Zamora	

Annex 3

Grouping		City	Tourist occupancy	Structural	Cultural	2014 index
Very low		Albacete	2.95	24.94	0.00	9.30
		Badajoz	17.15	6.19	2.50	8.61
		Lleida	23.12	2.52	19.97	15.20
		Alcalá de Henares	15.37	2.72	6.40	8.17
	Mean		**14.65**	**9.09**	**7.22**	**10.32**
Low		Vitoria-Gastéiz	22.62	15.22	13.94	17.26
		Zafra	0.00	44.91	19.99	21.63
		Burgos	27.91	26.91	21.29	25.37
		Cáceres	34.38	11.77	22.67	22.94
		Plasencia	22.86	8.09	31.06	20.67
		Jerez de la Frontera	34.03	18.63	23.83	25.50
		Ciudad Real	7.56	14.75	50.73	24.35
		Huesca	10.89	21.28	29.30	20.49
		León	30.70	25.49	16.00	24.07
		Ponferrada	6.58	43.96	20.79	23.78
		Logroño	46.20	4.08	7.53	19.27
		Pamplona/Iruña	28.70	15.82	12.39	18.97
		Ourense	17.90	15.75	18.83	17.49
		Oviedo	12.01	25.88	33.04	23.64
		Benavente	22.35	0.00	41.64	21.33
	Mean		**21.65**	**19.50**	**24.20**	**21.78**
Medium		Ávila	18.08	34.83	51.81	34.91
		Mérida	44.98	24.84	32.35	34.06
		Arcos de la Frontera	1.67	42.30	72.99	38.99

(continued)

Grouping		City	Tourist occupancy	Structural	Cultural	2014 index
		Santiago de Compostela	34.30	65.10	6.41	35.27
		Cuenca	23.08	35.31	45.36	34.59
		Granada	100.00	1.92	18.68	40.20
		Úbeda	22.31	27.22	53.65	34.40
		Ciudad Rodrigo	4.93	54.84	40.65	33.47
		Salamanca	52.53	14.59	27.90	31.68
		Soria	4.07	49.87	64.33	39.42
		Teruel	28.49	50.00	33.36	37.28
	Mean		**30.40**	**36.44**	**40.68**	**35.84**
High		Jaca	73.39	100.00	1.71	58.36
		Antequera	35.85	21.34	100.00	52.40
		Ronda	75.33	43.88	34.74	51.32
		Segovia	46.46	30.41	59.96	45.61
		Toledo	76.67	15.07	83.95	58.56
		Zamora	29.71	20.16	86.21	45.36
	Mean		**56.23**	**38.48**	**61.09**	**51.94**

Annex 4

Grouping		City	Alienation of the local population	Cultural	Occupancy	2021 index
Very low		Vitoria-Gastéiz	8.73	19.56	60.04	29.44
		Albacete	7.84	14.63	19.03	13.84
		Badajoz	9.16	16.26	43.46	22.96
		Ciudad Real	16.87	60.83	0.00	25.90
		Logroño	26.44	20.17	37.25	27.95
		Ourense	11.59	31.13	53.62	32.11
	Mean		**13.44**	**27.09**	**35.57**	**25.37**
Low		Ávila	43.18	72.78	3.95	39.97
		Burgos	36.19	25.62	66.85	42.88
		Cáceres	27.67	43.19	60.24	43.70
		Santiago de Compostela	77.55	0.00	40.00	39.18
		Lleida	13.91	26.14	89.65	43.23
		Pamplona/Iruña	14.47	18.53	83.03	38.68
		Oviedo	33.61	47.42	45.11	42.05
		Salamanca	46.03	29.32	53.92	43.09
		Teruel	36.99	47.38	21.94	35.44
	Mean		**36.62**	**34.49**	**51.63**	**40.91**
Medium		Mérida	39.56	47.09	76.46	54.37
		Jerez de la Frontera	30.18	36.53	87.03	51.25

(continued)

Grouping		City	Alienation of the local population	Cultural	Occupancy	2021 index
		Cuenca	43.45	68.33	39.60	50.46
		Granada	68.46	16.67	82.50	55.88
		León	27.44	23.31	100.00	50.25
		Ronda	100.00	21.72	21.37	47.70
		Palencia	0.00	87.05	54.09	47.04
		Soria	22.12	78.29	60.43	53.61
	Mean		**41.40**	**47.37**	**65.18**	**51.32**
High		Segovia	80.59	83.91	98.10	87.53
		Toledo	57.46	87.95	60.17	68.53
		Zamora	28.74	100.00	74.81	67.85
	Mean		**55.60**	**90.62**	**77.69**	**74.64**

References

Ali, R. (2016). Exploring the coming perils of overtourism. *Skift*. Retrieved July 7, 2019.

Ardura, Á., & Sorando, D. (2016). La destrucción creativa de las ciudades. *Catalunya Economica*, (533), 22–23.

Barrado-Timón, D. (2010). Gran ciudad y turismo en la transición postindustrial: nuevos y viejos procesos, nuevas y viejas teorías. *El ejemplo del área metropolitana de Madrid. Scripta Nova, XIV* (317).

Barrado-Timón, D. A., & Hidalgo-Giralt, C. (2019). Golden hordes or mere barbarians? discourses on tourism, touristification, and tourismophobia in Madrid's Lavapiés neighborhood. *Boletín de la Asociación de Geógrafos Españoles*, (83). https://doi.org/10.21138/bage.2824

Barrado-Timón, D., Palacios, A., & Hidalgo-Giralt, C. (2020). Medium and small cities, culture and the economy of culture. A review of the approach to the case of Spain in light of international scientific scholarship. *Sustainability, 12*(18), 7321. https://doi.org/10.3390/su12187321

Barrado-Timón, D., Escalona-Orcao, A., Hidalgo-Giralt, C., Vázquez-Varela, C., & Cea-D'Ancona, F. (2022). A local and territorial approach to the operation of the culture economy in medium and small cities: Case study of the Spanish cities of Mérida and Cuenca. *Boletín de la Asociación de Geógrafos Españoles*, (92). https://doi.org/10.21138/bage.3185

Blanco-Romero, A., Blàzquez-Salom, M., Morell, M., & Fletcher, R. (2019). Not tourism-phobia but urban-philia: Understanding stakeholders' perceptions of urban touristification. *Boletín De La Asociación De Geógrafos Españoles*, (83). https://doi.org/10.21138/bage.2834

Bland, J. M., & Altman, D. G. (1997). Statistics notes: Cronbach's alpha. *BMJ, 314*(7080), 572. https://doi.org/10.1136/bmj.314.7080.572

Bonilla, J., & Bonilla, L. (2008). La capacidad de carga turística: Revisión crítica de un instrumento de medida de sostenibilidad. *El Periplo Sustentable, 15*, 123–150.

Butler, R. W. (1974). The social implications of tourist developments. *Annals of Tourism Research, 2*(2), 100–111. https://doi.org/10.1016/0160-7383(74)90025-5

Cape Town Declaration on Responsible Tourism. (2002). Cape town: International conference on responsible tourism in destinations.

Clancy, M. (2019). Overtourism and resistance: Today's anti-tourist movement in context. In H. Pechlaner et al. (Eds.), *Overtourism* (pp. 14–24). Routledge.

Clawson, M., & Knetsch, J. L. (2011). *Economics of outdoor recreation*. RFF Press.

Cohen, E. (1988). Authenticity and commoditization in tourism. *Annals of Tourism Research, 15*(3), 371–386. https://doi.org/10.1016/0160-7383(88)90028-X

Colomb, C., & Novy, J. (Eds.). (2016). *Protest and resistance in the tourist city*. Routledge.
de la Calle-Vaquero, M. (2019). Touristification of urban centres: An attempt to clarify the debate. *Boletín de la Asociación de Geógrafos Españoles, 83*(2829), 1–40. https://doi.org/10.21138/bage.2829
de la Calle-Vaquero, M., & García-Hernández, M. (2020). Turismo, patrimonio y centros urbanos: ¿hacia un nuevo ciclo de turistifcación?. In G. X. Pons, A. Blanco-Romero, R. Navalón-García, L. Troitiño-Torralba, y M. Blázquez-Salom (Eds.), *Sostenibilidad Turística: overtourism vs undertourism*. Mon. Soc. Hist. Nat. Balears, 31, 135–146. ISBN 978-84-09-22881-2. Palma (Illes Balears).
de la Calle-Vaquero, M., Santos-Solla, X., Velasco-González, M., & García Hernández, M. (2019). Presentación del número monográfico. *Boletín De La Asociación De Geógrafos Españoles*, (83). Recuperado a partir de https://bage.age-geografia.es/ojs/index.php/bage/article/view/2882
de la Calle-Vaquero, M., Mendoza, S., Ferreiro, E., & García, M. (2020). El urbanismo como instrumento de contención de la actividad turística en los centros urbanos: alcance y limitaciones. In G. X. Pons, A. Blanco-Romero, R. Navalón-García, L. Troitiño-Torralba, y M. Blázquez-Salom (Eds.), *Sostenibilidad Turística: overtourism vs undertourism*. Mon. Soc. Hist. Nat. Balears, 31, 121–133. ISBN 978-84-09-22881-2. Palma (Illes Balears).
Forster, J. (1964). The sociological consequences of tourism. *International Journal of Comparative Sociology, 5*(2), 217–227. https://doi.org/10.1163/156854264X00207
Garay, L., Morales, S., & Wilson, J. (2020). Tweeting the right to the city: Digital protest and resistance surrounding the Airbnb effect. *Scandinavian Journal of Hospitality and Tourism, 20*(3), 246–267. https://doi.org/10.1080/15022250.2020.1772867
Garay-Tamajón, L., Lladós-Masllorens, J., Meseguer-Artola, A., & Morales-Pérez, S. (2022). Analyzing the influence of short-term rental platforms on housing affordability in global urban destination neighborhoods. *Tourism and Hospitality Research, 99*, pages 146735842110575.
García Hernández, M. (2000). Turismo y medio ambiente en ciudades históricas. de la capacidad de acogida turística a la gestión de los flujos de visitantes. *Anales de Geografía de la Universidad Complutense de Madrid, 20*, 131–148.
García Hernández, M., & de la Calle-Vaquero, M. (2012). Capacidad de carga en grandes recursos turístico-culturales. *Anales de Geografía de la Universidad Complutense de Madrid, 32*(2), 253–274. https://doi.org/10.5209/rev_AGUC.2012.v32.n2.39720
García Hernández, M., Ivars-Baidal, J., & Mendoza, S. (2019). Overtourism in urban destinations: The myth of smart solutions. *Boletín de la Asociación de Geógrafos Españoles, 83*(2830), 1–38. https://doi.org/10.21138/bage.2830
Gil, J. (2021). Turistificación, rentas inmobiliarias y acumulación de capital a través de Airbnb. *Cuadernos Geográficos, 60*(1), 95–117.
Gil, J., & Sequera, J. (2018). Expansión de la ciudad turística y nuevas resistencias. El caso de Airbnb en Madrid. *EMPIRIA. Revista de Metodología de las Ciencias Sociales, 41*, 15–32.
Goodwin, H. (2017). *The challenge of overtourism* (Working Paper 4) (pp. 1–19). Responsible Tourism Partnership.
Gretzel, U. (2019). The role of social media in creating and addressing overtourism. In R. Dodds & R. Butler (Eds.), *Overtourism: Issues, realities and solutions* (pp. 62–75). De Gruyter Oldenbourg.
Guitart, N., & Serrat, R. (2020). A methodological proposal for public responses to overtourism and its integration in urban destinations' policies. In G. X. Pons, A. Blanco-Romero, R. Navalón-García, L. Troitiño-Torralba, & M. Blázquez-Salom (Eds.), *Sostenibilidad Turística: overtourism vs undertourism*. Mon. Soc. Hist. Nat. Balears, 31: 265–281. ISBN 978-84-09-22881-2. Palma (Illes Balears).
Gutiérrez Puebla, J., García Palomares, J. C., & Salas, M. (2016). Big (geo) data en ciencias sociales: Retos y oportunidades. *Revista de Estudios Andaluces, 33*(1), 1–23. https://doi.org/10.12795/rea.2016.i33

Hidalgo, C., & Maene, O. (2017). The nature of Spain's international cultural tourism throughout the economic crisis (2008–2016): A macroeconomic analysis of tourist arrivals and spending. *Economies, 5*(3), 32. https://doi.org/10.3390/economies5030032

Hidalgo, C., Palacios, A., & Barrado, D. (2020). El comportamiento del turismo internacional con motivación cultural. El caso de las ciudades medias y pequeñas española ¿overtourism? ¿undertourism? In G. X. Pons, A. Blanco-Romero, R. Navalón-García, L. Troitiño-Torralba, y M. Blázquez-Salom (Eds.), *Sostenibilidad Turística: overtourism vs undertourism*. Mon. Soc. Hist. Nat. Balears, 31: 391–403. ISBN 978-84-09-22881-2. Palma (Illes Balears).

Judd, D. R. (2003). El turismo urbano y la geografía de la ciudad. *EURE (Santiago), 29*(87), 51–62. https://doi.org/10.4067/S0250-71612003008700004

Judd, D. R., & Fainstein, S. S. (1999). *The tourist city*. Yale University Press.

Koens, K., Postma, A., & Papp, B. (2018). Is overtourism overused? Understanding the impact of tourism in a city context. *Sustainability, 10*(12), 4384. https://doi.org/10.3390/su10124384

López Palomeque, F. (1995). La estrategia del turismo metropolitano: el caso de Barcelona. *Estudios Turísticos, 126*, 119–142.

López, L., & Pazos Otón, M. (2009). La imagen de Santiago de Compostela en el mercado turístico italiano. *Cuadernos de Turismo, 24*, 65–89. Retrieved from https://revistas.um.es/turismo/article/view/92661

López-Villanueva, C., & Crespi-Vallbona, M. (2021). Gentrificación y turistificación: dinámicas y estrategias en Barcelona. Encrucijadas: Revista Crítica de. *Ciencias Sociales, 21*, 7.

Manero Miguel, F. (2011). Patrimonio cultural y transformación urbana: ordenación espacial y proyección turística de los yacimientos arqueopaleontológicos de Atapuerca (Burgos). *Cuadernos de Turismo, 27*, 617–640.

Manly, B., & Navarro, J. (2016). *Multivariate statistical methods: A primer*. Chapman and Hal/CRC.

Martínez-Caldentey, M. A., Murray, I., & Blázquez-Salom, M. (2020). En la ciudad de Madrid todos los caminos conducen a Airbnb. *Investigaciones Turísticas, 19*, 1–27. https://doi.org/10.14198/INTURI2020.19.01

Milano, C. (2017). Turismofobia: cuando el turismo entra en la agenda de los movimientos sociales. *Marea urbana, 1*, 5–8.

Milano, C. (2018). Overtourism, malestar social y turismofobia. Un debate controvertido. *PASOS Revista de Turismo y Patrimonio Cultural, 16*(3), 551–564.

Miriam-Hermi, Z. (2019). Gentrificación y turismo urbano. ¿Cómo se articulan? Ar@cne. *Revista electrónica de recursos en internet sobre Geografía y Ciencias Sociales, 230*, 1–32.

Molina Collado, A., Martín-Consuegra, D., Esteban Talaya, A., & Díaz Sánchez, E. (2007). Segmentación de la demanda turística: un análisis aplicado a un destino de turismo cultural. *Revista de Análisis Turístico, 4*, 36–48.

Navarro, F., & Capote, A. (2021). ¿Overtourism en la ciudad de Granada? *Cuadernos Geográficos, 60*(1), 35–53. https://doi.org/10.30827/cuadgeo.v60i1.13717

Palacios-García, A., Hidalgo Giralt, C., & Narváez-Flores, C. (2020a). Transformaciones en los equipamientos culturales vinculadas a la gentrificación y la turistificación: Diferencias y similitudes entre ciudades grandes y medias. Desafíos y oportunidades de un mundo en transición: Una interpretación desde la Geografía / coord. por Jaime Escribano Pizarro, María Pilar Peñarrubia Zaragoza, José Javier Serrano Lara, Sabina Asins Velis; Joaquín Farinós i Dasí (dir.), 2020, ISBN 978-84-9133-305-0, págs. 649–662.

Palacios-García, A., Hidalgo-Giralt, C., & Narváez, C. (2020b). Transformaciones en los equipamientos culturales vinculadas a la gentrificación y la turistificación. Diferencias y similitudes entre ciudades grandes y medias. In J. Farinós, J. (Coord.), *Desafíos y oportunidades de un mundo en transición. Una interpretación desde la Geografía* (pp. 649–662). Universitat de Valencia y Tirant lo Blanch.

Palacios-García, A., Vallina-Rodríguez, A., Hidalgo-Giralt, C., & Villa, O. (2021). La gentrificación y los equipamientos culturales de Lavapiés: ¿merece la pena ser un barrio «cool»? In J. García (Ed.), *Libro de resúmenes de los trabajos del XXVII Congreso de la Asociación Española de Geografía* (pp. 483–484). AGE.

Pareto Boada, P., Torres-Delgado, A., Cerdan Schwitzguébel, A., & Elorrieta Sanz, B. (2020). Sostenibilidad social del turismo en Barcelona: un análisis geoespacial. In G.X. Pons, A. Blanco-Romero, R. Navalón-García, L. Troitiño-Torralba, y M. Blázquez-Salom (Eds.), *Sostenibilidad Turística: overtourism vs undertourism*. Mon. Soc. Hist. Nat. Balears, 31: 249–264. ISBN 978-84-09-22881-2. Palma (Illes Balears).

Piñeira, M. J., Fernández Tabales, A., y Mínguez, C. (2020). Vulnerabilidad y turistificación ¿quiénes son los perdedores del centro urbano? In G.X. Pons, A. Blanco-Romero, R. Navalón-García, L. Troitiño-Torralba, y M. Blázquez-Salom (Eds.), *Sostenibilidad Turística: overtourism vs undertourism*. Mon. Soc. Hist. Nat. Balears, 31: 83–98. ISBN ISBN 978-84-09-22881-2. Palma (Illes Balears).

Pons, G. X., Blanco-Romero, A., Navalón-García, R., Troitiño-Torralba, L., & Blázquez-Salom, M. (Eds.). (2020). *Sostenibilidad Turística: overtourism vs undertourism*. Mon. Soc. Hist. Nat. Balears, 31: 610 pp. ISBN 978-84-09-22881-2. Palma (Illes Balears).

Quebec Declaration on Ecotourism. (2002). Quebec: World ecotourism summit.

Romero-Padilla, Y., Cerezo-Medina, A., Navarro-Jurado, E., Romero-Martínez, J. M., & Guevara-Plaza, A. (2019). Conflicts in the tourist city from the perspective of local social movements. *Boletín De La Asociación De Geógrafos Españoles*, (83). https://doi.org/10.21138/bage.2837

Seraphin, H., Sheeran, P., & Pilato, M. (2018). Over-tourism and the fall of Venice as a destination. *Journal of Destination Marketing & Management, 9*, 374–376. https://doi.org/10.1016/j.jdmm.2018.01.011

Swarbrooke, J. (1999). *Sustainable tourism management*. Cab International.

Troitiño Vinuesa, M. A. (1995). El turismo en las ciudades históricas. *Polígonos, 5*, 49–65.

Troitiño Vinuesa, M. A., & Troitiño Torralba, L. (2009). Turismo y patrimonio en Castilla y León: las Ciudades Patrimonio de la Humanidad (Ávila, Salamanca y Segovia) como destinos turísticos de referencia. *Polígonos, 19*, 145–178.

Valenzuela Rubio, M. (1992). Turismo y gran ciudad, una opción de futuro para las metrópolis postindustriales. *Revista valenciana d'estudis autonòmics, 13*, 103–138.

Valero-Escandell, J. R., y García-Tortosa, F. (2020). Overtourism urbano y cambios demográficos. In G. X. Pons, A. Blanco-Romero, R. Navalón-García, L. TroitiñoTorralba, y M. Blázquez-Salom (Eds.), *Sostenibilidad Turística: overtourism vs undertourism*. Mon. Soc. Hist. Nat. Balears, 31: 519–530. ISBN 978-84-09-22881-2. Palma (Illes Balears).

Velasco-González, M., Herrero-López, R., & López-Sánchez, E. (2019). El orden del caos: la decisión del gobierno ante el problema del impacto del turismo en el centro urbano de Madrid. *Boletín De La Asociación De Geógrafos Españoles*, (83). https://doi.org/10.21138/bage.2832

WCED, U. (1987). *Our common future*. UN.

Wilson, J., Garay-Tamajon, L., & Morales-Perez, S. (2021). Politicising platform-mediated tourism rentals in the digital sphere: Airbnb in Madrid and Barcelona. *Journal of Sustainable Tourism, 1–22*, 1080–1101. https://doi.org/10.1080/09669582.2020.1866585

World Tourism Organization. (2018). *«Overtourism»?. Understanding and managing urban tourism growth beyond perceptions*. UNWTO. Recuperado de. https://doi.org/10.18111/9789284419999

World Travel & Tourism Council. (2017). *Coping with success: Managing overcrowding in tourism destinations*. McKinsey & Company.

Zárate Martín, M. A. (1995). Imagen y potencial turístico de un Centro Histórico singular, Toledo. In M. Valenzuela Rubio (Coord.), *Los turismos de interior* (pp. 289–304). Ed. Universidad Autónoma de Madrid.

Carmen Hidalgo-Giralt holds a European PhD in Geography from the Universidad Autónoma de Madrid (UAM) within the framework of the Doctoral Program "Territory, Environment and Society" taught by the Department of Geography of the UAM. She also has a Bachelor's Degree in Art History, a Diploma in Tourism and a Degree in Geography and History (UNED). She develops three main lines of research: cultural heritage and urban tourism; industrial heritage; and the use of new technologies in teaching. She develops part of her research within the Research Group "URByTUR" coordinated by Professor Manuel Valenzuela Rubio.

Antonio Palacios-García holds a PhD in Geography from the Universidad Autónoma de Madrid (UAM) within the framework of the Doctoral Program with Mention of Quality "Territory, Environment and Society" taught by the Department of Geography of the UAM. He also holds a Master's Degree in Urban Planning, He develops three main lines of research, two of them in urban geography (urban inequality and environmental justice and urbanism and housing), and another in Geography of tourism (cultural heritage and urban tourism). He develops part of his research within different research groups: "URByTUR", "GEOTAPLAN" and "Society, Landscape and Heritage.

Diego A. Barrado-Timón is Professor of Human Geography at the Universidad Autónoma de Madrid (UAM). He holds a degree and PhD in Geography from UAM and a degree in Social and Cultural Anthropology and a degree in English Cultural Studies. He has focused much of his work in the field of Geography of Tourism, mainly in relation to tourism development and the management and planning of tourism space. His other main field of scientific and teaching specialization is urban studies, mainly in aspects related to tourism, culture, heritage and cultural industry.

Francisca Cea-D'Ancona Graduated in Economics and Business Administration in 1983 from the Complutense University of Madrid. Doctorate: Doctoral thesis "Los sistemas de previsión social en España un tratamiento multivariante" (1987) with the qualification of outstanding Cum laude by the Universidad Autónoma de Madrid. She has the following lines of research: Models and theories of economic development; Big Data; Application of statistics to accounting; and Statistical treatment to data analysis.

Chapter 9
Overtourism in Mature Coastal Destinations on the Spanish North Atlantic Coast

Carlos Alberto Patiño-Romarís, Rubén Camilo Lois-González, and Breixo Martins Rodal

9.1 Introduction

In recent decades we have witnessed tourism's continuous growth on a global scale. Travelling has gone from being a right to a necessity. The rapid increase in tourist flows has therefore created problems of overload and has posed major challenges for destination management. In addition, the new tourist profile is characterised, among other trends, by an increase in the demand for environmental quality. There has been a growth in demand, linked to an unwise production of tourist spaces lacking all kinds of territorial planning. In other words, the production of tourist spaces has basically met the impulses of demand and the lack of sustainability criteria (Vera Rebollo & Monfort, 1994). In response to these dynamics, territorial tourism management currently faces significant challenges to provide sustainable activity and practices that minimise any adverse effects to its development at a territorial level. Among these challenges, we find problems linked to overcrowding, noise, rising home rental prices, etc. (UNWTO, 2018). The term overtourism arises and is directly related to the existence of these unsustainable effects of tourism development. However, within the academic literature we find different stances when addressing its definition (Blázquez-Salom et al., 2021). These are destinations where either the host population or the demand feel that visitors are overcrowding the space and that

C. A. Patiño-Romarís (✉) · R. C. Lois-González
Universidade de Santiago de Compostela, Santiago, Spain
e-mail: carlosalberto.patino@uvigo.es; rubencamilo.lois@usc.es

B. Martins Rodal
Universidade de Vigo, Vigo, Spain
e-mail: breixo.martins@uvigo.es

A. Blanco-Romero, M. Blázquez-Salom (eds.), *Spanish Tourism Geographies*, Geographies of Tourism and Global Change,
https://doi.org/10.1007/978-3-031-39780-6_9

the quality of life or the experience has deteriorated in an unacceptable way (Milano et al., 2018; UNWTO, 2018). Indeed, the perception of overcrowding can affect the satisfaction of visitors (Jin et al., 2016; Usher & Gómez, 2017; Peeters et al., 2018), especially in the case of sun and sand destinations (Alegre & Garau, 2010). In addition, this overfrequency generates an overload on infrastructures and negatively affects nature, culture and local heritage (Dichter & Gloria, 2017). Benner (2019) points out that overtourism is also associated with damage to the social, cultural or environmental fabric of the destination. Koens et al. (2018) highlight that overtourism describes a problem that is multidimensional and complex. Overtourism or tourist saturation thus becomes a threat to all agents who participate directly and indirectly in the tourism machinery: public administration, the private sector, travellers and residents (Milano, 2017). Agents who will have to face the challenges associated with overtourism: alienated residents, a degraded tourist experience, overloaded infrastructure, damage to nature or threats to culture and heritage (Peeters et al., 2018).

Another key element to consider within overtourism is its "perception". In this sense, we must be aware of the existence of individuals in a tourist destination with different readings of the same reality. Therefore, we find different types of overtourism; and consequently, the problems, indicators and solutions to this phenomenon must be specific to the destination in question (Blázquez-Salom et al., 2021).

Overtourism requires territorial strategic actions adapted to the characteristics of a given destination (Milano et al., 2018). A problem that has to be managed within the tourist destinations, since the quality holiday experience has an integral character and is not viable in a deteriorated space. This is why the growing competitive weakness of some destinations is linked to the loss of quality and the indifferenciation of experiences offered, in addition to the excessive tourist pressure on the natural and cultural resources that should enhance the destination's attractiveness. Territories have a limited capacity to accommodate human pressure and, beyond these limits, there can be increasing degradation processes that devalue the resources of the destinations and jeopardise their own tourist viability in the medium term (EXCELTUR, 2003).

We are facing a global scenario in the tourism sector marked by the need for structural change. A need for change motivated by the transformations that have occurred both in the configuration of the products offered and in the demand's tastes. The competitiveness of the tourism sector is closely linked to its sustainability, as the quality of destinations depends to a large extent on their natural and cultural environment, as well as on the attitudes of the local community (EU, 2017).

On another level, within this tourism market, tourism developed in coastal spaces continues to play a hegemonic role, although it requires transformations, adjustments and adaptations. A reality that has led to the reformulation of mature coastal destinations to give greater relevance to sustainability. These destinations enjoy a high level of resilience and their degradation is not a fait accompli, since any relevant problem can be structurally corrected with the appropriate territorial reorganisation strategies. A reorganisation process in which the municipal public administration must play an important role. A local administration that must exercise leadership in the formulation and implementation of territorial tourism regeneration strategies (Koens et al., 2018), where the focus of the debate lies on defining the limits and intensities of the intervention. A much-needed leadership of the municipal administration focused on bringing together both public and private agents towards this strategy of reorganisation and tourism differentiation.

In this direction, the first thing to determine is whether the destination has exceeded its physical or infrastructural, economic, social, cultural or environmental carrying capacity which can signify the beginning of the destination's decline. It is at this time, when public managers must promote lines of action to avoid the loss of attractiveness for tourist demand. We are talking about the adoption of a set of measures aimed at successfully overcoming this situation of decline. Lines of action established in order to face complex challenges. A set of actions aimed at regenerating, redefining, reorganising, restructuring and, ultimately, renewing the tourist destination. They are also aimed at maintaining adequate levels of competitiveness in the market, where the image, as an engine of demand, cannot be overlooked. Mature destinations enjoy a consolidated image within the market, which is an important strength when facing a loss of competitiveness. Therefore, reorganisation strategies should remain faithful to the established traditional image, while introducing new attributes that revive and prepare it for the coming years (Santana-Talavera et al., 2010).

Based on this approach, this study focuses on the case studies of the main Spanish Atlantic coastal destinations (San Sebastian, Santander, Llanes, etc.), paying particular attention to the case of the Sanxenxo municipality in the province of Pontevedra, where uncontrolled urban development has territorially determined the prevailing tourism model. An urban development that has shaped heterogeneous and dynamic urban and peri-urban coastal spaces with defined tourist functionality. That is, specialised territorial areas with unique forms of urbanisation unlike traditional urban spaces (Mullins, 1994; Antón, 1998). Coastal sectors that are formed as complex territorial systems where any intervention process requires a contextualisation in accordance with the processes that determined their origin.

Sanxenxo, a destination where the management of urban growth for tourist use is a key element to ensure that future development is sustainable, so that the carrying capacity in certain areas of its territory is not exceeded in the summer period and the loss of landscape identity is minimised. The spatial subdivision of these geographical areas will be essential in the process of tourist renovation and restructuring. A process of singularisation that enables specific management adapted to the tourist territorial organisation of each individualised coastal sector.

The renewal/restructuring of Sanxenxo involves a planning exercise. Planning that determines the territorial action strategic lines directed to pursue a desirable competitiveness. For this it is essential to have a territorial analysis that provides the necessary knowledge for the design of the most appropriate strategy.

9.2 A Review of Academic Literature on Tourist Destinations on the Spanish North Atlantic Coast

The main tourist destinations today are inevitably affected by the recent profound transformations in tourism globally (overtourism, COVID-19, etc.) and especially in terms of the evolution of some of the basic parameters that measure its sustainability. Indeed, coastal tourism has left an unquestionable imprint on the landscape, as directly responsible for a patent territorial transformation and the evident environmental degradation resulting from the commodification of the territory. In this section we assess the main features of the academic research that has addressed the analysis of the main tourist destinations on the north Atlantic coast of Spain (Fig. 9.1). A qualitative analysis that aims to capture the strategies of land occupation and the patterns of spatial organisation that have shaped them.

With few exceptions, coastal tourism along the North Atlantic coast of Spain has not been subject to a systematic analysis as a tourist space, despite the evidence of hospitality and summer holiday activities since the second half of the nineteenth

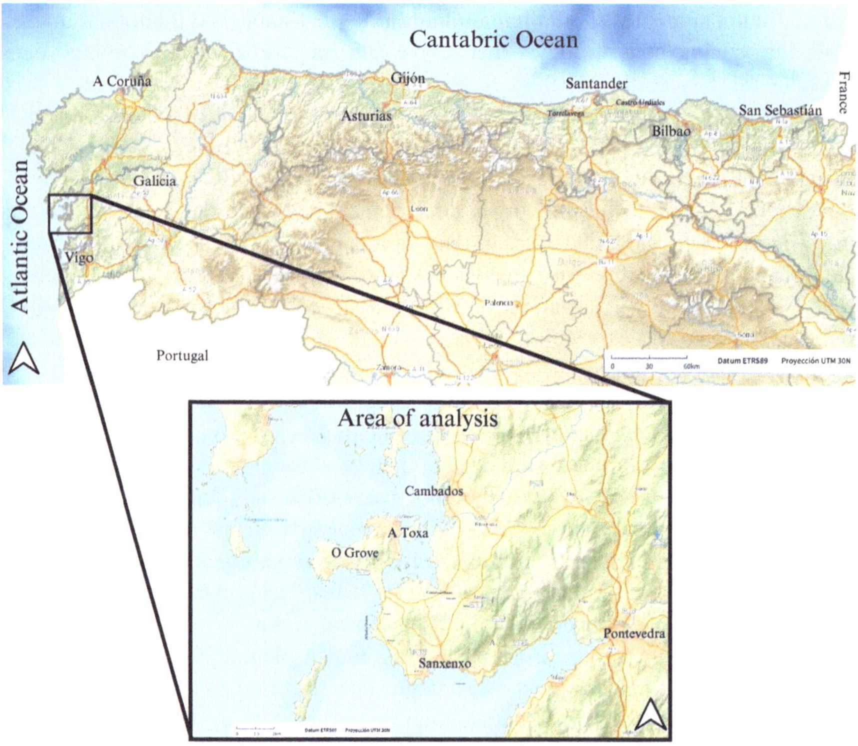

Fig. 9.1 General situation map and area of analysis. (Source: Own elaboration)

century (Barral, 2012; Beascoechea Gangoiti, 2002, 2017; Cava Mesa, 2009; Gil de Arriba, 1990, 1994, 2007), and the emergence of an elite tourism space linked to the figure of King Alfonso XIII (Gil de Arriba & Larrinaga, 2020). This type of leisure was exclusive to a minority (the country's political, economic and cultural elites) and a clear sign of prestige (Artola Blanco, 2015). The Basque Country and Cantabria became the preferred areas for this type of demand. This is the time of sea bathing with the epicentre in the beaches of San Sebastian and Santander. The presence of this elite contributed to create an image of an elitist, distinguished and elegant destination, (Gil de Arriba & Larrinaga, 2020: 27; Delgado Viñas, 2019: 293) that to some extent survives today. The derived urbanisation represents a specialised version of the bourgeois city, characterised by the large landscaped houses, the wide use of public spaces, such as the promenade, luxury facilities, etc. (Beascoechea Gangoiti, 2017). This first stage of tourism development on the North Atlantic coast is identified by the construction of the first spas (Leonardo Martín & Pozueta Echavarri, 1984: 95).

The beginnings of tourism development in the city of Santander and in other towns along the Cantabrian coast has been addressed in several studies. Among which we mention: Gil de Arriba (1990, 1992), García Cuadrillero (1996) and Delgado Viñas (2011, 2015). Pozueta Echavarri (1980) and Sazatornil Ruiz (1996) have analysed the case of Santander. However, the urban model par excellence linking tourism and urban planning is the case of San Sebastian (Beascoechea Gangoiti, 2017: 5). Walton (2002, 2013) and Larrinaga (2005) have studied the close intertwining between tourism promotion and the associated economic interests, with the planning and development of different urban projects in the city of San Sebastian.

The rest of the Spanish North Atlantic coastal sector has not been subject, with a few exceptions, to a systematic analysis as a tourist space, although there is evidence of activities related to services aimed at visitors in different nuclei from the second half of the nineteenth century (Álvarez Quintana, 1995; Barral, 2012; Cava Mesa, 2009; Gil de Arriba, 1990, 1994, 2007). In fact, between the 1890s and World War I, the network of maritime seaside resorts extended to Asturias (Gijón) and Galicia (A Coruña) (Lindoso-Tato & Vilar-Rodríguez, 2018: 218). Roza Candás (1995) studied the system of seaside resorts of Asturias in the period 1848–1935.

Galicia is no stranger to these trends, as it occupied relevant positions in the Spanish tourism ranking during this period. Within coastal tourism, the A Toxa spa, with its luxury hotel complex and clients with high purchasing power, stood out (Patiño Romarís, 2002), as well as the port cities of Vigo and A Coruña at the urban level (Lindoso-Tato & Vilar-Rodríguez, 2018).

From the Second World War there was a substantial change in coastal tourism: Hygienism gave way to new fashions, such as tanning (...) that began to promote the warm beaches of the European Mediterranean (Larrinaga, 2015). Spain became one of the main sun and sand destinations, and by the middle of the twentieth century there had already been a real meridionalisation of tourism, along with its democratisation or popularisation.

This was a process characterised by the expansion of tourist-type urbanisation in large sectors of the Spanish coast. This phenomenon has been studied mainly on the Mediterranean coast; however, it has also been widespread on the North Atlantic coast since the 1960s.

Tourism real estate growth has had important economic, urban and environmental consequences. The ways in which the urban and residential expansion process have been carried out has had important effects on the landscape that, "(...) on many occasions, goes as far as the production of new standardised landscapes, depersonalised products of real estate speculation, that often result in the alteration and dismantling of landscapes built by the social communities that have occupied and benefited from the territory over the centuries" (Delgado Viñas, 2008: 18).

In the case of Cantabria, there are important changes in the functions of tourist nuclei, which gradually become secondary residence places for inhabitants of nearby urban areas (Castro Urdiales and Laredo in relation to Bilbao, Suances to Torrelavega) or, even, in the case of Santander, becoming an area of permanent residence (Leonardo Martín & Pozueta Echavarri, 1984: 95).

Territorial and urban planning regulates the planning, intensity and uses of land and, therefore, conditions the spatial distribution and intensity of tourism use (Leonardo Martín & Pozueta Echavarri, 1984: 98). However, municipal planning results in contrasting landscapes depending on the territorial objectives of the plans and regulations (Rodríguez Gutiérrez et al., 2010).

In Cantabria, Castro Urdiales has been "(...) a summer resort for ever wider classes of population" (Delgado Viñas, 2011: 259) since the 1960s and 1970s. From this period, together with the traditional villas and farms, blocks of holiday apartments began to be built. A remarkable growth of the secondary housing stock that was closely related to the "(...) population coming (...) from the Basque provinces and, above all, from the metropolitan area of the Bilbao estuary" (Delgado Viñas, 2011: 261). A residential tourism development carried out on many occasions outside the current urban planning system (Delgado Viñas, 2011).

In recent decades the Asturian coast has increased the number of tourist places, especially in the eastern sector, "(...) supported by the improved accessibility, the proximity of very contrasting and well-connected landscapes, the work to promote and improve the offer and its diversification, adopting new trends (...)" of the market (Rodríguez Gutiérrez, 2016: 130).

Llanes is one of the best examples of a "sun and sand" destination on the Cantabrian coast (Sevilla Álvarez & Rodríguez Pérez, 2019). Gijón was a medium size industrial city in the early 1980s tourism was chosen by the administrations as a strategic line of territorial development. There was a real socio-economic transformation of the city from industry to tourism and services. A reality that has resulted in important physical transformations to the urban landscape such as the creation of important nautical facilities in the port (Alvargonzález Rodríguez, 1989). Álvarez Díaz and Valdés Peláez (2016) have carried out a study of interest which examines the sustainability of tourism development in Gijón through the proposal of a set of territorial-based indicators. García Quirós (2013) has analysed the actions undertaken since the 1990s in the western coastal area of Asturias to try to minimise the effects of the decline of its traditional economic, fishing and industrial activity (especially shipbuilding) in favour of tourism.

The Galician coast was traditionally far from the residential urban development linked to the "sun and sand" tourism product, except for very specific areas (Patiño Romarís, 2001; Santos Iglesias, 2002; Penas Murias, 1987), a circumstance that made it quite unique within Spain. In fact, due to its territorial enclave, the Galician coastline came out almost intact from the real estate fever of the 1980s and 1990s that devastated other areas of the coastline. Galicia was a very difficult area to access with long travel times. This situation of pseudo-isolation continued until the completion of the motorways which connected it to the inland of the country in 2002. This process of growth of secondary residences resulted in a deterioration of the traditional landscape and an excessive concentration of the space built in specific places, with a good tourism image and acceptable accessibility (Patiño-Romarís, 2016).

The resulting territorial tourism development model responds to a supply system that is poorly integrated in commercial markets and that is spatially materialised in extensive occupation models, with a predominance of secondary real estate developments and weighed down by a strong seasonality (Patiño Romarís, 2020: 173). The high temporary concentration translates into the tourist demand concentrated in the summer holiday months. Regarding the spatial concentration of the destination, it also has negative consequences, which are basically the loss of service quality and the deterioration of the physical environment that, in turn, makes the place less attractive (Patiño Romarís, 2021: 252 and 253).

The strategies implemented in the main tourist destinations on the Spanish North Atlantic coast are diverse. In fact, we find lines of action with a marked continuity with respect to past models, essentially characterised by the relevance of both the real estate activity and the residential-tourism offer. On the other hand, there are those strategies of rearrangement/repositioning and territorial renewal/requalification that introduce dynamics of change supported by new paradigms within the tourism market (Williams & Montanari, 1995; Vera Rebollo & Baños Castiñeira, 2010). Within these we find lines of action aimed at the modernisation of tourism establishments, the incorporation of new products in the destinations, the introduction of measures aimed at increasing sustainability, or the implementation of new management formulas.

In this complex territorial context, Sanxenxo in the words of Lois González (2020: 330) is a "(...) coastal destination that exemplifies the sun and sand model". A destination with the "(...) current issue, which some authors call touristification and others massification of the destination (...)". "Tourism that encompasses traffic congestion, both road and pedestrian; massive occupation of accommodation by tourists, a tendency towards crowds of people and the expulsion of citizens (...). Although these problems are limited in space (...) and time (July and August as the maximum exponents), they present a challenge that, if not managed properly, will get worse".

9.3 Methodology and Objective of the Study

The methodology used for this study has been based on both qualitative and quantitative techniques with the analysis of both primary and secondary data. With respect to the former, participant observation and semi-structured interviews have been prioritised; with respect to the latter, statistical sources from public agencies and institutional reports have been considered.

Within the methodology used to obtain information regarding the current dynamics, semi-structured interviews were conducted with various actors operating in Sanxenxo (the chosen destination due to its importance as the first tourist centre of the Spanish North Atlantic coast by volume of supply, visitors and its reputation). A panel of 11 experts were interviewed, including stakeholders from the following groups: public administrations, tourism service companies, business groups, local travel agencies and people who own tourism products.

Semi-structured interviews, with a level of freedom for both the interviewee and the dialogue, guide the flow of the conversation around a series of topics so that the results are comparable. To this end, a questionnaire was used to ensure that all the planned topics were covered (Trillo Santamaría & Paül Carril, 2016). The interviews addressed two issues: (a) tourism planning, management and development and (b) future of the destination. The dialogues were held in April 2022. Interviews lasting between one to one and a half hours were recorded with the consent of the interlocutor and subsequently transcribed. The interview data were structured through open coding, which has enabled us to analyse, reduce and abstract the data collected in a scientifically ordered manner. The purpose was to know the perception that agents have of the territorial tourism reality in Sanxenxo. The analysis focused on the search for common elements in the discourse of the different agents, on the causal attributions and the explanatory mechanisms used.

The quantitative techniques focusing on the management of pressure indicators have helped to establish the territorial reality of tourism activity in Sanxenxo by using a system of indicators based on statistical data from the Galician Tourism Agency (accommodation offer) and the INE (Population and Housing Census). The indicator system is intended to provide information of interest that is comprehensible.

The objective is to elaborate a diagnosis, an x-ray of the activity in the municipality that highlights the importance of tourism, placing it as an element of social debate and openly exposing its fit in the territory. This will consist of an initial diagnosis of the main macromagnitudes that contextualise the evolution and current situation as well as an initial territorial analysis that will form the basis of the final diagnosis. The tourist information system implemented is based on the European Tourism Indicators System (ETIS) adapted to the particularities of a Galician coastal destination (EU, 2017).

9.4 Initial Diagnosis of the Territorial Tourism Reality of Sanxenxo

Sanxenxo is the main Galician coastal tourist destination. The activity, according to data from EpData (2020), generates more than 85% of the local GDP and more than 70% of direct employment. According to the latest report of the Ministry of Finance and Public Administration (2013), the number of residents in Sanxenxo multiplies by 7.5 times in summer and is a destination likely to be classified as mature with evident symptoms of exhaustion. In fact, the main tourist product of Sanxenxo is the "sun and sand" (Patiño-Romarís, 2016). A product that has been deteriorating as a result of the proliferation of construction sites along its coastline. A tourism development model where excellence in landscape management is secondary.

The process of tourism functionalisation in Sanxenxo's space at first took place in areas near the beaches (Silgar, Canelas, Paxariñas, etc.)and later in the fertile agricultural landscapes and the urban property heritage of the village. The production of tourism spaces currently translates into a highly urbanised and overly densified waterfront in certain sections, generating landscape and economic contrasts between the coastal space and the interior. A residential tourism development resulting from the absence of integrated coastal management policies and the systematic non-compliance with urban planning discipline.

According to the data of the Tourism Agency of Galicia, in 2019 Sanxenxo was the Galician destination with the biggest offer of regulated tourist accommodation bed places[1] (23,641 bed places, 13.09% of Galicia), surpassing a tourist centre like Santiago de Compostela (14,007 bed places). It should also be noted that it has the largest number of hotel bed places (7906 bed places). However, at the aggregate level, supply exceeds demand, given that in relation to visitor entries and overnight stays the offer is relatively underutilised. An obvious fact when analysing occupancy rates. From a dynamic perspective, the supply of hotel bed places in Sanxenxo experienced a remarkable progression in the period 2003–2019 (335.45%).

The monthly evolution of the offer of bed places in hotel establishments has a marked seasonality. In fact, according to data from the EOH (Hotel Occupancy Survey, INE), the operating offer of the month of December (1033 bed places) represents only 11.41% of that corresponding to August (9051 bed places). A seasonality that is linked to the existence of small one- and two-star establishments that are incorporated into the market exclusively in the high season. A first reading of these data shows that in relative terms there is a relative oversupply for much of the year in Sanxenxo.

[1] This includes hotels, guest houses, tourist shelters, rural tourism, tourist camps, tourist apartments and holiday rentals (VUT).

Also, taking into account the values of secondary residences included in the 2011 Population and Housing Census (INE), we developed a series of pressure indicators to analyse the territorial dimension of the residential tourism offer in the area of Sanxenxo. The residential offer of Sanxenxo responds to a marked residential tourist approach (44,236 bed places, 4.15% of the total Galician coast). An approach that is evident in the 0.62 of its indicator "bed places in residences per de facto population". (Table 9.1).

Sanxenxo has a total of 67,887 accommodation bed places. The housing supply pressure indicators record the values of 3.91 bed places/inhabitant and 1505.92 bed places/km^2 (Table 9.2). In short, a predominantly residential offer is available. However, the regulated offer is large within the Galician context. Based on the accommodation data, it is the "sun and sand" tourist destination par excellence of the Galician coast.

The initial analysis of the pressure values of the accommodation supply in Sanxenxo shows a strong tourist functionalisation within the space and a maximum carrying capacity in the summer months of greater demand, which are the months where overtourism processes are observed. We do not believe that Sanxenxo as a tourist destination is in its decline, however we do believe that strategic measures are needed to competitively reposition it in the market.

Table 9.1 Residential tourist offer in the municipality of Sanxenxo[a]. Year 2011

	A	B	C	D	E
Sanxenxo	11,059	245.21	0.64	189.50	44,236
Pontevedra Estuary	32,865	114.63	0.20	55.14	131,460
Galician coastline	266,312	54.28	0.17	44.28	1,065,248

Source: INE: Municipal population register (2019) and Population and Housing Census (2011). Surface data km^2 IGN (National Geographic Institute)

[a]Indicators: (A) Number of secondary residences. We consider secondary residence those included in the Population and Housing Census as "non-main (secondary+empty)"; (B) Density (secondary residence per Km^2); (C) Secondary residences/Population in fact; (D) T.F.R. = Rs × 100/Rp; where: TFR is the Residential Function Rate; Rs is the number of secondary residences; Rp is the number of main residences; (E) Offer of secondary residences bed places. Offer of bed places in secondary residences = number of secondary residences × 4

Table 9.2 Total tourist accommodation offer and pressure indicators in the municipality of Sanxenxo[a]. Year 2019

	A	B	C	D	E
Sanxenxo	23,641	44,236	67,887	3.91	1505.92

Source: INE: Municipal population register (2019) and Population and Housing Census (2011). Surface data km^2 IGN (National Geographic Institute). Tourism of Galicia

[a]Indicators: (A) Total number of bed places regulated offer; (B) Offer of secondary residence bed places. Offer of bed places in secondary residences = number of secondary residences × 4.; (C) Total tourist accommodation bed places; (D) Total tourist accommodation bed places/de facto population; (E) Density (Total tourist accommodation bed places per Km^2)

9.4.1 *Interpretation of the Results Obtained from Semi-structured Interviews*

9.4.1.1 Tourism Planning, Management and Development in Sanxenxo, According to the Sector Stakeholders

As already mentioned, in order to know the problems of this mature destination first-hand, an amplified example of what happens in other important tourist locations on the North Atlantic coast and at risk of being affected by overtourism, leading actors in the Sanxenxo tourism sector have been consulted. All the interviewees share the opinion that Sanxenxo is a consolidated sun and sand destination with obvious seasonality problems. “It is a seasonal tourism. Summer tourism, beach tourism” (E9, Portonovo Fishermen’s Association-Guide). An endogenous tourist development that began in the decade of the 1950s. “It has been a development based on family sacrifice, not studied, not controlled, not monitored by the administrations” (E5-Hotel Businessman). In recent years there has been a leap in quality (E4-Restaurant Owner), “(. . .) it has been professionalised a lot thanks to the second and third generations (of the pioneering entrepreneurs) education and training” (E10, Sanxenxo Tourism Officer).

Sanxenxo is a consolidated national destination (E8, Hotel Manager). One aspect to be improved is its international character. “There are areas of Europe that do not know us and we do not expect them” (E4, Restaurant Owner). “The main strength is the sun and the beach” (E10, Sanxenxo Tourism Officer). “Today sun and sand is our great product (. . .), a fundamental pillar” (E5-Hotel Businessman). “The sun and sand product” (. . .) is so consolidated (. . .) that other products are overlooked. It’s all secondary. “Even in winter people come to enjoy the beaches (. . .)” (E10, Sanxenxo Tourism Officer).

However, it is also important to note a general consensus in highlighting the existence of a complementary offer as a strength of Sanxenxo, based essentially on gastronomy and nautical equipment (E11, CN Travel Tour Operator-Director). In fact, following the inauguration of the Juan Carlos I marina in 2005 “(. . .) tourism improved a lot” (E3, Bar Owner). In this direction, “they are trying to do things so that (the tourist season) lasts longer (. . .)” (E9, Portonovo Fishermen’s Association-Tourist Guide). By way of example, marine tourism is being promoted with an offer of products with a marked seasonality (E9, Portonovo Fishermen’s Association-Tourist Guide).

However, we find ourselves with more critical positions: “I think it should be a little more diversified because it is very focused on sun and sand tourism” (E9, Portonovo Fishermen’s Association-Tourist Guide). While “(. . .) we have a lot of things we can offer, we continue to offer the same as many years ago (. . .)” (E1, Hotel Businessman). “(. . .) we would like to have some new ingredients, like a park or some attraction to deseasonalise (. . .)” (E2, Interrias Tour Operator Commercial Director). “We have plenty to offer and for the customer to say I’m going to an area of the world that is a paradise where you can experience fantastic things” (E8, Hotel Manager).

Also, there is a broad consensus in the panel of interviewees in highlighting the progressive application of ICT (Information and Communication Technologies) in the process of commercialisation of the offer as the most relevant change within the tourism sector in Sanxenxo in the last decade. "(...) Social networks in particular, are a weapon and a fundamental support (...)" (E4, Restaurant Owner). On the other hand, within the tourism sector of Sanxenxo, there have been new emerging trends in demand:

- The "(...) customer increasingly knows more, is more demanding" (E3, Bar Owner).
- "(...) we have to be very observant of new trends, very responsive to possible new customer demands and try to adapt the offer (...). We, for example, believe that the customer demands more in terms of sustainability, digitalisation. (...), mainly in the younger audience" (E2, Interrias Tour Operator Commercial Director).
- "Changes in the length of holidays. Now there is no longer the tourist that stays for two weeks, a month (they almost don't exist)" (E10, Sanxenxo Tourism Officer).
- The "(...) season has been greatly extended. It has not been seasonally adjusted but has expanded" (E10, Sanxenxo Tourism Officer).
- "More travellers with an increasing purchasing power (...), there are people who return so there must be some high degree of satisfaction" (E2, Interrias Tour Operator Commercial Director). In the case of the demand for nautical facilities at the marina, there has been a move from a "(...) medium-high category customer that we had in the early 2000s to a, so to speak, high or very high category customer. The demand we have right now is from customers with very high purchasing power, very large boats (...)" (E7, Nauta Sanxenxo SL Boatswain).

The tourist development in the municipality of Sanxenxo has been out of all urban control until the approval on 27 February 2003 of its PGOM.[2] "We had a time where a lot was built and bad, as usual, in similar places. Many constructions of that time would not have been done today" (E10, Sanxenxo Tourism Officer). "Areas that have been degraded due to the urban development have lost their charm" (E11, CN Travel Tour Operator-Director). "The arrival of PGOM has stopped all that" (E5-Hotel Businessman). A normative document that did not entail a substantial change in the tourism development model implemented so far. The transformation that the waterfront of the village of Sanxenxo has undergone is specially worrying. "The development of the urban part in Sanxenxo was perhaps too aggressive (...). It is not normal (...) that there is a line of buildings along the waterfront when a few years ago there were small villas with their plots of land" (E1- Hotel Businessman). "I remember Sanxenxo with low rise houses 20 or 30 years ago. That can no longer be changed. What we must try to avoid is to continue constructing those types of buildings". (E6, President of the Yacht Club). "It is clear that it is not nice, there

[2]General Plan Municipal Planning.

had to be a more orderly growth and I think it was not like that. I think it was built too quickly" (E3, Bar Owner).

Among the interviewed tourism agents from Sanxenxo, and as an example of the destinations of the Spanish North Atlantic, we have seen a clear commitment to the need to orient the tourism model towards sustainability. "Before it was impossible to demolish a building and today the city council is considering demolishing buildings. That we consider demolishing buildings due to ugliness is something (...)" (E5- Hotel Businessman). "We have to make a huge effort to safeguard the natural environment, the landscapes, the mountains (...), it has to be our commitment" (E8, Hotel Manager). "We have to support the conservation of all the elements that we have and that show who we are to provide our sector with a fundamental future capacity" (E4, Restaurant Owner). In this sense, within the interviews there has been support for greater rigor in the application of the current legislation and territorial action strategies have been provided in order to improve the landscape and the environment within the municipality. "I think that if the PGOM is executed and if we really follow the legislation, I would be very pleased. Between PGOM, Coastal Law, River Law... the well applied legislation is capable of paralysing outrageous projects" (E5- Hotel Businessman). "There has been a lack of urban and tourism planning since tourism was seen as the main source of income" (E11, CN Travel Tour Operator-Director). "As a Tourism officer, I think that at some point we will have to put a limit on the number of bed places, tourism residences, etc. I think that somehow, they will have to slow it down in some way because it is going to result in a loss of quality" (E10, Sanxenxo Tourism Officer).

The existence of problems related to overtourism (overcrowding problems, noise, loss of quality in the experience, etc.) have also been subjected to comments in the interviews carried out.

"We have 17,000 inhabitants registered in the municipality, and in summer we reach almost 100,000 or 150,000. It's too much (...)" (E4, Restaurant Owner). "There are places where people are very concentrated" (E10, Sanxenxo Tourism Officer). "(...) Absolutely everything collapses, restaurants, cafes, supermarkets, parking lots, the beach (...) (E9, Portonovo Fishermen's Association-Tourist Guide)". The carrying capacity is largely exceeded in "Sanxenxo urban centre (...), the centre exceeds its carrying capacity because (...), there are too many secondary residences (...) and the secondary residence nowadays is often turned into a holiday rental" (E1- Hotel Businessman). The "(...) centre of Sanxenxo is massively overcrowded" (E2, Interrias Tour Operator Commercial Director). In the summer months "(...) we are overwhelmed by the number of people (...), we run out of parking spaces, supermarkets run out of fruit and vegetables before 18:00 in the afternoon. It's just that season. The rest of the year we don't have that. For eight months we are, so to speak, calmer. We don't have that overload" (E9, Portonovo Fishermen's Association-Tourist Guide).

In response to these problems linked to the tourism overload in certain enclaves of Sanxenxo, the panel of experts interviewed have made concrete proposals for action: "(...) I think it would be better to spread people throughout the year and not have everyone in the summer months" (E9, Portonovo Fishermen's Association-Tourist

Guide). "It's not easy, because of how it's built, because of how it's made. We should have done it sooner, we should have done infrastructure works, the issue of parking is complicated (. . .). Maybe pedestrianise a part. Because of how it is located and where it is located, as well as the streets and the buildings, it is very difficult to give a solution to the urban centre" (E1- Hotel Businessman). "A bypass road is a priority. The accesses in summer collapse. (. . .) The solution is to divert traffic that does not have to enter Sanxenxo. (. . .) Another project would be to bury the Sanxenxo promenade. That all traffic is underground and the walk is pedestrianised (E11, CN Travel Tour Operator-Director)". The municipal administration implemented measures with relative success: "(. . .) During COVID we tried to create deterrence zones so that people would go to lesser-known areas. It didn't work out badly, but last year it was completely ignored. We tried to organise trails, inland routes, etc. But it is difficult, because people like the same thing. The actions we tried to decentralise did not work badly but, of course, it was a during the pandemic" (E10, Sanxenxo Tourism Officer).

The interviewees have shown a diversity of opinions regarding the tourism strategies implemented by the Xunta de Galicia as well as by the municipal administration. A fundamental criticism is the lack of coordination between the different public administrations. "(. . .) I am not talking about the Municipality not being able to work with the Provincial Council or the Provincial Council not being able to work with the Xunta. It's just that the Xunta can't work with the Xunta. Tourism is horizontal, the different departments should work together" (E5- Hotel Businessman). On the other hand, among the private agents interviewed, their lack of participation in the design of public strategies is evident: "(. . .) they have to listen to us. (. . .)" (E11, CN Travel Tour Operator-Director). Within the field of municipal management, the lack of a territorial tourism development plan is highlighted: "Having a real tourism plan (. . .). Years ago we created a company to encourage planning, today that company is dedicated to hiring lifeguards and making t-shirts. We need to get back to planning (. . .). A master plan managed by the public and private sectors is essential" (E5- Hotel Businessman). However, favourable comments by interviewees are common:

- "Xunta and Sanxenxo City Council are very involved" (E7, Nauta Sanxenxo SL Boatswain).
- "It is clear that more can always be done". "A very important job is being done by Rías Baixas tourism, with promotion outside the Autonomous Region and even outside (. . .) outside Spain". "(. . .) if we look into specific issues, the management of the pandemic was not that bad, we could have done a little better, in the sense that knowing certain measures more in advance could have helped to minimise costs or to do things differently". (E8, Hotel Manager).
- The "(. . .) tourist voucher Quédate en Galicia (Stay in Galicia). That I think is a very good option for the people in Galicia to stay here and spend money here" (E9, Portonovo Fishermen's Association-Tourist Guide).

From the Local Government level, two major difficulties stand out in order to implement an effective tourism policy. The problem of financing tourist municipalities: "The municipality does not reach 20,000 inhabitants, but in summer it becomes the fourth biggest city in Galicia. But what the city receives is for the 17,000 inhabitants." "Everything is spent on beach cleaning and other similar things". In addition to the private sector agents, add a lack of collaboration: In the low season "(. . .) we find that the hotels are closed. We want to do events, but the hotels are closed, they do not read the emails, they are on vacation etc. There should be better public-private cooperation" (E10, Sanxenxo Tourism Officer). And within the lines of municipal tourism action, the initiatives aimed at alleviating the problem of seasonality stand out: "In the low season we are trying to have international tourists. (. . .). If you want to work in winter you have to target international tourism (. . .). I think that with international tourism we could extend the season" (E10, Sanxenxo Tourism Officer).

The possibility of applying a municipal tourist tax has received mixed opinions. "As a Tourism officer I don't think it's bad, but politicians don't like the measure. I don't think anything will happen" (E10, Sanxenxo Tourism Officer). We also find those clearly opposed: "(. . .) I am totally against it. In the end, the tourist tax is a tax that is invented by the Administrations to raise money for themselves, but they do not return it" (E1- Hotel Entrepreneur) to the tourism sector. This aspect is the key when it comes to having a favourable opinion by the different interviewees: "If it is used for something other than tourism, I would not agree. But if it helps tourism in Sanxenxo, I would support it one hundred percent" (E3, Bar Owner). "I am in favour of the tourist tax, if it is used exclusively for the promotion and defence of the tourism sector" (E5- Hotel Businessman).

The main challenges that Sanxenxo as a destination must face within an increasingly competitive market are essentially:

- Diversify the economic structure of the municipality: "Sanxenxo should not depend exclusively on tourism. I think it's very complicated to have a population that gets used to living from the work of three months a year. It is extremely dangerous" (E5- Hotel Businessman).
- Encourage cooperation between public and private actors (E5- Hotel Entrepreneur).
- Improve the quality of the product offered: Commit to "(. . .) quality and not quantity" (E7, Nauta Sanxenxo SL Boatswain), "strive for the quality that we offer. (. . .) if we cannot deal with more travellers we must not do so, but those who come must be well looked after" (E2, Interrias Tour Operator Commercial Director). "Continue working on the improvement of quality in both the public and private sectors" (E10, Sanxenxo Tourism Officer). At this point, two strategic actions are proposed; firstly, a higher implementation of quality systems to manage the offer (E8, Hotel Manager). And, secondly, to promote more quality in hotel facilities (E4, Restaurant Owner).
- The education and training of human capital linked to the tourism sector: "It is one of the main challenges. (. . .) the issue of professionalism. (. . .) the creation of a Hospitality Training Centre has been proposed. I think it's a very good idea (. . .)" (E8, Hotel Manager).

- Fight against the seasonality of supply and demand: "Try to deseasonalise the destination. I think there's a lot of talk about this but nothing has been done. In the end, the few things that are done are on a very personal level. This is one of the challenges of the future" (E11, CN Travel Tour Operator Director). Among the actions proposed is: to increase "(. . .) tourism promotion in the low season" (E10, Sanxenxo Tourism Officer).

9.4.1.2 The Future of the Destination

Participants expressed the belief that the future prospects for the sector are positive, as well as for the North Atlantic coast as a whole. "We have more and more tourists in Sanxenxo. Each time we are chosen by customers looking for different types of holidays, we are able to attract smaller niches to deseasonalise, such as golf (. . .). We see that there is a future as long as we are able to maintain the current level of quality (. . .) on offer" (E2, Comercial Interrrias Tour Operator-Director). "Many tourists are coming for the first time, and it is a challenge to develop their loyalty. I am concerned about the carrying capacity issue. At some point they will have to take action on the matter. To grow in quality is a challenge that at some point we will have to deal with" (E10, Sanxenxo Tourism Officer).

Finally, we highlight the demand for "(. . .) a plan that makes tourism socially profitable. We need to analyse why we are a destination for small entrepreneurs and why large chains have not come. If we do not analyse it, it means we are not interested" (E5- Hotel Businessman).

9.5 Conclusion

The spontaneity and lack of planning criteria for tourism development in Sanxenxo and other locations on the Spanish North Atlantic coast (Ribadesella, Castro Urdiales, etc.) explain situations of a waste of resources, a lack of rationality in land use and a lack of references for the development of interventions, both public and private, in addition to environmental impacts and poor integration of actions within a coherent framework in the implementation of tourism (Vera et al., 2011). Therefore, we find ourselves with the need to deploy a comprehensible and effective strategy to ensure the sustainable development of tourism within these destinations, which follow the same model as Sanxenxo, where tourism managers must work on the design of lines of action aimed at avoiding overtourism or mitigating its effects. In this direction, we believe that the process of renewal on a consolidated destination should have an impact on the quality of the offer, based on the enhancement of its territorial components from the principle of sustainability and of those products that, for the moment, have not received enough attention in their design.

Additionally, it is necessary to take the local community into account when developing the process of implementation for a territorial action strategy. To this end, it is essential that this process is developed within the framework of a true governance model. The municipal public administration must lead the renewal process, establishing a system of participation that should encourage cooperation between different public and private actors.

Also, the Smart Tourist Destinations Network provides technological solutions for the management of tourist overload in certain public spaces. Sanxenxo and other similar destinations have signed a collaboration agreement with SEGITTUR and the Ministry of Tourism to implement a Pilot Project of Sustainable Tourism Indicators. The objective is the development of an intelligent tourist destination as a tool to improve competitiveness, the residents quality of life and satisfaction with the tourist experience (Boes et al., 2015; Buhalis & Amaranggana, 2013). Open/big data analysis is a very useful tool to improve tourism management in a destination in terms of flow management. However, the potential of technological solutions and their supposed efficiency cannot be a pretext to avoid more far-reaching measures within a destination governance model such as limiting the growth of tourism supply and demand, and even slowing down growth (García-Hernández et al., 2019).

Finally, we must be aware that the maturity of the most prominent destinations on the North Atlantic coast must not imply a territorial strategy of reorganisation/renewal that results in a total break with the current model. The municipal administration must support a progressive reinvention of the tourism system, modifying or promoting products that can grant a territorial competitiveness bonus. Provide this set of destinations that we have exemplified with Sanxenxo with a wide tourist offer, one that is better regarded, sustainable and with added value, less seasonality and public-private cooperation that enhances its image in the tourism market. A quality sun and sand product and a real flagship, rich in green, cultural, sports and gastronomic alternatives.

References

Alegre, J., & Garau, J. (2010). Tourist satisfaction and dissatisfaction. *Annals of Tourism Research, 37*(1), 52–73. https://doi.org/10.1016/j.annals.2009.07.001.

Álvarez Díaz, R., & Valdés Peláez, L. (2016). Selección e indicadores para la evaluación del desarrollo sostenible de un destino turístico. Aplicación al municipio de Gijón. *ROTUR. Revista de Ocio y Turismo, 11*, 12–21. https://doi.org/10.17979/rotur.2016.11.1.1777

Álvarez Quintana, C. (1995). *Baños de mar en Ribadesella 1890–1936. Urbanismo, arquitectura y sociedad.* Asociación Cultural Amigos de Ribadesella.

Alvargonzález Rodríguez, R. M. (1989). Metamorfosis portuaria: el nuevo puerto deportivo de Gijón, notas para una tipología de los espacios de ocio litorales. *Ería, 18*, 77–84. https://doi.org/10.17811/er.0.1989.77-84

Antón, S. (1998). La urbanización turística. De la conquista del viaje a la reestructuración de la ciudad turística. *Documents d'Anàlisi Geogràfica, 32*, 17–43.

Artola Blanco, M. (2015). *El fin de la clase ociosa*. Alianza.

Barral, M. (2012). Casas de baños y balnearios en el desarrollo del turismo en Galicia: el caso de Vigo y Mondariz. In T. M. Ortega & M. A. Del Arco (Eds.), *Claves del mundo contemporáneo. Actas del XI Congreso de la Asociación de Historia Contemporánea* (pp. 1–8). Granada.

Beascoechea Gangoiti, J. M. (2002). Veraneo y urbanización en la costa cantábrica durante el siglo XIX: Las playas del Abra de Bilbao. *Historia Contemporánea, 25*, 181–202.

Beascoechea Gangoiti, J. M. (2017). De la estancia de baños al veraneo de clase media: el cambio de modelo urbano en Plentzia, Bizkaia. *Scripta Nova, 21* (568). https://bit.ly/2VaVUrg

Benner, M. (2019). From overtourism to sustainability: A research agenda for qualitative e-tourism development in the Adriatic. *MPRA Paper*, 92213. https://mpra.ub.uni-muenchen.de/92213.

Blázquez-Salom, M., Cladera, M, & Sard, M. (2021). Identifying the sustainability indicators of overtourism and undertourism in Majorca. *Journal of Sustainable Tourism*. https://doi.org/10.1080/09669582.2021.1942478

Boes, K., Buhalis, D., & Inversini, A. (2015). Conceptualising smart tourism destination dimensions. In I. Tussyadiah & A. Inversini (Eds.), *Information and communication technologies in tourism 2015. Proceedings of the international conference in Lugano, Switzerland, February 3–6, 2015* (pp. 391–403). Springer. https://doi.org/10.1007/978-3-319-14343-9_29

Buhalis, D., & Amaranggana, A. (2013). Smart tourism destinations. In Z. Xiang & I. Tussyadiah (Eds.), *Information and communication Technologies in Tourism 2014* (pp. 553–564). Springer. http://link.springer.com/10.1007/978-3-319-03973-2_40

Cava Mesa, M. J. (2009). Balnearios y baños de ola en Bizkaia. *Itsas Memoria, Revista de Estudios Marítimos del País Vasco, 6*, 369–380.

Delgado Viñas, C. (2008). Vivienda secundaria y turismo residencial como agentes de urbanización y segregación territorial en Cantabria. *Scripta Nova. Revista Electrónica de Geografía y Ciencias sociales, XII* (261). http://www.ub.es/geocrit/sn/sn-269.htm

Delgado Viñas, C. (2011). Castro Urdiales (Cantabria), de "villa marinera" a ciudad de servicios. La transformación urbanística de una "ciudad de frontera". *Ería, 86*, 237–270.

Delgado Viñas, C. (2015). Estructura y forma de la ciudad a través de la cartografía histórica. Castro Urdiales, Cantabria (1800–1960). *Investigaciones Geográficas, 63*, 17–32.

Delgado Viñas, C. (2019). Evolución urbanística de una ciudad portuaria burguesa (Santander, 1750–1941). *Ería: Revista cuatrimestral de geografía, 39*(3), 285–330.

Dichter, A., & Gloria, G. M. (2017). *Coping with success. Managning destinations* (Pdf file). McKinsey & Company and World Travel & Tourism Council. https://www.wttc.org/-/media/files/reports/policy-research/coping-with-success%2D%2D-managing-overcrowding-in-tourism-destinations-2017.pdf

EpData. (2020). *Sanxenxo - Datos, gráficos y estadísticas sobre el municipio*. https://www.epdata.es/datos/datosgraficos-estadisticas-municipio/52/sanxenxo/6757

European Union. (2017). *El Sistema Europeo de Indicadores Turísticos. Herramienta del ETIS para la gestión de destinos sostenibles*. Oficina de Publicaciones de la Unión Europea.

EXCELTUR. (2003). *Plan Renove de destinos turísticos españoles. Documento de Bases. Hacia un nuevo paradigma en el modelo de gestión de los destinos del litoral mediterráneo y las islas: Los Proyectos de Reconversión Integral de Destinos Turísticos (PRIDET)*. Área de Estudios e Investigaciones de Exceltur (Alianza para la Excelencia Turística).

García Cuadrillero, M. (1996). *Mar, veraneo y urbanización en Laredo. 1850–1964*. Universidad de Cantabria.

García Quirós, R. M. (2013). Diseño y turismo en la costa occidental asturiana, del Eo al Nalón. *Liño, 19*, 95–106. https://reunido.uniovi.es/index.php/RAHA/article/view/9935

García-Hernández, M., Ivars-Baidal, J., & Mendoza de Miguel, S. (2019). Overtourism in urban destinations: Themyth of smart solutions. *Boletín de la Asociación de Geógrafos Españoles, 83*(2830), 1–38. https://doi.org/10.21138/bage.2830

Gil de Arriba, C. (1990). La génesis de las actividades de ocio en Cantabria. Estudio del caso de Comillas. *Treballs de Geografía, 43*, 111–118.

Gil de Arriba, C. (1992). *Casas para baños de ola y balnearios marítimos en el litoral montañés, 1868–1936*. Universidad de Cantabria/Fundación Marcelino Botín.

Gil de Arriba, C. (1994). La práctica social de los baños de mar. Establecimientos balnearios y actividades de ocio en Cantabria (1868–1936). *Documents d'Anàlisi Geogràfica, 25*, 79–97.

Gil de Arriba, C. (2007). Las playas y el fenómeno urbano. In J. L. Estrada Llaquet, R. Bruttomesso, J. Alemany Llovera, & M. Velasco Cabeza (Eds.), *Playas urbanas. IV Curso Internacional de Relaciones Puerto-Ciudad* (pp. 33–48). Autoridad Portuaria de Santander. Navalia.

Gil de Arriba, C., & Larrinaga, C. (2020). La cornisa cantábrica como región turística en las primeras décadas del siglo XX (1902–1931). *Investigaciones de Historia Económica- Economic History Research, 17*, 26–36. https://doi.org/10.33231/j.ihe.2020.04.001

Jin, Q., Hu, H., & Kavan, P. (2016). Factors influencing perceived crowding of tourists and sustainable tourism destination management. *Sustainability, 8*(10), 976. https://doi.org/10.3390/su8100976

Koens, K., Postma, A., & Papp, B. (2018). Is overtourism overused? Understanding the impact of tourism in a city context. *Sustainability, 10*(12), 1–15. https://doi.org/10.3390/su10124384

Larrinaga, C. (2005). A century of tourism in northern Spain: The development of high-quality provision between 1815 and 1914. In J. K. Walton (Ed.), *Histories of tourism: Representation, identity and conflict* (pp. 88–103). Channel View.

Larrinaga, C. (2015). De las playas frías a las playas templadas: la popularización del turismo de ola en España en el siglo XX. *Cuadernos de Historia Contemporánea, 37*, 67–87. https://doi.org/10.5209/rev_CHCO.2015.v37.50987

Leonardo Martín, J., & Pozueta Echavarri, J. (1984). Transformación del litoral e intervención pública el caso de Cantabria. *Ciudad y territorio: Revista de ciencia urbana, 62*, 93–99.

Lindoso-Tato, E., & Vilar-Rodríguez, M. (2018). El despertar del turismo urbano en España: Baños de mar, ocio y urbanismo en A Coruña, 1900–1935. *Cuadernos de Turismo, 42*, 213–236. https://doi.org/10.6018/turismo.42.09

Lois González, R. C. (2020). Galicia: as singularidades xeográficas do territorio para o turismo. In R. Villares, J.-F. Botrel, & Rivalan Guégo, C. (Coords.). *Galicia-Bretaña: Olladas comparadas* (pp. 315–335). Consello da Cultura Galega.

Milano, C. (2017). *Overtourism y Turismofobia: Tendencias globales y contextos locales*. Ostelea School of Tourism & Hospitality.

Milano, C., Cheer, J. M., & Novelli, M. (2018). *Overtourism: A growing global problem*. https://theconversation.com/overtourism-a-growing-global-problem-100029

Mullins, P. (1994). Class relations and tourism urbanization: The regeneration of the petite bourgeoisie and the emergence of a new urban form. *International Journal of Urban and Regional Research, 18*(4), 591–608.

Patiño Romarís, C. A. (2001). *Espacios turísticos e de ocio nas Rías Baixas: Unha análise a diferentes escalas*. Direción Xeral para o Turismo, Consellería de Cultura, Comunicación Social e Turismo, Xunta de Galicia.

Patiño Romarís, C. A. (2002). *A illa da Toxa. O planeamento urbanístico no proceso de construcción e transformación dun producto turístico litoral*. Servicio de Publicacións, Deputación Provincial de Pontevedra.

Patiño Romarís, C. A. (2020). Análisis del modelo turístico del litoral gallego desde el enfoque de la oferta. *Minius, 25*, 173–202.

Patiño Romarís, C. A. (2021). Factores determinantes de la demanda del producto turístico litoral gallego. *Minius, 26*, 221–256.

Patiño-Romarís, C. A. (2016). *Espazos turísticos e de ocio no litoral galego: Unha análise a diferentes escalas*. Doctoral thesis, Universidade de Santiago de Compostela.

Peeters, P., Gössling, S., Klijs, J., Milano, C., Novelli, M., Dijkmans, C., Eijgelaar, E., Hartman, S., Heslinga, J., Isaac, R., Mitas, O., Moretti, S., Nawijn, J., Papp, B., & Postma, A. (2018). *Research for TRAN committee - Overtourism: Impact and possible policy responses*. European Parliament, Policy Department for Structural and Cohesion Policies.

Penas Murias, M. V. (1987). *El Área de Oleiros-Sada: Un espacio de Ocio en la periferia de La Coruña*. Excma. Diputación Provincial de La Coruña.

Pozueta Echavarri, J. (1980). *El proceso de urbanización turística. La producción del Sardinero*. Doctoral thesis, Universidad de Santander.

Rodríguez Gutiérrez, F. (2016). Las actividades económicas y la organización funcional del territorio litoral de Asturias. *Anales de Geografía de la Universidad Complutense, 36*(1), 109–147. https://doi.org/10.5209/rev_AGUC.2016.v36.n1.52716

Rodríguez Gutiérrez, F., Fernández Prieto, J. A., Blanco Fernández, J., & Rodríguez Cienfuegos, H. (2010). La Marina asturlucense: Experiencia de gobernanza en un territorio muy complejo. *Cuadernos de geografía, 88*, 179–198.

Roza Candás, M. (1995). *Los balnearios marítimos en Asturias (1848–1935)*. Fundación Alvargonzález.

Santana-Talavera, A., Jonay Rodríguez Darias, A., Díaz Rodríguez, P., & Ramos Domínguez, A. M. (2010). Innovación con compromisos. Retos en la renovación de la imagen en destinos turísticos maduros (Fuerteventura, Islas Canarias). In R. Hernández Martín & Agustín Santana-Talavera, A. (Coords.), *Destinos turísticos maduros ante el cambio. Reflexiones desde Canarias* (pp. 137–160). Instituto Universitario de Ciencias Políticas y Sociales, Universidad de La Laguna.

Santos Iglesias, M. V. (2002). *El turismo en el Val Miñor. Transformaciones espaciales y socioeconómicas en el suroeste de la provincia de Pontevedra*. Degree memory (Unpublished). Santiago de Compostela: Departamento de Xeografía Aplicada, Facultade de Xerografía e Historia, Universidade de Santiago de Compostela.

Sazatornil Ruiz, L. (1996). *Arquitectura y desarrollo urbano en Cantabria en el siglo XIX*. Universidad de Cantabria.

Sevilla Álvarez, J., & Rodríguez Pérez, C. (2019). Relevancia y desafíos del paisaje en la formación de una imagen turística de prestigio: la costa de Llanes (Asturias). *Cuadernos de Turismo, 3*, 435–469. https://doi.org/10.6018/turismo.43.17

Trillo Santamaría, J. M., & Paül Carril, V. (2016). Transboundary protected areas as ideal tools? Analyzing the Gerês-Xurés transboundary biosphere reserve. *Land Use Policy, 52*, 454–463.

UNWTO. (2018). *¿Exceso de turismo? Entender y Gestionar el Crecimiento del Turismo Urbano más allá de las Perspectivas. Resumen Ejecutivo*. OMT.

Usher, L. E., & Gómez, E. (2017). Managing stoke: Crowding, conflicts, and coping among Virginia Beach surfers. *Journal of Park and Recreation Administration, 35*(2), 9–24. https://doi.org/10.18666/JPRA-2017-V35-I2-7596

Vera Rebollo, J. F., & (Coord.), López, F., Marchena, M., & Antón, S. (2011). *Análisis territorial del turismo y planificación de destinos turísticos*. Tirant lo Blanch.

Vera Rebollo, J. F., & Baños Castiñeira, C. J. (2010). Renovación y reestructuración de los destinos turísticos consolidados del litoral: Las prácticas recreativas en la evolución del espacio turístico. *Boletín de la Asociación de Geógrafos Españoles, 53*, 329–353.

Vera Rebollo, J. F., & Monfort, V. M. (1994). Agotamiento de modelos turísticos clásicos. Una estrategia territorial para la cualificación: la experiencia de la Comunidad Valenciana. *Estudios Turísticos, 123*, 17–45.

Walton, J. K. (2002). Planning and seaside tourism: San Sebastián, 1863–1936. *Planning Perspectives, 17*, 1–20.

Walton, J. K. (2013). Another face of 'mass tourism': San Sebastián and Spanish beach resorts under Franco, 1936–1975. *Urban History, 40*(3), 483–506.

Williams, A., & Montanari, A. (1995). Introduction. Tourism and economic restructuring in Europe. In A. Montanari & A. Williams (Eds.), *European tourism: Regions, spaces and restructuring* (pp. 1–15). Willey.

Carlos Alberto Patiño-Romarís. Was awarded a PhD for a dissertation entitled: "Tourist and leisure spaces in the Rías Baixas: An analysis at different scales", in Applied geography by the University of Santiago of Compostela. He currently an professor in the Department of History, Art and Geography of the Uni-versity of Vigo. His work experience has focused on consulting, specifically in the field of Market-ing and Strategic Tourism Planning. In the academic field, he has published books, book chapters and articles on the subject of territorial marketing, local de-velopment policies, planning and management in the field of tourism.

Rubén Camilo Lois-González is a professor of Geography at the University of Santiago de Compostela, in the Department of Geography. He was published numerous articles on urban and cultural Geography, tourist analysis referred to historic cities, cultural itineraries and pilgrimage routes. He has been Director General for Tourism position of the Autonomous Galician Government (2005–2009). Besides, he has also been Director of the Euro-regional Study Centre (CEER), an organism for scientific cooperation between the universities in Galicia and the northern region of Portugal (2013–2017). Currently he is Vice President of the International Geographical Union (UGI/IGU), Head of Geography Department, coordinator of the Competitive Reference Group ANTE (Territorial Analysis) and member of the Scientific committee of the University Institute for Studies and Development of Galicia (IDEGA) of the USC.

Breixo Martins Rodal holds a PhD from the University of Santiago de Compostela. So far, he has focused his research on the Rías Baixas. It is also the main coastal tourist destination in Galicia. He has investigated this tourist activity in the town of Sanxenxo, the main tourist destination of the region. He has been a lecturer and researcher at the University of Vigo in the Faculty of Tourism. He has conducted research and participated in international projects related to tourism research and territorial development. He has worked in international research institutions analysing tourism and territorial behaviour in the EU.

Chapter 10
Governing the Social Production of Spain's Sun and Sand Tourist Resorts

Macià Blázquez-Salom and Ivan Murray

10.1 Introduction

Sun and sand tourist resorts are on of the key spatial products of recent urban development processes in Spain. Analyses of their development and use have been conducted in the field of tourism geography and in many other scientific disciplines. Among the multitude of possible approaches, it is interesting to explore the processes that led to their creation through the social production of space, their integration in global value chains and trade networks (Lee, 2010), and their incorporation in consumption by the popular classes. Who promotes them and who benefits from them are issues that are explored in studies of the agency of institutions and social agents through analyses of capital flows, power, forms of social organization, and geographically and historically established regulatory mechanisms (Bianchi, 2009; Mosedale, 2011; Murray, 2012). Successive capitalist crises have triggered the changes that they have undergone, leading to fluctuations in their production capacity, real estate earning capacity and financial profitability (Murray et al., 2017). Without relinquishing their role as spaces of production for the provision of services, in most cases, the value of their built environment attracts investment capital and new functions, consolidated through spatial tourism planning.

Butler's tourist area life cycle (1980), Agarwal's restructuring theory (2002), Gormsen's model of tourism development for the tourist peripheries (1997) and Brouder and Eriksson's evolutionary tourism geography (2012) all make theoretical headway in explaining changes in the tourist industry and in tourism spaces. Orthodox theories are inspired by Schumpeterian notions of "innovation" and Rostow's concept of economic growth stages (1960), while critics base their

M. Blázquez-Salom (✉) · I. Murray
Departament de Geografia, Univesitat de les Illes Balears, Palma, Spain
e-mail: mblazquez@uib.cat

A. Blanco-Romero, M. Blázquez-Salom (eds.), *Spanish Tourism Geographies*, Geographies of Tourism and Global Change,
https://doi.org/10.1007/978-3-031-39780-6_10

arguments on critical theories, like the world systems theory or uneven geographical development. Adaptive strategies founded on the dominant theories are put forward for coastal mass tourism destinations, based on the notion of boosting their competitiveness by confiding in growth and innovation through renewal, differentiation, digitalization and improvements at an environmental or cultural level or in the quality of life of the resident population (Anton-Clavé, 2012). Capital's role in the social production of tourism spaces is less widely analysed (Britton, 1991). It is not just a question of studying their production, but also their destruction and post-crisis restructuring. Crises are an inherent part of capitalism (Harvey, 2014) and they are overcome by making spatiotemporal adjustments, aimed at tackling the over-accumulation of capital and at restoring the rate of profit (Murray et al., 2017).

This chapter conducts a review of advances in the field of tourism geography in Spain in studies of sun and sand resorts and the agency of those responsible for their production and restructuring within the context of capitalism's changing regulatory mechanisms. The need to fill this gap has been pointed out in environmental (Pallicer & Blázquez-Salom, 2016) and libertarian circles (Casamajor, 2015), and this chapter aspires to complement these studies by tying them in with social theories on the social production of space (Lefebvre, 1974), the political economy of tourism (Britton, 1991; Bianchi, 2018) and social agency (Bourdieu, 1972). Because Spain is a host society, Spanish academic literature shows a greater concern for the spatial, environmental and socio-economic conditions of destinations than Anglo-Saxon studies, which take a tourist-based approach. As a result, first, we will explore studies of the social production and restructuring of space, mainly in reference to the built environment. Then we will turn our attention to studies of the role of the different agents involved in touristification processes; primarily the State, closely followed by building developers, hotel owners and financial bodies. We conclude with an evaluation of environmental and social aspects of recent transformative processes in a context of recurrent environmental, social and economic crises. The research approach is that of a theory-based critical essay, seeking evidence of the agents involved in touristification processes and their influences within and on social structures.

10.2 The Sun and Sand Tourism and Residential Mix

The built environments that stem from tourism development share very little in common with conventional urban spaces, characterized by a greater heterogeneity, better planning and more social interaction (Capel, 1975). Tourist resorts have been defined as a mosaic: they lack the identity of a real town, with no community centre, infrastructure or services typical of conventional towns or cities (Donaire & Mundet, 2002). This type of built environment arises when priority is given to tourism demands, and these places are hence highly seasonal in nature. At an economic level, the production of tourism space is based on the "fixing" of capital into the built environment–through the construction, purchase and sale of property for tourist use–in order to provide accommodation services (tourist beds or residential accommodation) and complementary services (such as restaurants and leisure businesses).

The social production of space is the outcome of a set of relations and forms of production, confrontation and domination by social agents (Lefebvre, 1974). These forms of production, along with forms of consumption and distribution, are determined by stages in the history of capitalism. The said social structures condition and, in turn, are dependent on the agency of individuals. Although agency is the capacity of individuals to act independently, this agency is limited by structures, and both exert a mutual influence on one another (Bourdieu, 1972).

As for the development of sun and sand tourist resorts in Spain, the highpoint was in the 1960s during the period of developmentalism and the Franco regime's adaptation to the country's tourism and real estate functions (Gaviria, 1974; Murray, 2015). By way of an example, Manuel Fraga was a driving force behind this process through the various different official state posts that he held (Pack, 2006). Developmentalism was fuelled by the growth mantra (Rist, 2002), the expansion of surplus value (Arrighi, 1996), the Cold War and global access to hydrocarbon fuel deposits (Murray, 2015). This period has been named the 30 Glorious Years due to the acceleration of the Fordist economy, in tandem with the consolidation of the welfare state (López & Rodríguez, 2010).

The reproduction of capital is based, first and foremost, on its spatial expansion, with the incorporation of new cheaper inputs (mainly manpower, food, energy and raw materials); and, secondly, on the intensification of its appropriation through improved capitalization and improvements in social and technological innovations (Moore, 2015). These dynamics were the guiding forces behind sun and sand development processes, absorbing capital in order to "fix" it in space and time (Harvey, 1982) and helping to overcome the contradictions of capitalism (Fletcher, 2011). The production of coastal tourism spaces went hand in hand with growing uneven geographical development, with maximum benefit being made of the peripheries of the world capitalist system (Wallerstein, 1991), leading tourist destinations to become known as the pleasure peripheries (Turner & Ash, 1975).

Systemic cycles of accumulation determine these spatial production, destruction and restructuring processes (Rullan, 2008). This change in the built environment is sustained by capital switching, aimed at generating capital gains through its utilization for tourism, real estate and financial purposes (Christophers, 2011). Spain joined in the process of post-war developmentalism from the 1960s, following the Franco autocracy. Sun and sand tourism development is a key spatial symbol of Fordism. During this early stage, this development process was funded by the landowning class, the industrial and commercial bourgeoisie (at a local and international level and also through the repatriation of capital by emigrants), smugglers and international tour operators (Gaviria, 1974, 1975; Bray & Raitz, 2001; Pack, 2006; Vallejo, 2015). The Franco regime promoted and legitimized this process through legislation, planning and subsidies to foster tourism. Through to the mid 1970s, the expansion of the hotel sector in Spain was mainly concentrated along the Mediterranean coast and in the Balearic Islands, followed later by the Canary Islands. Shortly after the approval of the 1959 Stabilization Plan, Jorge Vila Fradera–one of the promoters of tourism and a figure with close ties with the regime–stated that "the development of hotels in Spain during the last decade has been so swift that few examples of industrial growth can be compared with it" (1961:85). From the mid 1960s, there was a dizzying rise in coastal tourism, particularly in the Balearic Islands, which accounted for more than one out of every four Spanish hotel beds in 1984 (López-Palomeque, 1988; Yrigoy, 2014).

For the sun and sand sector, tourism development implied the consumption of tourism by the growing middle classes of industrial economies (Callejo et al., 2005). The construction of tourist apartments skyrocketed in areas like the coast of Malaga or the Balearic Islands. These apartments were not governed by any specific regulations or controls, and there was a strong speculative element, with apartments even being resold before they were built (Vallejo, 2015). Toward the end of 1970s, in some cases, like the Costa del Sol (Malaga), a substantial socio-spatial change could be observed, characterized by the purchase of housing by foreign investors (Jurdao, 1979). The commodification of leisure became yet another distinguishing social hallmark and tourism came to acquire a symbolic value, like a fetish that organizes meaning and the desire to enjoy things for reasons that go beyond our material needs (MacCannell, 1976). Access to tourism consumption fuelled the illusion of belonging to the middle classes (Rodríguez, 2022). Tourist perceptions were based on images inspired by the movies (Brotons Capó et al., 2016) or by festivals, aimed at creating a brand identity. Spain's flourishing period of mass tourism culminated in the **crisis** of the early 1970s due to the drop in the rate of profits and "inflated expectations" reflected by growing movements in defence of employment and civil rights, feminism, pacifism and environmentalism (Tello, 2005).

The 1970s Fordist crisis or crisis in regulated capitalism led to a slow-down in the construction of tourism accommodation, impacting on the production of tourism space along the Spanish coast. The emergence from the Fordist crisis, through the deployment of neoliberal capital, gave rise to restructuring processes and an extended range of real estate products, in particular housing estates of second homes. 1980s tourism policies were characterized by State government's inaction. In the words of Bote and Marchena (1996), the 1982–1989 period was governed by the motto "the best tourism policy is a non-existent one". These crisis years led to the deterioration of the built space in many coastal areas and the destruction of their value. By helping to relaunch the neoliberal order through new more flexible regulations that fostered investment, public and private sector partnerships and interterritorial competitiveness, hence generating a "favourable business climate", the path was cleared for a future growth cycle of capital investment in tourism. In a nutshell, in coastal areas, the urban entrepreneurialism analysed by Harvey (1989) took the form of tourism entrepreneurialism. To understand the magnitude of the political change that took place during the change of cycle, it is also important to take into account Spain's integration into the European Union as well as the process of global financialization (López & Rodríguez, 2010; Charnock et al., 2014).

One spatial indication of the new system of neoliberal regulation was the spread of housing developments for residential and tourist use, with the term an "urbanizing tsunami" being coined to describe it (Fernández-Durán, 2006). The real estate bubble was fuelled by investment and borrowing, tied in with the attraction of capital and foreign migrants (Huete, 2009; Hof & Blázquez-Salom, 2013). The construction of hotels and housing estates was accompanied by megaprojects of all kinds, including transport infrastructure and urban supply networks, yacht marinas, golf courses, theme parks, mega events and more, with a strong speculative element and public financial support (Baños-Castiñeira, 2012; Romero et al., 2018; Villar & Fernández, 2013; Anton-Clavé, 2005; del Romero Renau, 2010).

10.3 The Role of the State in the Social Production of Sun and Sand Tourist Resorts

The State regulates spatial and tourism planning through planning and legislation. Although Spain has had framework legislation in the field of urban planning since 1956, State support for tourism development led to the approval of specific legislation that gave priority to it: Spanish Act 197/1963 of December 28th on Centres and Areas of National Interest to Tourism, approved when Manuel Fraga was the Spanish Minister for Tourism. This legal figure was applied on 78 occasions between 1964 and 1975, earmarking over 22,000 hectares for the development of almost 890,000 beds (Galiana & Barrado-Timón, 2006). This illustrates the State's role of in promoting urban development through expansionist spatial planning aimed at attracting investment and tourists, added to economic measures like the devaluation of the peseta (Esteve & Fuentes, 2000; Murray, 2015).

The Costa del Sol is one of Spain's most prominent destinations for sun and sand tourism. Its tourist resorts were developed during the Franco regime through the aforementioned Spanish Act on Centres and Areas of National Interest to Tourism, promoted by business entrepreneurs with close links with the dictatorship. By way of an example, local 'feudal' figures, such as Hohenlohe, Meliá and Banús in Marbella, Girón in Fuengirola, and Bolín in Benalmádena, have been named as driving forces behind the development of tourism on the Costa del Sol (Fernández-Carrión, 2005). The last of them, Luís Antonio Bolín Bidwell, assisted Franco in his coup d'état by acting as an intermediary in payments by the Majorcan financier Joan March. Bolín was one of the ideologists behind Spain's touristification. He was the Director of the National Tourism Board responsible for creating a series of 'War Routes' (1938–1939) and Director General for Tourism (1938–1953) (Larrinaga, 2021). The Bolín brothers, Luís Antonio and Enrique, and their relatives combined their political posts with business activities on the Costa del Sol and in the Gibraltar tax haven. From 1966 to 2007, Enrique Bolín Pérez-Argemi held office almost uninterruptedly as mayor of Benalmádena until he was disqualified by the courts for granting illegal planning permits.

In the aftermath of the 1970s crisis, the adoption of a neoliberal system of accumulation coincided in Spain with the democratic transition, the country's integration into the European Union and, latterly, with the introduction of the euro. In this context of monetary unification and the single currency, Spanish economic policy fostered the "urbanizing tsunami" through legislation on land use (described as "all developable") by the Conservative government of José M. Aznar (1996–2004) (Rullan, 1999; Naredo & Montiel, 2010). This new regulatory framework simplified the requirements for the construction of new housing estates in a context of economic euphoria, with low interest rates and huge financial facilities. The flourishing urban development and tourism sectors helped to offset deficits in the Spanish balance of payments through financial and real estate capital flows and the trade surplus in tourism (Naredo, 2006; Murray & Blázquez-Salom, 2009).

Within the framework of the tourism and real estate boom, measures were introduced by the Spanish State in the exercise of its regulatory functions to accommodate growing social demands and to mitigate the lack of sustainability, particularly in coastal tourist areas, with the adoption of Local Agenda 21s (Blázquez-Salom, 2001; Brunet Estarellas et al., 2005). Nonetheless, numerous cases of political and business corruption came to the surface involving failure to comply with spatial planning regulations (Agudo & Iglesias, 2007; Jerez et al., 2012; Romero et al., 2018). In response to the progressive physical deterioration of sun and sand tourist resorts and their declining profitability, the State organized investment into the built environment through new strategic planning instruments: Tourism Excellence Plans, with 84 of them from 1992; Tourism Dynamization Plans, 97 from 1996; and Holistic Spanish Tourism Quality Plans (2000–2006) (López-Palomeque et al., 2022; Beas, 2012). For instance, the Tourism Excellence Plans were used to fund the freeing up of space in urban areas through the demolition of obsolete hotels, in addition to improvements to public roads and investment into public services (Donaire & Mundet, 2002; Simancas, 2015; Blázquez-Salom & Yrigoy, 2016). In some cases, as an exceptional measure closely tied in with the intensity of local social protests, the State introduced legislation to try and mitigate excesses in urban development through containment measures (Rullan, 2011), the protection and management of visits to natural spaces (Llausàs et al., 2019), and even by reversing their development by restoring coastal areas to their natural state (Quintana et al., 2018).

After the systemic crisis and bursting of the financial bubble in 2008, capital investors managed to bring about legislative changes so that new capital gains could be reaped from mature tourist destinations. If we focus on the effects of this legislation on the built environment, they consist of increases in plot ratios and volumetric parameters, changes of use, the fragmentation of hotel ownership (condohotels), and the possibility of tourist shows being held on beaches (Vera & Rodríguez, 2012; Blázquez-Salom et al., 2023). Legislation was also passed to facilitate the financialization of real estate, with assets like hotels passing into new hands and growing amounts of property being owned by real estate investment funds (Rodríguez & Espinoza, 2018; Gabarre, 2019). The public authorities of Spain's islands and some cities (like Barcelona) imposed restrictions on new tourism accommodation through public management of licenses, leading to their appreciation in value and the consolidation of monopoly ownership in the hands of local elites (Bianchi, 2004; Blázquez-Salom et al., 2023). These measures gave rise to growing real estate trade in hotels and licenses for hotel beds, based on the effectiveness of the resulting "cap and trade" system (in the same vein as greenhouse gas emissions trading). Leaving aside the typical mantra of sustainable development, these changes were often whitewashed by justifying them with motivations like improved digitalization, reduced seasonality, circularity, safety, accessibility, etc. (Ivars-Baidal & Vera-Rebollo, 2019; López-Palomeque et al., 2022; Blázquez-Salom et al., 2023). By way of an example, the notion of tourist congestion has even been used to foster real estate development, as illustrated by Spanish Act 197/1963 on Centres and Areas of National Interest for Tourism, which promoted the creation of tourist beds. The act opens with the sentence "The big increase in tourist flows has given rise to congestion and strain upon some areas and towns in Spain, brought about by the number of visitors, giving rise to the need to solve certain problems with urgency". Even after the COVID-19 health crisis, touristification was resorted to once again as a solution, with the State coming to the tourist sector's rescue (Méndez, 2020) and greater financialization of the hotel sector (Murray & Cañada, 2021).

10.4 The Development and Funding of Hotel Chains

In studies of the political economy of tourism, the role of multinational companies in the tourist industry has been explored (Britton, 1991; Bianchi, 2018), with this analysis also extending to Spain (Murray, 2012, 2015). Sun and sand tourist resorts are inextricably linked to the makeup of Spain's most powerful multinational hotel chains (Buades, 2006). Tourism entrepreneurs have been seen as a new social group that precipitates structural change (Amer, 2006). EXCELTUR, a Spanish lobby created in 2002, acts as an umbrella for Spain's main multinational companies, and its presidents have mostly been hoteliers: Gabriel and Sebastián Escarrer (Meliá Hotels), Simón Pedro Barceló (Barceló Corporación), Amancio López (Hotusa), Fernando Conte (Orizonia) and José Mª Rossell (Hoteles Playa). These multinationals take advantage of their salient characteristics: (1) their large size, which enables them to negotiate conditions with the States where they plan to invest, such as tax savings, the provision of public land and infrastructure, and favourable labour or environmental regulations; (2) their integration processes, often encompassing related tourism services (intermediation, travel, accommodation etc.) in a built environment where they participate in its urban development; (3) their high capacity for the creation of employment and for intra-group commercial operations; and (4) the worldwide commercial strength of their brand image in the repatriation of profits (*leakage*), in benefitting from the legal immunity afforded by tax havens in order to evade paying tax, in the attraction of investment (through the sale of shares, franchises, temporary corporate agreements etc.) and in receiving official aid for development (Artigues & Blázquez-Salom, 2019).

The Franco dictatorship contributed to the creation of clientelist business networks, encouraging power and influence through the requirement that coastal land had to be owned by Spanish citizens, with financial support for their businesses or the granting of licenses at the discretion of the authorities. Revolving doors in politics and local capital switching from other business sectors contributed to the emergence of Spanish hotel business groups (Meliá, Barceló, Riu or Iberostar, which continue to head the world ranks of Spanish hotel chains in terms of the number of hotel beds). Their links with northern European tour operators ensured funding through guaranteed occupancy agreements which would later mature into more solid links; for instance, between Riu or the Matutes hotel group and TUI, between Barceló and Thompson Holiday or First Choice Holidays, and between Meliá or Iberostar and Neckermann und Reisen. Big Spanish financial institutions have contributed to the development of hotel chains, such as Grupo March, which owns shareholdings in the same companies as the hotel chain Iberostar.

Today's recurrent crises are intensifying the concentration of capital and hotel financialization processes (Casamajor, 2015) through the sale of the built environment to investment funds, with Blackstone now being one of the main hotel owners in Spain (Murray & Cañada, 2021). Hence capital gains are harvested in the secondary circuit of capital, while ongoing management of hotels is ensured through rental, management or franchise agreements (Blázquez-Salom, et al., 2011; Yrigoy, 2016). The State has supported this through regulatory changes, such as the legalization of public listed real estate investment companies (SOCIMI in Spanish), the relaxation of urban planning and labour requirements, rescues through Official Credit Institute guarantees, temporary redundancy plans, and the investment of European recovery funds to tackle the economic and social consequences of the pandemic (Murray & Cañada, 2021). Meanwhile, business groups use holdings or front companies to mitigate their tax burdens by registering in tax havens. This is a practice adopted by Spanish island-based hotel chains like Melía, Riu and Martinón (Villarino & Pascual, 06/04/2016).

Through their hegemonic role in the hotel industry in Spain, they can continue to reactivate their rate of profit and trigger the required conditions for the accumulation of capital. They continue to use the State as an instrument in this, fixing capital in the built environment in order to boost the standard of the services they offer and their RevPar (the primary circuit of capital accumulation), while also attracting speculative capital interested in real estate investment (the secondary circuit of capital accumulation) (Blázquez-Salom et al., 2023). By way of an example, Formentor Hotel (a famous hotel built in 1929 in a secluded area of Pollença, Mallorca) was sold in the midst of the COVID-19 pandemic by the Barceló hotel chain (which had bought it in 2006 for 36 million euros) to the Andorra investment company Emin Capital (for 165 million euros), which carried out building work of questionable legality to run it with the Four Seasons hotel chain (Murray & Cañada, 2021; Hosteltur, 16/06/2006).

10.5 Tourism and Socio-environmental Decline

Hall and Page (2012) list the aspects of global change that are most closely associated with tourism: urban development, energy consumption by the transport sector (particularly in relation to the greenhouse gas emissions that contribute to global warming and extreme meteorological phenomena), reduced biodiversity, the spread of illness and disease (like the COVID-19 pandemic, as a result of hyper-mobility) and water shortages (also extensible to other resources or inputs). Added to this, the tourist industry also plays an influential role in the decline of certain social and environmental conditions, mainly associated with labour conditions, the gender gap (Cañada, 2015; Yrigoy & Cañada, 2019) and greater social inequality (Martínez-Caldentey & Murray, 2019). The complexity of this last problem has been exacerbated by touristification and the commodification of housing and public spaces to the detriment of the right to the city (Blanco-Romero et al., 2019).

The built environment of sun and sand tourist resorts is ageing, but it still attracts capital investors. The adoption of certain features of a multifunctional urban space are promoted (Pié, 2005; Anton-Clavé, 2012), combining permanent accommodation and conventional urban functions with tourism and leisure-related ones (Rovira, 2008; Rovira & Anton-Clavé, 2014). Hotels are refurbished to upgrade them (Blázquez-Salom et al., 2023) or to convert them into high-class accommodation for residential use, pensioners, expats or digital nomads (López-Palomeque et al., 2022). Let us not forget that sun and sand tourist resorts share some features of the centres of power of today's globalized world, chosen by the financial elite for the quality of the environment and existence of exclusive meeting points (Méndez, 2018). Spurred on by utopian libertarian ideas, the fantasy of an escape to an idyllic coastal retreat has no geographical bars thanks to global connectivity (Blázquez-Salom, 2022). Thus it is that over 60% of the world's tax havens are islands, immune to democratic interference or bothersome clashes between social classes (Campling & Colás, 2021).

Recurrent environmental and social crises lead to conflicts and disputes over the monopolization of the world's remaining resources. This prompts elite groups to seek refuge in places like sun and sand tourist resorts, pushing out less well-off tourists with the aid of neoliberal planning. Tourism gentrification flies in the face of the per-capita limitations on energy consumption and materials that are needed for the planned, democratic, fairer distribution of wellbeing, as advocated in degrowth theories (Blázquez-Salom et al., in press). By continuing to promote social tourism and proximity tourism at sun and sand resorts, this will contribute to the existence of a less vulnerable, more supportive social economy.

Acknowledgements This publication is part of the research project "Overtourism in Spanish Coastal Destinations. Tourism Degrowth Strategies", grant number RTI2018-094844-B-C31 funded by MCIN/AEI/ 10.13039/501100011033 and by "ERDF A way of making Europe".

References

Agarwal, S. (2002). Restructuring seaside tourism: The Resort Lifecyle. *Annals of Tourism Research, 29*(1), 25–55. https://doi.org/10.1016/S0160-7383(01)00002-0

Agudo, J., & Iglesias, F. (2007). *Urbanismo y democracia. Alternativas para evitar la corrupción*. Madrid.

Amer, J. (2006). *Turisme i política. L'empresariat hoteler de Mallorca*. Documenta Balear.

Anton-Clavé, S. (2005). *Parques temáticos: más allá del ocio*. Ariel.

Anton-Clavé, S. (2012). Rethinking mass tourism, space and place. In J. Wilson (Ed.), *The routledge handbook of tourism geographies*. Routledge.

Arrighi, G. (1996). *The long twentieth century: Money, power, and the origins of our times*. Verso.

Artigues, A. A., & Blázquez-Salom, M. (2019). Empresas multinacionales turísticas. In E. Cañada & I. Murray (Eds.), *Turistificación global. Perspectivas críticas en turismo* (pp. 199–224). Barcelona.

Baños-Castiñeira, C. J. (2012). El parque temático *Terra Mítica* (Benidorm). El fracaso de un proyecto estratégico en la renovación de un destino consolidado. In J. F. Vera & I. Rodríguez-Sánchez (Eds.), *Renovación y reestructuración de destinos turísticos en áreas costeras. Marco de análisis, procesos, instrumentos y realidades* (pp. 297–322). Publicacions de la Universitat de València.

Beas, L. (2012). Los planes de excelencia turística en España (1992–2006). Una apuesta por la revitalización de los destinos. Revisión de su implantación e impactos. *Scripta Nova, XVI*(411), 1–12.

Bianchi, R. V. (2004). Tourism restructuring and the politics of sustainability: a critical view from the European periphery (the Canary Islands). *Journal of Sustainable Tourism, 12*(6), 495–529. https://doi.org/10.1080/09669580408667251

Bianchi, R. V. (2009). The 'critical turn' in Tourism Studies: a radical critique. *Tourism Geographies, 11*(4), 484–504. https://doi.org/10.1080/14616680903262653

Bianchi, R. V. (2018). The political economy of tourism development: A critical review. *Annals of Tourism Research, 70*, 88–102. https://doi.org/10.1016/j.annals.2017.08.005

Blanco-Romero, A., Blázquez-Salom, M., Morell, M., & Fletcher, R. (2019). Not tourism-phobia but urban-philia: understanding stakeholders' perceptions of urban touristification. *Boletín de la Asociación de Geógrafos Españoles, 83*(2834), 1–30. https://doi.org/10.21138/bage.2834

Blázquez-Salom, M. (2001). Auditorías ambientales de destinos turísticos. Diagnosis territorial para el desarrollo de Agendas 21 locales. *Cuadernos de Turismo, 8*, 39–59.

Blázquez-Salom, M. (2022). A gold cage. From mass tourism to a safe haven. *Geographical Sciences, 76*(3), 157–177. https://doi.org/10.20630/chirikagaku.76.3_157

Blázquez-Salom, M., & Yrigoy, I. (2016). Caso 1. La planificación del turismo en áreas litorales en España. In M. Simancas (coord.), *La planificación y gestión territorial del turismo* (pp. 175–203). Síntesis.

Blázquez-Salom, M., Cañada, E. & Murray, I. (2011). Búnker playa-sol. Conflictos derivados de la construcción de enclaves de capital transnacional turístico español en El Caribe y Centroamérica. *Scripta Nova, XV*(368), 1–14.

Blázquez-Salom, M., Murray, I., & Artigues-Bonet, A. A. (2023). Hegemonía hotelera. El rol del Estado en la reactivación de la tasa de beneficio hotelera tras la crisis de 2008. *Investigaciones turísticas, 25*, 220–249. https://doi.org/10.14198/INTURI.22122

Blázquez-Salom, M., Murray, I., Fletcher, R., Sekulova, F., Blanco-Romero, A., & Cañada, E. (in press). Tourism and degrowth: Beyond the capitalist growth imperative. In Wilson, J. & Müller, D. (Eds.): New Routledge handbook of tourism geographies. Routledge.

Bote, V, & Marchena, M (1996). Política turística. In A. Pedreño (dir). *Introducción a la Economía del Turismo en España* (pp. 295–326). Civitas.

Bourdieu, P. (1972). *Outline of a theory of practice*. Cambridge University Press.

Bray, R., & Raitz, V. (2001). *Flight to the sun. The story of the holiday revolution*. Continuum.

Britton, S. (1991). Tourism, capital and place: towards a critical geography of tourism. *Environmental and Planning D: Society and Space, 9*(4), 451–478.

Brotons Capó, M. M., Murray-Mas, I., & Blázquez-Salom, M. (2016). Viaje de ida y vuelta, al mito. La contribución del cine a la formación de la iconografía turística de Mallorca. *Anales de Geografía de la Universidad Complutense, 36*(2), 203–236. https://doi.org/10.5209/AGUC.53583

Brouder, P., & Eriksson, R. H. (2012). Tourism evolution: on the synergies of tourism studies and evolutionary economic geography. *Annals of Tourism Research, 43*, 370–389.

Brunet Estarellas, P. J., Almeida García, F., Coll López, M., & Monteserín Abella, O. (2005). Agenda 21: Subsidiariedad y Cooperación a favor del Desarrollo Territorial Sostenible. *Boletín de la Asociación de Geógrafos Españoles, 39*, 423–446. https://bage.age-geografia.es/ojs/index.php/bage/article/view/511

Buades, J. (2006). *Exportando paraísos. La colonización turística del planeta*. La Lucerna.

Butler, R. (1980). The concept of a tourist area cycle of evolution: implications for management of resources. *Canadian Geographer, 24*(11), 5–12.

Callejo, J., Gutiérrez, J., & Viedma, A. (2005). Proceso de constitución de España en una sociedad turística. *Política y Sociedad, 42*(1), 151–168.

Campling, L., & Colás, A. (2021). *Capitalism and the sea*. Verso.

Cañada, E. (2015). *Las que limpian los hoteles. Historias ocultas de precariedad laboral*. Icaria Editorial.

Capel, H. (1975). La definición de lo urbano. *Estudios Geográficos, 138–139*, 265–301.

Casamajor, E. (2015). Pomes podrides? Els negocis de la corrupció i de la indústria turístico-immobiliària. *Tot inclòs, danys i conseqüències del turisme a les nostres illes*. Available online: https://totinclos.noblogs.org/corrupcio/pomes-podrides/ (consulted 23th September 2022).

Charnock, G., Purcell, T., & Ribera-Fumaz, R. (2014). *The Limits to Capital in Spain*. Palgrave Macmillan.

Christophers, B. (2011). Revisiting the urbanization of capital. *Annals of the Association of American Geographers, 101*, 1347–1364.

del Romero Renau, L.. (2010). Dos décadas de urbanismo-espectáculo en España: los grandes eventos como motor de cambio urbano. *Boletín de la Asociación de Geógrafos Españoles, 53*. Recuperado a partir de https://bage.age-geografia.es/ojs/index.php/bage/article/view/1203

Donaire, J. A., & Mundet, L. (2002). Estrategias de reconversion de los municipios litorales catalanes. In P. Pumares, M. A. Asensio, & F. Fernández (coord.). *Turismo y transformaciones urbanas en el siglo XXI* (pp. 39–49). Universidad de Almería.

Esteve, R., & Fuentes, R. (2000). *Economía, historia e instituciones del turismo en España*. Pirámide.

Fernández-Carrión, M.-H. (2005). Empresarios del turismo en la Costa del Sol. In *VIII Congreso de la Asociación de Historia Económica. Universidad de A Coruña, Santiago de Compostela y Vigo, (13–16 de septiembre de 2005)*.

Fernández-Durán, R. (2006). *El tsunami urbanitzador espanyol y mundial*. Virus Editorial.

Fletcher, R. (2011). Sustaining tourism, sustaining capitalism? The tourism industry's role in global capitalist expansion. *Tourism Geographies, 13*(3), 443–461.

Gabarre, M. (2019). *Tocar fondo. La mano invisible detrás de la subida del alquiler*. Traficantes de Sueños.

Galiana, L., & Barrado-Timón, D. A. (2006). Los Centros de Interés Turístico Nacional y el despegue del turismo de masas en España. *Investigaciones turísticas, 39*, 73–94.

Gaviria, M. (1974). *España a Go-Go. Turismo chárter y neocolonialismo del espacio*. Ediciones Turner.

Gaviria, M. (1975). *El escándalo de la "Court Line". (Bancarrota del turismo español)*. Editorial Cuadernos para el Diálogo.

Gormsen, E. (1997). The impact of tourism on coastal areas. *GeoJournal, 42*(1), 39–54.

Hall, C. M., & Page, S. J. (2012). From the geography of tourism to geographies of tourism. In J. Wilson (Ed.), *The Routledge handbook of tourism geographies*. Routledge.

Harvey, D. (1982). *The limits to capital*. Basil Blackwell.
Harvey, D. (1989). From Managerialism to Entrepreneurialism: The Transformation in Urban Governance in Late Capitalism. *Geografiska Annaler Series B, Human Geography, 71*(1), 3–17. https://doi.org/10.1080/04353684.1989.11879583
Harvey, D. (2014). *Seventeen contradictions and the end of capitalism*. Profile Books.
Hof, A., & Blázquez-Salom, M. (2013). The Linkages between Real Estate Tourism and Urban Sprawl in Majorca (Balearic Islands, Spain). *Land, 2*, 252–277. https://doi.org/10.3390/land2020252
Hosteltur. (16/06/2006). *Barceló se hace con el emblemático Hotel Formentor de Mallorca*. Available online https://www.hosteltur.com/35703_barcelo-se-hace-emblematico-hotel-formentor-mallorca.html (consulted 26/09/2022).
Huete, R. (2009). *Turistas que llegan para quedarse. Una explicación sociológica sobre la movilidad residencial*. Universidad de Alicante.
Ivars-Baidal, J. A., & Vera-Rebollo, J. F. (2019). Planificación turística en España. De los paradigmas tradicionales a los nuevos enfoques: planificación turística inteligente. *Boletín de la Asociación Española de Geografía, 82*, 1–31. https://doi.org/10.21138/bage.2765
Jerez, L. M., Martín, V. O., & Pérez, R. (2012). Aproximación a una geografía de la corrupción urbanística en España. *Ería, 87*, 5–18.
Jurdao, F. (1979). *España en venta: compra de suelos por extranjeros y colonización de campesinos en la Costa del Sol*. Ayuso.
Larrinaga, C. (Ed.). (2021). *Luís Bolín y el turismo en España entre 1928 y 1952*. Marcial Pons.
Lee, J. (2010). Global commodity chains and global value chains. In R. A. Denemark (Ed.), *The international studies encyclopedia* (pp. 2987–3006). Wiley – Blackwell.
Lefebvre, H. (1974). *Laproduction de l'espace*. Éditions Anthropos\Verso.
Ley 197/1963, *de 28 de diciembre, sobre Centros y Zonas de Interés Turístico Nacional* (Spanish Act 197/1963 of December 28th on Centres and Areas of National Interest to Tourism). BOE núm. 313, 31/12/1963, 18226–18230.
Llausàs, A., Vila-Subirós, J., Pueyo-Ros, J., & Fraguell, R. M. (2019). Carrying capacity as a tourism management strategy in a marine protected area: A political ecology analysis. *Conservation and Society, 17*(4), 366–376.
López, I., & Rodríguez, E. (2010). *Fin de ciclo: financiarización, territorio y sociedad de propietarios en la onda larga del capitalismo hispano (1959–2010)*. Traficantes de Sueños.
López-Palomeque, F. (1988). Geografía del turismo en España: una proximación a la distribución espacial de la demanda turística y de la oferta de alojamiento. *Documents d'Anàlisi Geogràfica, 13*, 35–64.
López-Palomeque, F., Vera-Rebollo, J. F., Torres-Delgado, A., & Ivars-Baidal, J. A. (2022). *El turismo, ¿fin de época?* Publicacions de la Universitat de València.
MacCannell, D. (1976). *The tourist. A new theory of the leisure class*. Schocken Books.
Martínez-Caldentey, M. A., & Murray, I. (2019). Crisis y Desarrollo Geográfico Desigual en la UE (2009–2015). *Revista de Geografía Norte Grande, 72*(163), 163–184. https://doi.org/10.4067/S0718-34022019000100163
Méndez, R. (2018). *La telaraña financiera. Una geografía de la financiarización y sus crisis*. RIL editores – Instituto de Estudios Urbanos y Territoriales UC.
Méndez, R. (2020). *Sitiados por la pandemia. Del colapso a la reconstrucción: apuntes geográficos*. Revives.
Moore, J. (2015). *Capitalism in the web of life. Ecology and the accumulation of capital*. Verso.
Mosedale, J. (Ed.). (2011). *Political economy of tourism. A critical perspective*. Routledge.
Murray, I. (2012). *Geografies del capitalisme balear: Poder, metabolisme socioeconòmic i petjada ecològica d'una superpotència turística*. [Geographies of balearic capitalism: Power, socioenvironmental metabolism and ecological footprint] (Unpublished doctoral dissertation). University of the Balearic Islands.
Murray, I. (2015). *Capitalismo y turismo en España. Del "milagro económico" a la "gran crisis"*. Alba Sud.

Murray, I., & Blázquez-Salom, M. (2009). El dinero, la aguja del tejido de la globalización capitalista. *Boletín de la Asociación de Geógrafos Españoles, 50*. Available online: https://bage.age-geografia.es/ojs/index.php/bage/article/view/1113
Murray, I., & Cañada, E. (2021). Turistificación confinada. In E. Cañada & I. Murray (Eds.), *#TourismPostCOVID19. Lockdown touristification* (pp. 14–80). Alba Sud.
Murray, I., Yrigoy, I., & Blázquez-Salom, M. (2017). El papel de las crisis en la producción, destrucción y reestructuración de los espacios turísticos. El caso de las Islas Baleares. *Investigaciones Turísticas, 13*, 1–29. https://doi.org/10.14461/INTURI2017.13.01
Naredo, J. M. (2006). *Raíces económicas del deterioro ecológico y social. Más allá de los dogmas*. Siglo XXI.
Naredo, J. M., & Montiel, A. (2010). *El modelo inmobiliario español y su culminación en el caso Valenciano*. Icaria.
Pack, S. D. (2006). *Tourism and Dictatorship: Europe's Peaceful Invasion of Franco's Spain*. Palgrave Macmillan.
Pallicer, A., & Blázquez-Salom, M. (2016). Turismo y caciquismo hotelero en las Baleares. La publicación Tot Inclòs y la quiebra del consenso social. *Ecología política, 52*, 88–92.
Pié, R. (2005). Ara toca fer ciutat. In *Debat Costa Brava. Congrés un futur sostenible* (pp. 50–87). Col·legi Oficial d'Arquitectes de Catalunya.
Quintana, X., Capellà, J., Colomer, À., Hors, X., Marí, M., & Pié, R. (2018). La restauració de la Pletera: crònica de la desurbanització. In *Llibre de la Festa Major de Torroella de Montgrí* (pp. 29–49). Available online: https://raco.cat/index.php/LlibreFestaMajor/article/view/341003
Rist, G. (2002). *El desarrollo: historia de una creencia occidental*. Los Libros de la Catarata.
Rodríguez, E. (2022). *El efecto clase media. Crítica y crisis de la paz social*. Traficantes de Sueños.
Rodríguez, R., & Espinoza, M. (2018). *De la especulación al derecho a la Vivienda. Más allá de las condiciones del modelo inmobiliario español*. Traficantes de Sueños.
Romero, J., Brandis, D., Delgado, C., García, J. L., Gómez Moreno, M. L., Olcina, J., Rullan, O., Vera, F., & Vicente, J. (2018). Aproximación a la Geografía del despilfarro en España: balance de las últimas dos décadas. *Boletín de la Asociación de Geógrafos Españoles, 77*, 1–51. https://doi.org/10.21138/bage.2533
Rostow, W. W. (1960). *The stages of economic growth: A non-communist manifesto*. Cambridge University Press.
Rovira, M. T. (2008). Consideraciones metodológicas para el estudio de los nuevos procesos territoriales y turísticos en destinos litorales consolidados. In J. A. Ivars & J. F. V. Rebollo (Eds.), *Espacios turísticos. Mercantilización, paisaje e identidad* (pp. 183–192). Universidad de Alicante.
Rovira, M. T., & Anton-Clavé, S. (2014). De destino a ciudad. La reformulación urbana de los destinos turísticos costeros maduros. El caso de la Costa Daurada central. *ACE: Architecture, City and Environment, 9*(25), 373–392. https://doi.org/10.5821/ace.9.25.3632
Rullan, O. (1999). La nueva Ley del Suelo de 1998 en el contexto del neoliberalismo postmoderno. *Investigaciones Geográficas, 22*, 5–22.
Rullan, O. (2008). Reconversión y crecimiento de las zonas turística. Del fordismo al postfordismo. In M. A. Troitiño, J. S. García Marchante, & M. García Hernández (coord.), *Destinos turísticos: viejos problemas, ¿nuevas soluciones?* (pp. 587–624). Ediciones de la Universidad de Castilla-La Mancha.
Rullan, O. (2011). La regulación del crecimiento urbanístico en el litoral mediterráneo español. *Ciudad y territorio. Estudios territoriales, XLIII*(168), 279–297.
Simancas, M. R. (2015). *La moratoria turística de Canarias. La reconversión de un destino turístico maduro desde la Ordenación del Territorio*. Servicio de Publicaciones, Universidad de La Laguna.
Tello. (2005). *La historia cuenta. Del crecimiento económico al desarrollo humano sostenible*. El Viejo Topo.
Turner, L., & Ash, J. (1975). *The Golden Hordes International tourism and the pleasure periphery*. Routledge.

Vallejo, R. (2015). ¿Bendición del cielo o plaga? El turismo en la España franquista, 1939–1975. *Cuadernos de Historia Contemporánea, 37*, 89–113.
Vera, J. F., & Rodríguez, I. (Eds.). (2012). *Renovación y reestructuración de destinos turísticos en áreas costeras. Marco de análisis, procesos, instrumentos y realidades*. Universitat de Valencia.
Vila Fradera, J. (1961). *Hoteles, hoy*. Editur.
Villar, A., & Fernández, A. (2013). Diagnóstico y perspectiva territorial del golf en Andalucía: entre la cualificación turística y el desarrollismo inmobiliario. *Boletín de la Asociación de Geógrafos Españoles, 62*, 357–378.
Villarino, Á., & Pascual, A. (06/04/2016). Las grandes hoteleras españolas Meliá, Riu y Martinón tuvieron sociedades en Panamá. *El Confidencial*. Available online: https://www.elconfidencial.com/economia/papeles-panama/2016-04-06/melia-escarrer-martinon-riu-hoteles_1179297/ (consulted 23th September 2022).
Wallerstein, I. (1991/1974). *El moderno sistema mundial*. Siglo XXI.
Yrigoy, I. (2014). The production of tourist spaces as a spatial fix. *Tourism Geographies, 16*(4), 636–652. https://doi.org/10.1080/14616688.2014.915876
Yrigoy, I. (2016). Financialization of hotel corporations in Spain. *Tourism Geographies. An International Journal of Tourism Space, Place and Environment, 18*(4), 399–421. https://doi.org/10.1080/14616688.2016.1198829
Yrigoy, I., & Cañada, E. (2019). Fixing creditor-debtors' tensions through labor devaluation. Insights from the Spanish hotel market. *Geoforum, 98*, 180–188. https://doi.org/10.1016/j.geoforum.2018.11.012

Macià Blázquez-Salom is a Professor in the Department of Geography at the University of the Balearic Islands. He teaches and researches on tourism, sustainability and land use planning. He has been a visiting scholar in universities of México (Toluca and La Paz), Nicaragua (UNAM), Dominican Republic (INCTEC), Austria (Salzburg), Germany (Rurh-Bochum, Sweden (Mid-Sweden) and the Netherlands (Wageningen). At a way to link activism and research within the framework of Radial Geography and Political Ecology, he collaborates with social movements in Spain, particularly in the Balearic Islands (https://www.gobmallorca.com/), but also in Latin America (https://www.albasud.org/).

Ivan Murray holds a PhD in Geography from the University of the Balearic Islands and an MSc in Environmental Sustainability from the University of Edinburgh. He is Associate Professor of geography at the University of the Balearic Islands. He is a member of the Research Group on Sustainability and Territory. His research merges political ecology, political economy and ecological economics of tourism. Hi has recently co-edited a boon on global touristification with Ernest Cañada. He is also involved in social movements

Chapter 11
Reinventing Destination Management and Planning: Taking Stock of a Decade of Smart Destinations Development in Spain

Josep A. Ivars-Baidal, Francisco Femenia-Serra, Marco A. Celdrán-Bernabeu, and David Giner-Sánchez

11.1 Smart Destinations as A Political Discourse: From Abstraction to Action

Smart destinations (SDs) have attracted massive attention in academic fora and have become the object of multiple theoretical reflections (e.g., Femenia-Serra et al., 2019; Gretzel et al., 2015a; Williams et al., 2020) that have progressively led to more empirical approaches in the past years, including multiple case studies (e.g., Boes et al., 2016; da Costa Liberato et al., 2018; Femenia-Serra & Ivars-Baidal, 2021). Smart destinations are argued to be a new destination management model aimed at achieving higher levels of sustainability, better governance and stronger innovation, thanks, among others, to data analysis and use of technologies (Ivars-Baidal et al., 2019; Soares et al., 2021). This is clearly linked to SDs connection to previous discourses focused on smart cities. Conceptually, smart cities rely on the implementation of a strong digital agenda and a wider use of technologies in the city operations, including the implementation of technological solutions related to energy efficiency, security and transport, among other. Smart cities have been the focus of a great academic debate, frequently based on their different dimensions, objectives

J. A. Ivars-Baidal (✉) · M. A. Celdrán-Bernabeu
Tourism Research Institute, University of Alicante, Alicante, Spain
e-mail: josep.ivars@ua.es; ma.celdran@ua.es

F. Femenia-Serra
Department of Geography, Complutense University of Madrid, Madrid, Spain
e-mail: ffemenia@ucm.es

D. Giner-Sánchez
Valencian Institute for Tourism Technologies (INVAT.TUR), Benidorm, Spain
e-mail: giner_dav@gva.es

A. Blanco-Romero, M. Blázquez-Salom (eds.), *Spanish Tourism Geographies*, Geographies of Tourism and Global Change,
https://doi.org/10.1007/978-3-031-39780-6_11

and implementation. The rather technology-oriented understanding of smart cities has received important criticisms by scholars, who argue that this concept responds to the interest of big tech companies and neo-liberal agendas (Greenfield, 2013; Townsend, 2013). Technocentric visions of smart cities are, however, challenged by more holistic perspectives, in which issues such as social and environmental sustainability are also considered (e.g., Caragliu et al., 2011; Giffinger et al., 2007).

SDs partly build on these ideas (Del Chiappa & Baggio, 2015), but also revolve around tourist experiences and governance of destinations (Gomis-López & González-Reverté, 2020). Thus, while some parallelisms can be found, SDs have created their own ideology and have adapted the principles of smart cities to the complex reality of tourism destinations, in which experiences, interaction and technology use are marked by different factors (Femenia-Serra et al., 2019). They incorporate the wider and long-term objectives of sustainability in their discourse, thus evolving from purely technological proposals to rather comprehensive ones, in which ethical principles, governance and planning are fundamental (Gretzel & Jamal, 2020).

This academic and theoretical reflection process around SDs is parallel to the consolidation of an active endorsement by some public institutions that have actively supported the discourse around smart destinations and the implementation of different initiatives. At the European level, efforts are concentrated in the European Commission's initiative called "European Capitals of Smart Tourism". This initiative is aimed at publicly recognising those destinations that are leading tourism transformation in four different axes, namely sustainability, cultural heritage and creativity, accessibility, and digitalisation (European Commission, 2022). The European Commission relies on experts who assess the candidates in each of those categories and decide overall and category winners. The Spanish cities of Málaga (2020) and Valencia (2022) have been recognized by the commission as winners together with other destinations in the continent. The winners of the contest are recognized in public events and awarded with promotional materials, which are expected to generate an impact in the public opinion. Beyond smart tourism capitals, the European Union has fostered a specific call for smart destinations, defined as an "initiative to support EU cities to facilitate access to tourism and hospitality products and services through technological innovation. EU cities will learn how to implement innovative digital solutions to make tourism sustainable and accessible, fully leveraging on their cultural heritage and creativity to improve the tourism experience. In particular, the project will focus on how EU cities can adopt data-driven approaches to become or improve as a smart tourism destination, and it will be characterised by the involvement of a large number of stakeholders, including as policy-makers, private sectors practitioners and academic researchers." (Smart Tourism Destinations, 2021).

At the national level, Spain is clearly the country that has invested more public resources in the development of smart tourism initiatives. SEGITTUR (*Sociedad Mercantil Estatal para la Gestión de la Innovación y las Tecnologías Turísticas*), the public company that promotes research and innovation in the Spanish tourism sector, has been designated as the principal actor in the creation of a strategy for the country's tourism destinations in their digital transformation. This organization coined the first definition of a smart destination a decade ago, arguing that a smart destination could be defined as "an innovative space, accessible to everyone, and consolidated on a cutting-edge technological infrastructure that guarantees the sustainable development of territories, facilitates the integration and interaction of visitors with the environment, and increases the quality of their experiences at the destination as well as residents' quality of life" (SEGITTUR, 2015, p. 104). The lines of action of SEGITTUR, which channels the mandates of the State Secretariat for Tourism and, therefore, of the Ministry of Industry, Trade and Tourism, are manifold. First, the organization frequently publishes documents aimed at facilitating the knowledge acquisition by destination management organizations (DMOs), in charge of the implementation of smart strategies. These documents include digital books and guidelines, best practices, case studies, list of resources for destinations, etc. Second, the institution has developed its own smart destination model (also referred as a *methodology*), based on five axes (technology; innovation; accessibility; sustainability; governance) and a set of indicators geared towards assessing the fulfilment of the established requirements in each of these five dimensions. Third, SEGITTUR has also created a national network of smart destinations (red DTI), which currently has 483 members in February 2022, including more than 300 destinations across the country, other public institutions and private companies and organizations (SEGITTUR, 2022). The network is inspired by previous initiatives, such as the national network of smart cities (RECI), and has become a critical link in the construction of the smart destination discourse and policies in the country. The destinations that become members of the network are compelled to apply the methodology described in the first place, going through an evaluation that consists in the assessment through the indicators. The SDs network provides the necessary tools and spaces for networking, benchmarking, exchange of solutions, services provision, access to data, training, and guidance by SEGITTUR experts. The alliance has become the main forum for destinations in the country, where issues are discussed, and agreements made.

This model and methodology have been referred as a remarkable example by many for other institutions, including the United Nations World Tourism Organization (UNWTO), the Organisation for Economic Co-operation and Development (OECD) and the World Travel & Tourism Council (WTTC). Membership also grants the different institutions (mainly DMOs and local councils) potential access to grants and other funding lines. This is the case of a recently approved subsidy (Order ICT/1527/2021, BOE-A-2022-483), consisting of six million of euros, aimed at modernising and developing technological solutions for tourism destinations belonging to the SDs network. These resources are accompanied by parallel lines for small and medium tourism companies, and a specific fund for academic research projects to be developed by universities. The provision of funds is derived from the line no.14 of the recovery, transformation and resilience plan approved by the Spanish government and funded by the European Union-NextGenerationEU programme. This line's objective is the modernisation and competitiveness improvement of the Spanish tourism sector (Gobierno de España, 2021). Beyond the Ministry of Industry, Trade and Tourism, a specific call for smart destinations by Red.es, a public institution within the Ministry of Economic Affairs and Digital Transformation, is also funding 25 projects across all the Spanish regions, co-funded by the European Regional Development Fund (Red.es, 2021). Moreover, there are other sources that have served destinations in their mission to transform their reality into a digital one. These include the also EU-co-funded EDUSI initiatives (strategies for a sustainable and integrated urban development).

However, initiatives are not only being developed at the national level, but also at the regional and local scales. The division of competences in Spain makes the regional scale (autonomous communities) a particularly relevant administration level in terms of tourism policies. Among all the regions, the Valencian Community (*Comunitat Valenciana*) is the one where smart destinations have become an important part of the political agenda. The region relies on the Valencian Institute for Tourism Technologies (Invat.tur) in terms of application of its own model of smart destination. This model involves a different, although partly shared, set of indicators compared to those of the central administration. Invat.tur has also promoted its own smart destinations network, composed by manifold tourist territories of the region. This institution also offers technical assistance and other resources for the regional destinations, which are at the same time integrated in the national network (Invat.tur, 2021).

At the local level, differences are remarkable between destinations. The idiosyncrasy of each destination is paramount in their understanding of the smart destination approach and their application of the discourse. Despite being part of the network, most Spanish small and medium destinations (in terms of tourist arrivals) are far from applying a comprehensive smart destination project (Ivars-baidal et al., 2022). Only a few cities are capable of developing strategies that are articulated with the rest of the administration departments (transport, security, parks, IT, etc.). Overall, the dependence of local administrations on external funding is almost complete, as the councils' budget for tourism departments are usually very limited. Therefore, destinations rely on the potential funding derived from the calls by the Spanish central administration, which are at the same time dependent on European funds most of the times. Hence, the local initiatives and the regional ones have become an integral part of an increasingly complex ecosystem devoted to the digitalisation of tourism destinations, in which the model of SEGITTUR and the Spanish network of smart destinations have become the central piece.

All in all, smart destinations have become a central piece within tourism policies in Spain, with a considerable amount of resources and efforts being devoted to the implementation of initiatives related to this discourse that has been constructed in the public, academic and private spheres. In this country, smart destinations initiatives are mainly the result of a public drive, which is distinctive of the country and different to the processes happening in other countries, in which this discourse has not permeated or is being addressed in a different way. Figure 11.1 summarizes the evolution of smart destination policy in Spain and its interconnection with European Union initiatives.

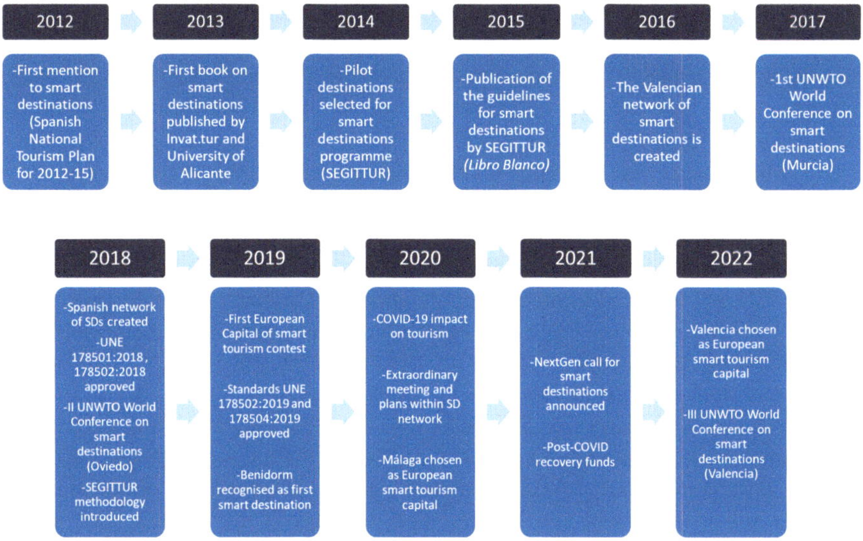

Fig. 11.1 Main milestones in smart tourism destinations development in Europe and Spain. (Own elaboration)

11.2 The Planning and Implementation of Smart Destinations: A Territorial Perspective

The development of smart destinations is embedded in a recent theoretical-conceptual framework marked by two fundamental perspectives: the integration of smart destinations in the emerging smart tourism paradigm (Celdrán-Bernabeu et al., 2018; Gretzel et al., 2015a, b) and its interpretation as an extended conception of smart cities (Lamsfus et al., 2015; Xiang et al., 2021). The convergence of these two perspectives explains the growing interest in smart destinations. Recent bibliometric research (Bastidas-Manzano et al., 2021; Sustacha Melijosa et al., 2022) shows a significant increase in research on this topic since 2015, mainly in countries with a high degree of digitisation, especially in Europe and Asia, with a technological bias that is opening up to other fields of analysis.

Thus, the development of smart destinations is associated with innovation and competitiveness in both urban (Boes et al., 2016; Koo et al., 2016) and coastal environments (Sigalat-Signes et al., 2020), improvement of destination information systems (Buhalis & Amaranggana, 2014), knowledge transfer (Del Chiappa and Baggio (2015), improvement of the tourism experience (Neuhofer et al., 2014), and even a paradigm shift towards smart tourism design (Xiang et al., 2021). In this context, Soares et al. (2021) observe a growing linkage of the smart destination with an emerging approach to destination planning and management to the extent that, following Hall's (2008) interpretative framework of tourism planning, this approach implies significant changes in the underlying assumptions (digitalization processes, the emergence of the smart tourist or the management of tourism in a knowledge economy); the tourism planning problems (understanding new business models, avoiding digital gaps or the access and management of data for decision making); and the methods used (big data analytics, netnography, predictive analysis, etc.); while providing new ways of understanding the structure and functioning of destinations through smart destination models.

Like the smart city concept, smart destinations are also criticised for their technological, technocratic and urban biases (Gretzel, 2018); for an excessive political-institutional orientation (Mínguez & Ruiz, 2014); for the limitations for a real transformation towards a greater commitment to sustainability (Ivars-baidal et al., 2022) and the need to provide public value to the host community (Brandão Cavalheiro et al., 2020); as well as its vagueness and lack of theoretical and empirical clarity (Gelter et al., 2020).

In the evolution of this theoretical-conceptual framework, and its applied dimension, the smart destination initiatives developed in Spain have become an international benchmark. These initiatives deserve to be contextualised in the evolution of tourism policy in Spain. In this evolution, different approaches to destination planning and management can be identified, adapted to the political and socio-economic contexts and the dynamics of the tourism market. Ivars-Baidal and Vera-Rebollo (2019) explain the recent evolution of these approaches in different periods in which they coexist in a variable way according to the emphasis of tourism policies: the rise of strategic planning to regain competitiveness in the 1990s; the extension to quality improvement in the early 2000s; the search for a new tourism economy based on innovation, formalised in the national tourism plan, *Horizon 2020 (2008–12)*; and the commitment to smart destinations in the *Integral National Plan 2012–15* as a path to innovation in destination management and their digitalization, given the influence of information and communication technologies (ICTs) in the tourism sector. Subsequently, the national tourism strategy focuses on sustainability, described in the working document *General Guidelines for the Sustainable Tourism Strategy, Spain 2030* (2019), an orientation maintained in the COVID-19 crisis recovery plans, in which smart destinations continue to occupy a prominent place. Hence, Spain is taking on a pioneering role in the definition and development of the smart destination as a new destination management approach (OECD, 2018; Soares et al., 2021). SEGITTUR's approach embraces at least on a theoretical level, the fundamental challenges of destination management such as governance, sustainability and innovation. In terms of institutional communication, smart destinations are proving to be an attractive solution for innovation in destination management thanks to the intensive use of technology and data. However, at the implementation level, doubts arise about their real scope, starting with the difficulty of reconciling integrative approaches with fragmented administrative competences in different departments (land use planning, environmental management, etc.).

Smart destinations have undoubtedly become one of the buzzwords in Spanish tourism policy, but the implementation of actions around the SD has been the object of consistent technical design and monitoring by SEGITTUR and regional and local tourism bodies. The work in pilot destinations of different types facilitated the adaptation of the model to the needs and capacities of the destinations, although the urban bias of the smart approach emerged soon, and initiated the collective learning process that the development of the SD in Spain has become. The creation of specific indicator systems is the fundamental tool for SD management, as it translates the principles of this approach into measurable criteria and makes it possible to analyse both the current situation and the evolution of destinations (Ivars-Baidal et al., 2021). The application of indicators has also served to contrast the situation of destinations with respect to the smart destination model in territorial areas such as the province of Barcelona (González Reverté, 2018). In addition, indicators are the fundamental instrument of the SD standards developed in Spain by AENOR (the National Standardisation Agency) and SEGITTUR: UNE 178501: 2016 (Requirements for smart destinations management systems), revised in 2018, and UNE 178502:2018 (Indicators and tools for smart tourism destinations). In this sense, Spain applies to tourist destinations the path of standardisation that has been followed for smart cities internationally. However, no specific SD standards have been created internationally, although Spain is exporting this initiative to Latin American countries that are committed to the development of SD, such as Mexico and Colombia.

The implementation of the smart approach presents a duality: It can be oriented towards a destination planning and management process leading to the SD label, or to the implementation of a standard. The latter approach is the least widespread and its scope is limited to the body being certified, generally a destination management organization. The management process designed by SEGITTUR consists of two phases: (1) Diagnosis and planning, and (2) implementation and monitoring. The diagnosis, based on the review of 97 requirements and 261 indicators, serves as the basis for the definition of the strategy that leads to the implementation and monitoring of actions. To receive the SD label, destinations must have a degree of compliance with the requirements set by SEGITTUR of 80% or higher, a score currently only achieved by five destinations: Benidorm, Gijón, Málaga, Tenerife Island and Santander. However, the commitment to apply this process to evolve towards a SD allows the integration of the destination in the national network. The network aims at facilitating collaboration between institutions and companies, the development of advisory activities for destinations and the exchange of experiences and best practices. It is made up of public administrations, mostly at local level, and collaborating organisations, among which technology and tourism consultancy companies stand out.

The development of SD-related plans and projects promotes a collaborative approach characteristic of the triple helix (administration, business and research centres), also evident in smart city experiences. In a certain sense, the SD approach fosters a technical vision of management that introduces improvements to traditional processes but raises questions about how and by whom the destination strategy is constructed. Theoretically conceived as an integrative approach, SD-derived instruments cannot replace but rather complement the strategic, urban or environmental planning of the destination. This limitation must be recognised in view of the perception of the SD as a global solution and requires adequate coordination between the different planning processes taking place at the local level.

The analyses carried out on SDs with more consolidated strategies and projects, such as Benidorm, the first destination to be certified with the Spanish standard, show significant advances in destination marketing and management: an improvement in competitive intelligence thanks to data driven marketing and decision making; the improvement of municipal interdepartmental coordination; the promotion of innovation and public-private collaboration; and even a positive influence on the reputation of the destination (Femenia-Serra & Ivars-Baidal, 2021). The impact of the SD project cannot be separated from the territorial and tourist singularity of the destination, a context favourable to Benidorm due to its high degree of tourist specialisation, public leadership and a particularly proactive DMO. For this reason, analyses of SD initiatives in other territorial and tourism environments are necessary, an objective that could be dealt with in the framework of the SD network by exploiting the indicator system as a benchmarking tool.

The typological classification of the local destinations in the national network according to their geographical environment provides an interesting perspective for analysis. The map in Fig. 11.2 shows the destinations of the network according to their typology: coastal (municipalities traditionally considered to be "sun and sand" destinations), urban (destinations with a higher demographic range, both on the coast and inland, with the multi-product offer typical of cities) and rural (considering small rural municipalities with a population of less than 5000 inhabitants).

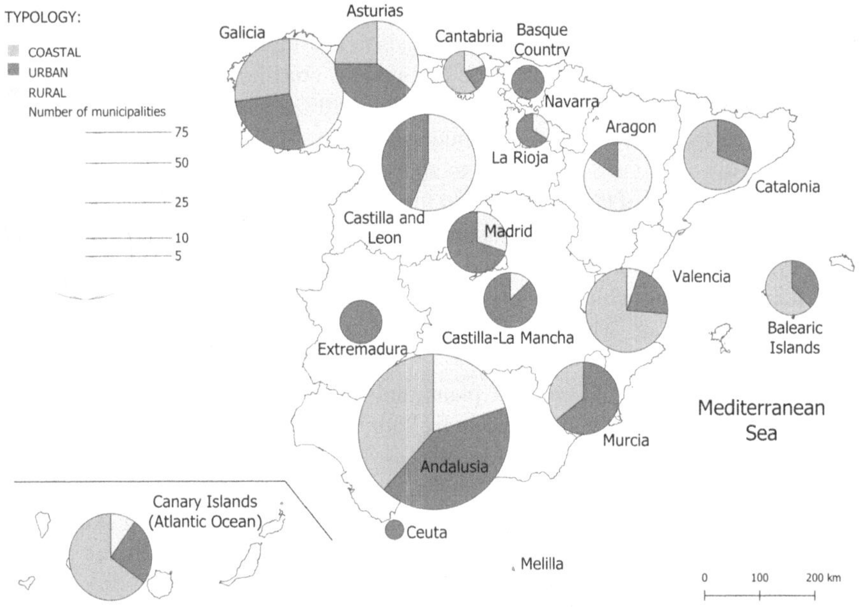

Fig. 11.2 Distribution of the municipalities integrating the SD network by geographical environment. Source: SEGITTUR (data as of February 2022)

At the national level, urban destinations predominate (40.7%), which reflects the urban bias of the smart approach, but the percentages of coastal (33.2%) and rural (26%) destinations are also high. This distribution can be viewed positively because the SD model is being extended to other territorial environments, but doubts emerge about the possibilities of implementing SD actions in municipalities with smaller demographic size and fewer resources and capacities, both in the public and private sphere, in coastal environments and, above all, in rural areas. In fact, the only five destinations in the network meeting more than 80% of the indicators are either purely urban (Santander and Gijón) or have an outstanding tourist specialisation and a significant urban range (Benidorm). The island of Tenerife is a singular case insofar as the indicators of all the municipalities that make up the island are evaluated, a procedure that may be methodologically inconsistent as it aggregates municipalities with different tourist, socio-economic and territorial realities, as well as an uneven evolution towards the SD model. At the regional level, the municipalities integrated in the network are unevenly distributed. Andalusia stands out, with 24% of the municipalities belonging to the network, together with regions such as Galicia, Castilla y León, Asturias, the Canary Islands and Valencia. These six regions account for 68.6% of the municipalities in the network. In the Mediterranean coastal regions there is a clear predominance of coastal and urban destinations, with the exception of Andalusia, which integrates a high percentage of small rural destinations. In inland regions such as Castilla y León and Aragón, rural municipalities are more prominent, while two regions, the Basque Country and Extremadura, only have urban destinations in the network.

Finally, the SD approach has proven its usefulness during the COVID-19 crisis and will presumably play a relevant role in the digital and ecological transition of Spanish tourism in accordance with the guidelines set by the European Commission for the post-COVID recovery. During the digital jump provoked by the pandemic, advances in digitization, enabled by the SD network, facilitated the development of action protocols for tourism businesses and services, including: the provision of information for businesses on regulations and aid programmes; the support for training and the development of new skills; the application of official protocols in public spaces of tourist interest; the permanent updating of information; the generation of new data sources adapted to the context and useful for forecasting the recovery of demand; the development of new metrics and indicators; and the communication plans adapted to the interruption of tourism activity.

11.3 The Role of Technology in Smart Destinations: Technological Solutions by Territorial Context

As highlighted in the preceding paragraphs, the SD approach is partly based on a strong technological component that reinforces aspects related to governance and the decision-making process, with the aim of improving destination management and the tourist experience. The seeking of higher levels of competitiveness through the application of technology encourages the development of smart destinations, an approach of tourism planning and management that promotes the measurement and analysis of data derived from tourism development in order to improve processes. In the present scenario, the use of technology applied to smart management is a key aspect in fostering the change of governance processes in the destination and thus enhancing the levels of tourism competitiveness through the promotion of sustainability and innovation in the destination.

As detailed by authors such as Buhalis and Sinarta (2019), technologies applied to smart management, has transformed value creation from a product-centric process to a consumer-centric approach. These technologies, also known as smart technologies, are also essential elements in allowing the development of smart ecosystems, thanks to the interconnection of all the stakeholders involved in the destination (tourism companies, market, tourist consumer, wider actor network). This is how co-creation processes appear as a result of the interaction of these agents with smart technologies. To this we must add the hyper-connection of tourists, destinations and companies, which generate data that represent new levels of intelligence for smart destinations (Gretzel et al., 2015a, b).

It is in this context, it is possible to generate more personalised tourism experiences, which also reinforce tourism governance processes, the core element of smart destinations as a result of the improvement of public-private collaboration, leadership and administrative coordination. A representative example can be found in Benidorm (Spain), where the implementation of the smart destination approach has been reflected in the creation in 2016 of the Smart Tourism Destination Management Unit (EGDTI); the application of the Smart Tourism Destination Management System (UNE 178501 Standard); the implementation of the Technical Office for Innovation and Intelligence (Smart Office DTI); the development of the Tourism Intelligence System focused on R&D&I activities, among other actions (SEGITTUR, 2022).

However, in the deployment of the smart destination model, major challenges and difficulties must be faced. One of the most relevant is that smart destination approach is not universal, it is not a standard model. On the contrary, it must be adapted and sized depending on the territorial context in which it is developed because of social, economic and technical differences. For this reason, smart destination strategies and projects must also be adapted to respond to the specific needs of each territorial area. As follows, we can identify three main territorial levels where the smart destination model and the associated technological solutions could be applied:

Coastal destinations. Coastal destinations with a high level of tourist arrivals and a high concentration of tourism supply. In these destinations, the use of technology in beach management (safety, monitoring of environmental parameters or improvement of public services) is worth highlighting. As an example, we can point out the case of the city of Barcelona, which has launched a pilot project in 2022, based on the use of drones, to improve the management of Barcelona's beaches and bathing areas and to estimate the capacity in real time (Barcelona City Council, 2022).

Urban, cultural and heritage destinations. Cities with a high volume of tourism supply and tourist arrivals, albeit with unequal levels of affluence and diverse product typology. Within this group, the historic and heritage cities are particularly noteworthy for their uniqueness (SEGITTUR, 2022b). Urban and cultural destinations are showing a recent commitment to the digitisation of their resources and tourism supply, integrating the same type of technological solutions as other types of destinations, highlighting their application in the recovery and preservation of heritage (SEGITTUR, 2022b). In this territorial context, we can highlight the case of the city of Granada, a World Heritage City, where a pilot project has been developed based on the installation of sensors to measure in real time the temperature and humidity of the Zirí Wall, with the aim of knowing the damage caused by environmental factors and which have motivated restoration interventions (Gutiérrez-Carrillo et al., 2021).

Rural destinations. Destinations with a smaller population and a more limited volume of tourism supply and tourist arrivals, poor connectivity and limited capacities. The use of technology in rural areas is being reflected in the application of Iot (Internet of Things). Specifically, the sensorisation of public spaces to improve basic services (lighting and waste management), as well as the installation of sensors to control gauging in natural spaces, among other applications. In this regard, the Intelligent Rural Territory project, developed by the Castilla y León region, stands out. A project developed within the framework of the Network of Digital Municipalities of Castilla y León, and co-financed by European Regional Development Fund (ERDF). The main objective of this project is to improve the efficiency of public services through the use of IoT, improve the quality of life of citizens in rural areas in a sustainable way, as well as to fix population by offering 4.0 services and creating new business opportunities for SMEs (JCyL, 2022).

Considering this territorial segmentation, Table 11.1 has been designed, containing a categorisation of the solutions applied in each of the previously described territorial contexts. This table refers to the analysis of the technological solutions proposed by both SEGITTUR and Invat-tur in their reports on the adaptation of the DTI model to different territorial tourism destinations. It includes technological solutions implemented by urban, sun and beach and rural destinations, including both destinations certified by SEGITTUR and others that have planning and management framed within the DTI model. These solutions are categorised by management areas related to digital marketing, sustainability and accessibility. These areas are strongly influenced by this new management approach and the recent evolution of the tourism scenario (Giner & Celdrán, 2021), although it is a systematisation of good practices whose application is not generalised to all destinations in the network.

It should be noted that the use of these technological solutions are associated with the different travel phases (before, during and after), each of them performing different functions and allowing the Destination Management Organisation (DMO) to acquire information on both tourist behaviour and tourist resource use for smart destination management.

Table 11.1 Smart solutions by management domain

Digital Marketing	Sustainability	Accessibility
Virtual tourism assistants (chatbots) based on Artificial Intelligence (AI). Big Data for destination analysis (supply, demand). Monitoring of visitor interests and needs. Brand reputation monitoring. 3D multimedia content via VR & AR. Gamification processes applied to the tourist experience. Web platforms for tourism marketing and promotion and SNNs. Promotion of events, resources and activities on social networks. Virtual tourist offices. Virtual tourist cards with contactless technologies. Tourism Intelligence Systems. Business intelligence technologies for monitoring the evolution of tourism industry. On-line booking of services. Digitalisation of tourist information and signage. Interconnection of emergency and public safety services.	Parking control by sensorisation (counting, tracking and monitoring of vehicles). Drones for security and surveillance of tourist areas. Monitoring of tourist flows and carrying capacity. Videometry for the control of capacity. Sensorisation and Wi-fi for the control of occupation density on beaches, natural areas and historical and cultural heritage. Sensorisation for real-time measurement of environmental parameters related to tourist activity at the destination (CO2 emissions according to transport, energy consumption, etc.) Sensorisation of tourist equipment to control use (signals, showers, etc.). Digitisation of cultural heritage for preservation and tourist promotion.	Destination apps for people with limited movement. Audio guides. Inclusive chatbots. Digitisation of tourist information and signage for people with intellectual disabilities, the elderly, etc. Accessible online tourist guides. Mobility and public transport management. Accessible navigation of web pages. Sensorisation of the management of parking spaces reserved for people with reduced mobility. Sensorisation of access to tourist resources and adapted spaces and equipment.

Own elaboration based on Segittur (2021, 2022a)

11.4 Discussion

The institutional commitment to the development of SDs in Spain represents a new approach to the planning and management of destinations better adapted to the digitisation of tourism and in favour of a greater degree of innovation based on an intensive use of technology and data. This new approach has provided a stimulus for more advanced tourism management as opposed to the crisis of traditional formulas (FEMP, 2008). However, the media appeal of the model, linked to the fascination with new technologies, and its supposed holistic nature (encompassing areas such as governance or sustainability), makes it advisable to delimit its real scope in the transformation of destinations. The design of a well-structured programme from the initial pilot destinations to the creation of a system of indicators, and the creation of the network, has generated an interesting learning process with a significant contribution to the improvement of destination management. Its effects are expected to increase with the progressive consolidation of the network.

Specific studies and catalogues of good practices show significant progress in digital marketing, competitive intelligence, development of innovative projects and integration of policies such as tourism and cultural heritage (SEGITTUR, 2022b). However, these developments, like the innovative solutions referred to in the previous section, are unevenly applied and therefore not widespread. The measures implemented in the SDs have increased their management efficiency and reinforced the need to move towards destinations with better governance, more sustainable and innovative. Nonetheless, from the point of view of governance, the SD model cannot be limited to a sophisticated technical process and must reinforce social participation in planning to the extent that the construction of the strategic vision of a destination requires a debate and a degree of social consensus that goes beyond any exclusively technical approach. In this sense, the adoption of the SD model does not necessarily imply a strategic reflection towards parameters of greater sustainability (González-Reverté, 2019), nor does it open the debate about the limitations on the growth of supply and even the de-growth of destinations (Blázquez-Salom et al., 2019). Likewise, the social dimension of sustainability must be strengthened to address relevant issues in destinations, such as the quality of tourism employment, the contribution of tourism to gentrification or the impact on the quality of life of residents (Ivars-baidal et al., 2022).

The smart approach has led to progress in the challenge of destination innovation, both in the development of the SD model and in projects financed by technology policy (Red.es), with local councils, technology companies and research centres as the main actors. Nevertheless, the integration of small and medium-sized tourism companies in smart initiatives is still limited, a fundamental aspect in order to evolve towards more connected destinations in terms of technology and the use of data to strengthen cooperation at a local level. Otherwise, there is a risk of dependence on technology companies and data providers. In the field of innovation, negative aspects are also visible, such as the adoption of mimetic initiatives, disconnected from the needs of the destinations, the legal and administrative complexity as a limiting factor, or the dependence on public funds and the difficulties of scalability of the projects.

Finally, the expansion of network members into territories other than cities is positive to overcome the urban bias of the smart approach and reduce the digital divide between rural and urban environments, but it also requires adapting the model to the singularities of less populated areas. This adaptation would entail a greater degree of cooperation between actors at different territorial scales, a route envisaged in the national network by incorporating provinces and associations, both of companies and municipalities. The evolution towards SD in sparsely populated environments requires a sub-regional scale of intervention to facilitate cooperation between different municipalities, as well as an adaptation of processes and indicator systems. SEGITTUR and Invat.tur are both working on this strategy. In supra-municipal areas with a certain degree of homogeneity and complementarity, the development of the model makes sense, but the declaration of larger and more diverse territories such as provinces as SD, as the case Tenerife, responds to an institutional perspective rather than to the logic of planning and management of destinations.

In a certain sense, the SDs are comparable to learning destinations (Cooper & Sheldon, 2010), and their capacity for transformation requires relatively long periods of time. Therefore, the period of ten years for their analysis might not be enough and presents several limitations. With this time perspective, it seems clear that the SDs are not a magic solution to the problems of destinations (García-Hernández et al., 2019), but they do represent a new way of evolving their planning and management This approach is undoubtedly more ambitious, although with a certain technical and technological bias which can generate different effects between two opposing poles, ranging from their usefulness in rethinking the strategy of destinations towards a real commitment to sustainability to the buzzword that protects and even aggravates "business as usual".

Acknowledgement This paper has received funding from the EU's programme HORIZON 2020 under the Grant Agreement no. 870753. Project title: Cities as mobility hubs: tackling social exclusion through 'smart' citizen engagement (SMARTDEST).

References

Barcelona City Council. (2022). *Barcelona chooses drone technology to improve its management of public spaces*. https://ajuntament.barcelona.cat/imi/en/noticia/barcelona-chooses-drone-technology-to-improve-its-management-of-public-spaces_1173724

Bastidas-Manzano, A. B., Sánchez-Fernández, J., & Casado-Aranda, L. A. (2021). The past, present, and future of smart tourism destinations: A bibliometric analysis. *Journal of Hospitality and Tourism Research, 45*(3), 529–552. https://doi.org/10.1177/1096348020967062

Blázquez-Salom, M., Blanco-Romero, A., Vera-Rebollo, F., & Ivars-Baidal, J. (2019). Territorial tourism planning in Spain: From boosterism to tourism degrowth? *Journal of Sustainable Tourism, 27*(12), 1764–1785. https://doi.org/10.1080/09669582.2019.1675073

Boes, K., Buhalis, D., & Inversini, A. (2016). Smart tourism destinations: ecosystems for tourism destination competitiveness. *International Journal of Tourism Cities, 2*(3), 108–124.

Brandão Cavalheiro, M., Antonio Joia, L., & Marcuzzo do Canto Cavalheiro, G. (2020). Towards a smart tourism destination development model: Promoting environmental, economic, socio-cultural and political values. *Tourism Planning & Development, 17*(3), 237–259. https://doi.org/10.1080/21568316.2019.1597763

Buhalis, D., & Amaranggana, A. (2014). Smart tourism destinations. In Z. Xiang & I. Tussyadiah (Eds.), *Information and communication technologies in tourism* (pp. 553–564). Springer.

Buhalis, D., & Sinarta, Y. (2019). Real-time co-creation and nowness service: Lessons from tourism and hospitality. *Journal of Travel & Tourism Marketing, 36*(5), 563–582. https://doi.org/10.1080/10548408.2019.1592059

Caragliu, A., Del Bo, C., & Nijkamp, P. (2011). Smart cities in Europe. *Journal Urban Technology, 18*(2), 65–82. https://doi.org/10.1080/10630732.2011.601117

Celdrán-Bernabeu, M. A., Mazón, J. N., Ivars-Baidal, J. A., & Vera-Rebollo, J. F. (2018). Smart tourism. A study on systematic mapping. *Cuadernos de Turismo, 41*, 655–658. https://doi.org/10.6018/turismo.41.326971

Cooper, C., & Sheldon, P. (2010). Knowledge management in tourism: From databases to learning destinations. In D. G. Pearce & R. Butler (Eds.), *Tourism research: A 20–20 vision* (pp. 215–227). Goodfellow Publishers.

da Costa Liberato, P. M., Alén-González, E., & de Azevedo Liberato, D. F. V. (2018). Digital Technology in a Smart Tourist Destination: The case of Porto. *Journal of Urban Technology, 25*(1), 75–97. https://doi.org/10.1080/10630732.2017.1413228

Del Chiappa, G., & Baggio, R. (2015). Knowledge transfer in smart tourism destinations: Analyzing the effects of a network structure. *Journal of Destination Marketing & Management, 4*(3), 145–150.

European Commission (2022). *Smart Tourism initiative*. https://smart-tourism-capital.ec.europa.eu/index_es

Femenia-Serra, F., & Ivars-Baidal, J. A. (2021). Do smart tourism Destinations really work? The case of Benidorm. *Asia Pacific Journal of Tourism Research, 26*(4), 365–384. https://doi.org/10.1080/10941665.2018.1561478

Femenia-Serra, F., Neuhofer, B., & Ivars-Baidal, J. A. (2019). Towards a conceptualisation of smart tourists and their role within the smart destination scenario. *The Service Industries Journal, 39*(2), 109–133. https://doi.org/10.1080/02642069.2018.1508458

FEMP. (2008). *Modelos de Gestión Turística Local. Principios y Prácticas*. Federación Española de Municipios y Provincias y Secretaría General de Turismo.

García-Hernández, M., Ivars-Baidal, J., & Mendoza de Miguel, S. (2019). Overtourism in urban destinations: The myth of smart solutions. *Boletín de la Asociación Española de Geografía, 83*, 10.21138/bage.2830.

Gelter, J., Lexhagen, M., & Fuchs, M. (2020). A meta-narrative analysis of smart tourism destinations: Implications for tourism destination management. *Current Issues in Tourism, 24*(20), 1–15. https://doi.org/10.1080/13683500.2020.1849048

Giffinger, R., Fertner, C., Kramar, H., Kalasek, R., Pichler-Milanovic, N., & Meijers, E. (2007). *Smart cities: Ranking of European medium-sized cities*. Centre of Regional Science, Vienna University of Technology.

Giner, D., & Celdrán, M. A. (2021). El escenario Smart y sus derivadas en la estrategia de marketing online de los destinos turísticos. El caso de la Comunitat Valenciana. *Investigaciones Regionales – Journal of Regional Research, 3*(51), 107–128. https://doi.org/10.38191/iirr-jorr.21.021

Gobierno de España. (2021). *Plan de recuperación, transformación y resistencia*. https://www.lamoncloa.gob.es/temas/fondos-recuperacion/Documents/16062021-Componente14.pdf.

Gomis-López, J. M., & González-Reverté, F. (2020). Smart tourism sustainability narratives in mature beach destinations. Contrasting the collective imaginary with reality. *Sustainability, 12*(12), 5083.

González Reverté, F. (coord.). (2018). *Les destinacions turístiques intel.ligents a les comarques de Barcelona. Diagnosi i propostes d'actuació*. Diputació de Barcelona. https://llibreria.diba.cat/cat/libro/les-destinacions-turistiques-intelligents-a-les-comarques-de-barcelona-diagnosi-i-propostes-d-actuacio_60589

González-Reverté, F. (2019). Building sustainable smart destinations: An approach based on the development of Spanish smart tourism plans. *Sustainability, 11*(23), 6874-1–6874-24. https://doi.org/10.3390/su11236874

Greenfield, A. (2013). *Against the smart city*. Do Publications.

Gretzel, U. (2018). From smart destinations to smart tourism regions. *Investigaciones Regionales-Journal of Regional Research, 42*, 171–184.

Gretzel, U., & Jamal, T. B. (2020). *Guiding principles for good governance of the smart destination*. Travel and Tourism Research Association International Conference: Advancing Tourism Research Globally. https://scholarworks.umass.edu/ttra/2020/research_papers/42

Gretzel, U., Sigala, M., Xiang, Z., & Koo, C. (2015a). Smart tourism: Foundations and developments. *Electronic Markets, 25*(3), 179–188.

Gretzel, U., Werthner, H., Koo, C., & Lamsfus, C. (2015b). Conceptual foundations for understanding smart tourism ecosystems. *Computers in Human Behavior, 50*, 558–563. https://doi.org/10.1016/j.chb.2015.03.043

Gutiérrez-Carrillo, M. L., Delgado, M. G., Ramos, J. S., Diaz, J. A., Cardiel, I. B., & Dominguez, S. A. (2021). Mitigating damage on heritage structures by continuous conservation using thermal real-time monitoring. Case study of Ziri Wall, city of Granada, Spain. *Journal of Cleaner Production, 296*, 126522. https://doi.org/10.1016/j.jclepro.2021.126522

Hall, C. M. (2008). *Tourism planning: Policies, processes and relationships* (2nd ed.). Pearson/Prentice Hall.

Invat.tur-Instituto Valenciano de Tecnologías Turísticas. (2021). *Red DTI-CV*. https://invattur.es/sobre-la-red-de-destinos-turisticos-inteligentes.html

Ivars-Baidal, J., & Vera-Rebollo, J. F. (2019). Planificación turística en España. De los paradigmas tradicionales a los nuevos enfoques: Planificación turística inteligente. *Boletín de la Asociación Española Geografía, 82*, 2765-1–2765-31. https://doi.org/10.21138/bage.2765

Ivars-Baidal, J. A., Celdrán-Bernabeu, M. A., Mazón, J.-N., & Perles-Ivars, Á. F. (2019). Smart destinations and the evolution of ICTs: A new scenario for destination management? *Current Issues in Tourism, 22*(13), 1581–1600. https://doi.org/10.1080/13683500.2017.1388771

Ivars-Baidal, J. A., Celdrán-Bernabeu, M. A., Femenia-Serra, F., Perles-Ribes, J. F., & Giner-Sánchez, D. (2021). Measuring the progress of smart destinations: The use of indicators as a management tool. *Journal of Destination Marketing &Management, 19*, 100531-1–100531-20. https://doi.org/10.1016/j.jdmm.2020.100531

Ivars-baidal, J. A., Vera-rebollo, J. F., Perles-ribes, J., Femenia-serra, F., & Celdrán-bernabeu, M. A. (2022). Sustainable tourism indicators: What 's new within the smart city/destination approach? *Journal of Sustainable Tourism, 31*, 1–24. https://doi.org/10.1080/09669582.2021.1876075

Junta de Castilla y León (JCyL). (2022). *Territorio Rural Inteligente*. https://rmd.jcyl.es/web/es/territorio-rural-inteligente.html

Koo, C., Shin, S., Gretzel, U., Canon, W., & Chung, N. (2016). Conceptualization of smart tourism destination competitiveness. *Asia Pacific Journal of Information Systems, 26*(4), 367–384. https://doi.org/10.14329/apjis.2016.26.4.561

Lamsfus, C., Martín, D., Alzua-Sorzabal, A., & Torres-Manzanera, E. (2015). Smart tourism Destinations: An extended conception of smart cities focusing on human mobility. In I. Tussyadiah & A. Inversini (Eds.), *Information and communication Technologies in Tourism 2015*. Springer.

Mínguez, M., & Ruiz, P. (2014). Los destinos turísticos inteligentes en España: ¿Un proyecto institucional o el futuro del sector?. In Fernández Tabales, A. y Navarro-Jurado, E. (Eds.). *Espacios turísticos e inteligencia territorial: Respuestas ante la crisis* (pp. 65–78). Universidades de Málaga y Sevilla. Red de Impresión.

Neuhofer, B., Buhalis, D., & y Ladkin, A. (2014). A typology of technology-enhanced tourism experiences. *International Journal of Tourism Research, 16*(4), 340–450. https://doi.org/10.1002/jtr.1958

OECD. (2018). *OECD tourism trends and policies*. https://doi.org/10.1787/tour-2012-en

Red.es. (2021). *Destinos Turísticos Inteligentes. Información General*. https://red.es/es/iniciativas/proyectos/destinos-turisticos-inteligentes/informacion-general

SEGITTUR. (2015). *Informe destinos turísticos inteligentes: construyendo el futuro.* http://www.SEGITTUR.es/opencms/export/sites/segitur/.content/galerias/descargas/proyectos/Libro-Blanco-Destinos-Tursticos-Inteligentes-construyendo-el-futuro.pdf

SEGITTUR. (2021). *Catálogo de soluciones tecnológicas para Destinos Turísticos Inteligentes.* https://www.SEGITTUR.es/blog/destinos-turisticos-inteligentes/catalogo_soluciones_tecnologicas_destinos_inteligentes/

SEGITTUR. (2022). *La Red de Destinos Turísticos Inteligentes celebra su tercer aniversario sumando 483 miembros.* https://www.SEGITTUR.es/sala-de-prensa/notas-de-prensa/la-red-de-destinos-turisticos-inteligentes-celebra-su-tercer-aniversario-sumando-483-miembros/

SEGITTUR. (2022a). *Guía de las actuaciones más destacadas en los destinos distinguidos con el reconocimiento de Destino Turístico Inteligente.* https://www.SEGITTUR.es/wp-content/uploads/2022/04/SEGITTUR_best-practices_ESP_2022_200422-OK.pdf

SEGITTUR. (2022b). *Guía de buenas prácticas en patrimonio cultural para Destinos Turísticos Inteligentes.* https://www.SEGITTUR.es/wp-content/uploads/2022/03/Guia-buenas-practicas-en-patrimonio-cultural.pdf

Sigalat-Signes, E., Calvo-Palomares, R., Roig-Merino, B., & García-Adán, I. (2020). Transition towards a tourist innovation model: The smart tourism destination: Reality or territorial marketing? *Journal of Innovation and Knowledge, 5*(2), 96–104. https://doi.org/10.1016/j.jik.2019.06.002

Smart Tourism Destinations. (2021). About us. https://smarttourismdestinations.eu/about-us/

Soares, J. C., Domareski Ruiz, T. C., & Ivars Baidal, J. A. (2021). Smart destinations: A new planning and management approach?, *Current Issues in Tourism.* https://doi.org/10.1080/13683500.2021.1991897

Sustacha Melijosa, I., Baños Pino, J. F., & Del Valle Tuero, E. (2022). Análisis de la investigación sobre destinos turísticos inteligentes mediante la visualización de redes bibliométricas. *Investigaciones Turísticas, 23*, 266–289. https://doi.org/10.14198/INTURI2022.23.12

Townsend, A. M. (2013). *Smart cities: Big data, civic hackers, and the quest for a new utopia.* W.W. Norton & Company.

Williams, A. M., Rodriguez, I., & Makkonen, T. (2020). Innovation and smart destinations: Critical insights. *Annals of Tourism Research, 83*, 102930. https://doi.org/10.1016/j.annals.2020.102930

Xiang, Z., Stienmetz, J., & Fesenmaier, D. R. (2021). Smart tourism design: Launching the annals of tourism research curated collection on designing tourism places. *Annals of Tourism Research, 86*, 103154. https://doi.org/10.1016/j.annals.2021.103154

Dr. Josep A. Ivars-Baidal is a Professor in Geography and Tourism at the University of Alicante (Spain). His research interests include tourism policy, planning and management of tourism destinations, as well as sustainability and innovation in the field of tourism. Along with his academic activities, he has also been involved in planning projects and has held different positions in public bodies devoted to tourism management.

Dr. Francisco Femenia-Serra is a Lecturer at Complutense University of Madrid, and a member of the research group "Turismo, Patrimonio y Desarrollo". His research focuses on tourism geographies, particularly on the planning, management and marketing of tourism destinations, smart destinations and smart cities, as well as on the intersection between places, technologies and humans in the travel context.

Dr. Marco A. Celdrán Bernabeu is a Research Assistant at the Tourism Research Institute (University of Alicante), a tourism multidisciplinary research centre where he works on projects associated with the Spanish National R&D&I Plan. Marco's current research focuses on smart tourism, smart destinations and technology influence on tourism. He combines his research activity with teaching tasks at the Department of Regional Geography (UA).

Dr. David Giner Sánchez is a project coordinator at the Valencian Institute for Tourism Technologies (INVAT.TUR) in Benidorm (Spain) and a teaching assistant at the Department of Regional Geography, University of Alicante. He conducts research on smart destinations, social media marketing, destination management and information and communication technologies.

Part II
Different Approaches

Chapter 12
Tourist Visitation and Beach Erosion: Analysis of the Evolution of the Coastline, Frequentation and Beach Area on the Island of Menorca (2001–2015) (Balearic Islands, Spain)

Guillem X. Pons, José Ángel Martín-Prieto, and David Carreras-Martí

12.1 Introduction

Climate change is having an undeniable influence on coastal areas. A substantial part of the coastlines around the world are suffering erosion due to the increase in storms (Amores et al., 2020) and floods (catastrophic punctual events), anthropic modifications of the environment that modify the coastal drift, a bad coastal management, over-frequency, as well as the rise in sea level (López-Dóriga et al., 2019; Nicholls & Cazenave, 2010). Given the constant changes in our coastal environments, Earth observation satellites and aerial photography analysis are a very useful tool to study their evolution.

For decades, coastal areas have experienced intense urbanization and strong population growth (Small & Nicholls, 2003). Although the different lengths of coastline analyzed imply a fractal dimension, it could conclude that the European Union has some 68,000 km of coastline, three times more than the United States. According to the European Environment Agency, almost half of the EU population lives less than 50 km from the sea; Furthermore, the coastal areas are the most popular holiday destination in Europe. Coastal tourist destinations have airports that increase their operations in summer. Mallorca is the third busiest airport in Spain, after Madrid and Barcelona, and in 2022 it had 28.5 million passengers (−3.9%

G. X. Pons (✉) · J. Á. Martín-Prieto
Universitat de les Illes Balears, Palma, Spain
e-mail: guillemx.pons@uib.es

D. Carreras-Martí
Observatori Socioambiental de Menorca (OBSAM), DCM i GXP, Institut Menorquí d'Estudis (IME), Mahón, Spain

A. Blanco-Romero, M. Blázquez-Salom (eds.), *Spanish Tourism Geographies*, Geographies of Tourism and Global Change,
https://doi.org/10.1007/978-3-031-39780-6_12

variation compared to 2019 in a pre-pandemic situation); Ibiza with 8.1 million and Menorca with 3.9 (11.6% more passengers compared to 2019) (data from the Ministry of Transport, Mobility and Urban Agenda).

The presence of beaches characterizes the five main world vacation destinations (Wolch & Zhang, 2004), due to their contribution as a primary resource to produce economic and social value (Houston, 2013), with high tourist profitability (Alves et al., 2014). The interest in the environmental management of tourist beaches has developed a large scientific corpus, based on the diagnosis of their deterioration (Roig-Munar & Martín-Prieto, 2005), to define procedures and tools to mitigate it (Yepes, 1998, 2007; Chen & Teng, 2016). Beaches are part of the public maritime and terrestrial domain in Spain, which ensures their free access, through state management (Torres-Alfonsea, 2010).

Around 10% of the coast of the Balearic Islands (161 km) are sandy accumulations. The coastal zone, and specifically the beaches, are a very dynamic environment with many physical processes that cause changes in the coastline, such as periodic flooding due to tides, sea level rise, erosion, or deposition. These processes play an important role in the change of the coastline and the development of the coastal landscape (Agulles et al., 2021). The coastline is considered one of the most dynamic processes in coastal areas.

The rate and extent of coastal erosion is expected to intensify as a result of sea level rise, but erosion trends cannot be easily predicted due to the interaction between various factors, such as the sediment balance and the hydrodynamics of the coast along with climatic variables and anthropic processes such as over-frequency (beach users). The anchoring of yachts discharges and climate change can alter, in the Balearic Islands, the main sediment factory, the *Posidonia oceanica* -seagrass meadows- (Jordà et al., 2012; Marbà et al., 2014; Grelaud & Ziveri, 2020). Anthropic processes also have a general consequence, such as the rise in sea level or the increasingly frequent episodes of extreme climatic events (precipitation, winds, waves... intense). In addition, man also manages and conserves these spaces, sometimes without knowing how this fragile and dynamic ecosystem works. In fact, by the end of the century, the average retreat of the coast in the Balearic Islands will oscillate between 9.19 and 11.7 m in a moderate scenario and between 33.3 and 35.4 m if extreme climatic conditions are considered (Agulles et al., 2021). This implies the total disappearance of some beaches. Considering the extreme scenario, between 72 and 314 of the 869 existing beaches will be completely flooded by the end of this century, with the disappearance of 80% of all the beaches (Agulles et al., 2021).

Considering the importance of this reduced space, the information on the evolution of the coastline and its trend is extremely valuable. Standardized and widely accepted methodology to analyze long-term shoreline change is essential to carry out an integrated approach to shoreline evolution and therefore support shoreline management. In addition, the availability of different databases that provide knowledge on the evolution of the coast are of key importance to support both management experts and users.

12.2 Background

The analysis of the number of visitors and beach surface have been dealt with in different publications. Most of them with a static conception of the beach, when this is a very dynamic and changing element over time.

There is no single definition or method for calculating beach carrying capacity. The World Tourism Organization proposes carrying capacity is the maximum number of people who can visit a tourist site at the same time, without damaging the physical, economic, and socio-cultural environment, and without unacceptably reducing the quality of the visitor experience. To better understand this carrying capacity proposal, we could break down this concept into different aspects: environmental (related to the minimum environmental impact), geographic (location, dimensions, typology, fragility, optimal surface for carrying out the activity), perceptual (satisfaction of the user and also residents) and economic (profitability of the activity). Each of them has an associated weight and relative importance depending on the place analyzed.

Botero et al. (2008), presents a methodology to measure the carrying capacity in tourist beaches, beyond the environmental component. He conceives the beach as a complex system in movement towards its sustainable development. Five beaches of the Colombian North Caribbean were selected and classified, through four tourist sub-uses: intensive, conservation, shared and ethnic. In them, they establish carrying capacity ranges, depending on their classification (urban A or natural B), determined by the average layout of the beach surface, between 10 and 45 m^2/visitor respectively. Several authors have proposed a similar methodology for European beaches, finding that a similar measurement of the physical load capacity in urban beaches of 4 m^2/user, considering that below these densities there would be overcrowded conditions (Yepes, 1998; Da Silva, 2002).

During the pandemic period, new carrying capacity limits for the beaches have been established, taking exclusively into account the frequentation and the current beach area through theoretical calculations in terms of m^2/user, in COVID-19 times. (Del Rosario Rodríguez & Pérez-Chacón, 2022), after different orders from the Government of Spain (Order SND/414/2020 (BOE 138), Order SND/440/2020 (BOE 146); Order SND/445/2020 (BOE 150)). Natural parks, beaches, coastal spaces and other natural areas of the Balearic Islands constitute an important tourist and recreational resource. The tourist and recreational pressure on them, especially the one concentrated during certain periods of the year, mainly in summer and part of the spring-autumn, generates natural and social impacts of various kinds, some important, that require urgent solutions or management.

There are different scientific contributions that deal with the evolution of the position of the coastline on various beaches of the Balearic Islands, in the long term (since 1956) and in the medium term (since 2002) (Gómez-Pujol et al., 2017a, b, 2018; Martín-Prieto et al., 2016, 2020, 2018a, 2018b; Tintoré et al., 2009). Regarding studies on carrying capacity, several works have been carried out since the beginning of the 2000s in which their analysis is considered a good coastal management tool (Roig-Munar, 2002). Subsequently, Roig-Munar and Martín-Prieto (2002) and Roig-Munar (2003) analyze the physical carrying capacity and the perceptual carrying capacity in the case of Menorcan beaches located in protected natural spaces on the northern and southern coast (Areas Natural Areas of Special Interest, hereinafter ANEI, Me-3 and Me-14. In both works, over-frequency of the five beaches that are in ANEI Me-14 is already observed, which leads to the degradation of these spaces. it becomes clear how the advertising of these spaces seems to be related to over-frequency. Roig-Munar et al. (2020) analyze the carrying capacity of a total of 16 Menorcan beaches between the years 2000 and 2017, stressing that the patterns of use and frequentation are a useful tool in the management and planning of the coast It also reaffirms, as in 2002, that the publicity of these spaces plays an important role.

Most of these works are physical descriptions and few analyzed or interpreted the results as a whole, especially those that incorporate anthropic factors (frequency, management...) in their analysis. In this regard, our working group is a pioneer in the integrated analysis of this loss -erosion- (or gain -accretion-) of beach sand.

12.3 Material and Method

12.3.1 *The Study Area: Menorca Biosphere Reserve*

UNESCO declared Menorca a biosphere reserve on October 7, 1993, in view of the high degree of compatibility achieved between the development of economic activities, the consumption of resources and the conservation of heritage and a landscape that has maintained, and continues to maintain today, an exceptional quality. Menorca is an intensely humanized territory, with a very rich traditional rural landscape. It is home to a remarkable diversity of Mediterranean habitats, in which animal and plant species exclusive to the island live, some of which are in danger of extinction.

Around the world there are more than 400 biosphere reserves where the concept of sustainable development is experimented with. They are places where human activity is carried out in a way that is compatible with the conservation of natural resources and cultural heritage. Menorca is part of the Spanish Network of Biosphere Reserves and maintains contacts with other biosphere reserves attached to UNESCO's Man and Biosphere Program (MaB Program). Menorca belongs to the World Network of Island and Coastal Biosphere Reserves (RMRBIC) was created in 2009, with the support of UNESCO, to promote sustainable development specifically on islands and coastal areas and promote adaptation strategies and mitigation against climate change.

With an approximate area of 700 km^2 and 200 km of coastline, Menorca is administratively divided into eight municipalities and is home to a registered population of 95,936 inhabitants (according to the National Institute of Statistics, year 2021). The preponderance of tourism, which in recent decades has become the island's main economic resource, explains the strong seasonality of human pressure, which occasionally hosts more than 160,000 people.

Menorca is the northernmost and easternmost of the Balearic Islands, with the highest rainfall (about 600 mm per year) and influenced by persistent winds. Elongated and slightly rectangular in shape, it stretches from east to west for about 53 km and presents a low-lying orography. Its maximum height is 358 m (El Toro).

The tourism sector is what drives a great part of Menorca's economy. Monitoring the movement of the coasts is key to understanding the evolution of coastal environments, as well as serving to demonstrate essential information to decision makers at the regional level. The tourist image of the island of Menorca is associated with the abundance of virgin beaches and coves that are distributed along the coastline (Fig. 12.1). This constitutes the main tourist resource of the island, which is reflected as a space that tries to satisfy the expectations of the tourist. The tourist offer, in addition to local demand, have ended up trivializing this natural resource, transforming the naturalness of the beaches at the mercy of the temporary, subjective and short-term needs of market demand (Roig-Munar et al., 2017).

The problem arises when over-frequency occurs in any of these spaces, concentrated in high season, so the pressure exerted is very high, generating environmental and social impacts (waste generation, habitat fragmentation, erosion, etc.).

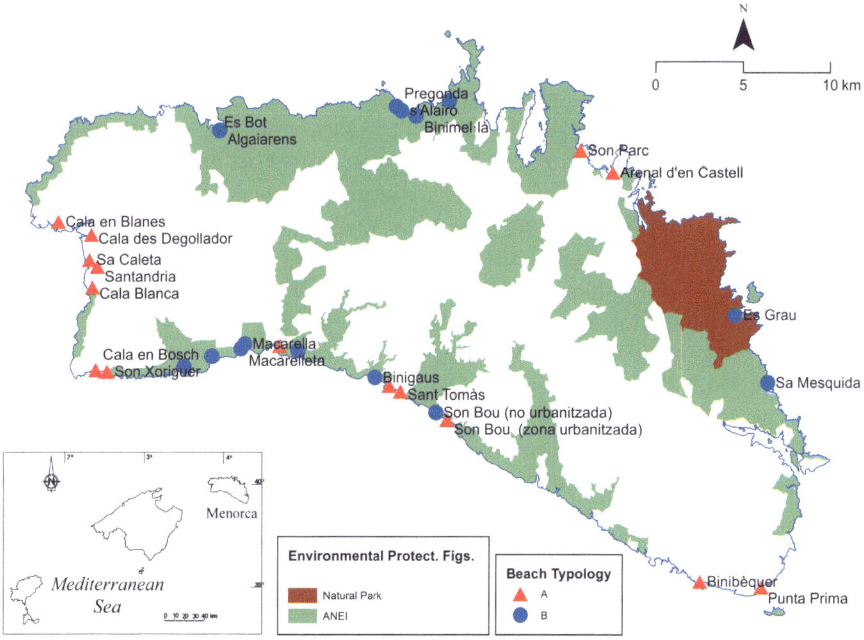

Fig. 12.1 Location of the beaches under study, typology and environmental protection areas

From the point of view of environmental protection, Law 1/1991 on Natural Spaces (LEN) developed the figure of Natural Areas of Special Interest (ANEI) to exclude 43% of the territory and 23% of the coastline from pressure planning (Fig. 12.1). This urban protection led to its tourist advertising. The increase in tourist frequentation was addressed, at the beginning of the 2000s, through management measures for its public use by the Beach Service of the Menorca Island Council. For this purpose, a classification of beaches was carried out for differentiated management measures based on the conservation and/or natural values of each one of them (Juaneda & Roig, 2002; Roig-Munar, 2003). This classification distinguishes three types of beaches:

Type A: located in urban or tourist areas, with beach services and vehicle accessibility on the beach,
Type B: located in ANEI, with high frequentation rates, close vehicle accessibility and no services,
Type C: located in ANEI, with low frequentation rates and with exclusively pedestrian accessibility.

For this reason and to improve its management, it is necessary to combine all the available information. In this sense, the availability of databases referring to the historical evolution of the coastline, the evolution of the beach surface and the frequentation of users constitute very important tools to know how it has developed throughout this period of study, which covers from 2001 to 2015, the impact of users for a total of 30 beaches on the island.

Apart from the evolution of the jure population, another indicator is introduced to measure the de facto population of the island: the Daily Human Pressure (PHD) indicator. According to the definition of OBSAM (www.obsam.cat), this indicator counts all the people in Menorca and does a daily follow-up, adding and subtracting from the total population the people who enter and leave daily through the ports and the airport of Minorca. The PHD gives us information about the actual number of people on the island at any given time. This maximum number of people occurred in 2015 with 202,743. However, during the last 4 years the maximum annual values of the PHD present similar values, always above 200,000 people, with a continuous increase since 2009 (Florit et al., 2016). Although the PHD shows a stabilization of the maximum population around 200,000 people in recent years, the total number of tourists arriving on the island of Menorca each year has grown year after year since 2010 with 944,799 tourists, reaching 1,178,729 in 2015. Therefore, we have an increase in the number of users from the beginning of the century until 2013, the year in which the numbers stabilize. Regarding the evolution of the population of Menorca, during the study period of this work, it has increased from just over 75,000 in 2001 to 92,348 inhabitants in 2015 (Florit et al., 2016), that is, an increase of 18.5%, with a continued increase until 2013 with 95,183 inhabitants. As of this year the curve begins to descend until the year 2015.

Another piece of data that provides information on mobility is the Average Daily Intensity (IMD). This indicator shows the evolution of traffic intensity based on the average number of daily vehicles that pass through the main gauging stations on the island (Florit et al., 2016). The IMD shows a constant growth from 2001 to 2007 with 10,846 vehicles, that decrease slightly until 2015, and with numbers similar to 2001 with 10,030 vehicles.

12.4 Methodology

In this work, the evolution of the coastline of 52 beaches in Menorca for a period of time between 1956 and 2015 (Martín-Prieto et al., 2020) has been analyzed, and 30 of them have been selected because they have good frequentation data.

The objective of this work is focused on determining if there is any relationship between the number of users and the evolution of both the coastline and the surface of a total of 30 selected beaches, distributed along the entire coastline of Menorca, 50% of which are classified as type A and the other 50% as type B (Fig. 12.1 and Table 12.1).

Table 12.1 NSM (M) data, in yellow erosion, orange above average; in green accretionary beaches. In red beaches exceeding 100% of the carrying capacity; in blue between 75 and 100% (Source: Obsam)

	Beach Tyipology	NSM	Average Beach Surface	Mx. Count Users 2001	Surf. mx. (m^2/us) 2001	Carr. Cap. % 2001	Mx. Count Users 2002	Superf. màx. (m^2/us) 2002	Carr. Cap. % 2002	Mx. Count Users 2006	Superf. màx. (m2/us) 2006	Carr. Cap. % 2006	Mx. Count Users 2008	Superf. màx. (m2/us) 2008	Carr. Cap. % 2008	Mx. Count Users 2015	Superf. màx. (m2/us) 2015	Carr. Cap. % 2015
Arenal de son Saura	A	-22.00	15042	865	20.2	28.8	1,625	9.8	50.9	923	15.5	32.4	1,470	8.7	57.2			
Arenal d'en Castell	A	-7.05	14931	2,410	6.3	80.7				1,419	10.6	47.4	1,328	11.1	44.9	1483	11.9	42.1
Cala Binibequer	A	-3.83	3517	555	7.8	78.9	686	4.3	116.9	693	4.6	107.6	692	5.6	88.9			
Cala Blanca	A	-52.74	2031	365	6.6	89.9	280	8.3	60.4	377	5.6	88.6	400	4.4	112.9			
Cala de Santandria	A	-7.08	1918	185	7.9	48.2	340	4.3	116.0	130	16.0	31.2	310	6.9	72.3			
Cala en Blanes	A	-7.92	1267	256	5.3	101.1	210	5.6	89.9	260	4.8	104.4	300	3.7	136.4			
Cala en Bosch	A	-13.33	6139	550	10.9	44.0	350	17.1	20.3	300	16.7	20.0	1,121	6.3	91.4			
Son Bou (urb.)	A	-7.32	35999	3,129	10.5	43.5	2,429	13.6	36.8	2,493	16.8	29.8	2,432	16.5	30.3			
Platja de Binicodrell	A	-4.11	4874	917	2.0	94.1	510	9.7	51.6	900	7.2	69.9				571	10.2	49.0
Platja de Punta Prima	A	13.40	7865	972	9.5	61.8	952	8.6	58.5	854	8.0	62.5	866	8.7	57.7	615	12.6	39.7
Platja de Sant Tomàs	A	-1.96	13987	1,350	10.1	48.3	1,085	11.8	42.3	1,300	11.6	43.1				950	14.5	34.4
Platja de so na Parets	A	0.55	4672	600	7.0	64.2				1,350	3.9	129.1	1,258	4.5	111.0			
Platja Gran	A	-6.54	1093							226	5.9	84.6	140	9.1	55.0			
Cala Galdana	A	-9.07	17600	2,800	5.6	79.5	2,970	5.4	92.6	1,800	11.1	45.0						
Sa Caleta	A	-27.42	1216	210	6.2	86.3				169	6.7	74.4						
Average		**-10.4**	**8811**	**1083**	**8.3**	**67.8**	**1040**	**8.9**	**67.7**	**886**	**9.7**	**65.3**	**938**	**7.7**	**78.3**	**905**	**12**	**41**
Llevant de son Saura	B	-29.18	10843	1,497	6.0	207.1	661	15.5	96.8	592	20.2	74.4	643	19.0	79.1			
Cala d'Algaiarens	B	2.70	6404	450	14.3	105.4	550	12.7	118.4	634	8.1	184.7	655	11.4	131.2			
Cala en Turqueta	B	-13.44	2341	503	4.4	322.3	570	3.6	421.0	320	6.7	223.7	526	4.9	305.7	288	9.4	159.3
Cala Macarella	B	-9.67	5373	539	11.4	150.5	648	7.6	197.1	610	9.0	167.1	615	8.9	168.6	1416	4.5	336.9
Cala Macarelleta	B	-4.99	1110	155	7.6	209.4	320	3.2	471.3	230	5.4	277.3	288	4.0	375.1	421	2.8	531.8
Cala Mesquida	B	-2.86	8565	396	23.7	69.4	355	25.4	59.1	336	28.3	53.0				385	19.7	76.3
Cala Mitjana	B	-10.30	3959	319	13.6	120.9				515	9.1	165.1	568	7.7	194.3			
Cala Pregonda	B	2.80	5898	233	22.2	59.3	210	28.7	52.3	289	20.8	72.2	198	31.0	48.4			
Son Bou (no urb.)	B	-2.97	28287	832	30.4	44.1	646	39.8	37.7	565	59.3	25.3	665	47.0	31.9			
Platja de Binigaus	B	-1.29	12131	768	6.8	95.0	723	13.8	108.8	1,280	13.3	112.9	934	16.7	89.8	753	17.9	83.7
Platja de Binimela	B	-1.58	6779	356	19.7	78.8	220	33.3	45.1	125	49.0	30.6	215	31.6	47.5			
Platja de Cavalleria	B	3.34	8353	649	8.0	116.5	520	11.6	129.6	796	7.5	198.9	520	11.8	127.0			
Platja de s'Alairo	B	8.38	2259	178	10.3	118.2	288	5.7	261.4	490	6.3	237.2	198	13.1	114.8			
Platja des Bot	B	-0.52	5339	130	41.5	36.5	127	34.4	43.7	235	24.3	61.7	374	17.7	84.9			
Platja des Grau	B	12.97	8339	430	14.9	77.4	490	15.8	94.7	418	16.2	92.3						
Average		**-3.1**	**7732**	**496**	**15.6**	**120.7**	**452**	**17.9**	**152.6**	**496**	**18.9**	**131.8**	**492**	**17**	**138**	**653**	**11**	**238**

The working hypothesis considers whether the processes of erosion/accretion of the beaches and their decrease/increase in surface area are linked to frequentation and exceeding the physical load capacity (5 m^2/user on urban beaches, A, and 15 m^2/user in the natural ones, B, close to the minimum with respect to the standards) of each one of the chosen beaches.

The frequentation data was obtained by the Menorca Island Council. The counts were conducted on the beach. We counted all people of any age on the sand and in the water, also people standing in the shade of the pine trees adjacent to the beach, and those who are temporarily walking on the adjacent rocks. People located on yachts anchored in the cove were not included in the count.

The counts are carried out 2 days each month (one every fortnight, in which one is intended to be carried out on a working day and the other on a holiday) during the months of July, August and September, considering the weeks of maximum presence of tourists on the island, since it is desired to find the maximum number of users for each beach. The counts are conducted at 12:00, 14:00 and 17:00, recording the data on previously established cards. For subsequent analysis, we used two values: the maximum number of users during the three summer months, and the maximum value of users detected during the month of August. It works with maximum values.

The counting methodology is always carried out in the same way, with small variations depending on the surface of the beach. It starts at the same point and progresses progressively along the beach.

Although we try to ensure that the weather conditions are homogeneous every day, and optimal so that the use of the beach is maximum.

12.4.1 Data and Statistical Tools: Digital Shoreline Analysis System

In order to systematically measure the variations of the coastline over a long period of time, a methodology developed through a Geographic Information System (GIS) has been established based on the use of orthophotos obtained from the IDE Menorca (Infraestructura de Dades Espacials de Menorca, www.cartografia.cime.es), corresponding to the years 2001, 2002, 2006, 2007, 2008, 2012 and 2015. Once the different coastlines have been digitized, they have been analyzed with the help of the Digital Shoreline Analysis System tool (Thieler et al., 2009), which allows the calculation of the rate of change of the coast based on the distance between two lines and the elapsed time. In this way, the tool offers several fundamental statistics and for this work the Net Shoreline Movement (NSM) has been used, which shows the distance in metres between the oldest and most recent shorelines.

The calculation of the surface of the beaches has been carried out by means of a GIS from the orthophotos indicated in the previous paragraph, together with the field work. From the sandy surface, that suitable for placing a towel is taken, called resting surface, which is defined as the beach surface not occupied by vegetation without dune morphologies, excluding wetlands or creek mouths, no counting the area of transit or access to the beach and the sea batting area on the first coastline (Pérez et al., 2018). Instead, the adjacent areas occupied by pine trees are included, where people are accustomed to placing in shade. It is important to keep in mind that beaches are not static systems and that their surface may undergo significant changes within the same year or over the years, depending on storms, floodings or aspects related to management.

To calculate the carrying capacity, the concept of physical carrying capacity has been used, which is defined as the maximum number of users that a given space can support indefinitely, without causing significant negative impacts, that is, that the available beach area per person is lower than that considered optimal for each type of beach. These reference values are 5 m^2/pers. for beaches of type A and 15 m^2/pers. for type B beaches. For the user count, the data provided by the Menorca Socio-environmental Observatory (OBSAM, www.obsam.cat) is used based on the user counts on the beach. This methodology has been followed continuously since the year 2000, established by the OBSAM and the Beach Cleaning Service of the Menorca Island Council (Carreras et al., 2018). The counts are made 2 days (1 day during the week and 1 day on the weekend) during the month of August at 11:00, 14:00 and 17:00, looking for the weeks with the maximum presence of tourists on the island, always using the maximum value detected.

Finally, an Exploratory Data Analysis was carried out by applying clustering techniques. Firstly, data on beach surface, carrying capacity and maximum headcount of people for the years 2001, 2002, 2006, 2008 and 2015 were standardized, missing data were estimated and similarity measures were subsequently applied (Euclidean distance) and the formation of hierarchical groups by the Ward agglomeration method. The purpose of CA is to form homogeneous groups or clusters based on the similarities or similarities between them.

12.5 Results

The overall result for the 52 beaches was a mean retreat (MSL) of −4.8 m for all the beaches on the island, at an average of −0.1 m/year (Martín-Prieto et al., 2020). From this study, 30 beaches have been analyzed, whose average retreat from the coastline is −6,8 m (Fig. 12.1 and Table 12.1).

Table 12.1 shows the data for each beach based on the maximum count for each year sampled, the beach surface and the carrying capacity between 2001 and 2015, together with the NSM values in addition to the typology of each beach.

It can be observed that the typology A beaches have an average shoreline retreat of −10.4 m, with four beaches with values above the average and where only two of them showing accretion. However, only 4 beaches have ever exceeded 100% of their carrying capacity, for a total of 9 occasions, being 2006 and 2008 the years with the highest number of times. When the occupancy is between 75 and 100% of the carrying capacity, the number of beaches is 9 (for a total of 12 occasions). The average surface of this typology is 8.811 m^2.

With regard to the beaches of typology B, 10 of them show shoreline retreat (5 of which above average), twice of accretionary beaches than type A beaches, with an average retreat (NSM) of −3.1 m. The total beaches of this typology that exceeds 100% its carrying capacity is 9, for a total of 34 observations (twice the beaches and 4 times more than on the beaches type A), distributed homogeneously throughout the sampling years. Cala d'Algaiarens, Cala en Turqueta, Macarella, Macarelleta, Platja de Cavalleria and Platja de s'Alairó are particularly noteworthy, that overcome carrying capacity in all the sampled years. When beach occupation is between 75 and 100% of the carrying capacity, the number of beaches is 6 (for a total of 10 occasions). The average surface of this typology is 7.732 m^2, slightly lower than the beach typology A (Table 12.1).

12.6 Spatial Analysis

A hierarchical agglomerative clustering of the 30 beaches is carried out using the Euclidean Distance through the Ward's Method. From this statistical analysis a dendrogram is obtained (Fig. 12.2) that groups the different observations into four groups or clusters.

Cluster 1 consists of 5 beaches which are characterized by the fact that they have the largest surface area (with an average surface of 17.969 m^2), with an average of 2.840 users. All beaches are erosive with an average setback of −8,6 m, the highest of all groups. All are typology A, except the beach of Son Bou non urbanized. The average of the set does not exceed 50% of the carrying capacity, with a surface available per user around 17,7 m^2.

Cluster 2 is formed by 11 units with an average surface of 3.905 m^2 (the lowest), where the average carrying capacity is 546 users. 7 beaches exceed 100% of their capacity for a total of 10 occasions. All except Es Grau are of typology A. The average retreat of the coastline (NSM) is −8,7 m (despite that the beaches des Grau and Punta Prima grow up to 12,9 and 13,4 m respectively between 2001–2015), with the beaches with the greatest coastline retreats being sa Caleta (−27,4 m) and Cala in Bosch with −13,3 m. The average carrying capacity is around 76%, with a surface available per user over 8,6 m^2.

Cluster 3 is formed by only one beach, urbanized Son Bou beach, which can be considered an outlier. It is the beach with the largest surface area on the island with 36,000 m^2 and a carrying capacity of 7200 users.

Cluster 4 is formed by a total of 13 units, all classified as typology B, with an average surface of 6.104 m^2. The average retreat of the coastline is −4,3 m, the lowest of

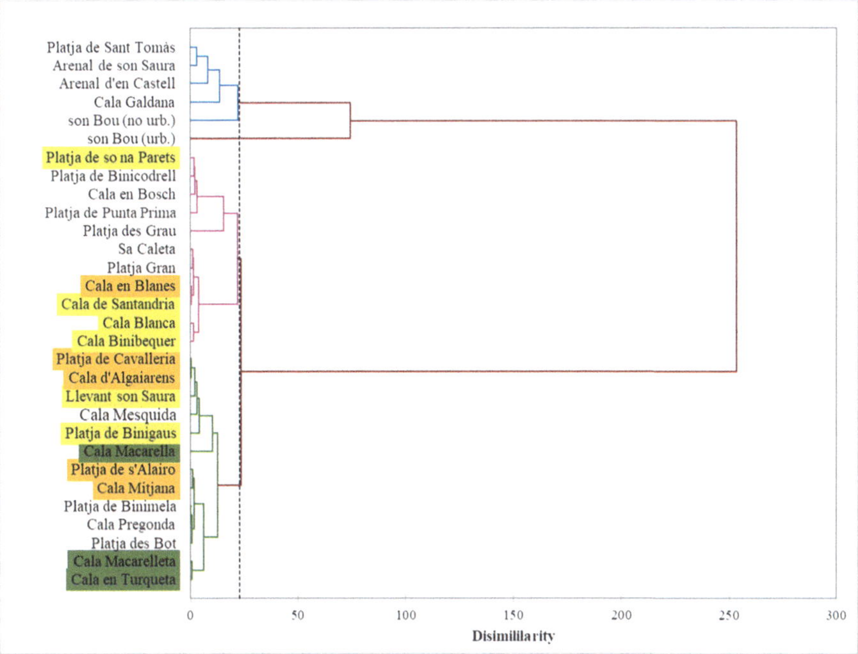

Fig. 12.2 Ward's Dendrogram by grouping. In yellow beaches where the CC has been overcome 1–2 occasions; In orange beaches where the CC has been overcome 3–4 occasions and in green beaches where the CC has been overcome 5 occasions

all. This group stands out because in it are the beaches (70%) where the carrying capacity has been exceeded in greatest number of occasions. This cluster also highlights because it concentrates those most frequented beaches, exceeding even 400% of their carrying capacity, as is the case of Macarelleta or Cala in Turqueta, with an average number of 506 users. The other beaches with a high frequency are Macarella and Cavalleria. The average of the set, has a carrying capacity around 178%, with an available surface per user over 14 m^2, at the limit of its capacity (which for typology B is 15 m^2). All beaches show a shoreline retreat, except the beaches of Algaiarens, Cavalleria, S'Alairó and Pregonda.

Figure 12.3 shows the relationship between erosion-accretion (NMS), the difference in available beach surface and the evolution of the carrying capacity between 2001 and 2015. Two groupings of beaches can be seen in the graph. Those located in the blue circle, which represents those where the beach surface difference between 2002 and 2015 has increased (except es Grau Beach). The opposite occurs with the beaches located inside the red circle, where the beach surface decreases and the carrying capacity increases. Cala Macarella, Macarelleta and Cala Mitjana are the beaches where the carrying capacity increases significantly.

Fig. 12.3 Scatter plot comparing NSM versus difference in beach area between 2001 and 2015 and the evolution of carrying capacity. The blue circle shows beaches where the surface area increases and the red circle shows beaches where the surface area decreases and the carrying capacity increases. The size of the dots is proportional to the increase in carrying capacity

12.7 Discussion

From the results, a certain relationship can be seen in the initial hypothesis in which the aim was to determine whether the processes of erosion/accretion of the beaches and their decrease/increase in surface are linked to frequentation and physical carrying capacity of 30 beaches on the island of Menorca. In other words, a priori, the higher the erosion rate, the smaller the surface of the beach, so the surface used for resting should also decrease.

In general, it should be noted that the trend in the number of users since 2006 has increased and that there is a predisposition to concentrate on certain beaches.

Based on the results (Fig. 12.2 and Table 12.1) it becomes clear that the type B beaches (located in ANEI, high rates of frequentation, relatively close road accessibility and no services, whose comfort surface is 15 m^2/user), with an average setback of the coastline of −3,1 m, three times less than type A beaches (located in urban or tourist areas, with beach services and road accessibility at beachfront, whose comfort surface is 5 m^2/user) have higher rates of frequentation and therefore have a higher carrying capacity exceedance.

One of the factors that explain this situation is the beach surface. Thus, the larger the surface, the lower the carrying capacity indices and vice versa, i. e. the smaller the surface and the higher the carrying capacity indices. In group 4 (Fig. 12.2) there are the most over-frequented beaches of the island, Cala Macarella, Macarelleta and Cala Turqueta, which exceed 100% by far (Table 12.1) and to a lesser extent the beaches of Algaiarens, Cavalleria and S'Alairó. I the case of Cala Turqueta, the beach surface increases, as well as coastline grows, which differs from the data of Cala Macarella and Macarelleta.

The reasons that encourage tourists to visit Menorca are mainly sun and beach in 51% of cases, according to a survey carried out in 2016 (Pérez et al., 2018), along with tranquillity and landscape (Florit et al., 2016). The explanation of the advertising-visitation relationship is convincing, especially for Cala Macarella, Macarelleta and Cala Turqueta. This fact was highlighted in a study carried out by Pérez-López and Roig-Munar in 2013, where they remark that the 8 most visited beaches on the island were among the 10 most advertized beaches (80% of wich are beaches of type B).

These data matches with the results of Table 12.1, where beaches cited before exceed 100% in all the years analyzed, with a range between 105,4 and 531,8%; in contrast to type A beaches, where the highest figure is 136,4% in 2008. These data also coincide with the car parks, where the most frequented beaches were already full between 09:00 and 11:00 (Florit et al., 2016).

The consequence of this over-frequency on type B beaches is that after a certain density, users look for places outside the resting area of the beach, invading the adjacent environments and deteriorating nearby ecosystems such as dune and wooded areas, where erosive impacts have already been observed.

On larger beaches, although erosion reduces the total area available per user, the area used is not affected since the strip of beach used will be displaced landward following the receding shoreline.

In terms of tourism planning, it is important to be aware of and anticipate this process of over-frequentation and decrease in rest area per user to avoid user dissatisfaction.

12.8 Conclusions

The beaches and calas of Menorca are one of the island's main tourist attractions. That is why the over-frequentation of some of these spaces gives rise to a feeling of overcrowding and saturation, which leads some users to invade neighbouring environments such as dunes, dune systems, etc., deteriorating these ecosystems.

However, the carrying capacity presents tolerable values, being within the recommended limit of m^2/user, except on some beaches where it is widely exceeded.

Based on the study of the historical evolution of the coastline between 2001 and 2015, the evolution of the beach surface and the analysis of the carrying capacity of 30 beaches, the degree of relationship between these three variables has been analyzed. Among the conclusions of this work, it should be noted that the retreat of the coastline is not decisive in the evolution of the beach surface, but there is a relationship between the two. In terms of beach surface, the larger the surface, the lower the carrying capacity and the less pressure on these spaces. Thus, the beaches with the largest surface area, regardless of their typological classification in A or B, are those which suffer the least overloading of the carrying capacity and the least frequentation. On the other hand, those with a smaller surface have their carrying capacity exceeded. To this must be added that some coastal areas are heavily advertised, leading to over-frequentation and saturation.

The typology of sun and beach tourist destination is dependent on the conservation of the latter resource, which is experiencing a growing decline, as has been demonstrated in the case of the Menorcan beaches analyzed. The tourist repercussion of this deterioration are greater on the most publicised and popular beaches, which are also the smallest and most natural. It has been demonstrated that the carrying capacity has been exceeded, in terms of desirable occupancy ratios (15 m^2/user), on natural beaches (typology B), which are the most erosive. Therefore, and as a final conclusion, the collection of data on the geomorphological and biogeographic evolution and the knowledge of the carrying capacity of the beaches and calas of tourist destinations, such as Menorca, should be a tool for the manager to avoid major problems.

Acknowledgements This work is a contribution to the research project financed with FEDER funds from the Ministry of Science, Innovation and Universities and State Research Agency entitled: "Overtourism in Spanish Coastal Destinations. Tourism Degrowth Strategies" (RTI2018-094844-B-C31), https://overtourism-degrowth.uib.eu/. We must thank Miguel McMinn for his review of the English text. Finally, thanks to the Menorcan Institute of Studies and the Natural History Society of the Balearic Islands for helping us carry out this study.

References

Agulles, M., Jordà, G., & Lionello, P. (2021). Flooding of sandy beaches in a changing climate. The case of the Balearic Islands (NW Mediterranean). *Frontiers in Marine Science, 8*(760725), 1–15.

Alves, B., Benavente, J., & Ferreira, O. (2014). Beach users' profile, perceptions and willingness to pay for beach management in Cadiz (SW Spain). *Journal of Coastal Research, 70*, 521–526.

Amores, A., Marcos, M., Carrió, D. S., & Gómez-Pujol, L. (2020). Coastal impacts of Storm Gloria (January 2020) over the North-Western Mediterranean. *Natural Hazards and Earth System Sciences, 20*, 1955–1968.

Botero, C., Hurtado, Y., González, J., Ojeda, M., & Díaz, L. H. (2008). Metodología de cálculo de la capacidad de carga turística como herramienta para la gestión ambiental y su aplicación en cinco playas del caribe norte colombiano. *Gestión y ambiente, 11*(3), 109–122.

Carreras, L., Galarza, N. Blanco, A., & de Pablo, F. (2018). *Servei d'informadors de la Reserva de Biosfera de Menorca. Any 2018.* Informe tècnic 02/2018. Departament de Medi Ambient Reserva de Biosfera-Consell Insular de Menorca. Observatori Socioambiental de Menorca-Institut Menorquí d'Estudis. Maó, 167 pp.

Chen, C., & Teng, N. (2016). Management priorities and carrying capacity at a high-use beach from tourists' perspectives: A way towards sustainable beach tourism. *Marine Policy, 74*, 213–219.

Da Silva, C. P. (2002). Beach carrying capacity assessment: How important is it? North Ireland. *Journal of Coastal Research, 36 SI*, 190–197.

Gobierno de España. Orden SND/414/2020, (BOE 138), de 16 de mayo, para la flexibilización de determinadas restricciones de ámbito nacional establecidas tras la declaración del estado de alarma en aplicación de la fase 2 del Plan para la transición hacia una nueva normalidad.

Gobierno de España. Orden SND/440/2020 (BOE 146), de 24 de mayo, por la que se modifican diversas órdenes para una mejor gestión de la crisis sanitaria ocasionada por el COVID-19 en aplicación del Plan para la transición hacia una nueva normalidad.

Gobierno de España. Orden SND/445/2020 (BOE 150), de 26 de mayo, por la que se modifica la Orden SND/271/2020, de 19 de marzo, por la que se establecen instrucciones sobre la gestión de residuos en la situación de crisis sanitaria ocasionada por el COVID-19 y la Orden SND/414/2020, de 16 de mayo.

Del Rosario Rodríguez, E., & Pérez-Chacón, E. (2022). Aforo de la playa de Las Canteras (Gran Canaria, Islas Canarias) en tiempos Covid-19. *XI Jornadas de Geomorfología Litoral, Galicia, 2022*, 399–403.

Florit, A., Canals, A., & Carreras, D. (2016). *Estudi de l'afluència de persones a les platges de Menorca (2000-2016).* Observatori Socioambiental de Menorca (OBSAM). Institut Menorquí d'Estudis. Maó, 139 pags.

Gómez-Pujol, L., Compa, M., Orfila, A., Álvarez Ellacuría, A., Balaguer, P., Roig, F. X., Fornós, J. J., & Tintoré, J. (2017a). Las playas de Menorca: naturaleza y distribución. *Monografies de la Societat d'Història Natural de les Illes Balears, 25*, 67–85.

Gómez-Pujol, L., Orfila, A., & Álvarez Ellacuría, A. (2017b). Evolución y dinámica de una playa encajada de bloques y gravas en la costa NW de Mallorca (Cala Deià, Mallorca). *Geotemas, 17*, 111–114.

Gómez-Pujol, L., Orfila, A., Morales-Márquez, V., Compa, M., Pereda, L., Fornós, J. J., & Tintoré, J. (2018). Beach systems of Balearic Islands: Nature, distribution and processes. In J. A. Morales (Ed.), *A: The Spanish coastal systems: Dynamic processes, sediments and management* (pp. 269–287). Springer.

Grelaud, M., & Ziveri, P. (2020). The generation of marine litter in Mediterranean Island beaches as an effect of tourism and its mitigation. *Scientific Reports, 10*(20326), 1–11.

Houston, J. R. (2013). The economic value of beaches: A 2013 update. *Shore & Beach, 81*(4), 4–11.

Jordà, G., Marbà, N., & Duarte, C. M. (2012). Mediterranean seagrass vulnerable to regional climate warming. *Nature Climate Change, 2*(11), 1–4.

Juaneda, J., & Roig, F. X. (2002). El pla de neteja integral de l'illa de Menorca com a eina de gestió ambiental. In M. Blàzquez & M. Cors (Eds.), *Geografía y Territorio, el papel del geografo a escala local* (pp. 43–48). Universitat de les Illes Balears, Palma de Mallorca.

López-Dóriga, U., Jiménez, J. A., Valdemoro, H. I., & Nicholls, R. J. (2019). Impact of sea-level rise on the tourist-carrying capacity of Catalan beaches. *Ocean and Coastal Management, 170*, 40–50.

Marbà, N., Díaz-Almela, E., & Duarte, C. M. (2014). Mediterranean seagrass (*Posidonia oceanica*) loss between 1842 and 2009. *Biological Conservation, 176*, 183–190. https://doi.org/10.1016/j.biocon.2014.05.024

Martín-Prieto, J. Á., Roig, F. X., Rodríguez Perea, A., Mir Gual, M., Pons, G. X., & Gelabert, B. (2016). La erosión histórica de la playa de Sa Ràpita (S. Mallorca). *Investigaciones Geográficas, 66*, 135–154.

Martín-Prieto, J. Á., Almaraz, A., Pons, G. X., & Carreras, D. (2018a). Evolución histórica de la línea de costa de playas de Menorca (Illes Balears): 1956–2015. A: Pons, G. X. et al. (Eds.). *Llibre de ponències i resums de les VII Jornades de Medi Ambient de les Illes Balears*. Societat d'Història Natural de les Balears (SHNB); Universitat de les Illes Balears.

Martín-Prieto, J. A., Roig, F. X., Rodríguez-Perea, A., Pons, G. X., Mir-Gual, M., & Gelabert, B. (2018b). Análisis de la evolución histórica de la línea de costa de la playa de Es Trenc (S. de Mallorca): causas y consecuencias. *GeoFocus, 21*, 187–214.

Martín-Prieto, J. Á., Carreras, D., Pons, G. X., & Almaraz, A. (2020). Shoreline historical evolution (1956–2015) of beaches of Menorca (Balearic Islands). In G. Malvárez & F. Navas (Eds.), *Global coastal issues of 2020. Journal of Coastal Research, SI* (Vol. 95, pp. 563–567).

Nicholls, R. J., & Cazenave, A. (2010). Sea-level rise and its impacts on coastal zones. *Science, 328*, 1517–1520.

Pérez, M., Pons, M., Gallofré, A., & Carreras, D. (2018). *Sistema d'Indicadors Turístics de Menorca 2018* (p. 406). Observatori Socioambiental de Menorca, Institut Menorquí d'Estudis.

Pérez-López, M., & Roig-Munar, F. X. (2013). Análisis de la relación entre la publicitación, la visitación y la ocupación de los sistemas litorales arenosos de Menorca (Islas Baleares). In A. Rodríguez-Perea, G. X. Pons, F. X. Roig-Munar, J. Á. Martín-Prieto, M. Mir-Gual, & J. A. Cabrera (Eds.), *La gestión integrada de playas y dunas: experiencias en Lationoamérica y Europa* (Mon. Soc. Hist. Nat. Balears) (Vol. 19, pp. 393–401).

Roig-Munar, F. X. (2002). Anàlisis de la capacitat de càrrega de les platges situades a l'Àrea Natural d'Especial Interès Me-3 de l'illa de Menorca. *Geografía y Territorio, el papel del geógrafo en la escala local*. Ed. Blázquez, M. et al, 279–285.

Roig-Munar, F. X. (2003). Análisis de la relación entre capacidad de carga física y capacidad de carga perceptual en playas naturales de la isla de Menorca. *Investigaciones Geográficas, 31*, 107–118.

Roig-Munar, F. X., & Martín-Prieto, J. A. (2002). Valoración de la capacidad de carga física y perceptual en playas situadas en espacios naturales protegidos. La Geografía y la Gestión del Turismo. *Actas VIII Coloquio de Geografía del Turismo, ocio y recreación* (pp. 343–351). Universidad de Santiago de Compostela.

Roig-Munar, F. X., & Martín-Prieto, J. A. (2005). Efectos de la retirada de bermas vegetales de Posidonia oceanica sobre playas de las islas Baleares: consecuencias de la presión turística. *Investigaciones Turísticas, 57*, 40–52.

Roig-Munar, F. X., Martín-Prieto, J. A., Rodríguez-Perea, A., & Balaguer, P. (2017). Gestión litoral en Menorca: un modelo basado en criterios geomorfológicos. In L. Gómez-Pujol & G. X. Pons (Eds.), *Geomorfología litoral de Menorca: dinámica, evolución y prácticas de gestión* (Mon. Soc. Hist. Nat. Balears) (Vol. 25, pp. 191–212).

Roig-Munar, F. X., Pintó, J., García-Lozano, C., Martín-Prieto, J. A., & Rodríguez-Perea, A. (2020). Análisis de los patrones de uso y frecuentación (2000-2017) en las playas de la isla de Menorca (Islas Baleares). *Cuadernos Geográficos, 59*(1), 171–195.
Small, C., & Nicholls, R. J. (2003). A global analysis of human settlement in coastal zones. *Journal of Coastal Research, 19*(3), 584–599.
Thieler, E. R., Himmelstoss, E. A., Zichichi, J. L., & Ergul, A. (2009). *Digital Shoreline Analysis System (DSAS) version 4.0. An ArcGIS extension for calculating shoreline change.* U.-S. Geological Survey Open-File Report 2008–1278.
Tintoré, J., Medina, R., Gómez-Pujol, L., Orfila, A., & Vizoso, G. (2009). Integrated and interdisciplinary scientific approach to coastal management. *Ocean and Coastal Management, 52*(10), 493–505. www.obsam.cat. Obsevatori Sociambiental de Menorca (OBSAM)
Torres-Alfonsea, F. J. (2010). Cuarenta años de leyes de costas en España (1969-2009). *Investigaciones geográficas, 52*, 167–198.
Wolch, J., & Zhang, J. (2004). Beach recreation, cultural diversity and attitudes toward nature. *Journal of Leisure Research, 36*(3), 414–443.
Yepes, V. (1998). Planificación y gestión turística de playas. *Actas del Curso sobre gestión Integral de la Costa.*
Yepes, V. (2007). Gestión del uso y explotación de las playas. *Cuadernos de Turismo, 19*, 241–254.

Guillem X. Pons is a biologist and tenured professor at the Department of Geography at University of the Balearic Islands (physical geography knowledge area). From 2017 to 2023 he has been the head of the Natural Sciences section of the Menorcan Institute of Studies (IME). He is a member of the board of the Natural History Society of the Balearic Islands (SHNB, www.shnb.org). He is a member of the Spanish Association of Geography (AGE) and the Board of Publications (BAGE). He dedicates his teaching and research tasks in biogeography, especially insular biogeography, and to the management and planning of natural areas, including beach-dune systems, their historical evolution, erosion processes, effects of climate change and tourist frequentation.

José Ángel Martín-Prieto, geographer, specialist in Geographic Information Systems (GIS). He has carried out a multitude of works related to coastal geomorphology, especially with sandy coasts. He is linked to the University of the Balearic Islands in different research groups and has worked on different projects, the latest Overtourism in Spanish Coastal Destinations. Tourism Degrowth Strategies" (RTI2018–094844-B-C31), https://overtourism-degrowth.uib.eu/.

David Carreras-Martí is a biologist, Director of OBSAM (Socio-environmental Observatory of Menorca). He is a member of the Menorcan Institute of Studies (IME) dependent on the Island Council of Menorca (CIMe). Among the tasks that OBSAM develops are the dynamics that occur on the coastline, such as the historical evolution of the coastline. As an observatory, and the fact that Menorca is a biosphere reserve, a fragile ecosystem, they analyze multiple environmental and social parameters of the island.

Chapter 13
The Relationship Between Water and Tourism in the Spanish Mediterranean: An Efficient Management?

Rubén Villar-Navascués, Carlos J. Baños Castiñeira, Jorge Olcina Cantos, and María Hernández Hernández

13.1 Introduction

Since the end of the Second World War, tourism has become one of the economic activities and sectors that has best identified the major changes occurring in contemporary societies throughout the world. From the political, business and scientific dimensions, this growth has traditionally been measured in economic terms from a quantitative approach. It is evaluated based on the number of trips and tourists and economic exchanges (GDP, share in international trade, etc.). This perspective of quantitative development also has a geographic reading, as the evolution of mass tourism has come about through the constant incorporation of regions and countries into the global tourism system, both as inbound and outbound markets, starting with countries in the central areas of the developed world (North America, Western Europe) then with the introduction of the peripheries closest to these central areas (Mediterranean, Eastern Europe, Caribbean) and, finally, from 1990, with the incorporation of all regions of the world. This geographic extension of tourism has traditionally been understood as economic progress for countries where tourism is conceived as a "passport for development" (De Kadt, 1979).

Among the natural resources necessary for tourism development, water has played a strategic role as it is fundamental for the drinking water supply, the

R. Villar-Navascués (✉)
Department of Geography, Complutense University of Madrid, Madrid, Spain
e-mail: ruvillar@ucm.es

C. J. Baños Castiñeira · J. Olcina Cantos · M. Hernández Hernández
Department of Regional Geographic Analysis and Physical Geography, University of Alicante, Alicante, Spain
e-mail: carlos.banos@ua.es; jorge.olcina@ua.es; maria.hernandez@ua.es

A. Blanco-Romero, M. Blázquez-Salom (eds.), *Spanish Tourism Geographies*, Geographies of Tourism and Global Change,
https://doi.org/10.1007/978-3-031-39780-6_13

functioning of certain recreational and leisure facilities (golf, water parks) or as a defining factor of the dominant landscape in certain natural spaces (Gössling et al., 2012). The development of tourism activities and infrastructures is directly associated with the increase in water consumption, which has given rise to the construction of major hydraulic infrastructures to increase the available resources. In addition to the direct consumption of tourists and the water needs of the recreational activities and facilities, water is also incorporated into the productive cycle of different inputs necessary for tourism production, such as construction materials, fuels and food (Gössling et al., 2012; Chapagain & Hoekstra, 2008). It is, therefore, one of the most critical and scarce resources for the tourism sector (UNWTO, 2003).

The pressure on water resources has been intensified by the fact that some of the large sun and beach tourism regions, such as the Mediterranean basin, also have arid and semi-arid climates (Olcina & Miró, 2017). Furthermore, the tourism activity, particularly residential tourism, has led to a change in the urban and demographic models of the populations of the European Mediterranean coast. Since the 1960s and 1970s, these dynamics have given rise to a growing demand for urban and tourism uses on the Mediterranean coast, particularly during the summer, when there is a lower availability of water. Currently, the Mediterranean receives approximately a third of international tourist arrivals and this figure is expected to increase in the future (Fosse et al., 2021). This would increase the area's vulnerability to situations of drought and could aggravate scenarios of water scarcity. Taking into account that water is an essential factor for the development of the tourism activity, it is crucial to analyse how climate change affects the availability of resources and how this can limit the development or even the maintenance of this sector. The sixth report of the IPCC confirms the decreasing trend in rainfall at the end of the last century and the increase in temperature in the Mediterranean region, generating not only a loss in climatic comfort but also an increase in water evaporation. Moreover, it also indicates an increase of extreme hydrological events (floods and droughts) which will condition water and territorial planning over the coming decades (IPCC, 2021, 2022). These effects have been apparent since the beginning of this century on the Spanish Mediterranean coast and have generated the development of actions in the tourism sector aimed at improving the efficiency of water and energy management and the adaptation of tourist facilities to the afore-mentioned effects of climate change. In short, the tourist destinations of the Spanish Mediterranean coast are faced with a decisive challenge in the next few years related to the activation of polices and specific actions to mitigate and adapt to climate change (Romero & Olcina, 2021).

These future challenges will take place in a territory, the Mediterranean coast, which for decades has needed to increase the availability of water in order to maintain urban and tourism growth and to adapt its hydraulic infrastructures to the seasonal use of water derived from the tourism activity. On the one hand, the increase in the available water has mainly been generated through the diversification of the sources of supply, incorporating non-conventional resources such as reclaimed water and desalinated seawater. On the other hand, the resizing of the hydraulic infrastructures in order to manage the consumption peaks occurring during the summer due to the tourism activity has required large investments so as to adapt, for example, the capacity of the wastewater treatment stations (Milano et al., 2019). In some cases, in order to avoid a shortage of supply during episodes of drought and to guarantee the availability of water for urban and tourism uses, a series of water governance agreements has been formed between urban users and irrigators for the exchange of regenerated water for drinking water (Rico-Amorós et al., 2013; Ricart et al., 2020). Furthermore, progress has been made in the incorporation of reclaimed water in some urban supply systems, such as that of the city of Alicante, where, from the beginning of the twenty-first century, a water supply network was constructed for reclaimed water so that this complementary source could be applied in non-drinking uses, such as street cleaning or the watering of public gardens and single-family homes (Morote & Hernández, 2018).

However, although the policies aimed at increasing the available resources (supply management policies) have been the traditional response to water scarcity in the coastal tourist areas (Vila et al., 2018), in tourist destinations demand management policies have also been developed to reduce water consumption. This group of measures includes tariff policies, awareness raising campaigns and technological solutions to promote an efficient use of water. For example, the measures related to environmental awareness raising have been developed for decades in western countries and have also been applied to the tourism sector (Hjalager, 1996; Niezgoda, 2011) in the form of campaigns that promote water saving aimed at the general population. Hotels and tourism companies have also conducted such campaigns among their customers and workers (Gabarda-Mallorquí et al., 2018; Rico et al., 2020). However, the principal stakeholders of the tourist areas on the Mediterranean coast all indicate that the reduction of the operating costs due to the decrease in water and energy consumption (Warren & Becken, 2017) and the reduction in the losses of water in the network (Vila et al., 2018) are some of the main indicators that guide their actions in relation to the sustainable management of water resources. For these reasons, the demand management policies have focused on promoting the improvement in efficiency in water use, such as the introduction of savings devices in homes and tourist accommodation (Gabarda-Mallorquí et al., 2017; Rico et al., 2020), and the management of the supply service through the improvement of hydraulic efficiency (Rico-Amorós et al., 2013). These policies respond to different objectives which are mutually exclusive. On the one hand, they seek to reduce as much as possible the environmental impact of the tourism activity and promote policies aimed at guaranteeing environmental sustainability, reducing the volume of water demanded. On the other hand, they seek to increase the availability of water to enable the continued growth of the tourism activity through a greater capacity of accommodation or the supply of new services which are intensive in the use of water.

In order to determine the impact that these measures have on the sustainable management of water resources, this chapter addresses the relationship existing between the policies of water efficiency and consumption in coastal tourism destinations of the Spanish Mediterranean coast. This analysis will take a multiscale perspective in order to identify the dynamics and factors that can explain the relationship between water efficiency and consumption in scientific publications specialised in the area of study indicated. Therefore, two levels of application of these policies and actions of saving and water efficiency will be distinguished. First, from a regional and local point of view, the measures carried out on the water supply systems to improve the efficiency in the use of water will be evaluated. We will differentiate between the different territorial models of implementing the tourism activity and how these measures have affected the water consumption trends. Second, the analysis will focus on the impact of the efficient use and water saving measures implemented in the tourist establishments, particularly hotels and tourist apartments.

To achieve the aforementioned objectives, different sources have been consulted. In general, a literature search has been conducted to identify publications that have analyzed water efficiency and consumption policies in coastal tourist destinations with particular attention to the factors that affect it. This has been carried out from the search for the following keywords: hydraulic efficiency, water consumption, urban model, network losses, water price, saving measures and Mediterranean. For the analysis of the water efficient use measures implemented in tourist establishments (mainly hotels), the keywords used have been: tourist accommodation, hotels, water saving devices, outdoor use and indoor use. Furthermore, official documents and plans of the tourism departments of the Autonomous Communities of the Spanish Mediterranean coast and of tourist destinations have been consulted in relation to the development of sustainability actions and adaptation to climate change, which have water as a key element of policy action.

13.2 Efficiency Policies in the Management of Water Resources: The Local Scale

The increase in the efficiency of urban water supply management has been identified as one of the principal factors explaining the decrease in water consumption in the majority of the urban areas of the developed world since the beginning of the twenty-first century. This trend grew stronger after the economic crisis of 2008 (Baldino & Saurí, 2018). In the case of the Mediterranean region, the development of an efficient water use management has been particularly relevant as a measure to respond to the scarcity of the resource, especially in semi-arid regions where drought has historically been a cause of insecurity in water supply (Anton Clavé et al., 2011; Martínez-Ibarra, 2015; Rico-Amorós et al., 2013; Rico et al., 2020). In many cases, the situations of drought have constituted a catalyst for the implementation of measures to increase water supply, the interconnection of basins and the establishment of resource exchange agreements between different users (Martínez-Ibarra, 2015). However, they have also acted as a driver of structural changes in the adoption of water saving measures (Dinarès & Saurí, 2015), and of the development of policies to increase efficiency in water use (Freire-González & Puig-Ventosa, 2014).

Water policies on a local scale have been implemented mostly by the local governments and the water cycle management companies. These policies have been aimed at reducing the leaks in the water distribution network through the improvement of hydraulic efficiency or hydraulic technical performance (HTP). The reduction in the losses of water in the distribution network has contributed, therefore, to the decrease in the volume of non-revenue water (NRW). According to the Spanish Association of Water Supply and Wastewater (AEAS), the percentage of NRW in urban networks decreased from 29% to 22% between 1998 and 2018 (AEAS & AGA, 2019). This evolution is the result of the improvement in the methods to detect and repair leaks, the sectorisation of the supply network, the renovation of pipes, the introduction of geographic information systems, the incorporation of remote control and monitoring and the development of remote reading (March et al., 2017), and fraud control (Morote & Hernández, 2018). However, the reduction in the volume of NRW depends on the level of investment made in the supply system on a local level and has an asymptotic progression. In other words, there is an economic-technical limit at which the investment necessary to reduce NRW does not justify the water savings generated (Álvarez et al., 2014). The existing literature finds that the level of NRW depends on several characteristics of the supply system, such as its physical conditions or management. First, the ageing of the water distribution network is the principal cause of leaks in the network. This is due both to the deterioration derived from its use and the materials used, as the pipes were formerly made from iron and lead, more vulnerable to corrosion, while today more resistant materials such as ductile case, steel and polyethylene are used (Martínez-Espiñeira et al., 2017).

Second, the urban model will also have a major influence on the level of efficiency of the management of the water supply service, as the management of the services in densely populated areas with higher levels of water efficiency will be less complex and require a lower investment (Picazo-Tadeo et al., 2009). Different studies have noted this circumstance, indicating the greater water efficiency of concentrated urban models with respect to disperse systems, both from the point of view of water supply management and the use of water made by the different types of tourist accommodation which predominate in each model (Rico, 2007; Rico-Amorós et al., 2009; Hof & Schmitt, 2011; Morote et al., 2017; Navarro et al., 2020). Cities and high density tourism models favour economies of scale in the use and management of water, giving rise to a greater efficiency in the management of the service, with a lower volume of NRW and a lower water consumption by tourist establishments mainly consisting of hotels and tourist apartments (Stoker et al., 2019). In these tourist destinations which are well inserted in the global tourism markets, the season is becoming longer, reducing seasonality and enabling an optimisation of water infrastructures, although in parallel, giving rise to a greater consumption of water in absolute terms. On the other hand, in municipalities with a clear tourism-housing function, the disperse or low density urban model based on single-family homes with outdoor water uses (gardens and swimming pools) generate a higher level of water consumption and a higher volume of leaks and losses of water in the network, due to its longer length and number of connections (Rico, 2007; Morote et al., 2017).

Third, the availability of water resources or the financial capacity of the local governments are also principal factors that explain the development of measures to improve the efficiency in water management on a local and regional scale. Some studies find that the greater the scarcity of the resource, the greater the concern for the maintenance of the network, which usually coincides with more complex water supply systems and with higher extraction and treatment costs (González-Gómez et al., 2012). Furthermore, the high maintenance and repair cost of the network means that, in response to situations of budget difficulties and municipal debt, the concession of the service for its management by an external company is related to a more efficient management (González-Gómez et al., 2012). However, there is no clear scientific evidence that indicates the superiority of either the public or private ownership of the water concessionaire companies in terms of efficiency in the management of the urban supply service (García-Rubio et al., 2010; Picazo-Tadeo et al., 2009; Martínez-Espiñeira et al., 2017). Finally, other authors indicate that there is a direct relationship between the price of water and the efficiency of the network, as in those municipalities where the actual costs of the service are reflected in the invoice, the performance of the network is greater (Cabrera et al., 1999). This is based on the economic analysis of the actions by the system manager, because if the price of water does not cover the renovation or repair costs of the network, it is likely that these measures are not taken as they would not be profitable. To do this, the need for greater investment and the difficulty in managing the supply service in disperse urban models leads to the managers of these systems to increase the fixed part of the water rate to compensate the losses produced by the high level of NRW (Kanakoudis et al., 2015).

The application of these measures is reflected in a significant improvement of the HTP. For example, the town of Benidorm, the leading tourist destination of the western Mediterranean in terms of hotel beds, is located between the cities with the highest water efficiency in Spain, at around 95%. This high level of efficiency has been possible thanks to the improvements in the control and search for leaks in the network, favoured by a high density urban model (Olcina Cantos et al., 2016). In many tourist destinations, the gradual reduction in the losses in the supply network enabled a continuous increase in the tourism activity until the emergence of Covid-19, without the need to increase the water supply. This counterintuitive result is usually referred to in the study of resource management as the "Jevons paradox", which establishes that the increase in the efficiency of production or management of a resource can generate an increase in its consumption, which would question the overall view that links technical progress with environmental sustainability (Polimeni et al., 2008).

Within the scientific literature on tourism there are a significant number of studies emerging on how technological improvements can generate different rebound effects on financial plans or on the consumption of materials, resources and energy (Hall et al., 2015) and the limitations of technology as a factor of reducing consumption (Hall, 2010). This is evident, for example, in the consumption of energy in tourist transport (Gascón, 2019) and, in relation to this, in the increase of greenhouse gas emissions (Bows et al., 2009; Scott et al., 2010). With respect to water, the relationship between the increase in efficiency in watering and the increase in water consumption has been identified, as the reduction in the marginal cost of water has led to the increase in the irrigated area or the introduction of new crops which require more water (Sears et al., 2018). Furthermore, the water savings gained by the increase in the efficiency of irrigation have been used in some cases to enable an increase in urban uses, particularly in agro-tourism contexts, such as the majority of the Mediterranean coast (Sanchis Ibor et al., 2016). It has also been shown that the decrease in the production costs of desalinated water, and, therefore, its retail price, could lead to an increase in its consumption (Meerganz & Moreau, 2007). With respect to tourism, some studies have evaluated how the increase in hydraulic performance or the efficiency of water management has occurred in parallel with an increase in water consumption on a municipal level with the increase in the availability of water and the possibility of growth in the tourism activity (Villar-Navascués et al., 2020). This is the case of different municipalities of the Spanish Mediterranean coast located in the Marina Alta district (Alicante), the Balearic Islands or the Costa del Sol (Málaga), characterised by the development of a distinctive tourism-residential territorial model, with the proliferation of single-family holiday homes with a garden and swimming pool (Villar-Navascués, 2019). In these cases the increase in the efficiency of the management of the service has enabled an increase in the availability of water and growth in the tourism activity. Therefore, despite the direct environmental benefits generated by these technical improvements in the management of the water resources, the increase in the efficiency of the supply does not necessarily lead to a net reduction in water consumption, but can have a rebound effect, producing the opposite impact.

13.3 Efficiency Policies in the Management of Water Resources: Tourist Accommodation

In parallel with the measures that have led to a more efficient management of water resources on an urban level, a group of policies and actions have been developed for the efficient use and consumption saving of water in tourist accommodation, particularly in hotels, which are usually indicated as the large water consumers in the tourism sector (Gabarda-Mallorquí et al., 2017; Rico et al., 2020; Mendoza et al., 2022). In Spain, from the mid-1990s, the installation of water saving devices in rooms and communal areas of hotels was common during hotel renovation processes, particularly in taps, shower heads and toilets with dosage devices and flow timers and policies for informing clients about water saving practices (Olcina Cantos et al., 2016). Similarly, some hotels transfer these environmental awareness policies to the staff of the hotel, establishing savings objectives and internal training for an efficient use of water (Olcina Cantos & Vera-Rebollo, 2016). These improvements have also been implemented in the general services of the hotels (kitchens, restaurants, laundry) with the installation of efficient tap services and water and energy-saving electrical appliances (dishwashers, washing machines). To a lesser extent, combined systems for saving water and energy in heating water have been installed (Yoon et al., 2022). In outdoor uses, autochthonous xerophyte species can be used for ornamental plants, which are resistant to drought and are irrigated through localised irrigation systems (Rico et al., 2020). On the other hand, swimming pools, commonly found in all sun and beach hotels, on the whole have a closed circuit filtering system and are covered in the low season, preventing the evaporation of the water. Among other actions, with the aim of monitoring water consumption for different areas or services offered by the hotel, some have installed several meters. Furthermore, in tourist apartments, other measures aimed at water saving include the installation of individual meters in each property of old buildings that formerly had a single communal meter, which has led to a more responsible consumption of the resource (Olcina Cantos et al., 2016).

Most authors indicate that all of these efficient use and water-saving measures have contributed to the reduction in the consumption of water in hotels (Gabarda-Mallorquí & Ribas Palom, 2016), and the reduction of operating costs, including energy costs associated with sanitary hot water (Rico et al., 2020). However, the impact of these measures to reduce water consumption differs depending on the hotel category. Tourists staying in a higher category hotel (four and five stars) are those who consume the most water due to the greater supply of services that use water, such as spas, jacuzzis and even small water parks and golf courses (Deyà Tortella & Tirado, 2011; Gössling et al., 2012; Dinarès & Saurí, 2015; Gabarda-Mallorquí et al., 2017; Rico-Amorós et al., 2009; Rico et al., 2020). Therefore, the commitment to an increase in efficiency in water use of hotels does not necessarily lead to a net reduction in water consumption, as, in many cases, the measures to reduce water and energy consumption are implemented within the framework of a strategy to improve the quality of the service, customer loyalty and the brand image, both on an individual level and as part of policies of the large hotel chains, in order to obtain environmental certifications (Deyà Tortella & Tirado, 2011; Olcina Cantos & Vera-Rebollo, 2016). On the other hand, water-saving measures are more frequently implemented by higher category hotels, due to their greater financial capacity to make investments and benefit from economies of scale, the existence of a wider variety of services and facilities that make an intensive use of water and their greater environmental impacts, which generate a higher level of social and political pressure to adopt pro-environmental measures (Dinarès & Saurí, 2015). However, it is difficult to generalise in this respect, as, sometimes in higher category hotels affiliated to large international chains, the compliance with quality standards (such as cleaning or maintaining services such as spas or private jacuzzis in the rooms), contradicts the implementation of water-saving measures (Deyà Tortella & Tirado, 2011).

However, the promotion of other measures, such as the installation of grey water re-use systems, which, when used for toilet discharge can reduce the water consumption of each tourist by 25% (March et al., 2004) or rainwater harvesting, are usually ruled out due to their technical complexity. Some hotel directors indicate that these measures should be contemplated during the hotel construction process (Yoon et al., 2022). They also point to economic reasons for ruling out their incorporation. They argue that, in some cases, water represents a small fraction of the operating costs in hotels, which discourages the adoption of savings measures (Deyà Tortella & Tirado, 2011; Olcina Cantos et al., 2016). However, some studies have estimated that the period of amortisation of the grey water re-use systems in hotels fluctuated between three and seven years, depending on the size of the hotel and the water treatment capacity (Atanasova et al., 2017). And in cases of success, we can observe that the benefits are both direct, related to a significant reduction in water consumption and the associated costs, and indirect, through the improvement of customer satisfaction and loyalty (Ruiz-Rosa et al., 2017). In any event, the managers usually opt for other less complex measures of efficient use and water saving from a technical and economic point of view, with a low initial investment, as there is a lack of incentives for promoting other types of actions (Tirado et al., 2019; Torres-Bagur et al., 2019). However, these limitations may be irrelevant if, for example, they generate restrictions in the use of water (Tirado et al., 2006), or the price of water exceeds certain thresholds. Although, in general, the price of water is not a determining factor to reduce water consumption in hotels (Deyà-Tortella et al., 2016, 2019), in some cases where rigorous tariff reforms have been implemented, there has been a greater propensity to introduce water saving measures in tourist establishments (Razumova et al., 2016).

Finally, the promotion of pro-environmental behaviours related to the use and consumption of water among tourists is another line of action in tourist accommodation aimed at a more efficient use of resources. However, the establishments and tourists are usually reticent to apply economic measures, such as supplementary payments aimed at saving water (Casado-Díaz et al., 2020a). Studies that have addressed the behaviour of tourists with respect to water consumption have found that there is no causal relationship between environmental awareness raising and pro-environmental behaviour (Gabarda-Mallorquí et al., 2018). In general, tourists who stay in hotels have a high level of awareness regarding environmental issues, such as water scarcity or climate change (Torres-Bagur & Pavón Gamero, 2021). However, their water use habits in the hotel largely respond to hedonistic motivations related to pleasure and enjoyment (Rodriguez-Sanchez et al., 2020). Nevertheless, there are factors that distinguish between different types of tourist in relation to their environmental behaviour. For example, the duration of the stay (Gabarda-Mallorquí et al., 2021), age and the female sex (Gabarda-Mallorquí et al., 2018), the attachment to the place of the destination (Casado-Díaz et al., 2020b) or the type of tourist accommodation in which the tourists stay, either a campsite or a rural house (Torres-Bagur et al., 2020) have a positive influence on greater water savings. On the contrary, tourists from non-Mediterranean countries (Torres-Bagur & Pavón Gamero, 2021) or on trips for business, events or urban/cultural reasons (Torres-Bagur et al., 2020) have been related to less responsible water consumption practices.

13.4 Conclusions

The increase in water consumption for tourism uses has led to the development of initiatives and studies aimed at improving the efficiency of supply on a city scale, adapting the tariff structure to the costs of operating the service, incorporating non-conventional resources or modifying the consumption habits of tourists through awareness raising policies (Baños et al., 2019). However, in order to guarantee the environmental sustainability of tourist destinations, the adoption of technological improvements in the management of the service on an urban level is insufficient. In many cases, this increase in efficiency and water saving generated is taken advantage of in order to increase residential or tourism supply. The development of demand management measures, an improved planned management of drought situations and the promotion of tourism models that consume less resources through urban and territorial planning are some of the alternatives with respect to water consumption that should accompany the measures to increase water efficiency and reduce NRW. Similarly, issues related to hydrological planning, the ordering of the territory and natural limits to tourism development should be considered in response to a scenario of climate change and a reduction in the availability of water.

In this way, tourist destinations should accelerate their adaptation to climate change and the more intense and frequent droughts based on the elaboration of adaptation plans which, in terms of water, could imply the construction of sustainable urban drainage systems, rainwater storage tanks and the application of treated wastewater for non-potable uses (CEDEX, 2017). Furthermore, these plans should consider different water demand scenarios for urban-tourism uses, as the increase in the summer temperature and the emergence of new tourist demands for this type of holiday could generate a deseasonalisation of the sun and beach tourism in the Mediterranean and an increase in tourism activity in the spring and autumn (Freire-González & Puig-Ventosa, 2014). In short, the climate change scenario in the Spanish Mediterranean tourism towns should lead to the definitive commitment to a rational territorial and urban planning and the abandonment of urban growth practices that generate serious environmental, territorial and environmental impacts in Spain (Burriel, 2014). However, some recent regulatory change processes taking place in certain Autonomous Regions based on a relaxation of the environmental requisites in territorial and urban planning, justified by constituting an accelerated way out of the economic crisis caused by the Covid-19 pandemic (Sáez et al., 2020), go against the logic that the actions to adapt to climate change in tourist areas should follow.

Furthermore, the increase in the price of energy and the investments necessary to maintain the level of efficiency of the network and comply with the principle of cost recovery required by the European Water Framework Directive are some factors that indicate a foreseeable increase in the price of water, particularly in the coastal towns, where the reduction in water consumption due to the halt in the tourism activity and the mobility restrictions related to Covid-19 has been particularly relevant.

This increase in the price of water could be the catalyst for a renewed promotion of water savings measures and the efficient use of water on all levels, including the tourism sector. However, the economic impact of the pandemic on this sector means that we can expect that the priority measures adopted will be those that require a low initial investment and a short amortisation period. Finally, future research should evaluate how the emergence of the Covid-19 pandemic has modified the behaviour of tourists and the choice of tourist destinations, as an increase in the demand for holiday homes and rural tourism at some distance from the traditional coastal tourism destination characterised by a greater massification has been observed (Agudo et al., 2021).

Acknowledgements This study is supported by the SIMTWIST project (ERA-NETWater JPI 2018) funded by the Spanish Ministry of Science and Innovation (PCI2019-103395) and the project "Overtourism in Spanish Coastal Destinations. Tourism Degrowth Strategies" (RTI2018-094844-B-C31).

References

AEAS & AGA (Asociación Española de Abastecimientos de agua y Saneamiento y Asociación española de empresas Gestoras de los servicios de Agua urbana). (2019). *Hacia una financiación más eficiente de las infraestructuras del ciclo urbano del agua en España*. PricewaterhouseCoopers.

Agudo, L., García, R., & Moreno, D. (2021). Análisis de los cambios en el comportamiento de los consumidores turísticos en la Región de Murcia (España) tras la aparición de la COVID-19. *Revista de Ocio y Turismo (ROTUR), 16*(1), 1–13. https://doi.org/10.17979/rotur.2022.16.1.8746

Álvarez, M., Castellví, E., Monzó, M., & Verdú, C. (2014). *La eficiencia en los sistemas de distribución: revisión sobre la gestión del agua no registrada*. Fundación Aquae. Retrieved from https://www.fundacionaquae.org/aquae-papers-4-eficiencia-sistemas-distribucion/

Anton Clavé, S., Rullán Salamanca, O., & Vera, J. F. (2011). Mass tourism development on the Mediterranean coast. *Tourism Geographies, 13*(3), 495–501. https://doi.org/10.1080/14616688.2011.575171

Atanasova, N., Dalmau, M., Comas, J., Poch, M., Rpdríguez-Roda, I., & Buttiglieri, G. (2017). Optimized MBR for greywater reuse systems in hotel facilities. *Journal of Environmental Management, 193*, 503–511. https://doi.org/10.1016/j.jenvman.2017.02.041

Baldino, N., & Saurí, D. (2018). Characterizing the recent decline of water consumption in Italian cities. *Investigaciones Geográficas, 69*, 9–21. https://doi.org/10.14198/INGEO2018.69.01

Baños, C., Hernández, M., Rico, A. M., & Olcina, J. (2019). The Hydrosocial cycle in coastal tourist destinations in Alicante, Spain: Increasing resilience to drought. *Sustainability, 11*, 4494. https://doi.org/10.3390/su11164494

Bows, A., Anderson, K., & Peeters, P. (2009). Air transport, climate change and tourism. *Tourism and Hospitality Planning & Development, 6*(1), 7–20. https://doi.org/10.1080/14790530902847012

Burriel, E. L. (2014). El estallido de la burbuja inmobiliaria y sus efectos en el territorio. In J. M. Albertos & J. L. Sánchez (Coords.), *Geografía de la crisis económica en España* (pp. 101–140). Publicaciones de la Universidad de Valencia.

Cabrera, E., Almandoz, J., Arregui, F., & García-Serra, J. (1999). Auditoría de redes de distribución de agua. *Ingeniería del Agua, 6*(4), 291–303.

Casado-Díaz, A. B., Sellers-Rubio, R., Rodríguez-Sánchez, C., & Sancho-Esper, F. (2020a). Predictors of willingness to pay a price premium for hotels' water-saving initiatives. *Journal of Travel & Tourism Marketing, 37*(7), 773–784. https://doi.org/10.1080/10548408.2020.1812469

Casado-Díaz, A. B., Sancho-Esper, F., Rodríguez-Sánchez, C., & Sellers-Rubio, R. (2020b). Tourists' water conservation behavior in hotels: The role of gender. *Journal of Sustainable Tourism., 30*, 1518–1538. https://doi.org/10.1080/09669582.2020.1839758

CEDEX. (2017). *Evaluación del impacto del cambio climático en los recursos hídricos y sequías en España*. Centro de Estudios Hidrográficos.

Chapagain, A. K., & Hoekstra, A. Y. (2008). The global component of freshwater demand and supply: An assessment of virtual water flows between nations as a result of trade in agricultural and industrial products. *Water International, 33*(1), 19–32. https://doi.org/10.1080/02508060801927812

De Kadt, I. (1979). *Tourism: Passport to development? Perspectives on the social and cultural effects of tourism on developing countries*. Oxford University Press.

Deyà Tortella, B., & Tirado, D. (2011). Hotel water consumption at a seasonal mass tourist destination. The case of the Island of Mallorca. *Journal of Environmental Management, 92*(10), 2568–2579. https://doi.org/10.1016/j.jenvman.2011.05.024

Deyà-Tortella, B., Garcia, C., Nilsson, W., & Tirado, D. (2016). The effect of the water tariff structures on the water consumption in Mallorcan hotels. *Water Resources Research, 52*, 6386–6403. https://doi.org/10.1002/2016WR018621

Deyà-Tortella, B., Garcia, C., Nilsson, W., & Tirado, D. (2019). Hotel water demand: The impact of changing from linear to increasing block rates. *Water, 11*, 1604. https://doi.org/10.3390/w11081604

Dinarès, M., & Saurí, D. (2015). Water consumption patterns of hotels and their response to droughts and public concerns regarding water conservation: The case of the Barcelona hotel industry during the 2007–2008 episode. *Documents d'Anàlisi Geogràfica, 61*(3), 623–649. https://doi.org/10.5565/rev/dag.255

Fosse, J., Kosmas, I., & Gonzalez, A. (2021). *The future of Mediterranean tourism in a (post) covid world* (Eco-med briefing 01/21). Eco-union. https://doi.org/10.5281/zenodo.4616983

Freire-González, J., & Puig-Ventosa, I. (2014). Retos y oportunidades económicas de la adaptación al cambio climático. El caso de Cataluña. *Revista de estudios regionales, 99*, 47–74.

Gabarda-Mallorquí, A., & Ribas Palom, A. (2016). Understanding reductions in water consumption in tourist areas: A case study of the Costa Brava, Spain. *International Journal of Water Resources Development, 32*(6), 912–930. https://doi.org/10.1080/07900627.2016.1142861

Gabarda-Mallorquí, A., Garcia, X., & Ribas, A. (2017). Mass tourism and water efficiency in the hotel industry: A case study. *International Journal of Hospitality Management, 61*, 82–93. https://doi.org/10.1016/j.ijhm.2016.11.006

Gabarda-Mallorquí, A., Fraguell, R. M., & Ribas, A. (2018). Exploring environmental awareness and behavior among guests at hotels that apply water-saving measures. *Sustainability, 10*, 1305. https://doi.org/10.3390/su10051305

Gabarda-Mallorquí, A., Garcia, X., Fraguell, R. M., & Ribas, A. (2021). Are hotel stay characteristics influencing guests' environmental behaviour? Predicting water conservation habits. *Current Issues in Tourism, 24*(16), 2342–2356. https://doi.org/10.1080/13683500.2020.1829565

García-Rubio, M. A., González-Gómez, F., & Guardiola, J. (2010). Performance and ownership in the governance of urban water. *Proceedings of the Institution of Civil Engineers, 163*(1), 51–58. https://doi.org/10.1680/muen.2010.163.1.51

Gascón, J. (2019). Tourism as a right: A "frivolous claim" against degrowth? *Journal of Sustainable Tourism, 27*(12), 1825–1838. https://doi.org/10.1080/09669582.2019.1666858

González-Gómez, F., Martínez-Espiñeira, R., García-Valiñas, M. A., & García Rubio, M. A. (2012). Explanatory factors of urban water leakage rates in southern Spain. *Utilities Policy, 22*, 22–30. https://doi.org/10.1016/j.jup.2012.02.002

Gössling, S., Peeters, P., Hall, C. M., Ceron, J.-P., Dubois, G., Lehmann, L. V., & Scott, D. (2012). Tourism and water use: Supply, demand, and security. *Tourism Management, 33*(1), 1–15. https://doi.org/10.1016/j.tourman.2011.03.015

Hall, C. M. (2010). Changing paradigms and global change: From sustainable to steady-state tourism. *Tourism Recreation Research, 35*(2), 131–143. https://doi.org/10.1080/02508281.2010.11081629

Hall, C. M., Gossling, S., & Scott, D. (2015). Tourism and sustainability: Towards a green (er) tourism economy? In C. M. Hall, S. Gossling, & D. Scott (Eds.), *The Routledge handbook of tourism and sustainability* (pp. 490–519). Routledge.

Hjalager, A. M. (1996). Tourism and the environment: The innovation connection. *Journal of Sustainable Tourism, 4*(4), 201–218. https://doi.org/10.1080/09669589608667268

Hof, A., & Schmitt, T. (2011). Urban and tourist land use patterns and water consumption: Evidence from Mallorca, Balearic islands. *Land Use Policy, 28*(4), 792–804. https://doi.org/10.1016/j.landusepol.2011.01.007

Intergovernmental Panel on Climate Change (IPCC). (2021). *Climate change 2021. The Physical science basis.* Retrieved from https://www.ipcc.ch/report/ar6/wg1/downloads/report/IPCC_AR6_WGI_Full_Report.pdf

Intergovernmental Panel on Climate Change (IPCC). (2022). *Climate Change 2022: Impacts, adaptation and vulnerability*. Retrieved from https://www.ipcc.ch/report/ar6/wg2/

Kanakoudis, V., Tsitsifli, S., Samaras, P., & Zouboulis, A. (2015). Water pipe networks performance assessment: Benchmarking eight cases across the EU Mediterranean basin. *Water Quality Exposure and Health, 7*(1), 99–108. https://doi.org/10.1007/s12403-014-0113-y

March, J. G., Gual, M., & Orozco, F. (2004). Experiences on greywater re-use for toilet flushing in a hotel (Mallorca Island, Spain). *Desalination, 164*(3), 241–247. https://doi.org/10.1016/S0011-9164(04)00192-4

March, H., Morote, A. F., Rico, A. M., & Saurí, S. (2017). Household smart water metering in Spain: Insights from the experience of remote meter Reading in Alicante. *Sustainability, 9*(4), 582. https://doi.org/10.3390/su9040582

Martínez-Espiñeira, R., García-Rubio, M. A., & González-Gómez, F. (2017). Which factors, and to what extent, influence the condition of urban water distribution networks. An empirical analysis of the Spanish case. *Water Resources and Economics, 18*, 20–33. https://doi.org/10.1016/j.wre.2017.02.002

Martínez-Ibarra, E. (2015). Climate, water and tourism: Causes and effects of droughts associated with urban development and tourism in Benidorm (Spain). *International Journal of Biometeorology, 59*, 487–501. https://doi.org/10.1007/s00484-014-0851-3

Meerganz, G., & Moreau, V. (2007). Modelling of water energy systems. The case of desalination. *Energy, 32*(6), 1024–1031. https://doi.org/10.1016/j.energy.2006.10.006

Mendoza, E., Ferrero, G., Slokar, Y. M., Amores, X., Azzellino, A., & Buttiglieri, G. (2022). Water management practices in euro-mediterranean hotels and resorts. *International Journal of Water Resources Development*. https://doi.org/10.1080/07900627.2021.2015683

Milano, C., Novelli, M., & Cheer, J. M. (2019). Overtourism and degrowth: A social movements perspective. *Journal of Sustainable Tourism, 27*(12), 1857–1875. https://doi.org/10.1080/09669582.2019.1650054

Morote, A. F., & Hernández, M. (2018). Unauthorised domestic water consumption in the city of Alicante (Spain): A consideration of its causes and urban distribution (2005–2017). *Water, 10*(7), 851. https://doi.org/10.3390/w10070851

Morote, Á., Saurí, D., & Hernández, M. (2017). Residential tourism, swimming pools, and water demand in the western Mediterranean. *The Professional Geographer, 69*(1), 1–11. https://doi.org/10.1080/00330124.2015.1135403

Navarro, S., Estruch-Guitart, V., & García, C. (2020). Uso de indicadores causa-efecto para el diagnóstico de la sostenibilidad hídrica en las Islas Baleares (España). *Boletín de la Asociación de Geógrafos Españoles, 85*(2833), 1–48. https://doi.org/10.21138/bage.2833

Niezgoda, A. (2011). The role of environmental knowledge, attitudes and initiatives in the development of a tourism product. *Turyzm, 21*(1–2), 33–39. https://doi.org/10.2478/v10106-011-0004-6

Olcina Cantos, J., & Vera-Rebollo, J. F. (2016). Adaptación del sector turístico al cambio climático en España. La importantita de las acciones a escala local y en empresas turísticas. *Anales de Geografía de la Universidad Complutense, 36*(2), 321–352. https://doi.org/10.5209/AGUC.53588

Olcina Cantos, J., Baños Castiñeira, C. J., & Rico-Amorós, A. M. (2016). Medidas de adaptación al riesgo de sequía en el sector hotelero de Benidorm (Alicante, España). *Revista de Geografía Norte Grande, 65*, 129–153. https://doi.org/10.4067/S0718-34022016000300007

Olcina, J., & Miró, J. (2017). *Actividad turística y cambio climático en la Comunidad Valenciana.* Universidad de Alicante & Agència Valenciana del Turisme. https://doi.org/10.14198/2017-Actividad-Turistica-ComValenciana

Picazo-Tadeo, A. J., Sáez-Fernández, F. J., & González-Gómez, F. (2009). The role of environmental factors in water utilities' technical efficiency. Empirical evidence from Spanish companies. *Applied Economics, 41*(5), 615–628. https://doi.org/10.1080/00036840601007310

Polimeni, J., Mayumi, K., Giampietro, M., & Alcott, B. (2008). *The Jevons paradox and the myth of resource efficiency improvements.* Earthscan.

Razumova, M., Rey-Maquieira, J., & Lozano, J. (2016). The role of water tariffs as a determinant of water saving innovations in the hotel sector. *International Journal of Hospitality Management, 52*, 78–86. https://doi.org/10.1016/j.ijhm.2015.09.011

Ricart, S., Arahuetes, A., Villar, R., Rico-Amorós, A. M., & Berenguer, J. (2020). More water exchange, less water scarcity? Driving factors from conventional and reclaimed water swap between agricultural and urban–tourism activities in Alicante, Spain. *Urban Water Journal, 16*(10), 677–686. https://doi.org/10.1080/1573062X.2020.1726408

Rico, A. M. (2007). Tipologías de consumo de agua en abastecimientos urbano-turísticos de la Comunidad Valenciana. *Investigaciones Geográficas, 42*, 5–34. https://doi.org/10.14198/INGEO2007.42.01

Rico, A. M., Olcina, J., Baños, C., Garcia, X., & Saurí, D. (2020). Declining water consumption in the hotel industry of mass tourism resorts: Contrasting evidence for Benidorm, Spain. *Current Issues in Tourism, 23*(6), 770–783. https://doi.org/10.1080/13683500.2019.1589431

Rico-Amorós, A. M., Olcina-Cantos, J., & Saurí, D. (2009). Tourist land use patterns and water demand: Evidence from the Western Mediterranean. *Land Use Policy, 26*(2), 493–501. https://doi.org/10.1016/j.landusepol.2008.07.002

Rico-Amorós, A. M., Saurí, D., Olcina-Cantos, J., & Vera-Rebollo, J. F. (2013). Beyond megaprojects? Water alternatives for mass tourism in coastal Mediterranean Spain. *Water Resources Management, 27*, 553–565. https://doi.org/10.1007/s11269-012-0201-3

Rodriguez-Sanchez, C., Sancho-Esper, F., Casado-Diaz, A. B., & Sellers-Rubio, R. (2020). Understanding in-room water conservation behavior: The role of personal normative motives and hedonic motives in a mass tourism destination. *Journal of Destination Marketing & Management, 18*, 100496. https://doi.org/10.1016/j.jdmm.2020.100496

Romero, J., & Olcina, J. (2021). *Cambio climático en el Mediterráneo: Procesos, riesgos y políticas.* Tirant Humanidades.

Ruiz-Rosa, I., García-Rodríguez, F. J., & Santamarta-Cerezal, J. C. (2017). Redirecting hotel management towards greater efficiency in water consumption: A case study. *International Journal of Sustainable Development, 20*(3/4), 230. https://doi.org/10.1504/IJSD.2017.089995

Sáez, E., Vadillo, V., & Sánchez, E. (2020, May 16). El ladrillo se impone al medio ambiente en plena pandemia para combatir la crisis económica. *El País.* https://elpais.com/sociedad/2020-05-16/el-urbanismo-se-impone-al-medio-ambiente-en-plena-pandemia-para-combatir-la-crisis-economica.html

Sanchis Ibor, C., García Mollá, M., & Avellà Reus, L. (2016). Las políticas de implantación del riego localizado. Efectos en las entidades de riego de la Comunidad Valenciana. *Boletín de la Asociación de Geógrafos Españoles, 72*, 9–35. https://doi.org/10.21138/bage.2330

Scott, D., Peeters, P., & Gössling, S. (2010). Can tourism deliver its "aspirational" greenhouse gas emission reduction targets? *Journal of Sustainable Tourism, 18*(3), 393–408. https://doi.org/10.1080/09669581003653542

Sears, L., Caparelli, J., Lee, C., Pan, D., Strandberg, G., Vuu, L., & Lin Lawell, C. Y. C. (2018). Jevons' paradox and efficient irrigation technology. *Sustainability, 10*(5), 1590. https://doi.org/10.3390/su10051590

Stoker, P., Chang, H., Wentz, E., Crow-Miller, B., Jehle, G., & Bonnette, M. (2019). Building water-efficient cities. A comparative analysis of how the built environment influences water use in four Western U.S. Cities. *Journal of the American Planning Association, 85*(4), 511–524. https://doi.org/10.1080/01944363.2019.1638817

Tirado, D., Gómez, C. M., & Lozano, J. (2006). Efficiency improvements and water policy in the Balearic Islands: A general equilibrium approach. *Investigaciones Económicas, 30*(3), 441–463.

Tirado, D., Nilsson, W., Deyà-Tortella, B., & García, C. (2019). Implementation of water-saving measures in hotels in Mallorca. *Sustainability, 11*(23), 6880. https://doi.org/10.3390/su11236880

Torres-Bagur, M., & Pavón Gamero, D. (2021). Percepción del cambio climático y respuesta de los turistas ante los recursos hídricos: el caso de la cuenca del río Muga (Cataluña). *Boletín de la Asociación de Geógrafos Españoles, 89*. https://doi.org/10.21138/bage.3114

Torres-Bagur, M., Ribas, A., & Vila-Subirós, J. (2019). Incentives and barriers to water-saving measures in Hotels in the Mediterranean: A case study of the Muga River basin (Girona, Spain). *Sustainability, 11*, 3583. https://doi.org/10.3390/su11133583

Torres-Bagur, M., Ribas, A., & Vila-Subirós, J. (2020). Understanding the key factors that influence efficient water-saving practices among tourists: A Mediterranean case study. *Water, 12*(8), 2083. https://doi.org/10.3390/w12082083

UNWTO. (2003). *Djerba declaration on tourism and climate change*. Retrieved from http://sdt.unwto.org/sites/all/files/docpdf/decladjerbae.pdf

Vila, M., Afsordegan, A., Agell, N., Sanchez, M., & Costa, G. (2018). Influential factors in water planning for sustainable tourism destinations. *Journal of Sustainable Tourism, 26*(7), 1241–1256. https://doi.org/10.1080/09669582.2018.1433183

Villar-Navascués, R. A. (2019). Evolución reciente del consumo urbano de agua en el litoral mediterráneo (2007–2016). In J. Melgarejo (Ed.), *Congreso Nacional del Agua 2019: innovación y sostenibilidad* (pp. 1593–1605). University of Alicante. https://doi.org/10.14198/Congreso-Nacional-del-Agua-Orihuela-2019

Villar-Navascués, R. A., Baños, C., Hernández, M., & Olcina, J. (2020). ¿Es la eficiencia en la gestión del agua una coartada para el crecimiento del turismo? In G. X. Pons, A. Blanco-Romero, R. Navalón-García, L. Troitiño-Torralba, & M. Blázquez-Salom (Eds.), *Sostenibilidad Turística: overtourism vs undertourism*. Mon. Soc. Hist. Nat. Balears, 31, 109–120. Retrieved from http://hdl.handle.net/10045/109701

Warren, C., & Becken, S. (2017). Saving energy and water in tourist accommodation: A systematic literature review (1987–2015). *International Journal of Tourism Research, 19*(3), 289–303. https://doi.org/10.1002/jtr.2112

Yoon, H., Saurí, D., & Rico, A. (2022). The water-energy nexus in hotels and recreational activities of a mass tourism resort: The case of Benidorm. *Current Issues in Tourism, 25*(4), 592–610. https://doi.org/10.1080/13683500.2021.1893283

Rubén Villar-Navascués, PhD in Geography (University of Alicante, 2019) is Assistant Professor at the Complutense University of Madrid, Spain. His research has focused on issues related to urban, tourist and agricultural water uses and management, users' perception of unconventional water resources, hydrological and drought planning, and water governance. He has participated in five research projects at European, national, and regional level. Since 2019 he collaborates as a lecturer at the master's in planning and Management of Natural Risks. Likewise, he has published around 15 articles in JCR and Scopus journals.

Carlos J. Baños Castiñeira, PhD in Geography (University of Alicante, 2009) is Associate Professor at the Department of Regional Geographic Analysis and Physical Geography (University of Alicante, Spain). His research has focused on issues related to the study of the tourist-recreational system, the sustainable management of resources and the territorial allocations linked to the development of tourism. He is the author or co-author of several books and papers published in journals indexed in databases such as Web of Science or Scopus.

Jorge Olcina Cantos is Full Professor of Regional Geographic Analysis at University of Alicante, where teaches on Spatial Planning, Climatology and Natural Hazards. His research has been focused on various geographical issues (climate change, water resources, spatial planning, natural risk). Author or co-author of more than a hundred publications (scientific papers, book chapters and book editor). He has participated in several research projects on geographical and historical issues developed. Main-speaker at the International Year of Planet Earth (2008), declared by UNESCO. Visiting professor at universities in Spain, Europe and Latin America. Member of Editorial Board of various scientific journals on geographical and environmental issues. Main Investigator of the Competitive Research Group of the University of Alicante in "Climate and Spatial Planning", since 2002. President of Spanish Association of Geography (2017–21).

Maria Hernández Hernández, PhD in Geography (University of Alicante, 1996) is Full Professor at the Department of Regional Geographic Analysis and Physical Geography (University of Alicante, Spain). Her research has been focussed on issues related to the cultural value and the new functionality of cultural landscapes, agriculture and rural development, rural development and tourism policies, landscape planning, urban and agricultural water uses and management and natural hazards (flood and hydrological and drought planning). She has participated in 20 projects at European, national, and regional level. She is the author or co-author of several books and around 90 papers published in journals indexed in databases such as Web of Science or Scopus.

Chapter 14
The Territorial Quality Mark and Landscape as a Strategy for Inland Tourism

Paloma Ibarra Benlloch, Isabel Rabanaque-Hernández, Elena De Uña-Álvarez, and Montserrat Villarino-Pérez

14.1 Introduction

The relationship between territory, landscape, and tourism has been a research topic within the Spanish Geography. The specific issues discussed about it disclose the global trends in the study of the Tourism Geography, the Spanish institutional status of this discipline, and the interests as well as concerns of Spanish Geographers along time. The first objective of this chapter is to present Spanish Geography conceptual framework for touristic strategies based on the territorial quality and landscape. These strategies allowed to the reinforcement of the global position for the own territory and landscape inland Spain, as well as its controversy and current risks. Starting from the links between the heritage assets, landscape, and culture as an expression of the territorial identity, new horizons of the tourism sustainability and resilience at different levels were considered.

P. Ibarra Benlloch (✉) · I. Rabanaque-Hernández
Department of Geography and Territorial Planning, University of Zaragoza, Zaragoza, Spain
e-mail: pibarra@unizar.es; irabanaque@unizar.es

E. De Uña-Álvarez
Department of History, Art and Geography, Campus of Ourense-University of Vigo, Geaat Group, Vigo, Spain
e-mail: edeuna@uvigo.gal

M. Villarino-Pérez
Independent Researcher, Tudistar Group, Cerdanyola del Vallès, Spain
e-mail: montserrat.villarino@usc.es

A. Blanco-Romero, M. Blázquez-Salom (eds.), *Spanish Tourism Geographies*, Geographies of Tourism and Global Change,
https://doi.org/10.1007/978-3-031-39780-6_14

The interest and significance of these questions are closely related to the global-local interactions and feedbacks. In relation to the impact of Covid-19 crisis on tourism, the *World Tourism Organization* data (UNWTO, 2022a, b) reflect a slowly restart in Spain. The key indicators for inbound and outbound international tourism show that Spain, after it spent considerable effort to recover from the previous global crisis (2008–2009), is now well positioned in the ranking of destinations (whit 31.1 million of arrivals in 2021) although it is still far below 2019 data (whit 83.3 million of arrivals). Aggregating destinations' net sentiment scores, a social indicator that reflects the state and dynamic of destinations and travel brands e-reputation, has increased along 2021 (score value, 28) and 2022 (score value, 39) being higher than the average of Europe (score value, 17 in 2021, 34 in 2022). Considering most of tourism success initiatives in Spain (Observatorio del Turismo Rural and de Ecoturismo en España, 2021), resilient and inclusive, it should be noted the gaining importance of the inland tourism (nature-based, rural, and experiential). The singularity of their touristic products, anchored in the own heritage and culture, play a key role in the reactivation and recovery of the outdoor and slow tourism far from mass destinations; so that natural and rural territories, their landscapes, are becoming increasingly demanded for live well-being experiences. Thus, the second research objective is to analyse the Spanish academic knowledge, made by Spanish geographers, in order to foster its dissemination. It was focused on the values of territorial quality mark and landscape for tourism, development and sustainability of Spanish inland areas. For this purpose, an analysis of the indexed works in the main scientific database was developed.

To select the Spanish contributions, the chosen time period extends from the year 2000 to the year 2022, recording articles, chapter books, and books. A step-by-step methodology was applied: searching the indexed publications in the main global scientific databases (Web of Science, Scopus, and Scholar), and also carry out a search in the main Spanish academic journals. The results from search terms "territorial quality", "mark", "landscape", and "inland tourism" both in Spanish, English, and French languages, have been screened, tabulated and analysed in the framework of the knowledge on inland tourism. They provide an overview regarding the evolution of research topic, their main Spanish geographical areas of interest and the specific types of tourism in consideration. Due to the required extent of this chapter, only the significant publications will be cited in the text of the chapter.

Further, in the recent decades, the installation of large wind or photovoltaic farms in Spanish inland territories has been multiplied. This situation particularly affects territories in which the quality of the landscape acts as a key to achieve a sustainable development. The third research objective is to analyse this question, considering an inland territory where a development strategy (expressly supported by territorial quality and landscape) has been implemented, but is affected by large renewable energy projects. Thus, the strategy of the Matarraña Region (Teruel) was analysed, starting from a survey about the citizen perception of the changes in landscape and its role as a relevant factor for the choice of a tourist destination.

14.2 Inland Tourism, Territory, and Landscape from the Spanish Perspective

Since the late twentieth century, the consideration of the territories and landscapes diversity in Spain as touristic resources and attractors -particularly in the local dimension- has a relevant role (Antón i Clavé et al., 1996). The growth of the Spanish studies about the Geography of tourism along the 1990s decade reinforced their significance, including the analysis of specific products for the revitalization and sustainability of inland areas (García Hernández & de la Calle Vaquero, 2004). Already at the beginning of the new millennium there is an increased interest about the territorial quality and the landscape (Fernández Tabales et al., 2010) as elements which provide competitive advantages for tourist destinations. In fact, between the dominant research themes developed by Spanish geographers in this period (Antón Clavé, 2013) stand out the territorial analysis of tourism, the values assessment of inland natural-cultural resources as territorial heritage, their alternative uses face the sun and beach tourism, and their role in the emergence of new inland tourist modalities. These trends go hand in hand with the great importance of the European Landscape Convention (Council of Europe, 2000), ratified by the Government of Spain in 2008, that increased the consideration of the landscape in territorial strategies. The Spanish context also implies changes in the conceptual and operative definition of the tourist destinations (Ávila Bercial & Barrado Timón, 2005; López Palomeque & Cànoves Valiente, 2014) supported by the pillars of the quality, regarding territorial and environmental systems, as well as the experiential offer and the products innovation.

Current critical issues in the Spanish tourism research, namely: the need of a new culture of territory (Asociación de Geógrafos Españoles, 2006, 2008), the valuation of the territory and natural-cultural landscape for touristic destination in times of crisis (Blàzquez et al., 2016), the configuration of smart destinations (Ivars-Baidal & Vera Rebollo, 2019) and the transition from spatial-temporal models of overtourism to undertourism (Pons et al., 2020) are hot topics. These critical issues are allowing to the Spain's inland areas in the context of the evaluation of tourism sustainability (essentially depending on the local stakeholders set) considering the relationships to societal and environmental processes that affect visitors and residents. These questions are calling for the development of adaptive, sustainable, resilient and collaborative strategies for inland tourism, agree with the *Agenda for a Sustainable and Competitive European Tourism* (2007), the *World Charter for Sustainable Tourism + 20* (2015), and the UNWTO proposals (2021).

The total of Spanish contributions (2000–2022 period) from database and scholarly journals explored sum 128 publications (Fig. 14.1). By time sequences, the most prolific was between 2012 and 2017. Regarding the specific issues considered in these studies, more than 35% of them correspond to the analysis of the concept and pillars for definition of territorial quality mark together with territorial heritage, landscape and inland tourism. Other studies that represent more than 31% of the total are related to the tourist role and significance of natural landscape, cultural

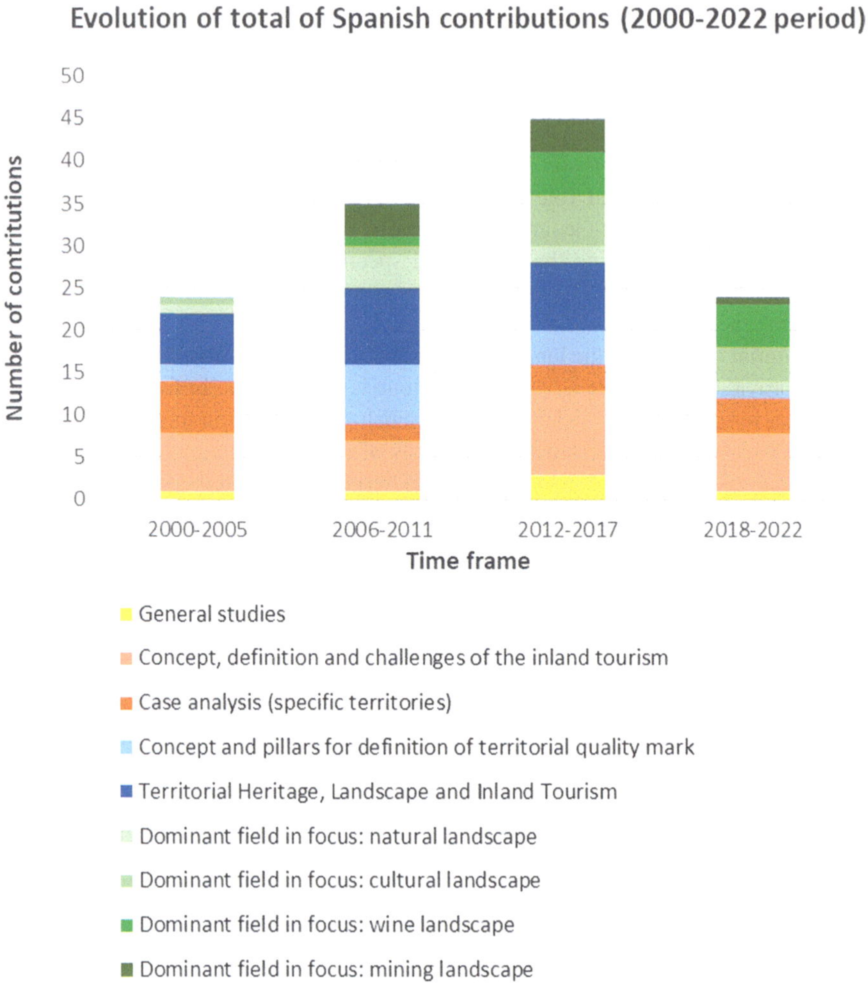

Fig. 14.1 Evolution of contributions from Spanish geographers. (Source: Own elaboration)

landscape, wine landscape and mining landscape for inland tourism. The contributions that refer to the concept, definition and challenges of the inland tourism together with case analysis (specific territories in Spain), account for 29% of the recorded publications.

Much of the contributions are research articles (67% of the total), mainly published in Spanish journals (Table 14.1) while there is an important presence of studies (20% of total articles) published in foreign indexed journals (e.g. *Geoforum*, *Tourism Geographies*, *Tourist Studies*, *The Tourist Review*, *Journal of Sustainable Tourism*, or *Journal of Heritage Tourism*). The first two Spanish journals that host a great number of contributions concentrate approximately 37% of the total of the research articles. On the contributions as books and chapter books it should be noted

Table 14.1 Top list of Spanish journals (2020–2022)

Spanish journal	% articles
Cuadernos de Turismo	23.80
Boletín de la Asociación de Geógrafos Españoles	13.09
Investigaciones Turísticas	7.14
Cuadernos Geográficos	5.95
Anales de Geografía de la Universidad Complutense	5.95
Scripta Nova	4.76
Estudios Geográficos	4.76
Documents d'Anàlisi Geográfica	4.76
Ería	3.57
Serie Geográfica	3.57
Total top list	77.48

Source: Own elaboration

Table 14.2 Top list of author's affiliations (2020–2022)

Authors affiliations	% total authors
University Autònoma of Barcelona	15.1
University of de Castilla-La Mancha	11.8
University of Santiago de Compostela	8.9
University of Alicante	8.1
University Autónoma of Madrid	5.9
University of Murcia	5.2
University of Vigo	5.2
University of Zaragoza	4.8
University of Gerona	3.7
University of Sevilla	3.7
University of Illes Balears	3.7
Total top list	76.1

Source: Own elaboration

the key role of the *Association of Spanish Geographers*, especially its *Geography of Tourism, Leisure and Recreation Group* that publishes regularly in open access the monographs whit the results of their Colloquiums (accounting for 35% of books and chapters found). Furthermore, these type of contributions have been published by Spanish editorials (44% of this category) as *Tirant Humanidades*, *Síntesis*, *Publications Services of the Universities of Valencia and Alicante*; and foreign editorials (21% of this category) as *Routledge*, *Emerald* or *IGI Global*. With regard to the affiliations of the authors, considering the universities or research centers to which they belong, the top list (Table 14.2) accounts for 76% of the contributions. Five universities (*Autònoma of Barcelona, Castilla-La Mancha, Santiago de Compostela, Alicante, and Autónoma of Madrid*) concentrate the most contributions, accounting for half of the total.

The searching results display a set of contributions about the evolution of the concept of "inland tourism" taking into account questions about local development, sustainability, slow tourism, smart destinations, under-tourism, and tourism de-growth; these contributions consider that territory and landscape are key resources due to their touristic values as well as their local authenticity/identity/intangible heritage linked to sociocultural conditions and processes drive by local actors along time. The development of rural tourism in Spain throughout the nineties identified inland tourism and rural tourism, both as a product for a tourist different from the traditional one: searching for an idyllic rural environment, with authenticity. The actions for local development were focused on this product, understood as a panacea for the development of rural territories. This process caused a very uneven development of the offer of rural tourism between inland areas, in quantity, quality and even in its own essence. Hence, the academic contributions address aspects such as the conceptualization (López Palomeque, 2008) because rural tourism grew in diversification as well as in intra and inter territorial typologies, with the negative consequences for its dissemination and commercialization (Cánoves Valiente et al., 2004; Herrera et al., 2005). The case analysis highlights the importance of different territorial resources in the development of rural tourism: eco-cultural (Plaza Gutiérrez, 2002), protected areas (Lozano de San Cleto et al., 2003; Muñoz Flores, 2008) or water resources (García González, 2004), among others. Territory and landscape cease to be seen only as a support to become a product, with new uses and new users (Antón Clavé, 2008). It is no longer possible to speak only of rural tourism but of inland tourism (Pillet Capdepón, 2012; Díez Santo, 2012; Cánoves Valiente et al., 2014, 2017) that is based on a territorial quality. In addition to natural and socio-economic resources, the activity of local or supra-local actors directly or indirectly involved is very important for the definition and mobilization of resources capable of attracting tourists, to build a collective vision of the tourist territory (De Uña Álvarez & Villarino Pérez, 2009) with construction and de-construction processes along time.

In this context, the structure, definition of resources, and territorial dynamics reflect a response to the main demand of tourism: activate the cognitive, emotional and recreational experience in the destination. This new requirement for tourism development determines the need of new methodologies allowed to create tourist products with activities integrated into the environment and that satisfies emerging demands. The application of product life cycle models to tourism (i.e. Cánoves Valiente et al., 2014) shows a phase of consolidation and even maturity in some territories. The diversification of inland tourism is already a fact and each territory tries to adapt to those modalities for which it is most suitable: Water tourism (e.g. De Uña-Álvarez, 2015), Tourism of natural/protected spaces (e.g. Serrano Gil & Aparicio Guerrero, 2017), Rural tourism (e.g. Villarino Pérez et al., 2009), Experiential Tourism (e.g. Rivera Mateos, 2015), Mountain Tourism (e.g. Cebrián Abellán & García Martínez, 2016), or Wine Tourism (e.g. Baraja et al., 2019; Plaza Gutiérrez & Fernández Álvarez, 2020). Some Spanish contributions show that in inland territories, innovative tourism products support local development by promoting other related activities and generating small tourism clusters (i.e. Cànoves Valiente et al., 2017). Thus it is essential to know the resilience of these groups. Regarding the territory-landscape-identity interactions, they imply a complex adaptive system dynamic. Its trajectories through different phases, from maintenance to renovation (or collapse), reveal different adaptive processes and changes. The possible states define open configurations, with feedback loops, with options for maintaining or reinforcing the territorial identity. Archetypal or metaphorical elements appear in most of them. There is a link between all processes: the identification and the definition of "one's own". Linked to the everyday and the extraordinary, it connects with the cultural heritage, the politics of the institutions and the condition of the social agents. Identifying one's own and elaborating a configuration of the collective territorial identity is an explicit action that continually modifies the ideas and strategies that include the territory and its landscape as the leitmotiv of tourist activity.

14.3 The Strategies Anchored in Territorial Quality and Landscape

The concept and pillars for the territorial quality mark as well as landscape in the general framework of Spanish inland tourism are displayed by other group of publications. The Spanish geographer's contributions show a concept of the territorial quality that involves their scientific, social, economic, cultural and educative values anchored to the local identity, not without controversy. Some studies (e.g. Vera Rebollo & Baños Castiñeira, 2004) highlight the increasing need of a sustainable tourism to attend environmental preservation, social equity, life quality and respect for local culture, as a driver of new leisure and recreational functions in Spanish inland territories. Other studies (e.g. Vaccaro & Beltrán, 2007) consider the redefinition of nature and culture in the globalization arena since 1970s, dominated by urban systems, a process leading to the modification of the inland touristic offer, now focused on urban cultural preferences. Without excluding the actions of public entities, local actors and agencies appear as key variables of this process and Spanish inland territories are defined as disempowered as demographic, socioeconomic, and cultural peripheries. Nonetheless, the available territorial resources are defined as pillars for a quality mark (e.g. Silva Pérez & Fernández Salinas, 2008), included in strategies of innovation for inland tourism.

The prominence of the territory and the configuration of its quality mark for tourism inland Spain is represented by contributions that address valuation of nature, heritage, landscape, and local culture accounting for adaptive or new strategies (Cebrián Abellán & García González, 2010; Hortelano Mínguez, 2015). The studies on the restructuring and renovation dynamics where inland tourism play an essential role (e.g. De Uña-Álvarez & Villarino-Pérez, 2017), reveal the importance of bottom-up processes in which it underlies -implicitly or explicitly- a collective definition of territorial identity that is embedded in place branding strategies (e.g. de San Eugenio-Vela & Barniol-Carcasona, 2015). In addition to fostering the local development particularly in multifunctional rural systems (e.g. Ruíz Pulpón & Cañizares Ruíz, 2022), those strategies strengthen the tangible and intangible values of territory and landscape for inland tourism.

The concept of territorial quality mark is clearly linked to landscape. Spanish geographers have published notable studies on the landscape, in coherence with the European Landscape Convention, to wit: their strengths and threats in the framework of territorial planning (i.e. Zoido, 2002; Nogué, 2010) or its concept as an own social and cultural construction (i.e. Martínez de Pisón, 2007; Mata Olmo, 2008) bringing out the significance of their narratives and collective imaginaries (i.e. Nogué, 2007) as well as how landscapes give rise to emotion, experience, and well-being (i.e. Puente Lozano, 2012). All these aspects undoubtedly influence the enhancement of landscape as a touristic resource, finding among the contributions on that subject (e.g. Espejo, 2011; Gómez Mendoza, 2013; Cañizares Ruiz, 2020) a concept of common heritage for Spanish landscapes which attract inland tourism but, conversely, are negatively affected by tourism. This recognition and rediscovery of landscape, defined as an identity sign and cultural legacy (Hernández Hernández, 2009), emphasizes the existence of threats and weaknesses related to the landscape consideration and uses in Spanish inland tourism. Such conflicts affect physical, environmental, and sociocultural character of landscape, including its visual imprint (de San Eugenio Vela et al., 2017) in the local creation of a territorial quality mark closely tied to places brand.

Dominant fields in those research that highlights the importance of the territorial quality and landscape in Spanish tourist strategies are represented by contributions on natural, wine, mining and rural areas in the Spanish hinterland. Its distribution regarding the main landscape character is quite balanced (Fig. 14.2). When analyzing those dominant research fields for the Spanish Autonomous Communities (Fig. 14.3) a remarkable variability is observed. The studies with wine related themes are presented in eight communities, followed by those related to the natural landscape represented in seven communities and those related to cultural landscape studies, which are reflected in six autonomous communities. Finally, mining studies are only present in three Autonomous Communities.

The inventory, characterization and assessment of the tourist potential that present geomorphological landscapes -considering the main type of visitors and sustainable management options- is analysed in the case of, for example, National Parks (Serrano & Gómez Trueba, 2011), Global Geoparks (Lagar Timón et al., 2013), Mountainous borders (De Uña-Álvarez et al., 2017), and Arid areas (Belmonte Serrato et al., 2019). All these studies point out the priority uses integrated in nature-based tourism and geo-tourism modalities, allowed to conservation, interpretation, revalorization, environmental education and sustainability of inland territories face the mass tourism developed in coastal territories; all of them take into account the current and possible future threats and impacts which could degrade the intrinsic and extrinsic values for tourism.

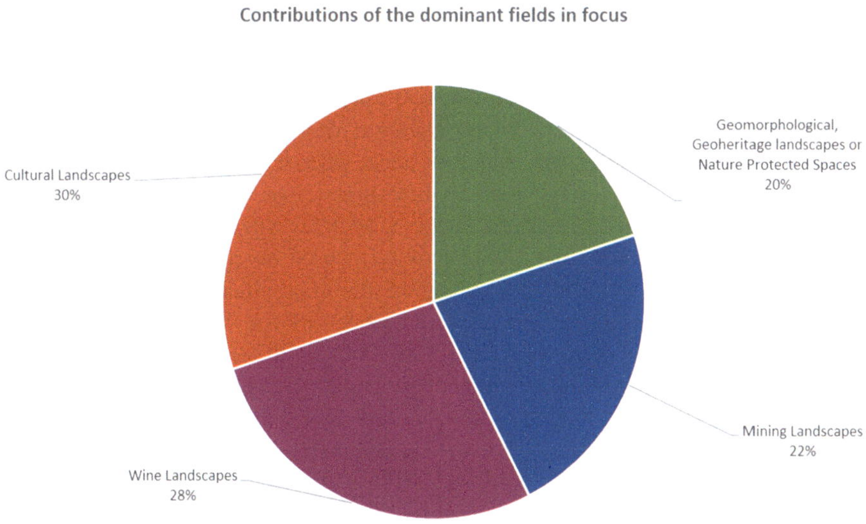

Fig. 14.2 Dominant fields in specific studies on territorial quality and landscape based-strategies for Spanish inland tourism. N = 40. (Source: Own elaboration)

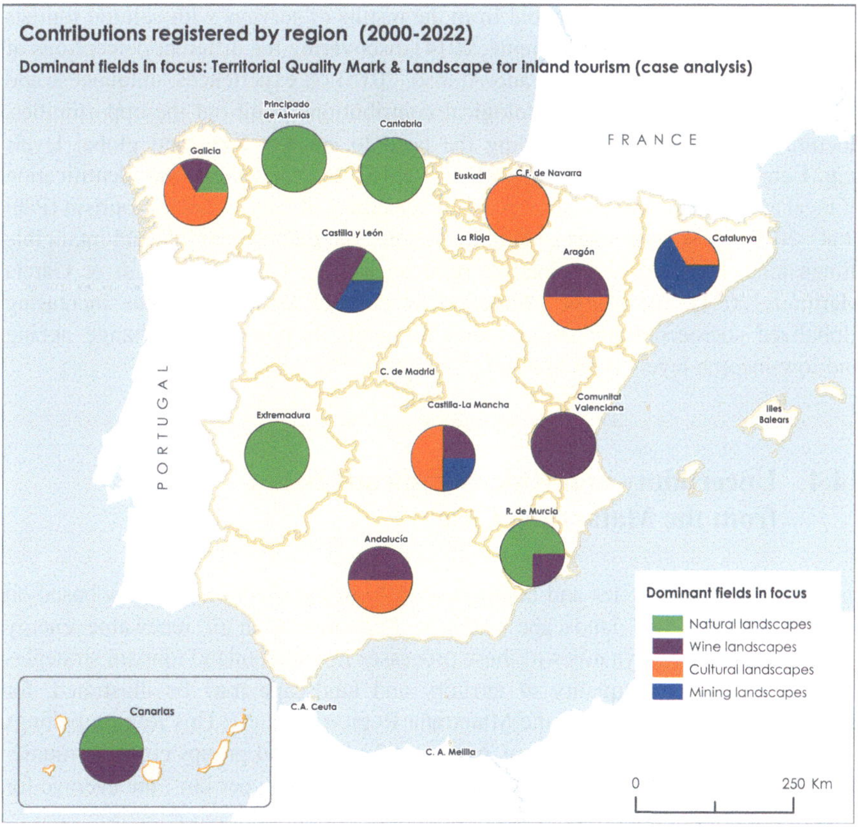

Fig. 14.3 Distribution of dominant focus in case studies on territorial quality and landscape based-strategies by Spanish Autonomous Communities. (Source: Own elaboration)

Besides this, the world of wine references has a singular significance in all the publications, due to the importance of vineyards and protected designations of origin for the Spanish inland areas. The wine-growing territories and their landscapes are one of the most considered tourist resources, anchored to territorial marks, looking at their socioeconomic and cultural values as the expression of own history and identity (Frutos Mejías et al., 2012; Martínez Puche & Morales Yago, 2016); indeed their values as heritage transcend the tourist activity focused on eno-tourism (Martínez Arnáiz & Molinero Hernando, 2019) being an important pillar for the local development in Spanish inland areas (De Uña-Álvarez & Villarino-Pérez, 2019) promoting their sustainability (Cañizares Ruiz & Ruiz Pulpón, 2020).

With regard to mining-industrial inland tourism, the Spanish contributions focus on the values of heritage assets as a powerful tool for the local development is less favoured (inland) territories (e.g. Valenzuela Rubio et al., 2008); and define new uses for these resources (Cueto Alonso, 2016) in the context of inland tourism (e.g. Díaz Soria & Romagosa Casals, 2016; Pardo Abad, 2017); the advantages and drawbacks of tourist uses also are considered (e.g. Llurdés Coit, 2017; Somoza-Medina & Monteserín-Abella, 2021).

Landscape is defined as an important factor of development in inland mountains through cultural tourism strategies (Panadero Moya et al., 2011). A great part of studies demonstrates this potential from the results of surveys with cultural tourists (e.g. Prat Forga & Cànoves Valiente, 2014) discovering the different perceptions of visitors and residents (e.g. Mercado Alonso, 2015) on experiences, uniqueness and quality. Theoretical and methodological contributions point out the opportunities, dilemmas and challenges regarding the need to connect local and global levels (e.g. Cebrián Abellán & García Martínez, 2016). In this context, the identification of local actors' narratives related to the enhancement of landscapes for tourism (Paül et al., 2019) show consensus and contradictions that include tangible and intangible dimensions. Particularly in Spanish rural areas (e.g. Cebrián Abellán & García Martínez, 2021), the tourist promotion of cultural landscapes faces increasing globalized socioeconomic requirements although its potential to change ageing and low-income levels is recognised.

14.4 Uncertainty and Risk, Significant Issues from the Matarraña Case

Increasingly controversies and conflicts arise from the clash of strategies based on territorial quality and landscape with the implementation of renewable energy macro-projects. The dynamics of these processes in which inland tourism strategies are supported by the quality of territory and landscape may be illustrated, for example, through the case of the Matarraña Region (Teruel). This Region belongs to the Autonomous Community of Aragon (933 km^2) and groups eighteen municipalities, with a very low population density (9 inhabitants per km^2, the average for

Aragon is 26 inhabitants per km^2). Its territorial singularity is the coexistence of mountain areas with a high forest richness and valley areas with dominant agricultural activity. Its landscapes, alive and lived, preserved both natural and cultural heritage values linked to territorial identity. Heavily affected by the recent pandemic, its attractiveness as an inland tourist destination is subject to the threat of the implementation of wind farms, which has generated an important social response, evident in the publication of manifestos such as the *Matarraña Landscape Conservation* and the mobilization of involved local actors (e.g. *Association of Entrepreneurs of Matarraña, Gent of Matarraña*).

In Matarraña, the efforts to promote tourism have been very important. The development of the tourism sector has had to do with the spectacular nature of the landscapes, together with a rich natural and cultural heritage, attracting an increasingly demanding inland tourism. Agricultural production in the region gives rise to a series of high-quality products: The *Teruel* Denomination of Origin ham, the olive oil attached to the Denomination of Origin of *Bajo Aragón*, the *Calanda* peach, wine, almonds, cheese, honey, lamb, marinades and artisan pastas. The tourist offer consists of 35 hotels, hostels or similar with 755 beds, 70 rural tourism homes, 3 campsites, 101 tourist apartments and 12 homes for tourist use, in total there are more than 2000 tourist beds registered. Distribution of economic activities by sectors shows a predominance of the services sector (62.3% of the activities), experiencing in the last ten years an increase of 40.4%. Within this framework the Matarraña territorial quality mark was born with the aim of defending and valuing those products, services, resources, and cultural manifestations, by entities and/or groups of interest that identify the territory and make it a singular. Based on the criteria of quality, social and environmental responsibility, economic development and always linked to territorial and cultural identity, this quality mark is a key instrument to become a development tool to achieve territorial balance and cohesion. It is part of the *European Territorial Quality Brand Network*. Despite the crisis, the increase in tourist activity in the region has taken off with the implementation of the territorial quality mark and the gradual certification of the different companies from 2012, the date on which the first certification was carried out (22 companies) in a formal act.

Regarding the role played by the landscape for inland tourism, its consideration has been decidedly driven by the signing of the *Landscape Charter of Matarraña* in 2010 after a long process of citizen participation (Comarca del Matarraña, 2010). It implies a voluntary agreement between the social actors of the territory, with the consensual specification of ten landscape quality objectives and proposed actions, aligned with the *European Landscape Convention* (CEP). The Region of Matarraña is the first to request the Government of Aragon to draw up the *Regional Landscape Maps* as a fundamental tool for precisely complying with the CEP; they are an exhaustive planning and management document for the regional development strategy. Thus, on the one hand, the aim is to promote the integration of the landscape in the different public policies that affect it (land planning, evolution of the agricultural and livestock sectors, tourism, conservation of natural and cultural heritage...); and, on the other hand, to integrate it in private activity through an awareness and training that encourage socioeconomic initiatives to value the landscape as a sign of identity.

Also noteworthy is the effort to value two key elements of the region's cultural landscape: dry stone and centuries-old olive trees. The Declaration of dry-stone constructions as World Heritage of Humanity is being promoted through the elaboration of an exhaustive inventory of them. However, one of the main problems is the intense process of abandonment that it has suffered. Apart from the undoubtedly important socio-economic repercussions, this abandonment supposes a loss of natural and cultural landscapes that are especially valuable.

The landscape with its natural and cultural heritage together with the territorial quality mark is integrated into the elaboration of a collective strategy by the social actors. In line with these guidelines, the *Strategic Development Program of the Region* involves in a proactive way both the different social actors who live in the region and the visitors or tourists. In connection with this approach, in November 2010 the *Matarraña Chair* was created at the University of Zaragoza with the aim of promoting the *Matarraña Territorial Quality Mark* and their landscapes, focusing on the creation of new knowledge and the development of practical training in two complementary areas (the landscape area and the food science area) to develop territorial quality charters. Likewise, as a milestone in this trajectory for the consolidation of inland tourism, the *Sustainable Matarraña Forum* emerges, organized by the *Matarraña Business Association* since November 2020 with the aim of debating the importance of territorial sustainability for the benefit of visitors and local population. In this scenario that calls for the operational definition of a sustainable and resilient inland tourism planning, the simulation of a future with wind turbines (Fig. 14.4) has been used to conduct a survey (Ibarra & Rabanaque, 2020) aimed at finding out the perception, ideas and opinions of the local and foreign population on the challenges of the future.

The perception of the landscape by the observer is an intangible, experiential and personal component that should never be ignored when considering a tourism product. The project for a cluster of wind farms in the provinces of *Teruel* and *Zaragoza* proposed the installation of 84 wind turbines with a rotor diameter of 170 m and a hub height of 115 m that would fully affect the *Matarraña* region. Its announcement during the public information phase caused a great commotion in the region, changing the scenario of the strategy for tourism based on the territorial quality mark and the landscape. A survey was carried out between May 13 and 30, 2020, in the midst of the confinement period, in online format (distributed through social networks and by email lists), anonymously and configured to be sent only once from the same device. The target sample of interest was the national population, interested in knowing the points of view of the local population, potential tourists and non-potential tourists.

The first section of the survey asked what the landscape suggests to them in its current state (Fig. 14.4a) in terms of the options it offers, sensations it evokes and its assessment, including the possible choice of this or similar landscape as a tourist destination. In the second section, the same questions were raised regarding the simulated scenarios with wind turbines (Fig. 14.4b). Finally, were discussed the opinions and perceptions regarding the socioeconomic and quality of life impacts that could arise from the installation of large wind turbines in these interior territories, where tourism emphasizing quality is a fundamental resource, as well as the type of planning required. The highest percentage of

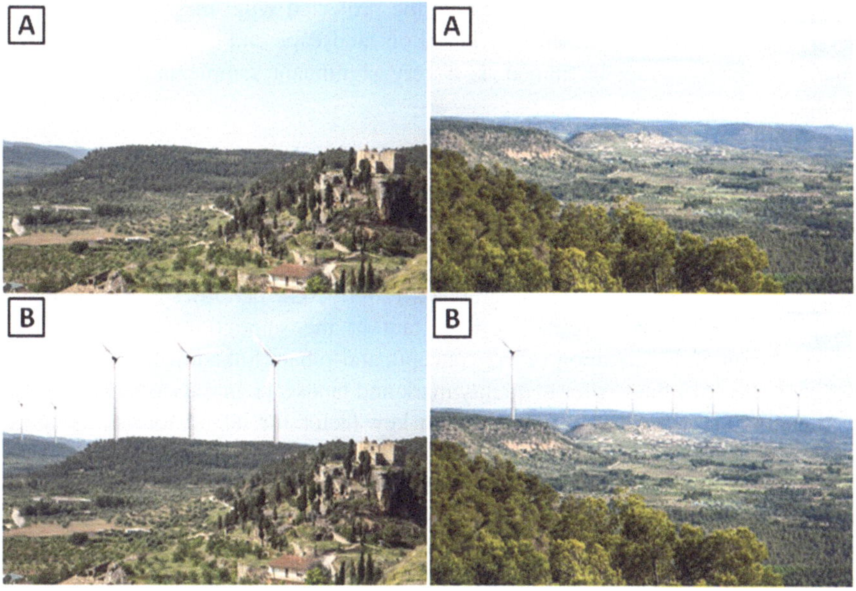

Fig. 14.4 Landscapes of the *Matarraña* region (**a**) and retouched images of landscapes with the incorporation of wind turbines (**b**). (Source: Own elaboration)

responses on a total of 1057 people from all over Spain corresponded to inhabitants from small municipalities (24%), a high proportion considering that the percentage of inhabitants residing in Spain for this type of municipality is 6%. In terms of age, the lowest weight of responses corresponds to the population over 65 years of age (5%), while this age group represents more than 20% of the total population of Spain as a whole. Both assessments demonstrate the interest aroused in the rural population and the inevitable bias of the older population in an online survey, without invalidating the validity of the survey.

The results of the survey showed that outdoor activities and enjoyment of heritage linked to tranquillity, escape and well-being were the primary factors considered when choosing a tourist destination like Matarraña, with the absence of major interventions in the landscape (73% of the responses). In addition, the 42.1% of Spanish people surveyed would choose an inland destination without wind turbines, while 33% do not know to what extent their presence would bother them and only 24.3% stated that their presence would not particularly affect them. The perception of the danger and the loss of environmental quality that the installation of wind turbines would entail was defined as a very severe (44%) and severe (20%) impact by the majority of those surveyed. This perception had a slightly greater weight in the group of foreigners (70% of responses) than in that of native people (63% of responses). In any case, 87% of those surveyed believed that the change in the landscape would cause a strong collision with the desirable strategy for promoting tourism and the future of territorial sustainability.

The strong collective reaction against the projected wind megaprojects and in defending the development strategy based on landscape and quality, in which the inland tourism product is framed, is a very significant sample of the collective thought of *Matarraña*.

14.5 Final Remarks

The contributions of Spanish geographers (2020–2022 period) show that inland tourism is becoming increasingly important at the same time that there is a reformulation and renovation of the concepts and pillars linked to its evolution in Spain. The role of the territorial quality mark and landscape in the strategies studied by Spanish Geography is postulated as a key factor for inland tourism, since it promotes the sustainability and resilience of less developed Spanish areas. Always from an integrated point of view, considering the perspective of its inhabitants as protagonists in the conformation of the tourist destination as well as the perspective of potential tourist, this relevance is reflected by the number of indexed and academic publications (articles, books, and book chapters) since the beginning of the present century.

The contribution from Spanish geographers also reveals the diversification of the focus and dominant research fields because inland tourism has multiple prisms, particularly linked to the particularities and singularities of the territories that host it. This imbrication in the territorial features and quality means that, on the one hand, it is a much more identified type of tourism with greater ties from the social, cultural and economic point of view with the area in which it takes place, but it also has greater threatens than conventional tourism since there is greater sensitivity and dependence on decisions made locally and globally that can alter landscapes and therefore territorial and tourist synergies. The point of view of Spanish geographers in this field, championed by milestones such as the European Landscape Convention, or the role of this tourism as the backbone of the territorial quality mark, makes it one of the most attractive research areas from the Spanish Geography.

The perception of the quality of the landscape can influence the choice as a tourist destination and can be affected by the deployment of large wind farms. These are issues that should be considered especially in inland areas where territorial quality is key to their development, which makes urgent a cross-cutting planning essential to foster the tourism sustainability and resilience.

Acknowledgements The development of this research corresponds to the project CSO2016-74861-R «Tourism of interior in Spain: challenges, valuation and strategies before changeable situations and of crisis to stimulate products and tourist destinations. Dynamics of cases» (2017–2019), funded by the Spanish Ministry of Economy and Competitiveness, in compliance with the call for «Research, Development and Innovation» program, targeting «Challenges of Society» (Principal researchers G. Cánoves and A. Blanco-Romero, Universitat Autònoma de Barcelona, Spain).

References

Antón Clavé, S. (2008). Turismo rural: del territorio a los productos. In F. Cebrián Abellán (Coord.), *Turismo rural y desarrollo local* (pp. 219–228). Ediciones de la Universidad de Castilla-La Mancha.

Antón Clavé, S. (2013). The geography of tourism in Spain: Institutionalization and internationalization. In J. Wilson & S. Antón Clavé (Eds.), *Geographies of tourism: European research perspectives* (pp. 151–177). Emerald. https://doi.org/10.1108/S1571-5043(2013)0000019009

Antón i Clavé, S., López Palomeque, F., Marchena Gómez, M. J., & Vera Rebollo, J. F. (1996). Tourism research in Spain: The contribution of geography (1960–1995). *The Tourist Review, 51*(1), 46–64. https://doi.org/10.1108/eb058217

Asociación de Geógrafos Españoles. (2006). *Por una nueva cultura del Territorio*. Retrieved from https://www.age-geografia.es/site/wp-content/uploads/2021/03/Manifiesto-por-una-nueva-cultura-del-territorio-y-adendas-2009-y-2018.pdf

Asociación de Geógrafos Españoles. (2008). Adenda. En defensa del territorio ante los nuevos retos del cambio global. 2018. Retrieved from https://www.age-geografia.es/site/adenda-2018-al-manifiesto-de-la-nueva-cultura-del-territorio/

Ávila Bercial, R., & Barrado Timón, D. A. (2005). Nuevas Tendencias en el desarrollo de destinos turísticos: marcos conceptuales y operativos para su planificación y gestión. *Cuadernos de Turismo, 15*(15), 27–44. https://doi.org/10.1097/JTO.0b013e3181dd1386

Baraja, E., Herrero, D., Martínez, M., & Plaza, J. I. (2019). Turismo y desarrollo vitivinícola en espacios de montaña con "alta densidad patrimonial". *Cuadernos de Turismo, 43*, 97–122. https://doi.org/10.6018/turismo.43.04

Belmonte Serrato, F., Romero Díaz, A., & Ballesteros Pelegrín, G. A. (2019). Los paisajes áridos abarrancados como recurso turístico. *Investigaciones Turísticas, 17*, 213–238. https://doi.org/10.14198/INTURI2019.17.10

Blàzquez, M., Mir-Gual, M., Murray, I., Pons, G. X., & (Eds.). (2016). *Turismo y crisis, turismo colaborativo y ecoturismo*. Monografies de la Societat d'Història Natural de les Balears. Retrieved from https://www.age-geografia.es/downloads/Monografies_SHNB_2016_vol023.pdf

Cañizares Ruiz, M. C. (2020). Procesos y retos en torno al patrimonio y a los paisajes culturales: Una reflexión teórica desde la geografía española. *Revista de Geografía Norte Grande, 76*, 189–212. https://doi.org/10.4067/S0718-34022020000200189

Cañizares Ruiz, M. C., & Ruiz Pulpón, A. R. (2020). Paisajes del viñedo, turismo y sostenibilidad: interrelaciones teóricas y aplicadas. *Investigaciones Geográficas, 74*, 9–28. https://doi.org/10.14198/INGEO2020.CRRP

Cánoves Valiente, G., Villarino, M., Priestley, G. K., & Blanco, A. (2004). Rural tourism in Spain: An analysis of recent evolution. *Geoforum, 35*(6), 755–769.

Cánoves Valiente, G., Villarino, M., Blanco-Romero, A., De Uña, E., & Espejo, C. (Eds.). (2014). *Turismo de interior: renovarse o morir. Estrategias y productos en Catalunya, Galicia y Murcia*. Universitat de València, Servei de Publicacions, Colección Desarrollo territorial (12).

Cànoves Valiente, G., Blanco-Romero, A., Prat Forga, J. M., & Villarino Pérez, M. (Coord.). (2017). *Turismo de interior en España: productos y dinámicas territoriales*. Universitat de València, Servei de Publicacions, Colección Desarrollo territorial (19).

Cebrián Abellán, F., & García González, J. A. (2010). Propuesta metodológica para la identificación, clasificación y puesta en valor de los recursos territoriales del turismo interior. la provincia de Albacete. *Boletín de la Asociación de Geógrafos Españoles, 54*, 361–383. Retrieved from Recuperado a partir de https://bage.age-geografia.es/ojs/index.php/bage/article/view/1293

Cebrián Abellán, F., & García Martínez, C. (2016). Uso y gestión del paisaje para la actividad turística en el medio rural: aproximación teórica y empírica en el sureste de Castilla-La Mancha (Albacete). *Boletín de la Asociación de Geógrafos Españoles, 72*, 381–407. https://doi.org/10.21138/bage.2345

Cebrián Abellán, F., & García Martínez, C. (2021). Landscape and tourism as tools for local development in Mid-Mountain rural areas in the Southeast of Spain (Castilla-La Mancha). *Land, 10*, 221. https://doi.org/10.3390/land10020221

Comarca del Matarraña. (2010). *Carta del Paisaje de la Comarca del Matarraña*. Pacto para la protección, ordenación, gestión y mejora del paisaje de la Comarca del Matarraña/Matarranya. Retrieved from http://aragonparticipa.aragon.es/sites/default/files/stories/Actualidad/noticias/enAragon/Carta_paisaje_Matarraa/carta_paisaje.pdf

Commission of the European Communities. (2007). *Agenda for a sustainable and competitive European tourism*. https://eur-lex.europa.eu/legal-content/EN/TXT/PDF/?uri=CELEX:52007DC0621&from=ES

Council of Europe. (2000). *The landscape convention* (European Treaty Series – No. 176). Retrieved from https://rm.coe.int/16807b6bc7

Cueto Alonso, G. J. (2016). Nuevos usos turísticos para el patrimonio minero en España. *Pasos, 14*(4), 1013–1026. Retrieved from http://www.pasosonline.org/Publicados/14416/PS416_15.pdf

de San Eugenio Vela, J., Nogué, J., & Govers, R. (2017). Visual landscape as a key element of place branding. *Journal of Place Management and Development, 10*(1), 23–44. https://doi.org/10.1108/JPMD-09-2016-0060

de San Eugenio-Vela, J., & Barniol-Carcasona, M. (2015). The relationship between rural branding and local development. A case study in the Catalonia's countryside: Territoris Serens (El Lluçanès). *Journal of Rural Studies, 37*, 108–119. https://doi.org/10.1016/j.jrurstud.2015.01.001

De Uña Álvarez, E., & Villarino Pérez, M. (2009). Territorio, turismo y desarrollo en la dimensión local. In Félix Pillet Capdepón, María del Carmen Cañizares Ruiz, Ángel Raúl Ruiz Pulpón (Coord.), *Geografía, territorio y paisaje: el estado de la cuestión* (pp. 207–221).

De Uña-Álvarez, E. (2015). Paisajes del Agua, territorio y turismo. En Grupo Geaat (Coord.), *Paisajes culturales del Agua* (pp. 21–28). Diputación Ourense.

De Uña-Álvarez, E., & Villarino-Pérez, M. (2017). Adaptive strategies for inland touristic destinations (Sil Canyon, Galicia, NW Spain). *Cuadernos de Turismo, 40*, 203–224. https://doi.org/10.6018/turismo.40.309671

De Uña-Álvarez, E., & Villarino-Pérez, M. (2019). Linking wine culture, identity,tourism and rural development in a denomination of origin territory (NW of Spain). *Cuadernos de Turismo, 44*, 93–110. https://doi.org/10.6018/turismo.44.404761

De Uña-Álvarez, E., Villarino-Pérez, M., & Cuquejo Bello, M. C. (2017). Territorio y Turismo a través de los actores locales. El caso del Larouco (Galicia, Noroeste de la Península Ibérica). *Anales de Geografía de la Universidad Complutense, 37*(2), 349–370. https://doi.org/10.5209/AGUC.57729

Díaz Soria, I., & Romagosa Casals, F. (2016). Patrimonio minero, paisaje y turismo de proximidad: explorando sinergias. El caso de Cardona (Barcelona). *Documents d'Anàlisi Geogràfica, 62*(1), 55–77. https://doi.org/10.5565/rev/dag.231

Díez Santo, D. (2012). Los turismos de interior: un enfoque desde la dimensión de las modalidades turístico-recreativas. *Documents d'Anàlisi Geogràfica, 58*(3), 373–396. https://doi.org/10.5565/rev/dag.16

Espejo, C. (2011). El paisaje como recurso turístico. In M. R. Simancas & A. Cortina (Coords.), *Retos y perspectivas de la gestión del paisaje de Canarias* (pp. 439–461). Gobierno de Canarias y Universidad Internacional Menéndez Pelayo.

Fernández Tabales, A., García Hernández, M., & Ivars Baidal, J. A. (Coords.). (2010). La investigación de la geografía del turismo en las comunidades autónomas españolas: orígenes, desarrollo y perspectivas de una disciplina en el horizonte de la Geografía. Asociación de Geógrafos Españoles.

Frutos Mejías, L. M., Castelló Puig, A., Hernández Navarro, M. L., Loscertales Palomar, B., Ruiz Budría, E., & Rubio Terrado, P. (2012). El espacio vitivinícola aragonés como paisaje cultural y factor de desarrollo. *Geographical, 62*, 5–34. https://doi.org/10.26754/ojs_geoph/geoph.201262846

García González, L. (2004). Agua y Turismo. Nuevos usos de los recursos hídricos en la Península Ibérica. Enfoque integral. *Boletín de la Asociación De Geógrafos Españoles, 37*, 239–255. https://bage.age-geografia.es/ojs/index.php/bage/article/view/1985

García Hernández, M., & de la Calle Vaquero, M. (2004). La investigación geográfica española en materia de turismo (1997–2004). *Anales de Geografía de la Universidad Complutense, 24*, 257–277. Retrieved from https://revistas.ucm.es/index.php/AGUC/article/view/AGUC0404110257A

Gómez Mendoza, J. (2013). Del patrimonio paisaje a los paisajes patrimonio. *Documents d'Anàlisi Geogràfica, 59*(1), 5–20. Retrieved from https://raco.cat/index.php/DocumentsAnalisi/article/view/262771

Hernández Hernández, M. (2009). El paisaje como seña de identidad territorial: valorización social y factor de desarrollo, ¿utopía o realidad? *Boletín de la Asociación De Geógrafos Españoles, 49*, 169–183. Retrieved from https://bage.age-geografia.es/ojs/index.php/bage/article/view/780

Herrera, L., Blanco, A., & Cánoves Valiente, G. (2005). Turismo rural en España; un análisis de la evolución en el contexto europeo. *Cuadernos de Geografía, 77*, 41–58.

Hortelano Mínguez, L. A. (2015). Patrimonio territorial como activo turístico en la «Raya» de Castilla y León con Portugal. *Cuadernos de Turismo, 36*, 247–268. https://doi.org/10.6018/turismo.36.230981

Ibarra, P., & Rabanaque, I. (2020). La marca de calidad territorial y el paisaje como estrategia para el turismo de interior en la comarca del Matarraña (Teruel). In G. X. Pons, A. Blanco-Romero, R. Navalón-García, L. Troitiño-Torralba, & M. Blázquez-Salom (Eds.), *Sostenibilidad Turística: Overtourism vs undertourism* (pp. 69–81). Monografies de la Societat d'Història Natural de les Balears. Retrieved from https://www.age-geografia-turismo.com/publicaciones/libros/actas-mao/

Ivars-Baidal, J. A., & Vera Rebollo, J. F. (2019). Planificación turística en España. De los paradigmas tradicionales a los nuevos enfoques: planificación turística inteligente. *Boletín de la Asociación de Geógrafos Españoles, 82*(2765), 1–31. https://doi.org/10.21138/bage.2765

Lagar Timón, D., García Marín, R., & Pulido Fernández, M. (2013). Caracterización del potencial turístico del Geoparque Villuercas-Ibores-Jara (Extremadura, España). *Investigaciones Turísticas, 5*, 136–162. https://doi.org/10.14198/INTURI2013.5.06

Llurdés Coit, J. C. (2017). El paisaje minero, ¿patrimonio o inconveniente paisajístico? El caso de la comarca del Bages (Barcelona). *Ería, 37*(2), 151–167. https://doi.org/10.17811/er.2.2017.151-167

López Palomeque, F. (2008). Delimitación conceptual y tipologías de turismo rural. En J. I. Pulido (Coord.), *El turismo rural*. Síntesis.

López Palomeque, F., & Cànoves Valiente, G. (Eds.). (2014). *Turismo y territorio. Innovación, renovación y desafíos*. Tirant Humanidades.

Lozano de San Cleto, M. J., González Cascón, V., & Martín Lou, M. A. (2003). Los espacios protegidos como factor de localización del turismo rural en Castilla y León. *Serie Geográfica, 11*, 177–190. http://hdl.handle.net/10017/1178

Martínez de Pisón, E. (2007). Paisaje, cultura y territorio. In J. Nogué (Ed.), *La construcción social del paisaje* (pp. 325–337). Biblioteca Nueva.

Martínez Puche, A., & Morales Yago, F. J. (2016). El vino, como recurso turístico para el fomento del desarrollo local. Una oportunidad para las comarcas del Vinalopó (Alicante) y el altiplano Yecla-Jumilla (Murcia). *Cuadernos de Turismo, 38*, 236–295. https://doi.org/10.6018/turismo.38.271451

Martínez Arnáiz, M., & Molinero Hernando, F. (2019). El valor patrimonial del paisaje como imagen de calidad en la estrategia comercial de los territorios vitivinícolas del Duero. *Cuadernos Geográficos, 58*(3), 169–194. https://doi.org/10.30827/cuadgeo.v58i3.8837

Mata Olmo, R. (2008). El paisaje, patrimonio y recurso para el desarrollo territorial sostenible. Conocimiento y acción pública. *Arbor, 184*(729), 155–172. https://doi.org/10.3989/arbor.2008.i729.168

Mercado Alonso, I. (2015). Percepción y valoración social de los paisajes disfrutados: aportaciones desde el visitante para una gestión sostenible de espacios turísticos rurales. El caso de la Sierra de Aracena (Huelva). *Investigaciones Turísticas, 9*, 160–183. https://doi.org/10.14198/INTURI2015.9.07

Muñoz Flores, J. C. (2008). El turismo en los espacios naturales protegidos españoles, algo más que una moda reciente. *Boletín de la Asociación De Geógrafos Españoles, 46*, 291–304. https://bage.age-geografia.es/ojs/index.php/bage/article/view/674/628

Nogué, J. (2007). Territorios sin discurso, paisajes sin imaginario. Retos y dilemas. *Ería, 73–74*, 373–382. Retrieved from https://www.unioviedo.es/reunido/index.php/RCG/article/view/1593/15038

Nogué, J. (2010). El paisaje en la ordenación del territorio. La experiencia del Observatorio del Paisaje de Cataluña. *Estudios Geográficos, 71*(269), 415–448. https://doi.org/10.3989/estgeogr.201014

Observatorio del Turismo Rural y de Ecoturismo en España. (2021). *Tendencias turísticas 2021*. https://www.escapadarural.com/blog/wpcontent/uploads/INFORME_TURISMO_RURAL_ECOTURISMO.pdf

Panadero Moya, M., García González, J. A., & Panadero Alarcón, J. D. (2011). Paisaje y turismo. El corredor bético de Alcaraz (Albacete). *Cuadernos de Turismo, 27*, 679–700. Retrieved from https://revistas.um.es/turismo/article/view/140161

Pardo Abad, C. J. (2017). The post-industrial landscapes of Riotinto and Almadén, Spain: Scenic value, heritage and sustainable tourism. *Journal of Heritage Tourism, 12*(4), 331–346. https://doi.org/10.1080/1743873X.2016.1187149

Paül, V., Trillo, J. M., & Haslam McKenzie, F. (2019). The invention of a mountain tourism destination: An exploration of Trevinca-A Veiga (Galicia, Spain). *Tourist Studies, 19*(3), 313–335. https://doi.org/10.1177/1468797619833364

Pillet Capdepón, F. (2012). El turismo de interior en la España peninsular: el patrimonio territorial como destino turístico. *Boletín de la Asociación De Geógrafos Españoles, 59*, 345–366. https://doi.org/10.21138/bage.1461

Plaza Gutiérrez, J. I. (2002). El turismo rural en territorios periféricos (el ejemplo de algunas comarcas del oeste castellano-leonés). *Investigaciones Geográficas, 27*, 83–106.

Plaza Gutiérrez, J. I., & Fernández Álvarez, R. (2020). Valor patrimonial de los paisajes del vino y su promoción en los valles y llanuras del este de Zamora. *Estudios Geográficos, 81*(289), e046. https://doi.org/10.3989/estgeogr.202060.060

Pons, G. X., Blanco-Romero, A., Navalón-García, R., Troitiño-Torralba, L., & Blázquez-Salom, M. (Eds.). (2020). *Sostenibilidad Turística: overtourism vs undertourism*. Monografies de la Societat d'Història Natural de les Balears. Retrieved from https://www.age-geografia-turismo.com/publicaciones/libros/actas-mao/

Prat Forga, J. M., & Cànoves Valiente, G. (2014). Integrated cultural tourism: New experiences in mountain areas. *Tourismos, 9*(2), 15–32.

Puente Lozano, P. (2012). El valor emocional de la experiencia paisajística. Querencias y paisajes afectivos. *Cuadernos Geográficos, 51*, 270–284. https://doi.org/10.30827/cuadgeo.v51i0.242

Rivera Mateos, M. (2015). Turismo experiencial y gestión estratégica de recursos patrimoniales: un estudio exploratorio de percepción de productos turísticos en las Sierras Subbéticas cordobesas (Andalucía). *Scripta Nova, XIX*, 511. http://www.ub.es/geocrit/sn/sn-511.pdf
Ruíz Pulpón, A. R., & Cañizares Ruíz, M. C. (2022). Intangible heritage and territorial identity in the multifunctional agrarian systems of vineyards in Castilla-La Mancha (Spain). *Land, 11*(2), 281. https://doi.org/10.3390/land11020281
Serrano Gil, Ó., & Aparicio Guerrero, A. E. (2017). La oferta turística vinculada a las áreas de influencia socioeconómica de los parques nacionales de Castilla-La Mancha: Tablas de Daimiel y Cabañeros. *Cuadernos de Turismo, 39*, 577–600. https://doi.org/10.6018/turismo.39.290711
Serrano, E., & Gómez Trueba, J. J. (2011). Environmental education and landscape leisure. Geotourist map and geomorphosites in the Picos de Europa National Park. *GeoJournal of Tourism and Geosites, 8*(2), 295–308.
Silva Pérez, R., & Fernández Salinas, V. (2008). El patrimonio y el territorio como activos para el desarrollo desde la perspectiva del ocio y del turismo. *Investigaciones Geográficas, 46*, 69–88. https://doi.org/10.14198/INGEO2008.46.04
Somoza-Medina, X., & Monteserín-Abella, O. (2021). The sustainability of industrial heritage tourism far from the axes of economic development in Europe: Two case studies. *Sustainability, 13*, 1077. https://doi.org/10.3390/su13031077
UNWTO. (2021). *A year in review: Tourism united, resilient and determined.* Retrieved from https://www.unwto.org/2021-a-year-in-review
UNWTO. (2022a). *International tourism and COVID-19.* Retrieved from https://www.unwto.org/international-tourism-and-covid-19
UNWTO. (2022b). *Tourism data dashboard.* https://www.unwto.org/unwto-tourism-dashboard
Vaccaro, I., & Beltrán, O. (2007). Consuming space, nature and culture: Patrimonial discussions in the hyper-modern era. *Tourism Geographies, 9*(3), 254–274. https://doi.org/10.1080/14616680701422715
Valenzuela Rubio, M., Palacios García, A. J., & Hidalgo Giralt, C. (2008). La valorización turística del patrimonio minero en entornos rurales desfavorecidos: actores y experiencias. *Cuadernos de Turismo, 22*, 231–260.
Vera Rebollo, J. F., & Baños Castiñeira, C. J. (2004). Turismo, territorio y medio ambiente. La necesaria sostenibilidad. *Papeles de Economía Española, 102*, 271–286.
Villarino Pérez, M., De Uña Álvarez, E., & Cánoves Valiente, G. (2009). Reflexiones sobre experiencias de diversificación del turismo rural en España. *Serie Geográfica, 15*, 67–78. http://hdl.handle.net/10017/13906
World Summit on Sustainable Tourism. (2015). *World Charter for Sustainable Tourism +20.* https://www.biospheretourism.com/en/blog/the-world-charter-for-sustainable-tourism-20/36
Zoido, F. (2002). El paisaje y su utilidad para la Ordenación del Territorio. In F. Zoido & C. Venegas (Coords.), *Paisaje y ordenación del territorio* (pp. 21–32). Junta de Andalucía.

Paloma Ibarra Benlloch is full professor of Geography at the University of Zaragoza, Spain (Department Geography and Land Management). Teacher in the degree of Geography and Land Management as well as the masters in Land and Environmental Planning. She is a member of the GEOFORETS Research Group (GEOenvironmental processes in FOREST areas) and a collaborating researcher of the TUDISTAR Research Group (Tourism and Social-Territorial Dynamics in Rural areas, Autonomous University of Barcelona, Spain). Her research interests are mainly related to the landscape, its knowledge, valuation and management for the territorial development of rural environments.

Isabel Rabanaque-Hernández is part-time professor of Geography at the University of Zaragoza, Spain (Department Geography and Land Management). Teacher in the degree of Geography and Land Management as well as the masters in Land and Environmental Planning and in Geographic Information Technologies. She is member of the GEOT Research Group (territorial planning) and a collaborating researcher of the TUDISTAR Research Group (Tourism and Social-Territorial Dynamics in Rural areas, Autonomous University of Barcelona, Spain).

Elena De Uña-Álvarez is full professor of Geography at the University of Vigo (Department of History, Art and Geography-Campus of Ourense, Spain). Teacher in the degree of Geography and History as well as the masters in Cultural Heritage and Teacher Training of the University of Vigo. She is a member of the GEAAT Research Group (Studies on Archaeology, Antiquity and Territory) and a collaborating researcher of the TUDISTAR Research Group (Tourism and Social-Territorial Dynamics in Rural areas, Autonomous University of Barcelona, Spain). Her research interests are mainly related to natural and cultural heritage management as key resources for territorial development.

Montserrat Villarino-Pérez is senior lecturer of Geography at the University of Santiago de Compostela. Teacher in human and economic geography and in specialization courses: Master's Degree in Education, Gender and Equality. She has done research in geography and gender and rural development, alone and with the Geography and Gender Research Group of the Universitat Autònoma de Barcelona. She has participated in Development cooperation projects in Latin America with Ecuador and research fellow in Argentina. Co-founder of the Interdisciplinary Center for Feminist Research and Gender Studies and Deputy Director of the Interdisciplinary Center for American Studies "Gumersindo Busto" (University of Santiago).

Chapter 15
Land Use in Spanish Coast: Tourism as a Driving Force of Landscape Change (1990–2018)

María-Dolores Pitarch-Garrido and Carmen Zornoza-Gallego

15.1 Introduction

The UNWOT (2005) considers that the coastal areas concentrate more than three quarters of the world's tourist offer, being the Mediterranean the main destination with more than one hundred million visitors per year. The so-called "sun and beach" model is the main offer of countries such as Spain where the two main resources, the good climate and an extensive coastline, are especially relevant.

Following UNWOT (2019), "coastal tourism refers to land-based tourism activities such as swimming, surfing, sunbathing and other coastal leisure, recreation and sports activities which take place on the shore of a sea, lake or river. Proximity to the coast is also a condition for services and facilities that support coastal tourism" (p. 46). This definition combines two elements: the activities carried out by the tourist and the type of territory or space where they take place. Given that it is the territory that supports tourist activity, among others, it is also the one that receives its impact, both positive and, above all, negative. Coastal areas are generally associated with mass tourism, linked in turn to overtourism, a problem that has occupied many pages in the conventional press and in scientific research over the last decades, particularly in the second decade of the twenty-first century, in the stage prior to the crisis associated with the Covid-19 pandemic. Even in the midst of the pandemic crisis, coastal destinations are proposing new ways to continue to host as many tourists as possible. In any case, overtourism is a problem that has an impact on the territory, on local society, on the very tourism businesses that are saturated, and on tourists who perceive overcrowding as a loss of quality (Lambin et al., 2001; Li et al., 2017; Greenpeace, 2016).

M.-D. Pitarch-Garrido (✉) · C. Zornoza-Gallego
Inter-University Institute for Local Development, University of Valencia, Valencia, Spain
e-mail: Maria.pitarch@uv.es; Carmen.zornoza@uv.es

A. Blanco-Romero, M. Blázquez-Salom (eds.), *Spanish Tourism Geographies*, Geographies of Tourism and Global Change,
https://doi.org/10.1007/978-3-031-39780-6_15

In this chapter we propose a complementary way of analysing saturation in coastal areas: through changes in land use. In this sense, coastal areas tend to be those that are home to the greatest number of people, economic activities and communication infrastructures. As Plieninger et al. (2016) point out, it is not common to analyse land use changes over wide spaces. Most studies focus on local, very specific areas. However, it is important to have a territorially broader view in order to know the true impact and magnitude of land artificialisation in recent decades. The artificialisation process is based on transforming agricultural or forest soil into soil that prevents infiltration. Moreover, diachronic analysis is also relevant, despite the problems linked to the comparison of results from different years. Satellite technology has improved in recent years, providing images with a detail that makes comparison with previous years difficult.

In addition to considering in this analysis the evolution of land uses, particularly artificialised land, it is important to turn our attention to the most current situation, i.e., the current picture of the Spanish coastal territory. The level of construction and sealing of natural soil is so high that it is necessary to take conservation and land-use planning measures as soon as possible, not only in Spain but in the coastal areas worldwide (Lambin et al., 2001). The implementation of sustainable development plans in Spain must take into account the rapid and efficient process of land use changes on the coast, especially in order to be able to face already evident problems such as those related to climate change (extreme rainfall events, storms, droughts, etc.) and social change (aging of the population, mobility problems, territorial imbalances in the supply of services to the population, etc.). The concentration of constructions on the coast, with an unequal demographic weight both in space and time (in tourist areas the population is concentrated only in a few months of the year), makes a conscious and long-term planning necessary.

An analysis of changes in land use along the Spanish coast, including the archipelagos, is proposed on the basis of satellite images provided by the CORINE Land Cover (CLC) project, which was set up with the aim of obtaining a common land use database for the European Union. In the case of Spain, CORINE was the first database to cover the entire national territory. There are currently 5 publications: CLC 1990, CLC 2000, CLC 2006, CLC 2012 and CLC 2018, so that the territorial evolution covers 28 years, until a relatively recent date, which provides us with a very broad panorama that complements other sources of information to analyse the phenomenon of massification or over-tourism in the Spanish coastal land.

15.2 Overtourism and Urban Development in Spanish Coast Line

There are different ways to approach the analysis of overtourism. Many proposals have been made to reduce the impact of the accumulation of tourists at certain times of the year, particularly in fragile environments such as the historic centres of some cities (see the example, repeated on many occasions, of Venice), in natural environments of great landscape value where the accumulation of people can aggravate their sustainability, and in coastal areas, which are particularly vulnerable at certain periods of the year, especially if we consider that these are territories that are strongly affected by the natural processes associated with climate change. According to UNWOT (2005), "almost all coastal destinations share the problems of control of the use and construction of the coast, the quality of sea water (or lakes), agglomerations in certain locations (beaches) in the high season, the creation of docks and piers, erosion, solid waste disposal on the beachfront, the identification and protection of species or fragile environments, and seasonal use." (p. 274). The first of these problems, the use and construction of the coast, is of great relevance in that it has been a habitual practice of tourist municipalities as a formula for creating tourist offers, fundamentally for accommodation, identifying real estate growth with tourist development, finally generating saturated spaces of low environmental and landscape quality, in addition to the enormous difficulties associated with the management of these spaces, which are empty for part of the year.

As noted by Romero Martínez et al. (2015), the forms of urban growth have had a special impact on coastal areas, many of which are strongly oriented towards leisure, sun and beach. In Spanish coastal areas there is a strong concentration of economic and demographic activities, often with urban functions. Some of these coastal territories are urban spaces that the authors call "ideal" produced by neoliberalism, which has turned them into very dynamic metropolitan areas, globally connected and with their own contemporary urban attributes, as is the case of the Costa del Sol in Malaga, with an enormous apparent vitality, even in times of crisis. However, this enormous urban growth, which has created an almost continuous conurbation along the entire Spanish coast, has imposed a generic city model, which has replaced a traditional and endogenous model, to meet the demand of foreigners and nationals (tourists), at a high environmental and social cost.

During the 80's and 90's of the last century, the Spanish coastline experienced an enormous growth of the urbanised surface, with the excuse of the development of a tourist and real estate vocation. Huge sums of capital were concentrated in the real estate sector, in a clear process of accumulation based on land rents. Spain's urban planning legislation in those years facilitated this practice, together with cheaper financing for real estate operations, feeding back into the accumulation process. More than 4.5 million homes were built in the Mediterranean coast in 10 years, between 1996 and 2007, until the real estate bubble burst in 2008.

One way of approaching the reality of tourist territories is through the quantification of tourists or accommodation places, although this may pose some problems (Simancas & Peñarrubia, 2019). As some authors point out, overtourism depends on the type of territory (for example, islands are much more fragile) and on its physical, ecological, economic and social capacity to assume the changes (Simancas et al., 2018).

According to some authors (Hof & Blázquez, 2013), since the late 1990s, Spain has become a country dependent on real estate investment, a fact that has favoured urban growth in coastal tourist areas, such as the Balearic Islands. Regional planning regulations have been adapted in order to attract capital flows to new real estate markets. The 2008 crisis led to a certain stagnation of urban growth, however, part of the solution to that crisis was to support and promote urban expansion once again. It seems clear that an important part of urban sprawl is related to neoliberal policies, as was the case with the legalisation of irregularly built holiday homes, even in protected areas (Blázquez et al., 2011).

Tourism is a product of the development of capitalism for many reasons (Britton, 1991). Tourism, in its broadest definition, i.e. all that individual and corporate tourism activity entails, is transformative of social and material spaces. When we talk about tourism, the meaning goes beyond the activity itself and even has an ideological impact on the perception of territory and development. Property development is an important part of tourism, because its impact on land use is very significant and, in most cases, irreversible. These dwellings may be second homes, i.e. occupied for only a few months or weeks a year, they may be used for residential tourism, especially by retirees from rich countries who are looking for better living conditions in other countries, or they may be short-term holiday homes or for workers in the sector. The reality is that construction means land occupation, artificialisation and the impossibility of alternative uses. The environmental impact is real and often very worrying, but the economic impact, especially in times of crisis and high unemployment, is sufficiently positive for responsible politicians to consider only short-term decisions.

In any case, the analysis of the supply of accommodation and its quantification complements other ways of measuring the tourist saturation of a territory, such as the study of artificial land or the evolution of construction. In coastal areas this is especially relevant, as has already been pointed out.

The countries of Mediterranean Europe have in common a process of intensifying economic dependence on their specialisation in tourism and construction (EEA, 2006). Urban growth and the development of infrastructures, particularly transport, characterise the mode of financial capitalism in these countries, among which Spain stands out, with its own model, as López and Rodríguez (2011) point out. These processes are not only occurring on the coast, but also in tourist cities, particularly in Eastern Europe, as in the case of tourist cities in Romania (Rusu et al., 2020) or the coastal area in Turkey (Hepcan et al., 2013).

Other regions of the world are also experiencing urban growth processes, particularly low-density urban growth, such as in the United States, particularly on the West Coast (Beyers & Nelson, 2000; Clark et al., 2009; Bryson & Wyckoff, 2010), or on the Australian coast (Gurran & Blakely, 2007). Other areas, such as rural or mountainous areas, have also experienced land transformation processes due to low-density urbanisation (Nepal & Jamal, 2011).

Analyses of the Spanish case are particularly abundant in both the national and international literature, which gives an idea of the importance of this case, or at least of the concern on the part of academia to analyse the relationship between urban growth and tourism. Most of the research is in the form of case studies and rarely has a general overview of coastal space been undertaken, as a fundamental example of the impact of tourism on the transformation of land and landscape.

There are several driving forces behind the increase in urbanised areas worldwide, such as demographic changes that have led to a concentration of the population in urban areas (starting with the rural exodus since the middle of the last century), the extension of the infrastructure network, the development of sun and beach tourism (which explains the massive occupation of the coastline and the growth of second homes), etc. Spain is the European country with the highest number of dwellings per inhabitant, which also explains, in part, the urban speculation and the real estate boom that characterised the first two decades of the twenty-first century.

The artificial surface has experienced unprecedented growth over the last 30 years. Analyses of the Mediterranean coastline highlight the phenomenon of expansion of peri-urban areas, such as the surface area of the metropolitan areas of Barcelona, Valencia and Malaga. In the Valencian Community, the urbanised surface area increased from 2.7% of the territory in 1990 to almost 5% in 2006, almost always at the cost of the loss of agricultural land. In Murcia, the growth of artificial uses has been the highest in Spain. The relatively irreversible nature of these changes and their location in particularly vulnerable areas, such as coastal zones and protected areas, can pose a considerable environmental problem (Turner et al., 2007; Barbero-Sierra et al., 2013; Plieninger et al., 2016; Fernández Nogueira & Corbelle Rico, 2017; Prieto López et al., 2021).

The environmental impact of the artificialisation of the Spanish coastline has been enormous. On the seafront, beaches have been altered by the construction of boardwalks that act as a barrier to the natural evolution of the beaches. This is in addition to the multiplication of breakwaters and sea walls. Few beaches remain as protected natural spaces and some of them, such as the Saler in Valencia or the Artola Fossil Dunes in Malaga, are threatened by new constructions at present. The traditional urban landscape has also been modified by new constructions, with few roots in local traditions and very similar to those of the rest of the coastal tourist areas of the world.

The diffuse and generic city, built during the periods of promotion of macro construction processes, first concentrated along the coast and then dispersed conurbation towards the interior, is the urbanisation model developed throughout the Spanish Mediterranean during the last 30 years (Guaita et al., 2008; Romero Martínez et al., 2015; Serra et al., 2014; Hof & Blázquez, 2013).

Spanish coastal areas have several advantages that position them in the international market: good communication and accessibility by air and road, a good climate, quality beaches and, what is more worrying with the consequences that this entails: a great potential for urban land for the coming years, if not on the seafront, then in the immediate hinterland. These characteristics, together with the country's own, among which stand out a wide experience as a tourist destination, legal security and the quality of services and facilities.

In short, coastal areas are particularly fragile from the point of view of sustainability, which justifies their interest for academic analysis. However, in Spain, in addition to being a special territory due to its physical peculiarities, these are also spaces with a defining characteristic: tourism is the activity that shapes their structure, explains the changes in land use and justifies their social and economic transformation over the last 5 years. These are territories with urban functions, but, above all, tourist functions. The evolution of land uses, particularly artificial uses, is associated with demographic, economic and social changes. In this chapter we will analyse changes in land use and their implications for the sustainability of coastal tourist areas in Spain.

15.3 Methodology

It is essential to know the basic characteristics of CLC in order to understand the type of results and the validity of the conclusions drawn from this source. The reference scale is 1:100,000, which means that the precision associated with this scale is 20 meters, resulting from multiplying the scale factor by the limit of visual perception (100,000 × 0.2 mm = 20 meters). In addition to this precision, it should be considered that the minimum polygon is set at 25 ha, so any area must have at least this area. This fact is very important for the study of developed areas, since there are a certain number of smaller areas that will not be represented.

The possibilities offered by CLC for the disaggregation of uses are wide, since it contains information on multiple typologies at different levels. The most general classification (1 digit) is divided into the following types:

1. Artificial surfaces
2. Agricultural areas
3. Forest and seminatural areas
4. Wetlands
5. Water bodies

Given that the object of study is basically the evolution of urbanisation on the coast, the work is done by breaking down Class 1 to obtain greater precision in these uses, while the others (Classes 2, 3, 4 and 5) remain at 1 digit. Class 1 is broken down as follows:

1.1.1. Continuous urban fabric
1.1.2. Discontinuous urban fabric
1.2. Industrial, commercial and transport: Industrial and commercial units (121), Road, rail and associated land (122), Port areas (123), Airports (124).
1.3. Mining, dump and construction sites: Mineral extraction sites (131), Dump sites (132), Construction sites (133)
1.4.1. Green Urban Areas
1.4.2. Sports and leisure facilities

The breakdown into 3 digits responds to the object of the study, in which knowing whether the type of artificial fabric is continuous or discontinuous is fundamental for recognising the pattern of occupation, relating it to the impact of tourism in these areas.

As a note, it is necessary to point out that, in the case of Spanish territory, there is another land use database called "*Sistema de Información de Ocupación del Suelo en España (SIOSE)*", whose definition is greater than that of CORINE. The main reason why on this occasion it has been decided to dispense with it and base the analysis on CORINE is that the time range of the latter is much greater, allowing the tourist phenomenon to be observed with greater perspective.

Following the methodology presented by Zornoza (2013), several strips are established from the coast line in order to be able to study the different territorial dynamics. Four study strips (bands) were established: Strip 1 (0–500 m); Strip 2 (500–1000 m); Strip 3 (1000–5000 m); Strip 4 (5000–10,000 m).

Bands 1 and 2, which correspond to those closest to the coast line, are of particular interest. It corresponds to the most fragile part of the ecosystem because the pressure generated by the union of the coastal and maritime system is more pronounced. It is also here that the massive urban developments associated with sun and beach tourism are concentrated. This combination of factors means that the possible problems are clearly represented here. Specifically, the 0–500 m strip encompasses the maritime-terrestrial public domain, so beaches, promenades or dune systems are part of its definition. It is interesting to make two differentiated strips to observe their differences, if any.

Fringes 3 and 4, which form the area from 1 km to 10 km, are also part of the coastal system and are not as exposed to coastal processes as the previous ones. The aim of this analysis is to recognise the extent to which they are affected by urbanisation processes and to compare the different situations.

In this analysis the *GIS* functions are used to delimit the indicated strips and to overview the phenomenon of the growth of urbanisation or artificialisation of land on the Spanish coast, in particular in the areas associated with sun and beach tourism.

To complement the information from land use, statistical information from the Spanish National Statistics Institute (INE) has also been used to highlight the tourism function of some particularly well-defined municipalities, which the INE defines as "tourist hotspots".

15.4 Global Changes of Land Use in Spanish Coast (1990–2018)

Spain is one of the countries where artificial surface has grown the most, together with Ireland and Portugal. The annual rate of artificialisation has been almost 2%, well above the European average of 0.68% (Guaita et al., 2008). Economic dependence on land-intensive sectors, such as construction, transportation and tourism, and the consolidation of a new city model, the dispersed city, are the main causes of this enormous growth. Although changes in land use towards greater urbanisation are general throughout the country, they have become more acute in the coastal regions (Fig. 15.1).

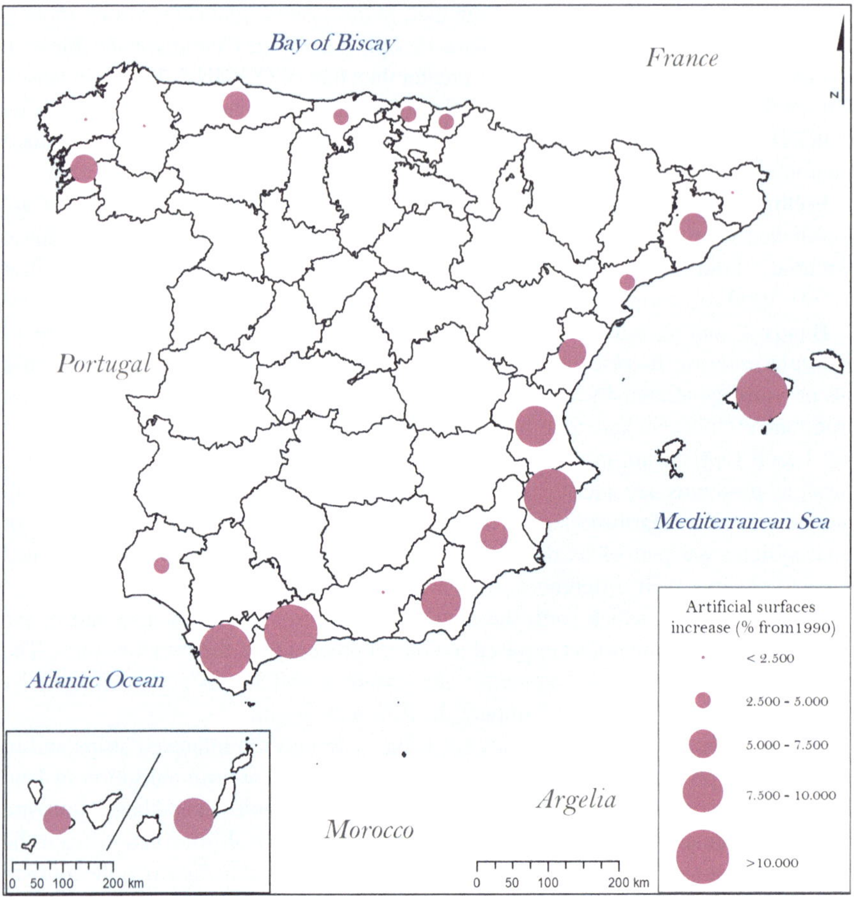

Fig. 15.1 Artificial surfaces increase, first 10 km zone from the coastline. (Own elaboration from data extracted from CLC 1990–2018)

Despite the fact that the coast is the most fragile natural area, it is also the most populated and active in the world and in Spain. Its high ecological value, both in aquatic and terrestrial areas, makes the impacts on this space especially relevant (Barragán, 1993). Conflicts derived from competition for land use, in which agriculture can hardly stop the push of urbanisation, water use, especially in the Mediterranean, where it is a scarce good and whose need increases with the concentration of activities in this space, the alteration of the landscape in areas where its value is directly related to economic benefit, etc. have over the years led to a process of flight forward that has provided solutions that are always linked to greater growth. That means an increase in engineering works and an increase in the artificialisation of land, or punctual solutions, repeated over time in a kind of fight by human being against nature, which is inexplicably expensive and ineffective, such as the enormous investments to regenerate beaches or the artificial extension of ports and promenades to gain space to the sea and continue to be productive spaces.

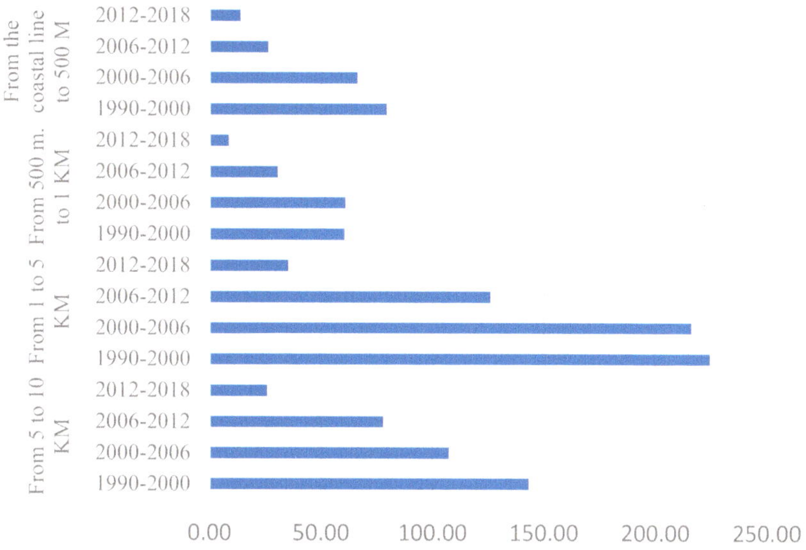

Fig. 15.2 Changes to artificial soil according to CLC in the first 10 km from the coast (in km^2). Total Spain. (Own elaboration from data extracted from CLC)

In Spain, according to the data analysed, between 1990 and 2008, 1302.4 km^2 have been transformed into artificial soil in the 10 km coastal strip. That is, this is in addition to what was previously artificialised. Most of this land has been transformed between 1990 and 2006, particularly in the first decade: between 1990 and 2000, 507.3 km^2 were transformed into artificial land, while between 2000 and 2006 this figure was 450.5 km^2. Between 2006 and 2018, the artificialisation of the soil of the coastal strip amounted to 344.6 km^2, i.e. lower than in the previous stages, despite a longer period. The financial and real estate crisis of 2008 paralysed construction and put a brake on the artificialisation of the land throughout the country. In the case of the coast, between 2006 and 2012 the highest figure of artificial land corresponds to the category of land under construction, 91 km^2, which shows the constructive inertia of the previous stage, a figure that is reduced to less than a sixth, 15 km^2 in the period 2012–2018. Next, we will approach the evolution of the process of artificialisation of the coast according to the different strips that we have considered (Fig. 15.2).

The initial situation, 1990, shows a coast (the first 10 kilometres) where forest and agricultural areas predominate, accounting for 52% and 38% of the land respectively. The artificial surface is 7%, a figure which may seem low in comparison with the previous ones, but which already represents a certain coverage especially in terms of continuous and discontinuous urbanisation, and to a lesser extent, industry and commerce, especially important in the first 5 km from the coast line.

It is clear that the Spanish coast is an overpopulated area where almost half of the population and more than 70% of tourism is concentrated. This coastal area barely accounts for 9% of the country's total surface area, yet it is home to 37% of the country's artificial surface area, which is evidence of its economic dynamism, i.e. its strength, but also its weakness, due to the enormous pressure to which this territory is subject. For more information, in 1990, 7% of the 10 km strip from the coast was artificial, a figure that rises to 11% in 2018. For the 1 km strip, the figure is even more striking: in 1990, 18% was artificial, while in 2018 it was 25%, which shows an increasingly strong saturation of the coastal soil.

In order to specify the analysis, we are going to focus on the changes from 1990 to 2018 that involve a transformation towards artificial surface. We will analyse the general figures for the whole of Spain, without losing sight of the fact that the transformations observed respond to different processes and not all of them are directly linked to the growth of tourist areas. This general vision brings us closer to the whole, although in certain areas of the coast, of which we will point out some examples, it is evident, due to the characteristics of the artificialisation process and the prevailing economic model, that the transformations respond to public and private decisions to create tourist accommodation offers in their different modalities, but, above all, real estate business.

In general terms, the coastal strip (10 km) is characterised by a change towards discontinuous urbanisation. Artificialisation has taken place on agriculture areas (61%) and forest areas (39%) (Fig. 15.3). In comparison, continuous urbanisation is scarce and only noticeable in the first part of the period under consideration, and the sites under construction are very important in the period of maximum urbanisation, before the 2008 crisis. Spaces transformed into industry, commerce and sports and leisure areas are also noteworthy in the first phase, decreasing, partly due to overcrowding, in the second part of the period considered.

If we consider only the changes to artificial use by stripes, that is, with the distance limits considered in this research, the result is relevant: the strip between 1 and 5 kilometres is the one that undergoes most of the changes, followed by the strip between 5 and 10 kilometres. In the 90 s, coinciding with the first real estate

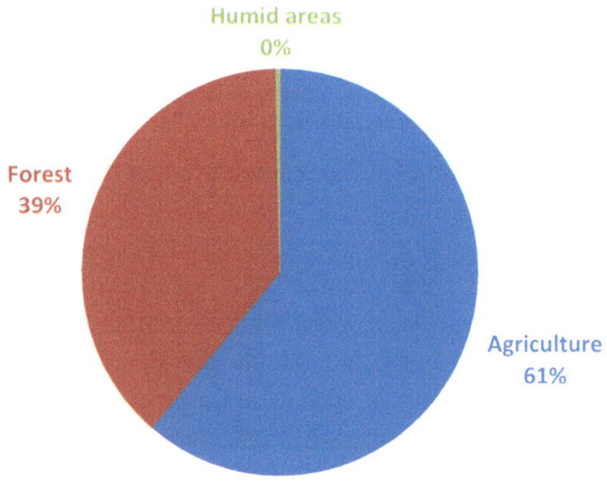

Fig. 15.3 Origin of the new artificial surfaces detected in 2018 in the 10 km strip from the coastline. (Own elaboration from data extracted from CLC)

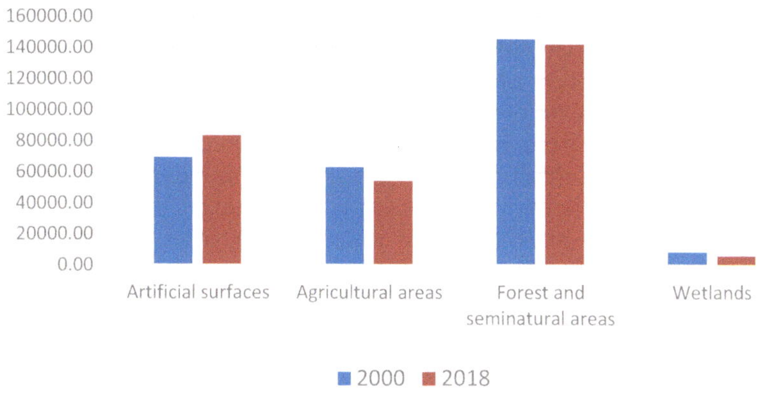

Fig. 15.4 Land use in the first 500 meters from the coastline, comparison between 2000 and 2018. (Own elaboration from data extracted from CLC)

boom, the transformation in the first 500 m from the coast line is really important, it is even higher than the second strip, between 500 meters and 1 kilometre. In this years, discontinuous urban growth, commercial growth and continuous urban development stand out, while in the second (2000–2006), and third period (2006–2012) spaces under construction (construction sites) stand out, typical of that moment of growth in the urbanisation process.

Finally, it is important to note the importance of the transformation of the area closest to the coastline, the first 500 metres of which are particularly fragile. Significant changes can be detected in this area, in particular the loss of natural land in favour of artificial land in general, with a great weight of urban and infrastructure uses (ports, roads, etc.) (Fig. 15.4). The loss of natural areas implies a reduction in the environmental quality and functionality of ecosystems, but also a reduction in the attractiveness of tourism, an activity that is highly dependent on environmental quality, especially in coastal areas.

15.5 Evolution of Artificial Surfaces by Provinces

The changes to artificial land are, as noted, an indication of economic activity, in particular real estate. In terms of tourism, we are mainly interested in the construction of housing and equipment related to improving access to the area. As an indicator of the relationship between land artificialisation, saturation and tourism, we propose the analysis of the coastal strip (10 km) at provincial level, i.e. to point out the changes towards the artificialisation of land in the coastal strip of each Spanish province.

With regard to the Autonomous Communities, it should be noted that Catalonia, Murcia, Valencia and Andalusia are the coastal regions that have experienced the greatest growth in artificial surfaces. These are joined by the archipelagos: the Balearic and Canary Islands, which have seen an increase of more than 30% in their artificial surface area over the last 30 years.

In terms of provinces, there are significant differences. In the northern provinces of Spain, the pressure of urbanisation and, above all, of infrastructure construction has been much less than in the Mediterranean. In these northern provinces most of the change in land use has been linked to large urban areas, whereas in the Mediterranean, tourism development has been fundamental to understanding these changes.

The percentage of artificial surface in the 10-kilometre strip depends on the pressure on the first kilometer of coastline, even more specifically, on the first 500 meters. The higher the pressure on the first kilometer, the more urbanisation (artificialisation of land) spreads inland, although the pressure decreases as we move away from the coast. However, the intensity is particularly noticeable in the 10-kilometre band in 2018 (Fig. 15.5). The provinces that have experienced the most significant changes are Alicante, Malaga, Cadiz and the Balearic Islands. Urban sprawl, which means artificialisation, is becoming increasingly significant for inland Majorca, where intensive tourist resorts had not previously been developed. (Hof & Blázquez, 2013). The six provinces with the greatest change towards artificial land in the stage considered account for more than 50% of the changes in the total of the first coastal strip (10 km), which is a very important weight.

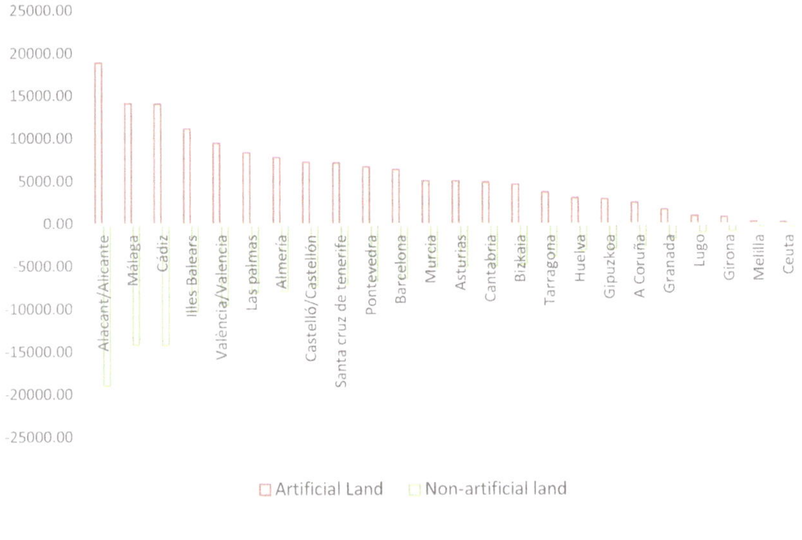

Fig. 15.5 Increase in artificial area and decrease in non-artificial area in the 10-kilometre coastal strip (1990–2018) (hectares). (Own elaboration from data extracted from CLC)

All of them are provinces with a strong growth in tourism. As an example, in Alicante, the increase of artificial surface on the coast is 3% per year in the last 28 years. In the case of the Majorcan coast, the increase of artificialisation in the same period is 2.25% per year. In both cases, these are very high figures, and in both cases the artificial surface is created at the cost of significant losses of agricultural surface. With regard to the type of artificial surface, in both cases the discontinuous urban surface stands out extraordinarily, followed by the continuous urban surface, but at a considerable distance, with similar figures to the surface destined for industry, commerce and transport, especially in 2018 and in Alicante. On the coast of Majorca, the discontinuous urban area has increased 5.3% per year, well above the Spanish average. In the case of the Costa Blanca, in Alicante, this increase is 3.3% per year.

15.6 Current Situation: Land Use in Coastal Tourist Municipalities (2018)

Tourism is the most important force for land use change in Spain, as evidenced by the fact that it is the provinces in which this activity linked to sun and beach is predominant that have undergone the greatest transformations, having experienced an artificialisation of the land. Tourism is of fundamental importance on the Spanish coast and is clearly a force for change in land use and landscape. Moreover, as it is a dynamic and constantly growing activity, it is a creator of employment and, therefore, tourist municipalities attract young national and international people in search of a job in the sector or in related sectors. Coastal tourism development in Spain presents a dual model from the point of view of accommodation: a residential model (second homes) and a hotel model (including all types of accommodation for tourists). Both cases involve growth of urbanised space, but the former much more so, especially linked to urban sprawl. According to Exceltur data, more than 80% of Spain's tourist accommodation capacity is concentrated on the coast. According to CORINE, more than 35% of the first kilometer of coastline is developed.

Tourist municipalities whose urban centres are not located by the sea have developed new areas of urbanisation next to the beach, which is the main attraction for national and international demand. The first line (500 metres) has the highest percentage of artificial surface area in Spain as a whole (Fig. 15.6), which decreases as the distance from the sea increases.

The expansion of the artificial surface, in particular of the built-up area, is most evident at the municipal level. The first coastal strip (10 km) of the most important tourist centers along the Spanish coast (60, according to the National Institute of Statistics) account for more than 30% of the artificial land of the total coastal strip (10 km) of Spain. This is a high percentage considering that in Spain there are 428 municipalities with a coastline, including large cities such as Barcelona, Valencia, Malaga, San Sebastian and Vigo, among others, which are accumulating urban

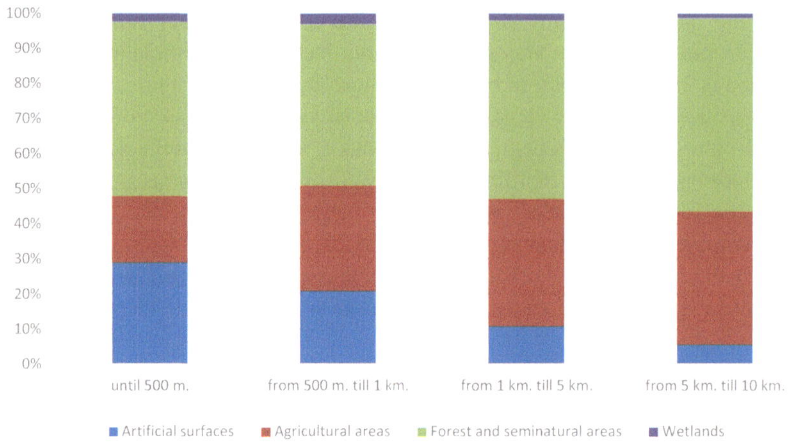

Fig. 15.6 Percentage of surface area by typology and strip from the coastline, year 2018. (Own elaboration from data extracted from CLC)

growth for reasons that are not exclusively tourist. If we eliminate the provincial capitals, the artificial land in 2018 of the tourist municipalities (42) amounts to 15.6%.

Some municipalities have a high percentage of developed or artificial land. This is the case of Cartagena (Table 15.1, Fig. 15.8a), where the area known as Manga del Mar Menor is completely occupied by housing and infrastructure constructions. This area, a 20 km long stretch, is the main driving force for sun and beach tourism in the Region of Murcia. However, in recent years, the massive occupation of this area has caused environmental problems, such as the disappearance of dunes and natural beaches, the drying up of lagoons, continuous dredging, the elimination of landscape values and the loss of biodiversity (Espejo Marín, 2011; Morales Yago, 2013). In Asturias, Llanes stands out as a traditional tourist destination, where despite the fact that all the beaches of the council, which are framed by the beaches of the Asturian Green Coast, which is considered a protected landscape, a significant urban development has been promoted since the end of the 1990s.

Always considering the coastline (10 km), it is worth noting the weight of the 20 tourist municipalities with the highest percentage of artificial land (Table 15.1), with the exception of the provincial capitals. As noted above, the Mediterranean destinations are the most prominent, although some destinations in northern Spain also appear among the first, such as Viveiro and Ribadeo, both in the province of Lugo. In both cases, there is a concentration of urban functions that go beyond the merely touristic and present an urban area on the seafront, a fact that does not occur in other municipalities where the orography defines a more inland and more dispersed settlement. On the other hand, Llanes, in the Principality of Asturias, is a council of great tourist attraction, as a result of which a notable urban growth has

Table 15.1 Residential land and artificial soil in the main tourist municipalities

Municipality	Percentage of residential land in the municipality out of the total residential land in the 10 km strip of the province	Percentage of the municipality's artificial soil out of the total artificial soil in the 10 km strip of the province
Cartagena	46,03	49,90
Marbella	20,46	16,86
Viveiro	20,83	16,28
Níjar	4,31	13,60
El Puerto de Santa María	12,43	12,84
Ribadeo	15,84	12,16
Roquetas de Mar	16,63	11,27
Lloret de Mar	11,38	10,40
Llanes	13,15	9,53
Calvià	9,52	9,15
Estepona	8,48	8,49
Algeciras	7,43	7,02
Elche/Elx	6,27	6,99
Dénia	6,27	5,80
Torrevieja	6,14	5,68
Cambrils	6,60	5,25
Gandia	5,55	4,81
Benalmádena	6,66	4,80
Salou	4,01	4,48
Benidorm	3,33	4,25

Own elaboration from data extracted from CLC

been generated in recent years. All the council's beaches, as part of the beaches of the Asturian Green Coast, are protected landscapes from an environmental point of view, so that the current possibilities for artificialisation of the land are very limited.

Most of the tourist municipalities with a high percentage of artificial land are located on the Mediterranean coast, with Cartagena, Marbella, Níjar, Puerto de Santamaría, Roquetas de Mar, etc. standing out. The Andalusian coast concentrates a high number of tourist municipalities where artificialisation of the land is very high. The paradigm is Marbella (Fig. 15.8b), which gives its name to this process, "marbellización de la costa", marked by urban speculation, landfills, promenades or a plan for marinas (Blanco Romero et al., 2021).

Other outstanding municipalities on the Mediterranean coast are Calvià, on the island of Mallorca (Fig. 15.7b), Lloret de Mar (Girona), Denia, Gandía and Benidorm (Fig. 15.7a), in the Valencian Community, among others (Table 15.1).

Finally, it should be noted that, in addition to the process of urban development over the last three decades, the municipalities along the Spanish coast have continued to grow in terms of accommodation places and the number of tourists each year.

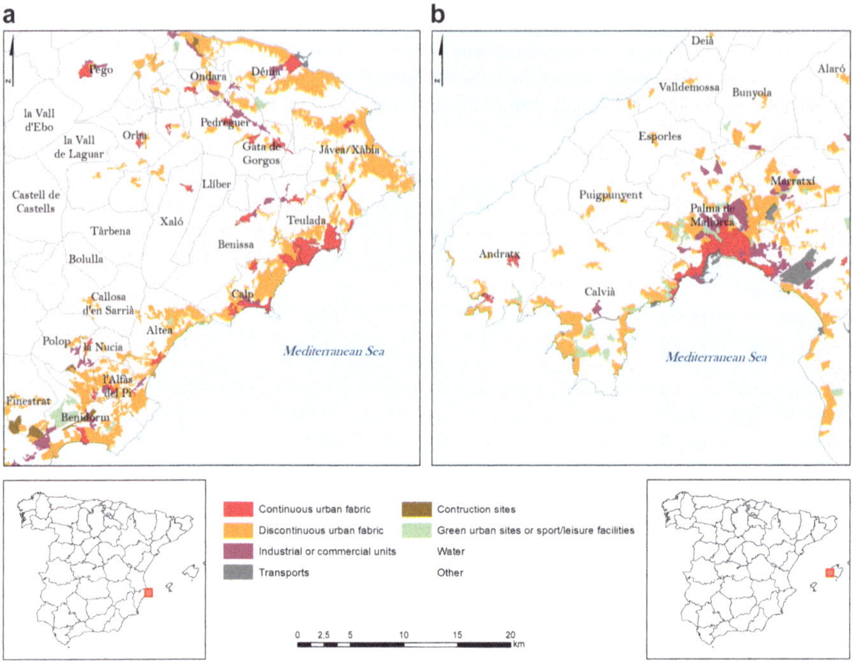

Fig. 15.7 Artificial land use 2018: Alicante and Majorca. (Own elaboration from data extracted from CLC)

Evidently, the increase in urbanisation and access infrastructures has facilitated the consolidation of these towns as important receiving centres. However, the dual offer of accommodation, the tourist accommodation (hotels, campsites, etc.) and second homes, has an obvious effect on local development: while the former generates employment and attracts population, the latter is at the root of the "ghost towns" that populate the Spanish coast during the winter.

15.7 Conclusion: Implications for the Sustainability of Coastal Tourism Territories

As indicated in the title of this communication, saturation of the coastline is neither a new nor an unforeseen phenomenon. On the contrary, there has been a process of artificialisation, accelerated between the late 1990s and the beginning of the twenty-first century, but which has not been reduced or reversed at any time or in any coastal territory. Tourism, or the excuse for it, has been one of the main architects of this transformation of the coastal space, and, therefore, of abusive urbanisation, in particular of the spaces most directly linked to the sea, that is, the first coastal strip, from the coastline to 10 km, or what is more, from the coastline to 1 km.

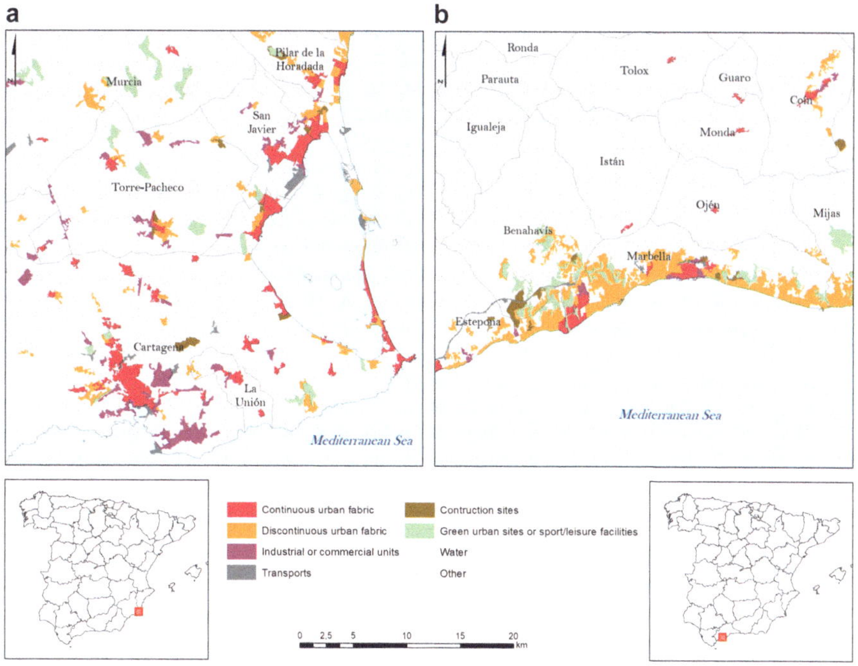

Fig. 15.8 Artificial land use 2018: Murcia and Malaga. (Own elaboration from data extracted from CLC)

With this first approach, we wanted to point out the magnitude of the phenomenon, as well as the need for a more protectionist planning. This phenomenon, which is not exclusive to Spain but common to the countries of the Mediterranean, shows the consequences that the generation of immediate benefits and the short-sighted visions of investors and governments have had in a particularly fragile space that is highly saturated by very diverse activities, to which tourism is added. The consequences of this process of saturation and artificialisation are unavoidable and impossible to reverse, which is why it will be necessary, looking ahead, to define the territorial model we want and to implement the necessary land planning actions. Throughout the years studied, it has been noted that, although there have been different visions according to the territories, it has not been possible to stop the process that seems to sweep away any attempt at soil conservation along the coastline. Although the crisis of 2018 meant a 100% slowdown, in the last period analysed (2012–18), a certain recovery of the growth rate of urbanisation is beginning to be seen.

More detailed studies will show the magnitude of this process at a regional and local level, as well as the existence of different growth models, which, linked to other data of a tourist nature, explain the evolution of the transformation and tourist processes in Spain.

References

Barbero-Sierra, C., Marques, M. J., & Ruíz-Pérez, M. (2013). The case of urban sprawl in Spain as an active and irreversible driving force for desertification. *Journal of Arid Environments, 90*, 95–102. https://doi.org/10.1016/j.jaridenv.2012.10.014

Barragán, J. M. (1993). *Ordenación, planificación y gestión del espacio litoral*. Oikos-Tau.

Beyers, W. B., & Nelson, P. B. (2000). Contemporary development forces in the nonmetropolitan West: New insights from rapidly growing communities. *Journal of Rural Studies, 16*(4), 459–474. https://doi.org/10.1007/1-4020-3902-6_13

Blanco Romero, A., Blázquez Salom, M., de la Calle Vaquero, M., Fernández Tabales, A., García Hernández, M., Lois González, R. C., Mínguez García, M. C., Navalón García, R., Navarro Jurado, E., & Troitiño Torralba, L. (2021). *Diccionario de Turismo* (389 pp). Ediciones Cátedra.

Blázquez, M., Murray, I., & Artigues, A. A. (2011). La balearización global: el capital turístico en la minoración e instrumentación del Estado. *Investigaciones Turísticas, 2*, 1–28.

Britton, S. (1991). Tourism, capital and place: towards a critical geography of tourism. *Environmental and Planning D: Society and Space, 9*(4), 451–478. https://doi.org/10.1068/d090451

Bryson, J., & Wyckoff, W. (2010). Rural gentrification and nature in the Old and New Wests. *Journal of Cultural Geography, 27*(1), 53–75. https://doi.org/10.1080/08873631003593232

Clark, J. K., McChesney, R., Munroe, D. K., & Irwin, E. G. (2009). Spatial characteristics of exurban settlement pattern in the United States. *Landscape and Urban Planning, 90*(3–4), 178–188. https://doi.org/10.1016/j.landurbplan.2008.11.002

Espejo Marín, C. (2011). Innovación para la competitividad turística en la Manga del Mar Menor (Murcia). *Cuadernos de Turismo de la Universidad de Murcia, 27*, 321–339.

European Environment Agency (EEA). (2006). *Urban Sprawl in Europe—The ignored challenge; EEA*. https://www.eea.europa.eu/publications/eea_report_2006_10

Fernández Nogueira, D., & Corbelle Rico, E. (2017). Cambios en los usos de suelo en la Península Ibérica: Un meta-análisis para el período, 1985–2015. *Biblio3W Revista Bibliográfica de Geografía y Ciencias Sociales*. Vol. XXII, n° 1.215 https://doi.org/10.1344/b3w.0.2017.26442

Greenpeace. (2016). *A toda Costa, Análisis de la evolución y estado de conservación de los bienes y servicios que proporcionan las costas*, in: https://es.greenpeace.org/es/wp-content/uploads/sites/3/2018/07/Informe-Ampliado.pdf

Guaita, N., López, I., & Prieto, F. (2008). Cambios de ocupación del suelo en España: implicaciones para la sostenibilidad. *Ciudad y Territorio Estudios Territoriales, 156*, 235–260.

Gurran, N., & Blakely, E. (2007). Suffer a sea change? Contrasting perspectives towards urban policy and migration in coastal Australia. *Australian Geographer, 38*(1), 113–131. https://doi.org/10.1080/00049180601175899

Hepcan, S., Hepcan, C. C., Kilicaslan, C., Ozkan, M. B., & Kocan, N. (2013). Analyzing landscape change and urban sprawl in a Mediterranean coastal landscape: a case study from Izmir, Turkey. *Journal of Coastal Research, 29*(2), 301–310. https://doi.org/10.2307/23353628

Hof, A., & Blázquez, M. (2013). The Linkages between Real Estate Tourism and Urban Sprawl in Majorca (Balearic Islands, Spain). *Land, 2*(2), 252–277. https://www.mdpi.com/2073-445X/2/2/252

Lambin, E. F., Turner, B. L., Geist, H. J., Agbola, S. B., Angelsen, A., Bruce, J. W., Coomes, O. T., Dirzo, R., Fischer, G., Carl, F., George, P. S., Homewood, K., Imbernon, J., Leemans, R., Li, X., Moran, E. F., Michael, M., Ramakrishnan, P. S., Richards, J. F., Skånes, H., Steffen, W., Stone, G. D., Svedin, U., Veldkamp, T. A., Vogel, C., & Xu, J. (2001). The causes of land-use and land-cover change: moving beyond the myths. *Global Environmental Change, 11*(4), 261–269. https://doi.org/10.1016/S0959-3780(01)00007-3

Li, L., Zhang, J., Nian, S., & Zhang, H. (2017). Tourists' perceptions of crowding, attractiveness, and satisfaction: A second-order structural model. *Asia Pacific Journal of Tourism Research, 22*, 1250–1260. https://doi.org/10.1080/10941665.2017.1391305

López, I., & Rodríguez, E. (2011). The Spanish model. *New Left Rev, 69*(3), 5–28.
Morales Yago, F. J. (2013). El impacto de la actividad turística sobre el paisaje de La Manga del Mar Menor (Murcia). *Estudios Geográficos, 74*(275), 523–556. https://doi.org/10.3989/estgeogr.201319
Nepal, S. K., & Jamal, T. B. (2011). Resort-induced changes in small mountain communities in British Columbia, Canada. *Mountain Research and Development, 31*(2), 89–101. https://doi.org/10.1659/MRD-JOURNAL-D-10-00095.1/
Plieninger, T., Draux, H., Fagerholm, N., Bieling, C., Bürgi, M., Kizos, T., Kuemmerle, T., Primdahl, J., & Verburg, P. H. (2016). The driving forces of landscape change in Europe: A systematic review of the evidence. *Land Use Policy, 57*, 204–214. https://doi.org/10.1016/j.landusepol.2016.04.040
Prieto López, A., Duarte-Abadía, B., & Boelens, R. (2021). Territory in conflict: Land dispossession, water grabbing and mobilization for environmental justice in southern Spain. *International Journal of Water Resources Development, 37*(6), 996–1020. https://doi.org/10.1080/07900627.2020.1854693
Romero Martínez, J. M., Romero Padilla, Y., & Navarro Jurado, E. (2015). Atributos urbanos contemporáneos del litoral mediterráneo en la crisis global: caso de la zona metropolitana de la Costa del Sol. *Scripta Nova. Revista Electrónica de Geografía y Ciencias Sociales, 19*, 500–526.
Rusu, R., Bodocan, V., & Man, T. (2020). Urban Sprawl and Its Impact on Urban Tourism in Romania. *Journal of Balkan and Near Eastern Studies, 22*(4), 521–533. https://doi.org/10.1080/19448953.2020.1775403
Serra, P., Vera, A., Tulla, A. F., & Salvati, L. (2014). Beyond urban–rural dichotomy: Exploring socioeconomic and land-use processes of change in Spain (1991–2011). *Applied Geography, 55*, 71–81. https://doi.org/10.1016/j.apgeog.2014.09.005
Simancas, M., & Peñarrubia, M. P. (2019). Analysis of the Accommodation Density in Coastal Tourism Areas of Insular Destinations from the Perspective of Overtourism. *Sustainability, 11*, 3031. https://doi.org/10.3390/su11113031
Simancas, M., García, J. I., Greifemberg, C., & Peñarrubia, M. P. (2018). Strategies to improve the quality and competitiveness of coastal tourism areas: Applying tourism standards. *Journal of Tourism Analysis: Revista de Análisis Turístico, 25*, 68–90. https://doi.org/10.1108/JTA-02-2018-0007
Turner, B. L., Lambin, E. F., & Reenberg, A. (2007). The emergence of land change science for global environmental change and sustainability. *Proceedings of the National Academy of Sciences of the United States of America, 104*(52), 20666–20671. http://www.jstor.org/stable/25450958
UNWOT. (2005). *Indicadores de Sostenibilidad para los Destinos Turísticos*. Guía Práctica, UNWTO.
UNWOT. (2019). *Tourism definitions*, https://www.e-unwto.org/doi/10.18111/9789284420858
Zornoza, C. (2013). *Crecimiento urbanístico en la zona costera de la Comunidad Valenciana (1987–2009). Análisis y perspectivas de futuro*. IIDL/PUV.

María-Dolores Pitarch-Garrido is Professor of Human Geography at the University of Valencia (Spain). She was director of the Inter-University Institute of Local Development (2018–2022). She has been visiting professor at the University of Trier, the London School of Economics and Massey University, among others. She has participated in research projects funded by EU Framework Programmes. She has carried out consultancy work on subjects related to tourism geography. She is currently working on aspects related to territorial tourism models, tourism sustainability indicators, among others, focusing her interest on the impact of these issues on people and territory.

Carmen Zornoza-Gallego is lecturer of Human Geography at the University of Valencia (Spain). She is a researcher of the group "Innovation and Local Development" from the University of València since 2015. She is a specialist on Cartography and GIS. She has participated in research projects funded by EU Framework Programmes. Her research is focused on mobility, urban structure, land use and the impact of climate change on tourism.

Chapter 16
Sustainable Tourism Indicators in Cities

Anna Torres-Delgado, Aurélie Cerdan Schwitzguébel, and Pol Pareto Boada

16.1 Introduction

Prior to 2019, global tourism grew dramatically due to widespread economic development, lower transport costs and a growing middle class in advanced and emerging economies. Cities became increasingly popular destinations, to the point that revenues generated by national and international tourism contributed significantly to the socio-economic and cultural development of many cities and their surroundings. However, this urban tourism growth, by its volume, created important challenges for cities in terms of sustainable tourism. Specifically, undesirable consequences were evident, linked to social and cultural disruption (space congestion, increased living costs and, in particular, housing, social inequality and vulnerability), and to environmental degradation (increased: energy and water consumption, air pollution and waste generation; and saturation of transport networks) (Martins, 2018). Also, for the local economy, the quality of tourism-related employment was questioned (Claveria González, 2009) and there were price increases for products and services in general (Sancho, 1998). Often, these various dynamics led to a loss of quality of life for the local community and, also, to deterioration in the visitors' experiences and satisfaction levels.

In 2020, the COVID-19 pandemic broke the tourism growth tendency and imposed a completely different scenario. Socioeconomic changes led by the pandemic have had a significant impact on tourism, including changes in: mobility,

A. Torres-Delgado (✉)
University of Surrey, Guildford, UK

University of Barcelona, Barcelona, Spain
e-mail: a.torres-delgado@surrey.ac.uk; annatorres@ub.edu

A. Cerdan Schwitzguébel · P. Pareto Boada
Independent Scholar, Barcelona, Spain

A. Blanco-Romero, M. Blázquez-Salom (eds.), *Spanish Tourism Geographies*, Geographies of Tourism and Global Change,
https://doi.org/10.1007/978-3-031-39780-6_16

socialisation and consumption patterns, leisure and work (Romagosa, 2020). Therefore, the pandemic has induced a crisis in the tourism industry that has implications (opportunities and threats) for the sustainability of the sector.

Sustainability (social, cultural, economic and environmental) has become a 'must' to guarantee the recovery of a better and more resilient model of tourism in cities (López Palomeque et al., 2021). Destination Management/Marketing Organisations (DMOs) need new terms of reference that go beyond destination promotion to increase visitor volume and expenditure, and must overcome sustainability rhetoric driven by short-term economic factors (Amore & Hall, 2016; Tervo-Kankare, 2011; Wyss et al., 2014). Going forward, DMOs must focus on planning and managing tourism as a vehicle to improve the quality of life of residents and visitors, and the sustainability of their cities. To meet these new requirements, tourist boards need more complete data on tourism impacts and new management skills to deliver a broader range of policy decisions. Recognising the changes required, several European cities are now giving fresh impetus to creating indicators to monitor sustainability. Indicators are particularly important because they are instruments that generate objective information on the processes in the destinations (Torres-Delgado & López Palomeque, 2014).

Although tourism is an information-intensive sector, it lacks adequate knowledge transfer among its participants (WTO, 2012a). There is, therefore, a clear need to develop indicators, articulated and interpreted under the paradigm of sustainability, that can systematically collect and share available information. This production of data, assimilation of information into knowledge and, subsequent, exchange of knowledge between stakeholders must contribute to a strategic change in the way tourism policy is conducted.

This chapter is devoted to a deeper consideration of the tourist sustainability indicators that shape the reality of urban environments. The aims are to: (i) identify specific components of urban environments in which tourism generates added, or different, pressures to those generated in other geographical environments (coastal, rural, and mountain) in relation to sustainability; (ii) gather relevant experience of the use of indicators relating to urban or tourism sustainability; (iii) exemplify the implementation of STIs in urban destinations through gathering several experiences in Spanish cities.

16.2 Sustainability and Tourism in the Urban Environment

Cities, and the urban environment by extension, have their own characteristics that give a certain uniqueness to the tourist practices that take place there. Urban tourist resources are unique, as many of them are part of the city's structure (markets, urban fabric, buildings, parks, squares etc.) and are accessible public amenities at no additional cost (Mathieson & Wall, 1982). Such attractions often involve intense and complicated interaction between the various parties involved in tourism, whose interests, motivations and sensitivities may be unalike (Timur & Getz, 2009). Good coexistence between tourists and citizens, for example, is one of the great challenges of urban tourism (Koens et al., 2018) and its management is not easy, as often both parties have similar patterns of urban consumption, which forces them to compete for space (Füller & Michel, 2014).

Cities have broad and diversified economic bases that enable them to have more independence when formulating sustainable tourism policies in comparison to other geographical environments that are more dependent on tourism (such as sun and beach destinations). Moreover, urban tourism tends to be less seasonal, based on resources and products that do not depend on climatic qualities (McKercher et al., 2015). However, cities already have considerable environmental conflicts and tourism intensifies these conflicts by increasing: population mobility (contribution to gas emissions); consumption of resources (water and energy); and waste generation (Miller et al., 2015).

Undoubtedly, the complexity of tourist flow, and the overlap of this with the city's multiple functions (Fernández-Güell & González-López, 2016), greatly complicates the management of impacts (Page & Hall, 2003), which tend to be highly subjective between stakeholders and variable over time. Therefore, measuring tourism sustainability in cities requires identifying the differential factors in which tourism exerts additional pressure on overall sustainability. A review of the literature on this topic identifies eight pressure points:

1. *The use of public space, infrastructure, and services.* The multifunctional nature of urban tourist spaces means that tourists and residents meet repeatedly and often in significant intensities. This causes congestion in public spaces (streets, squares or green areas) and in infrastructure and services (transport, housing, restaurants and trade), leading to competition for access to amenities, overcrowding of the environment and shifts away from the original functions of the spaces (García-Hernández et al., 2017; Blanco-Romero et al., 2018).
2. *The coexistence of tourists and residents.* Cities' functions and services are shared between the people who visit and those who live there (Ashworth & Page, 2011). When this interaction is intense, the local community's daily life is disrupted. Disrespectful tourist attitudes towards the host environment or society, when they occur, lead to tensions in the community (Zaar & Pontes da Fonseca, 2019).
3. *Air quality.* Mobility is an intrinsic condition of tourist practice and, for cities, the increase in transport provision, to accommodate the additional demand, increases one of the most severe environmental problems in the city: air pollution (WTO, 2004).
4. *The cityscape.* The urban landscape (city structure and buildings) is often one of the most important tourist attractions in cities. Tourism can act as a positive catalyst for the recovery of degraded or obsolete urban areas or, conversely, it can lead to overexploitation and trivialization or degradation of the urban landscape (WTO & World Tourism Cities Federation, 2018).
5. *The material and historical heritage.* The historical centres of cities are usually nuclei of heritage agglomeration and high tourist attraction. Tourism can contribute to the recovery, enhancement and conservation of material, historical heritage (WTO & World Tourism Cities Federation, 2018) but it can also threaten that

heritage if tourist flows are intense and highly concentrated in the space (Aranburu et al., 2016).

6. *Intangible cultural heritage*. Globalisation and the cosmopolitan nature of cities are intrinsic conditions that threaten the customs, traditions and identities of urban environments. In this sense, tourism can help to recover and revalue this intangible heritage or, conversely, to generate cultural trivialisation in destinations (WTO, 2012b).
7. *Safety*. Safety at a destination is an element that directly conditions the tourist experience and satisfaction levels. Cities tend to have higher crime rates than other, less populated areas. However, tourism potentially exacerbates illegal practices (such as drugs, prostitution, theft), which are often common in tourist areas (García-Hernández et al., 2017; Aris Anuar et al., 2012).
8. *The cost of living*. The administrative centrality of cities, combined with greater capital investment and wealth opportunities, condition greater economic development and population growth than is found in other environments. Inevitably, this means competition for space and resources, which translates into an increase in living costs. Tourism contributes to this competition, providing a profile of consumers who are willing to pay more, which leads to an increase in the prices of products and services in tourist areas. Moreover, it is necessary to consider the processes of replacement of residential housing in favour of tourist rent that, together with other processes of urban regeneration, contribute to gentrification and the emergence of speculative processes (Cócola Gant, 2016; Diaz-Parra & Jover, 2021).

16.3 Sustainable Tourism Indicators (STI) for Tourism or Cities

Sustainable tourism, as a concept, is meaningless without indicators and other monitoring tools to inform us about the impacts of activities and determine whether or not they are acceptable (Cool et al., 2001; Wheeller, 1993). Tourism policy needs objective data to measure changes in tourism development, to enable policy makers to anticipate, manage and prevent undesirable (or unsustainable) situations at their destination. Sustainability indicators are essential in monitoring sectoral development to facilitate the assessment of tourism policies and practices, measure sectoral performance and develop suitable strategies for a preferred future (Torres-Delgado & Saarinen, 2013; Castellani & Sala, 2010; Gahin et al., 2003). Moreover, sustainability indicators are useful tools to support knowledge transfer, as they generate objective, quantitative data that provide a fuller understanding of tourist phenomena in a spatial context (Blancas et al., 2010; Sánchez & Pulido, 2008).

Although there has been considerable research into, and application of, sustainability indicators to inform tourism policy, few studies have focused specifically on urban environments. The next three sections provide overviews of International, European and Spanish experience in the application of sustainable indicators, from an urban or tourist perspective.

16.4 International Experience

At the international level, the United Nations (UN) acts as a reference institution in sustainable development, currently under the guidance of the *2030 Agenda for Sustainable Development,* which includes 17 Sustainable Development Goals (SDGs). Goal 11 of the SDGs stands out, as it directly relates to the sustainability of cities: "*Goal 11: Make cities and human settlements inclusive, safe, resilient and sustainable*" (UN, 2015). Through the global Sustainable Development Solutions Network (SDSN), the UN has already proposed indicators to monitor this challenge but none of them relate specifically to tourism.

The WTO is responsible for translating the SDGs into the tourism sector and, with reference to Goal 11, the WTO states that tourism *"can advance urban infrastructure and accessibility, promote regeneration, and preserve cultural and natural heritage"* (WTO, 2022). Although WTO has extensive experience in producing documents to guide destinations in measuring sustainable tourism (see, for example, WTO, 1996, 2004), for its report to the UN Statistics Division for the Global SDG Indicators database, the WTO is only custodian of two indicators, "8.9.1 Tourism direct GDP as a proportion of total GDP and in growth rate" and "12.b.1 Number of sustainable tourism strategies or policies and implemented action plans, with agreed monitoring and evaluation tools". Neither of these two is related to tourism's contribution to urban sustainable development.

Currently, the WTO, in collaboration with the UN Statistics Division, is developing the Measuring the Sustainability of Tourism (MST) programme to measure the impacts and contributions of tourism on the economy, society and the environment in all countries of the world. Alongside this initiative, in 2004 WTO launched the International Network of Sustainable Tourism Observatories (INSTO), which urges destinations to measure sustainable development in tourism. Since its establishment, 32 observatories have joined the network including 4 in Spain (Barcelona, Mallorca, Canary Islands and Navarre). However, the INSTO doesn't define or endorse specific indicators; it points to generic issues that need to be monitored (such as seasonality, employment, economic benefits, governance, local satisfaction, energy/water/waste management).

Other well established, worldwide examples of monitoring sustainable development in tourism include the Global Destination Sustainability Movement (GDSM), the Global Sustainable Tourism Council (GSTC) and the Responsible Tourism Institute (RTI). The GDSM uses 70 indicators to measure the sustainability performance of destinations and provides a global index that enables benchmarking among members (https://www.gds.earth). In Spain, the index has only been applied to the cities of Barcelona, Bilbao and Valencia, by giving them a relatively low position in the ranking (far from the top 20). Conversely, the GSTC has developed a certification with criteria that outline principles and requirements to guide destinations and tourism businesses towards more sustainable schemes. It also includes performance indicators to measure compliance with the criteria (https://www.gstcouncil.org). Also, the RTI uses the Biosphere certification to certify the sustainable commitments of destinations, companies and others, and builds indicators to monitor the effectiveness of the proposed actions (https://www.responsibletourisminstitute.com). The Biosphere certification has been the most spread in Spain with more than 15 destinations certified in 2022.

No examples of measuring tourism sustainability through an urban lens have been identified. However, the *ICLEI – Local Governments for Sustainability* (founded in 1990 as the International Council for Local Environmental Initiatives) is worth mentioning. This is an international network of the governments of more than 2500 cities, towns and regions in 125+ countries worldwide. Members are committed to achieving low-carbon, resilient, biodiverse, resource-efficient and productive, healthy and happy cities with a green economy and smart infrastructure (https://iclei.org). ICLEI, through the World Council on City Data (WCCD), collects performance indicators of cities to monitor their evolution towards more sustainable scenarios. The indicators support the WCCD in certifying cities with the *ISO 37120: Sustainable development of communities* certificate. In the same vein, the *Sustainable Cities International* (SCI) programme has designed a set of common indicators that are used by various international cities to guide them towards urban sustainability (European Commission DG Environment, 2018).

16.5 European Experience

Sustainable development in Europe has been promoted mainly through the European Commission (EC) and the European Environment Agency (EEA). The need for indicators and tools for management and decision-making was repeatedly mentioned in institutional initiatives on European tourist policy (Torres-Delgado & López Palomeque, 2014). In response, the EC launched the European Tourist Indicator System (ETIS) in February 2013, which aimed to promote sustainable management at the destination level through a comprehensive and flexible system of indicators (EC, 2016). The ETIS toolkit identified 43 core indicators that covered the fundamental aspects of sustainability monitoring and allowed for comparison over time and benchmarking between destinations (see https://ec.europa.eu/growth/sectors/tourism/offer/sustainable/indicators_en). Additionally, the ETIS also suggested considering supplementary indicators to tailor the tool to destination peculiarities (such as urban areas). However, few examples were provided and destinations were asked to develop others on their own. After two pilot studies (with 10 Spanish destinations engaged), the EC ceased to support the ETIS in 2016 and transferred ownership of the tool to single destinations (Font et al., 2021). Since then, many destinations have adapted the ETIS to their needs and possibilities (see for example Torres-Delgado et al., 2021). This has enabled them to meet their aims and more easily collect data, although harmonisation between destinations has been lost and comparison is no longer possible.

Other important European tourism organisations have also developed indicators to measure sustainability in tourism. For example, the European Travel Commission

(ETC), within its benchmarking project, defined 12 STIs to facilitate the exchange of data among its member countries (see ETC, 2021). Also, European Cities Marketing (ECM), a non-profit, knowledge sharing network with a long tradition of benchmarking tourism in cities, has created a knowledge group on sustainability to propose common key performance indicators (KPIs) that go beyond the number of overnight stays. Another example is the Network of European Regions for Competitive and Sustainable Tourism (NECSTouR), which includes a working group on indicators designed to facilitate European regions to work together to optimise and improve the European system of statistics and measurement (see https://necstour.eu/strategic-lines/statistics-and-measurability).

The EEA has also considered tourism sustainability in cities but only from an environmental perspective. In fact, the EEA has been working, since 2000, to produce specific indicators as part of a global strategy to incorporate tourism into the European framework of sustainable development indicators. The proposed indicators pay particular attention to the polluting effects of transport and many of them are displayed in EUROSTAT.

16.6 Spanish Experience

The Ministry of the Environment (MMA) (now known as the Ministry for the Ecological Transition and Demographic Chal-lenge (MITECO)) was the leading agency in generating indicators of urban sustainability and sustainability in tourism in Spain. The OECD proposed, for Spain, a system of indicators to assess environmental sustainability in tourism, based on the EEA's sector-specific indicators. The Spanish Environmental Tourism Indicator System (SETIS) was adapted and published in 2003 by the MMA (Ministry of the Environment, 2003) and became a sector benchmark at the regional level. Later, in 2008, these indicators were aggregated into a single value known as the Sustainable Tourism Index (ST INDEX). This index enabled benchmarking of environmental sustainability in Spanish destinations (see Sánchez & Pulido, 2008). Later the SDGs defined a more holistic approach to sustainable tourism but, at the national level, only guidelines have been drafted to date, not measurement tools.

In Spain, the regional level has been prolific in developing Sustainable Tourism Indicators (STIs). For example, the Balearic Islands was one of the leading regions, producing in 2000 its own set of STIs (see CITTIB & University of the Balearic Islands, 2000). Although this set had an environmental bias, it was one of the first examples in Spain of sustainable indicators being used to monitor tourism development and its impact. The tool was designed to inform and guide future policies on tourism in the islands. Andalusia also made a significant contribution by building and implementing a set of indicators and synthetic indexes to assess, to monitor and evaluate results, and to rethink tourism policies when needed (see Junta de Andalucía, 2015). The Andalusian data system was presented in a friendly, intuitive way to facilitate the decision-making process, with results coloured in a "traffic lights" scheme to represent whether or not an indicator had evolved towards the objective in which it was framed. Also, the region of Barcelona built its own set of indicators based on the ETIS. Using this set, longitudinal data analysis on sustainable tourism performance was conducted in 27 municipalities and 11 counties (see Torres-Delgado et al., 2021). Later, these indicators were harmonised at the city level for Barcelona and were included in the Barcelona Tourism Observatory (see https://www.observatoriturisme.barcelona/en/sustainability-sits-otb-otb-sustainable-tourism-indicators-system).

Sustainability was not approached in any of these examples from an urban perspective - only from a tourism approach. However, a focus on sustainability in cities was provided by the MMA in coordination with the Barcelona Urban Ecology Agency (AEU-BCN). Jointly, they developed two models for quantifying urban sustainability: 1) The Municipal System of Sustainability Indicators, to measure the degree of sustainability of any municipality in Spain under the same criteria; and 2) The System of Indicators and Conditioners, a tool for large and medium-sized cities (with more than 50,000 inhabitants) that introduced urban specificities to select the best sustainability indicators, thus encompassing the complexities of cities (Ministry of the Environment and Fluvial and Marine Affairs, 2012). Although the general value of these tools is evident, the measurements do not, specifically, consider the tourism sector's impact on sustainability.

16.7 Examples of STIs Implemented in Spanish Cities

The following sections introduce examples of Spanish cities where STIs have been applied/tested. In general, cities tend to have well developed statistics on tourism, but very few of them refer to them as indicators for sustainable tourism. Several arguments about the usefulness of "urban-adapted" STIs to manage urban destinations can spur from the experiences presented.

16.8 Barcelona

Tourism is an economic activity with a long history in the city of Barcelona, with public tourism policies exercised for more than a century (Palou, 2016). In this sense, Barcelona's strategy is no different from other cities that seek to position themselves as popular destinations in the global, urban hierarchy, to attract both visitors and financial investments (Murray, 2014). However, in the last decades, a large increase in tourist influx has led to a broad citizen debate about the city's tourism model, being debated from the street to the institutions, passing through the media, national and international (Milano, 2017). At the same time, a counter-discourse emerged, from the neighbourhood organisation, that contributed new elements to the debate on social sustainability of tourism (Fernández & Pardo,

2016). In just a few years, Barcelona went from "coexisting" with tourism to "suffering" tourism (Elorrieta et al., 2022); the social conflict requires an effective tourism policy for the future, to manage the costs and distribute the benefits of tourism.

Accordingly, in 2019 researchers of the University of Barcelona built and implemented a set of indicators (see Table 16.1) for assessing the social impact of tourism the city of Barcelona (Pareto Boada et al., 2020). The research tested geospatial analysis of the indicator data as a mean to evaluate the impacts of tourism from a geographical perspective since there were significant differences in the economic, social, environmental and tourist realities among the districts that make up a city.

Table 16.1 Indicators for measuring tourism social sustainability in the city of Barcelona

Criteria	Indicators
1. Respect for the socio-cultural authenticity of the host communities	1.1. % of events held in the destination focused on traditional and local culture and/or heritage
	1.2. No. of categories of tourism resources
	1.3. % of cultural entities involved in tourism
	1.4. % of residents satisfied with the impact of tourism on the identity of the destination
2. Preservation of the social carrying capacity of the city	2.1. No. of tourist beds per 100 inhabitants
	2.2. No. of tourists per square kilometre
	2.3. Ratio between tourist and resident populations
3. Preservation of the functional diversity of the city	3.1. % of souvenir shops
	3.2. % of tourist shops
4. Citizen participation in tourism	4.1. % citizen associations of the overall engaging in the design of tourism strategies
5. Facilitated access to housing	5.1. Average house purchase price by average income
	5.2. Average house rental price by average income
	5.3. % change in house purchase price
	5.4. % change in house rental price
	5.5. % social housing
	5.6. No. of eviction orders
6. Accessibility for disabled people	6.1. % of hotels with adapted rooms for people with reduced mobility
	6.2. % of hotels with adapted rooms for people with visual disabilities
	6.3. % of hotels with adapted rooms for people with hearing disabilities
	6.4. % of tourist products and attractions accessible to people with some disability (motor, visual, auditory and cognitive)
	6.5. % of accessible public transport

(continued)

Table 16.1 (continued)

Criteria	Indicators
7. Mutually positive coexistence between residents and visitors	7.1. % of the residents that considers tourism important or beneficial for the city
	7.2. % of residents who consider that there are too many tourists in the city
	7.3. % of residents who believe that tourism causes price rises in their neighbourhood
	7.4. % change in population census
	7.5. % of residents annoyed with tourist apartments
8. Access to decent employment	8.1. Average annual salary of residents employed in tourism
	8.2. Ratio between the average salary of residents employed in tourism and residents employed in all sectors
	8.3. Ratio between the no. of temporary and permanent contracts
9. Promoted security	9.1. % of victimisation of overall
	9.2. No. of thefts
	9.3. % of the population who believe that tourism increase is the reason why the safety of their neighbourhood has worsened
	9.4. Opinion (tourist/resident) about the security at the destination
10. Guaranteed gender equality	10.1. % of women employed in the tourism sector
	10.2. % of tourism companies with leadership positions held by women
	10.3. Gender salary gap

Source: Based on Pareto Boada et al. (2020)

Based on the results provided by the indicators, various thematic maps were designed, of which a sample of maps are presented in the Fig. 16.1, showing the most significant indicators. This mapping allowed a deeper understanding of the level of social sustainability in the districts.

The indicators were a valuable instrument for identifying significant differences in the social sustainability of tourism between districts. Barcelona was shown as a polarised destination, with some intensely touristic and crowded districts (overtourism) and others little explored. The indicators and maps in this case study showed extraordinarily contrasting realities in the same tourist city. In general terms, the most touristic districts (Ciutat Vella and Eixample) recorded negative impacts on a significant number of social variables, mainly related to safety, tourist-resident coexistence, access to housing, commercial diversity and decent work. However, in parallel, these same districts recorded positive results in terms of diversification of tourism resources and products, and citizen participation and associations.

16.9 Gijón

The city of Gijón is a coastal historic city with a strong industrial background, located in the northern area of Asturias. Its industrial past implied an urbanism growth, especially explosive during the '60 s and '70 s and a consequently complicated reconversion to other activities that greatly influenced Gijón's urban planning (Saupin, 2011). The decade of the '80 s and '90 s of the past century were especially

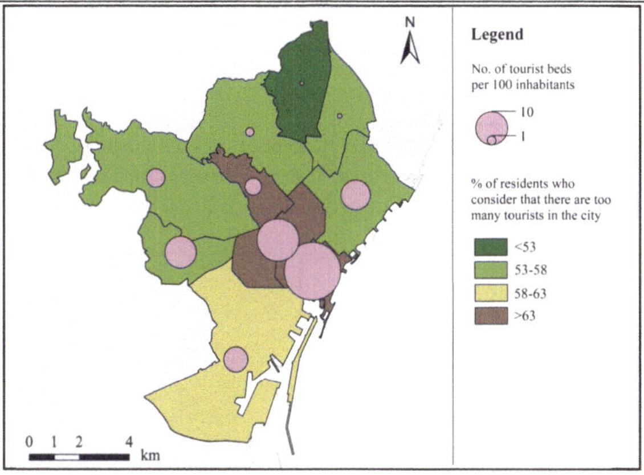

Example 1. Tourists/100 inhabitants and residents' negative perceptions of tourism (Criteria 2 and 7, see Table 1). Districts of Barcelona, 2017. Source: Based on data published by the Barcelona City Council (2017a & 2017b).

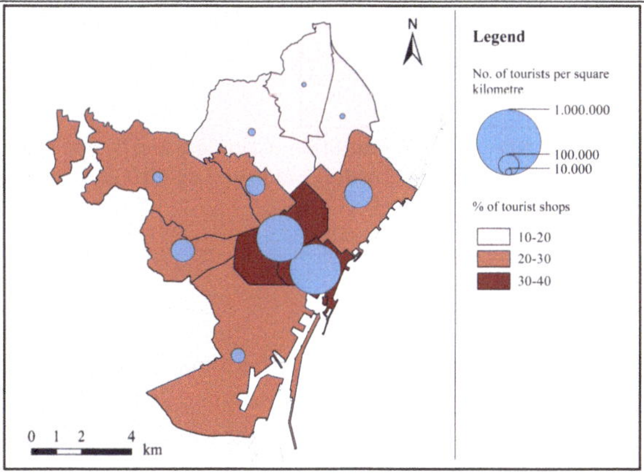

Example 2. Density of tourists and tourism-related shops (Criteria 2 and 3, see Table 1). Districts of Barcelona, 2016. Source: Based on data published by the Barcelona City Council (2016a & 2019).

Fig. 16.1 Geospatial representation of a sample of STIs implemented for Barcelona. Source: Based on Pareto Boada et al. (2020)

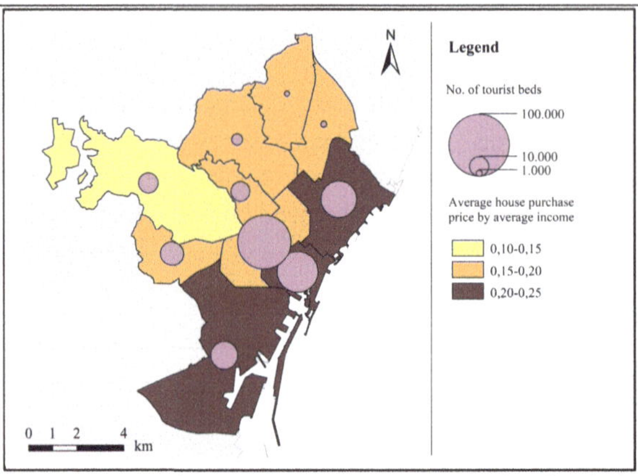

Example 3. Tourist places and house purchase prices in relation to average income (Criteria 2 and 5, see Table 1). Districts of Barcelona, 2017. Source: Based on data published by the Government of Catalonia (2013) & Barcelona City Council (2017c & 2019).

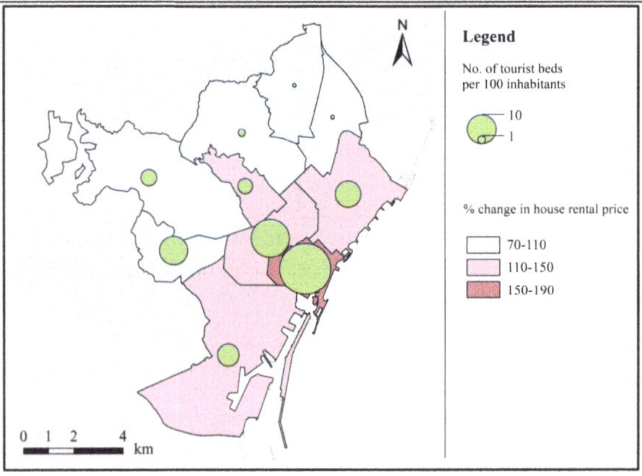

Example 4. Tourists/ 100 inhabitants and rental price variation (Criteria 2 and 5, see Table 1). Districts of Barcelona, 2000-2017. Source: Based on data published by the Government of Catalonia (2017) & Barcelona City Council (2017a).

Fig. 16.1 (continued)

active, as the city benefited from multiple public funds (National and European) oriented to the rehabilitation of historic neighbourhoods (Morales, 2011).

Aligned with this idea, the city reoriented its activity toward the tourism sector. Since then Gijón has experienced constant growth built upon its urban (Latorre & Solá, 2016) and cultural heritage (Saupin, 2011), receiving in 2019 a total of 1,126,945 tourists (InvesMark, 2019). In an attempt to improve its competitiveness and management, Gijón is reorienting its tourism development towards a more sustainable path. Namely, in 2013 the city received the Biosphere certification acknowledging its efforts toward a more sustainable tourism development (Fraiz Brea, 2018).

This interest in sustainability led to the proposal of a set of sustainability indicators to assess Gijón's sustainability level in 2016. The proposed system was rooted in the ETIS, the SETIS and the Asturias regional indicator system SITA (Indicator System of Tourism Information of Asturias). This system aimed to assess sustainability from a municipal, holistic and longitudinal perspective (using data from 2005 to 2014) (Álvarez Díaz & Valdés Peláez, 2016). The resulting set followed the triple-bottom-line approach common in ETIS and other mainstream systems, with a governance or management dimension, and economic, social and environmental dimensions. Such as, Gijón's STIs comprise indicators of tourist satisfaction, certification of tourism business, tourism flows, private economic profits, employment, social impact (tourist figures in relation to the number of residents), visits to the cultural heritage, waste and water management, energy consumption, and bath water quality. The results testified the tourism growth in the city in terms of visitors, but also in importance as an economic activity. However, the historic city centre registered an increased pressure of visitants which could lead to potential social and economic disruption.

16.10 Valencia

It was not until the late 1990s that Valencia became a conventional urban tourist destination. The city positioned itself as an international destination, following the example of other large Spanish cities (Marco-Serrano & Rausell-Köster, 2005), and it experienced a dramatic increase in visitor arrivals (Salom & Pitarch-Garrido, 2017). In short, Valencia went from being a great unknown in the international tourism scene to being a recognized destination (Visit Valencia Foundation Valencia tourist towards 2020).

However, beyond the positive figures for tourist arrivals and expenditure, the city must deal with the negative impacts caused by its rapid tourism growth. Aiming to develop tools to achieve a liveable city for all, the DMO has recently shifted towards a growing interest in sustainable development. Since 2016, sustainability has been essential to the destination's strategic plans (Visit València Foundation, 2022). Accordingly, the city has set the aim of becoming a neutral destination in emissions by 2025, and it has measured the carbon footprint generated by tourism activity.

The study has used a public-private collaboration model to assess the greenhouse gas sources and their effect in ten areas related to tourism: transport to and within the city, tourist accommodation, restaurants, leisure or retail offer oriented to tourism, visitors and cruise passengers, waste treatment, water management, and tourism support infrastructure. These ten areas cover the 3 scopes of the carbon footprint: 1) footprint by passenger transport to and from destination and domestic transportation, 2) indirect GHG emissions from energy consumption in tourism activities, and 3) other indirect emissions. As a result, Valencia determined that 81% of CO_2 tons attributed to tourism in 2019 corresponded to visitors' transportation to Valencia.

16.11 Zaragoza

Since the 2008 International Exposition, Zaragoza has invested heavily in tourism (Zaragoza Turismo, 2017). In parallel with the growth of tourist activity, the debate on sustainability has become necessary, as well as the need to set in precise way its objectives of sustainability (Navarro Jurado & Solis Becerra, 2009). In this sense, no system of indicators has been implemented at the institutional level in the city of Zaragoza for tourism sustainability measuring. However, an indicator system has been proposed by Aranda and Bosque (2014) that approaches tourism sustainability from the environmental point of view in the face of the need to make urban tourism more sustainable in Zaragoza.

Based on a review of experiences in environmental indicators related to tourism, they collected a total of 187 different indicators potentially applicable. Grounded on them, they made an adapted proposal considering the urban context's particularities. The STIs proposal included a total of 40 quantitative and qualitative environmental indicators divided into thirteen topics: tourism trend, tourism infrastructure, transport and mobility, noise, air, energy, water consumption, wastewater, urban solid waste, use of municipal land, information/social instruments, documentation on the functional structure of the municipal environment, and regulatory instruments. Although this proposal was not strictly applied, its applicability was tested as all the STIs were based on reliable data sources for the city of Zaragoza. Aranda and Bosque (2014) drafted a detailed process to be followed for implementing their proposal (forms, responsibilities, updating periods, data sources...), and even suggested how to create a Tourist Observatory to monitor the STIs.

16.12 Conclusions

Destinations are embedded in a particular social, economic, and territorial context which results in different needs, aims and expectations for sustainable tourism. The geographical condition of a destination determines its resources and attractions and, to some extent, its development model (Mathieson & Wall, 1982; Page & Hall, 2003). Cities, and urban environments in general, have specific qualities that lead to patterns of tourism consumption that are clearly different from rural, costal or mountain destinations. In particular, tourism challenges sustainability in the use of public spaces, infrastructure and services, tourist-resident coexistence, air quality, urban landscape, tangible and intangible heritage, safety and the cost of living. Thus, understanding how tourism differs and affects sustainability in different geographical contexts (coast, mountain, cities, and rural areas) is a challenge to design and adapt sustainable tourism strategies.

Indicators have typically been adopted to provide evidence on tourism impact at a destination level. Given their methodological simplicity, STIs have proved to be useful for decision-making as they enable specific impacts to be detected and practical solutions to be applied. However, indicators have often failed to strike a balance between context-dependent specificity and comprehensive relevance (Torres-Delgado & Saarinen, 2013). Many authors justify the need to monitor the sustainable development of tourism at the local level since that is where the tourism takes place (Torres-Delgado & López Palomeque, 2014; Vera & Ivars, 2003; Vila et al., 2010). Nevertheless, this local approach is problematic because STIs build on local data sources which are often limited, lacked comparability and fail in addressing the global-local nexus.

Conversely, experiences at regional level tend to rely on official statistics sources, such Eurostat (see Alfaro Navarro et al., 2020). Although this approach overcomes the challenges of data availability and comparability, it is controversial as hiding local context differences, needs and levels of tourism development. Besides, the regional level could be far away from the actual local government, thus STIs are not relevant for their purposes. These issues result in high flexibility and adaptability of the current STIs which often gather core indicators plus site- and destination-specific indicators (Biermann et al., 2017; European Commission, 2016; WTO, 2004).

The analysis conducted in this chapter has shown that no single, universally accepted system to measure urban sustainable tourism has been identified. Although urban tourism has been shown to have distinctive impacts on sustainability issues, to date the indicators for sustainable planning and management of urban destinations refer to sustainability in the city or sustainability in tourism, but not to both combined, i.e., to urban tourism sustainability. This gap in knowledge makes it difficult to adapt tourism policies when aiming to improve sustainability in urban destinations.

The implementation of STIs in urban destinations has been exemplified in experiences of several Spanish cities. These experiences have shown that indicators mostly respond to simple calculations. Indicators' simplicity is utterly needed to enable interpretation and communication (Torres-Delgado & Saarinen, 2013) and contributes to generating an objective and reasoned social debate about tourism incidence in the city. Besides, when STIs are periodically measured, they can also be used to monitor tourist sustainability in the city, as well as to evaluate the effectiveness and efficiency of the implemented policies (Castellani & Sala, 2010). However, future research lines must consider the urban quality of destinations as a distinctive element to be considered when selecting indicators. Also, exploring the use of STIs in evidence-informed policies needs more attention (Font et al., 2021). Data must be better captured and communicated broadly, to ensure future decisions are informed and knowledge is shared. If successfully achieved, this will enable more sustainable destination planning and management.

Acknowledgements This research has received funding from the European Union's Horizon 2020 research and innovation programme under the Marie Skłodowska-Curie grant agreement No 890281.

References

Alfaro Navarro, J. L., Andrés Martínez, M. E., & Mondéjar Jiménez, J. A. (2020). An approach to measuring sustainable tourism at the local level in Europe. *Current Issues in Tourism, 23*(4), 423–437.

Álvarez Díaz, R., & Valdés Peláez, L. (2016). Selección e indicadores para la evaluación del desarrollo sostenible de un destino turístico. Aplicación al municipio de Gijón. *ROTUR, Revista de Ocio y Turismo, 11*, 12–21.

Amore, A., & Hall, M. (2016). From governance to meta-governance in tourism? Re-incorporating politics, interests and values in the analysis of tourism governance. *Tourism Recreation Research, 41*(2), 109–122.

Aranburu, I., Plaza, B., & Esteban, M. (2016). Sustainable cultural tourism in urban destinations: Does space matter? *Sustainability (Switzerland), 8*(8), 1–14.

Aranda, S., & Bosque, R. (2014). Propuesta de indicadores ambientales para un turismo sostenible en la ciudad de Zaragoza. *Geographicalia, 66*, 99–132.

Aris Anuar, A. N., Noorbaizura Bookhari, S., & Azah Aziz, N. (2012). The effectiveness of Safe City Programme as safety basic in tourism industry: Case study in Putrajaya. *Procedia - Social and Behavioral Sciences, 42*, 477–485.

Ashworth, G., & Page, S. J. (2011). Urban tourism research: Recent progress and current paradoxes. *Tourism Management, 32*(1), 1–15.

Barcelona City Council. (2016). *Cens d'activitats econòmiques en planta baixa de la ciutat de Barcelona*. Barcelona City Council.

Barcelona City Council. (2017b). *Percepció del Turisme a Barcelona*. Barcelona City Council.

Barcelona City Council. (2017c). *Distribució territorial de la renda familiar disponible per càpita a Barcelona 2017*. Barcelona City Council.

Barcelona City Council. (2019). *Estadística i Difusió de Dades. Anuari estadístic de la ciutat de Barcelona 2019*. Barcelona City Council.

Barcelona Turisme. (2019). *Barcelona, por un turismo accesible*. Barcelona Turisme.

Biermann, F., Kanie, N., & Kim, R. E. (2017). Global governance by goal-setting: The novel approach of the UN sustainable development goals. *Current Opinion in Environmental Sustainability, 26–27*, 26–31.

Blancas, F. J., González, M., Lozano-Oyola, M., & Pérez, F. (2010). The assessment of sustainable tourism: Application to Spanish coastal destinations. *Ecological Indicators, 10*(2), 484–492.

Blanco-Romero, A., Blàzquez-Salom, M., & Cànoves, G. (2018). Barcelona, housing rent bubble in a tourist city: Social responses and local policies. *Sustainability (Switzerland), 10*(6), 1–19.

Castellani, V., & Sala, S. (2010). Sustainable performance index for tourism policy development. *Tourism Management, 31*(6), 871–880.

CITTIB & University of the Balearic Islands. (2000). *La mesura de la sostenibilitat del turisme a les Illes Balears. Indicadors de Sostenibilitat del Turisme a les Illes Balears*. CITTIB.

Claveria González, O. (2009). Dimensió econòmica del turisme. In F. L. Palomeque (Ed.), *Atles del Turisme a Catalunya. Mapa Nacional de l'Oferta i els Productes Turístics* (pp. 455–461). Generalitat de Catalunya.

Cócola Gant, A. (2016). Holiday rentals: The new gentrification battlefront. *Sociological Research Online, 21*(3), 112–110.

Cool, S. F., Moisey, R. N., & Nickerson, N. P. (2001). What should tourism sustain? The disconnect with industry perceptions of useful indicators. *Journal of Travel Research, 40*(2), 124–131.

Diaz-Parra, I., & Jover, J. (2021). Overtourism, place alienation and the right to the city: Insights from the historic Centre of Seville, Spain. *Journal of Sustainable Tourism, 29*(2/3), 158–175.

Elorrieta, B., Cerdan Schwitzguébel, A., & Torres-Delgado, A. (2022). From success to unrest: The social impacts of tourism in Barcelona. *International Journal of Tourism Cities, 8*, 675–702. https://doi.org/10.1108/IJTC-05-2021-0076

ETC. (2021). *Sustainable tourism implementation. A framework and toolkit to support national approaches*. ETC.
European Commission. (2016). The European tourism indicator system ETIS toolkit for sustainable destination management. Luxembourg. https://doi.org/10.2873/982144.
European Commission DG Environment. (2018). *Indicators for sustainable cities. In-depth report 12, produced for the European Commission DG Environment by the science communication unit*. UWE.
Fernández, H., & Pardo, D. (2016). La lucha por el decrecimiento turístico: El caso de Barcelona. *Ecología Política, 52*, 104–106.
Fernández-Güell, J. M., & González-López, J. (2016). Cities futures. A critical assessment of how future studies are applied to cities. *Foresight, 18*(5), 454–468.
Font, X., Torres-Delgado, A., Crabolu, G., Palomo Martinez, J., Kantenbacher, J., & Miller, G. (2021). The impact of sustainable tourism indicators on destination competitiveness: The European tourism indicator system. *Journal of Sustainable Tourism*. https://doi.org/10.1080/09669582.2021.1910281
Fraiz Brea, J. A. (2018). El turismo sostenible y sus derivadas. *Cuadernos de derecho local, 39*, 229–249.
Füller, H., & Michel, B. (2014). "Stop being a tourist!" New dynamics of urban tourism in Berlin-Kreuzberg. *International Journal of Urban and Regional Research, 38*(4), 1304–1318.
Gahin, R., Veleva, V., & Hart, M. (2003). Do indicators help create sustainable communities? *Local Environment, 8*(6), 661–666.
García-Hernández, M., de la Calle-Vaquero, M., & Yubero, C. (2017). Cultural heritage and urban tourism: Historic city centres under pressure. *Sustainability (Switzerland), 9*(8).
Government of Catalonia. (2013). *Compravendes d'habitatges registrades i el preu de venda a Catalunya*. Government of Catalonia.
Government of Catalonia. (2017). *Lloguers Barcelona per districtes i barris*. Government of Catalonia.
InvesMark. (2019). *El turismo en Gijón 2019*. Retrieved on 6th July 2022 from https://gijonturismoprofesional.es/es
Junta de Andalucía. (2015). *Sistema de Indicadores de Desarrollo Turístico Sostenible para Andalucía (Consejería)*. Consejería de Turismo, Comercio y Deporte, Junta de Andalucía.
Koens, K., Postma, A., & Papp, B. (2018). Is overtourism overused? Understanding the impact of tourism in a city context. *Sustainability (Switzerland), 10*(12), 1–15.
Latorre, J., & Solá, J. (2016). La apertura de Gijón al futuro: La revisión limitada del Catálogo Monumental. *VLC arquitectura, 3*(2), 149–162.
López Palomeque, F., Vera-Rebollo, J. F., Torres Delgado, A., & Ivars Baidal, J. A. (2021). *El turismo, ¿fin de época? Desafíos de España como destino turístico en un nuevo escenario*. Publicacions de la Universitat de València, Colección Desarrollo Territorial.
Marco-Serrano, F., & Rausell-Köster, P. (2005). Una aproximación al turismo urbano: el valor de la ciudad empaquetada en el caso de Valencia. *Revista de economía, sociedad, turismo y medio ambiente: RESTMA, 2*, 9–34.
Martins, M. (2018). Tourism planning and Tourismphobia: An analysis of the strategic tourism plan of Barcelona 2010–2015. *Journal of Tourism, Heritage & Services Marketing, 4*(1), 3–7.
Mathieson, A., & Wall, G. (1982). *Tourism, economic, physical and social impacts*. (9ª ed.). Longman Pub Group.
McKercher, B., Shoval, N., Park, E., & Kahani, A. (2015). The [limited] impact of weather on tourist behavior in an urban destination. *Journal of Travel Research, 54*(4), 442–455.
Milano, C. (2017). *Overtourism y Turismofobia. Tendencias globales y contextos locales*. Ostelea School of Tourism & Hospitality.
Miller, D., Merrilees, B., & Coghlan, A. (2015). Sustainable urban tourism: Understanding and developing visitor pro-environmental behaviours. *Journal of Sustainable Tourism, 23*(1), 26–46.

Ministry of the Environment. (2003). *Sistema español de indicadores ambientales de turismo*. Centro de Publicaciones, Secretaría General Técnica del Ministerio de Medio Ambiente.
Ministry of the Environment and Fluvial and Marine Environment. (2012). *Sistema de indicadores y condicionantes para ciudades grandes y medianas*. Retrieved on 4th April from http://www.upv.es/contenidos/CAMUNISO/info/U0722854.pdf
Morales, M. C. (2011). Marca Gijón: La imagen de la ciudad. *Liño: Revista anual de historia del arte, 13*, 123–135.
Murray, I. (2014). Bienvenidos a la fiesta: turistización planetaria y ciudades-espectáculo (y algo más). *Ecología Política, 47*, 87–91.
Navarro Jurado, E., & Solis Becerra, F. J. (2009). La importància dels indicadors de sostenibilitat en les destinacions turístiques. Aplicació d'indicadors de capacitat de càrrega a la Costa del Sol oriental. *Anuari de l'Agrupació Borrianenca de Cultura: revista de recerca humanística i científica, 20*, 31–44.
Page, S. J., & Hall, C. M. (2003). *Managing urban tourism*. Prentice Hall.
Palou, S. (2016). *Destinació BCN. Història del turisme a la ciutat de Barcelona*. Efadós – Ajuntament de Barcelona.
Pareto Boada, P., Torres-Delgado, A., Cerdan Schwitzguébel, A., & Elorrieta Sanz, B. (2020). Sostenibilidad social del turismo en Barcelona: un análisis geoespacial. In G. X. Pons, A. Blanco-Romero, R. Navalón-García, L. Troitiño-Torralba, & M. Blázquez-Salom (Eds.), *Sostenibilidad Turística: overtourism vs undertourism* (pp. 249–264). Palma (Illes Balears).
Romagosa, F. (2020). The COVID-19 crisis: Opportunities for sustainable and proximity tourism. *Tourism Geographies, 22*, 690–694. https://doi.org/10.1080/14616688.2020.1763447
Salom, J. S., & Pitarch-Garrido, M. D. (2017). Análisis del impacto en el turismo de la estrategia de desarrollo urbano basada en megaproyectos. El caso de la ciudad de Valencia. *Cuadernos de Turismo, 40*, 723–726.
Sánchez, M., & Pulido, J. I. (2008). *Medida de la sostenibilidad turística. Propuesta de un índice sintético*. Editorial universitaria Ramón Areces.
Sancho, A. (1998). *Introducción al turismo*. OMT.
Saupin, G. (2011). Gijón: From Asturian regional port and Industrial City to touristic and cultural Centre for the European Atlantic arc, from the nineteenth century to the present. In P. Borsay & J. K. Walton (Eds.), *Resorts and ports: European seaside towns since 1700* (pp. 167–182). Channel View Publications.
Tervo-Kankare, K. (2011). The consideration of climate change at the tourism destination level in Finland: Coordinated collaboration or talk about weather? *Tourism Planning & Development, 8*(4), 399–414.
Timur, S., & Getz, D. (2009). Sustainable tourism development: How do destination stakeholders perceive sustainable urban tourism? *Sustainable Development, 17*(4), 220–232.
Torres-Delgado, A., & López Palomeque, F. (2014). Measuring sustainable tourism at the municipal level. *Annals of Tourism Research, 49*, 122–137.
Torres-Delgado, A., & Saarinen, J. (2013). Using indicators to assess sustainable tourism development: A review. *Tourism Geographies: An International Journal of Tourism Space, Place and Environment, 16*, 31–47.
Torres-Delgado, A., López Palomeque, F., Elorrieta Sanz, B., & Font Urgell, X. (2021). Monitoring sustainable management in local tourist destinations: Performance, drivers and barriers. *Journal of Sustainable Tourism*. https://doi.org/10.1080/09669582.2021.1937190
UN. (2015). *Transforming our world: The 2030 agenda for sustainable development*. United Nations.
Vera, J. F., & Ivars, J. A. (2003). Measuring sustainability in a mass tourist destination: Pressures, perceptions and policy responses in Torrevieja, Spain. *Journal of Sustainable Tourism, 11*(2–3), 181–203.
Vila, M., Costa, G., & Rovira, X. (2010). The creation and use of scorecards in tourism planning: A Spanish example. *Tourism Management, 31*(2), 232–239.

Visit València Foundation. (2022). *VLC Visit València*. Retrieved on 8th July 2022 from https://fundacion.visitvalencia.com/
Wheeller, B. (1993). Sustaining the ego. *Journal of Sustainable Tourism, 1*(2), 121–129.
WTO. (1996). *What tourism managers need to know: A practical guide to the development and use of indicators of sustainable tourism*. UNWTO.
WTO. (2004). *Indicators of sustainable development for tourism destinations*. UNWTO.
WTO. (2012a). *Global report on City tourism – Cities 2012 project*. UNWTO.
WTO. (2012b). *Tourism and intangible cultural heritage*. UNWTO.
WTO. (2022). *SDG 11 – Sustainable Cities and Communities*. Tourism for SDGS. Retrieved on 29th March 2022 from https://tourism4sdgs.org/sdg-11-sustainable-cities-communities/
WTO & World Tourism Cities Federation. (2018). *UNWTO/WTCF City tourism performance research*. UNWTO.
Wyss, R., Abegg, B., & Luthe, T. (2014). Perceptions of climate change in a tourism governance context. *Tourism Management Perspectives, 11*, 69–76.
Zaar, M. H., & Pontes da Fonseca, M. A. (2019). Turismo y consumo del espacio urbano en Barcelona. Análisis de la relación residente-turista en el barrio de la Barceloneta. *Cuadernos de Turismo, 44*, 487–514.
Zaragoza Turismo. (2017). *Plan estratégico de Turismo de Zaragoza 2017–2021*. Retrieved on 8th July 2022 from https://www.zaragoza.es/sede/portal/turismo/

Dr. Anna Torres-Delgado is a Marie Skłodowska Curie Research Fellow at the University of Surrey (UK) and lecturer at the University of Barcelona in tourism and sustainability in planning and destination management. Her area of expertise is the development of sustainable tourism indicators and policy. Currently, she is leading an EU-funded project about governance of urban tourism by using sustainable tourism indicators to create evidence-informed policy.

Aurélie Cerdan Schwitzguébel is a PhD candidate at the University of Barcelona researching on the urban tourist destinations fields and specializing in the tourism-associated social impacts. She is graduated in Tourism Management and holds a Master in Management of Urban Tourist Destinations at the CETT-UB Tourism and Hospitality Education/Research Centre.

Pol Pareto Boada is graduated in Geography and holds a Master in Management of Urban Tourist Destinations at the CETT-UB Tourism and Hospitality Education/Research Centre.

Chapter 17
Touristification and Gentrification in Spain: Perspectives and Challenges for the Post-pandemic Era

Gustavo Javier Macías Mendoza and Antonio Paolo Russo

17.1 Introduction

The 2008 Financial Crisis has been a global event with glittering consequences in the economies of all the countries of the world. Its nature, originated in the real estate market and construction sector and its collapse, led to a series of dynamics that shaped the internal processes and development cycles of various nations. Many local economies found themselves at the sudden and urgent crossroads of changing their configuration.

A renewed boost to urban tourism has represented, for many cities of the world, a quick and significant way out of the crisis. This approach has been rapidly diffused in southern Europe and Latin America. Local administrations took advantage of the situation and undertook policies to enhance and brand the architectural and historical heritage of city centers, sometimes in deviation of preexisting urban planning schemes and regulation to facilitate the attraction and use of tourists. Housing prices had experienced a remarkable drop after the crisis, hitting Southern European countries with great intensity. Property owners and promotors saw this as an opportunity to earn high returns on their assets. A recent escalation of around a 50% since 2014 of housing prices has been observed according to a recent study from the Bank of Spain (Andújar Llosa et al., 2020). The combination of a strong

G. J. Macías Mendoza (✉)
Programme Tourism and Leisure, Rovira i Virgili University, Tarragona, Spain
e-mail: gustavojavier.macias@estudiants.urv.cat

A. P. Russo
Department of Geography, Rovira i Virgili University, Tarragona, Spain
e-mail: antonio.russo@urv.cat

A. Blanco-Romero, M. Blázquez-Salom (eds.), *Spanish Tourism Geographies*, Geographies of Tourism and Global Change,
https://doi.org/10.1007/978-3-031-39780-6_17

campaign carried out by municipal governments, a growth in real estate speculation coinciding with the increase of the stock of short-term accommodation, and innovations in areas of interest—the expansion of low-cost travel infrastructure and the concomitant diffusion of digital rental platforms like Airbnb—have arguably made it possible to abate the costs of city travel and exponentially develop the economic importance of tourism in countries most struck by the crisis, among which Spain.

The expansion of urban tourism meant, in turn, a change of the dynamics of development of the city. Historic centers were increasingly exposed to the flows and uses of tourists and immigrants, driving an accelerated metamorphosis of the urban landscape. The real estate market, local population dynamics, supply of services, commercial fabric: none of these issues was unscathed by this new reality. Experts observed this phenomenon and understood that what was happening was something like gentrification[1] but under a new context and driving novel dynamics. Debates around what could be understood as "tourist gentrification"[2] appeared in the urban and tourism studies literature. This production was increasingly fueled by the debate around what was beginning to be understood as the problem of "over-tourism", the emergence in the local population of "tourism-phobia", and other side effects caused by the intensification of tourism in urban fabrics.

In 2020, however, this growth in tourism was suddenly interrupted by the transformation of China's COVID-19 epidemic into a pandemic. Countries around the world closed their borders, interrupted their flights, and took quarantine decisions, locking the population in their homes. The constant and massive flow of people to urban centers came to a halt, bringing to standstill economies that, in 10 years, had learned that the key to their growth lay in tourism.

The current work is situated in a context in which the world tries, with greater or lesser success, to return to a supposed normalcy after the ravages caused by the pandemic. Gradually, cities around the world return to receive tourists. Businesses that closed have reopened, and those that went bust during the pandemic are likely to be replaced by new ones. There is, all over the world, and of course in Spain, a certain confidence that the situation will normalize; which is supported by the early data on tourism activity from the summer season of 2022.

However, the recovery of urban economies in the post-pandemic period must take in, as a starting point, an accurate and critical view of pre-pandemic trends. Tourist cities had become increasingly hostile to the life of the local inhabitants. They saw

[1]Term coined by Ruth Glass in 1964 that refers to the process by which traditional neighborhoods experience the arrival of social groups with a higher economic and cultural level, which transform them, increase the price of land, and end up expelling the most vulnerable traditional neighbors (Calo et al., 2021: 3).

[2]In 1994, Marie-Françoise Lanfant spoke of "touristification", a concept that expresses the process by which, in certain areas, tourism caused aspects such as commercial and residential offerings, infrastructures and services, or urban planning to be oriented to the demand and preference of tourists, to the detriment of the resident population.

their economic base diluted by the intensification of leisure and entertainment services designed for tourists. Access to housing had become increasingly difficult because of the growth of the short-term vacational market and the emergence of platforms such as Airbnb. Urban tourism has a double face. On the one hand, it helps to revitalize and rebuild cities, putting them in the global spotlight as attractive, dynamic and hospitable. On the other hand, it represents a risk factor in relation to social and cultural dynamics, or a factor of dissipation of their authenticity that could result, paradoxically, in the depletion of their tourist attractiveness.

In other words, the conception of the "tourist city" prior to the crisis caused by the pandemic is at risk of remaining as a short-term and exclusive model. In this sense, some authors (e.g. Guerra-Luzuriaga, 2020) introduced a "sustainable city" model as a benchmark for tourism-driven development trends: the idea that advocates promoting healthier urban spaces; inclusive for the disadvantaged and for the local small business; and protectors of heritage and local culture.

This work understands that the pandemic opened apolitical window to start thinking cities in this sense. For this reason, we propose to carry out, for the Spanish case, an analysis of the main challenges that the post-pandemic era opens to its tourist cities to improve their urban policies. To do this, we will carry out a general review of the Spanish bibliography that takes tourist gentrification or touristification as a theme and we will bring out the most representative approaches and subtopics. The idea is to be able to review the available studies on the problem in the pre-pandemic period and even during the pandemic. Then, we will delve into the debates around what is the correct approach when talking about the main causes of gentrification and how really is the link between this phenomenon and the growth of tourism in recent years. Behind this decision is the idea that these debates do not reach unanimous conclusions and that different authors consider different variables.

Later, we will delve into one of the most relevant issues about tourism in recent years. The social and cultural effects and the responses of different groups to tourist gentrification are determining factors. This is so because, as will be seen, urban policies are key in shaping the real estate market and the local economies that derive from tourism. Knowing the new political actors born of gentrification itself—that is, their interests, positions, and problems—will allow a clearer picture of what are the challenges and resistances that must be considered for the post-pandemic era.

Finally, and considering the findings obtained from previous research, we will specify a number of axes on which future policies that seek the reconfiguration of sustainable tourist cities should tackle. That is, decisions that advocate generating healthy and inclusive urban spaces for both visitors and long-term residents. Also, cities that can preserve that "authenticity" that makes them relevant in the tourism market and that, after all, transform them into the most precious commodity in an increasingly globalized, connected, and digital world.

17.2 Some Characteristics and Problems of the Spanish Tourism Model

17.2.1 Methodology

To reconstruct the situation of the study of tourist gentrification in Spain, a review of the available literature on the subject has been carried out. To do so, we have relied on other bibliographic studies, especially the comparative analysis carried out by Calo et al. (2021), which offers an excellent review not only of the most outstanding articles, but also offers a classification of the most relevant thematic blocks.

Although the classification of these authors gave us a comprehensive and in-depth look at the field of study, it has particularities that need to be highlighted. In the first place, their work was not aimed at capturing all the studies on the subject, but rather the relevant ones when making a classification. They have included and highlighted the relevant works of each thematic area that fit a specific geographic case study (Calo et al., 2021: 7). This does not represent a problem for the interests of the present work. On the other hand, the authors decided to exclude from their analysis those texts that address the problem only in a theoretical-conceptual way, that is, without extending empirically to reality of a particular territory. While this makes sense within the objectives of that study, in the present work it represents a key question. We consider that the works that are inserted in the debate surrounding the understanding of the dialectic between gentrification and tourism in Spain are important when it comes to understanding how the phenomenon is conceived in epistemological terms in this country's context. This fact will be relevant in the second part of this chapter, in which we will delve into specialized debates about the causes of gentrification, a debate that is far from unanimous in the literature.

For this same reason, our review has included conceptual works that address the problem of the dialectical relationship between gentrification and tourism. In turn, and to adapt the bibliographic review to its own objectives, works have been sought that introduce the contextual variable of the COVID-19 pandemic. For this latter objective, we have allowed ourselves to look outside the Spanish context, considering that studies from abroad can be useful to think about different edges that are opened because of the pandemic.

Therefore, we have carried out research supported by different online bibliographic search engines and repositories (Google Scholar, Academia.edu, Researchgate.net, Microsoft Academic, SCOPUS). Terms such as "gentrification", "tourism", "touristification", "Spain", "pandemic" and "Covid-19" have been crossed in various combinations. Although no specific time filters have been established, special attention has been paid to work carried out in the last decade, especially after the financial crisis of 2008. The growth of urban tourism in Spain, in southern European countries and in Latin America in the latest years produced, in turn, a great development of research on the subject.

17.3 The Study of Tourist Gentrification in Pre-pandemic Spain

The large amount of research available on the phenomenon of tourism in Spain is consistent with an expansion of this economic activity within the context of urbanity. In this sense, the rise to urban of tourism (especially after the financial crisis of 2008) as a strong alternative for economic growth, caused many major cities in the country to intensify their efforts in the form of public policies that boost the tourism potential of their cities urban spaces; especially in their historic centers, as in the case of Madrid, Barcelona, Seville, Palma, among others.

It is necessary at this point to introduce the special character of the Spanish tourism model and to situate the recent rise of urban tourism in that context. For this reason, we look at Murray (2015), who carried out an accurate study of the way in which tourism has become the main economic activity in Spain, tracing the systemic, political and historical origins of this trend. The author points out that "the triumph of the neoliberal dogma in the academic world buried a whole set of critical reflections on tourism capital" (p. 18). This critical approach that he adopts is one of his most important contributions: it is an alternative vision to studies focused solely on the positive side of tourism. The critical analysis of tourism has very little echo among tourism specialists, who are more willing to be complacent and useful to business initiatives and institutional powers. Few scientists have been interested in publishing critical, rigorous and independent studies on tourism in the Spanish State. In this sense, Murray's work is rather unique in highlighting the links between tourism development and the dictatorship, political corruption, labor and ecological deterioration, and financial and real estate speculation. The Spanish tourist "miracle" is, then, strongly linked to Francoism and the geographical expansion of the "Marca España" tourist model: the reconquest of America with its "modernization" that developed the real estate bubble that led to the global crisis, an research approach that has been initiated by Ivars-Baidal (2004) in his critical reading of the 'planning stages' in Spain following Getz's (1986) seminal model. Murray points out that, because of the real estate production initiated during the Franco regime, the "cement fever" reached national dimensions with the redefinition of Spanish capitalism. This, in turn, was reinforced by entry into the European Union and the promotion of neoliberal rules. The real estate bubble and its exponential growth, derived from financial liberalization and monopolistic rents caused by tourist specialization, originated at that time, in the second half of the twentieth century. An institutional scaffolding was developed there that promoted the processes of urban production of space, including urban planning at the pleasure of investors and the construction of transport mega-infrastructures subsidized by European funds. Real estate speculation spread until the arrival of the crisis, which favored criticism and social participation in the face of the devastation of the physical and social space.

Spain, in this way, went from being a peripheral and globalized country to a globalizing one, thanks to the process of modernization of capital under the protection of the neoliberal governments of the democratic regime. The transformation is manifested in the expansion of Spanish transnational capital, beginning with the great privatizations until what the author calls the recolonization of Latin America, through economic-financial arrangements of Spanish capital, fundamentally tourism. Murray (op.cit.) underlines the crucial role of Spanish tourists in the tourism-real estate explosion and the construction process of tourist industrial zones. He also highlights what he calls the "other accounts" of Spanish "tourism success": the biophysical accounts (the ecological footprint of the model) and the social accounts that warn of social underdevelopment. The ecological footprint is based on the supposed ideal of equitable distribution of the territory and its resources among the entire world population. This assumption implies the questioning of the appropriation and accumulation of wealth in the hands of a few, which is the situation they face. Tourism, at least international or long-distance, is an activity within the reach of a tiny fraction of the world's population, with immense energy requirements and a toxic load. The ruling classes, to resolve the contradictions of capital that led to the crisis, tried to recompose profit rates through the tourist exploitation of space. In this process, some urban spaces or cities such as Barcelona can be highlighted, which, taking tourism as an urban accumulation strategy, caused new socio-urban conflicts. This is how, beyond the optimistic narratives about the growth of capital, in this case of Spanish capital, the enormous ecological footprints are unmasked, and the conditions of social underdevelopment are made visible, which, for authors such as Murray, place Spanish society in disadvantageous positions of the rest of the European countries. These are key aspects that must be considered for further studies on the impacts of tourism in a context of globalization and generalized crises, such as that of Covid-19.

Murray's work allows contextualizing the accelerated gentrification that took place in recent years rooting in tourism development, and consistently with new efforts for urban revitalization. In this sense, understanding the meaning of these processes is deeply necessary. One of the most important definitions of gentrification is that made by the geographer Neil Smith in his book *The New Urban Frontier* (2012). The "frontier" appears to collect the meeting point between the wild and civilization, between us and them: thus, Smith affirms, a legitimizing discourse has been produced as a process of conquest. Gentrification is an example of a new way of producing spaces that develop unevenly through the actions of the owners of capital: banks, real estate developers, lenders and distribution chains; all this, sponsored to a greater or lesser extent by the State.

The concept of gentrification has its origins in the gentry, the new social class that emerged from the process of agricultural revolution in the United Kingdom. However, it is after the Second World War that it begins to be used to explain the urban development of industrial cities. The neoclassical theory of the urban development of cities and their configuration explained these phenomena through the income of the social groups they inhabited, that is, it is based on consumer preferences as the main explanation. However, Smith understands that the preferences of the producers cannot be ignored: builders, real estate agents, public agents. In a capitalist society, the land and the buildings built on it are commodities that change value according to the logic of the market, causing interest in investing or disinvesting in certain spaces, which leads to a certain configuration of the city. For Smith (2012), then, "A theory of gentrification needs to explain the historical process of devaluation of capital in inner cities and the precise way in which this devaluation generates the possibility of profitable reinvestment" (p. 116).

For Smith (2012), the main justification for the gentrification process lies in his theory of the potential difference in rent, which would be the difference between the rent that is estimated to be possible for a piece of land and the current capitalized rent of the land. If the current capitalized ground rent falls, developers can buy at low prices, pay the builders' costs and profit from the restoration. Thus, gentrification for Smith constitutes a moment back to the city, but a movement back to the city on the part of capital rather than the people.

It is important to note that the author focuses on the empirical analysis of urban processes in the West. By this I mean that his theory is constructed through experiences in North American and European cities.

The Hispanic bibliography has peculiarities that make it different from the Anglo-Saxon tradition of the study of gentrification. In general, Spanish research places great emphasis on the importance of urban policies. That is, the existence of institutional, social, and economic structures that respond idiosyncratically to Spanish particularities (Calo et al., 2021: 6). Of course, this responds to the fact that the material and historical conditions of Spain differ greatly from those of countries such as Great Britain, the United States or those of Northern Europe; therefore, the focus of the investigations, both methodologically and conceptually, is different. In this way, Spanish cities fall within a type of "semi-periphery", where other cities in Southern Europe can also enter, such as Lisbon or the Italian ones.

Of course, it is necessary to understand the process of tourist gentrification or touristification as a multidimensional process, in which various cultural, economic and political factors come into play. Thus, it is possible to see in research on the subject many different approaches that try to address it in a generalized way. It is necessary to detail some of the main approaches that the Hispanic literature on the subject has adopted. Doing so will allow us to identify, in turn, those problems that are key to understanding the future of the problem of tourism in the country.

One of the most complete investigations in recent years on the phenomenon of touristification has been conducted by Cañada and Murray (2019). *Turistificación global. Perspectivascríticasen turismo* is a compilation book with a plurality of approaches and theoretical frameworks, mainly from researchers in the Spanish-speaking world, which help to think critically about the current configuration of the tourist world. The objective is to question both the story and the academic production that tries to legitimize the tourism industry. Therefore, they focus on the contradictions and flaws in the functioning of the current state (p. 8).

In this approach to the world of tourism lies a fundamental concern for power and how it is configured. The main concern of Cañada and Murray (2019) is the inequality and exclusion generated in this global touristification process. In short, it is an analysis and reflection on the current tourism industry from a heterodox framework, since, for these authors, the existing gaps throughout the literature dedicated to the study of this phenomenon are the product of approached mainly from orthodox approaches with little incidence from critical perspectives.

Among these studies, interesting perspectives can be highlighted to understand the problem of tourism on a global scale. Fletcher (2016) shows that tourism works where it is established as a tool to propagate neoliberal logic. In this sense, neoliberal capitalism has set out to preserve the places that are in the best environmental condition to be able to sell them as precious exceptions. In turn, the book by Cañada and Murray (2019) includes relevant approaches to the paradigms of the present, such as feminist views on the impacts of touristification, understanding the need to approach the study of tourism from said feminist perspective, considering phenomena such as sex tourism, feminization of employment and gender-aware tourism policy.

In the fourth block of the book directed by Cañada and Murray (2019) is where there is a greater depth in the relationship between gentrification and touristification. There, several authors deal with the issue of urban spaces and the conflicts generated within them due to the growth of the tourism phenomenon. Cocola-Gant discusses tourist gentrification and explains how peripheral spaces with economies dependent on tourism and marked by job insecurity tend to be gentrified thanks to the purchasing power of tourists. On the other hand, he argues that there is a displacement through three spheres: the residential, the commercial and the symbolic. Marc Morell, for his part, discusses the multidimensional relationship that occurs between tourist saturation and the income differential. By feeding on gentrification and class struggle, they become the fuel that capital needs for its reproduction. In turn, Javier Gil focuses on the case of Airbnb as a paradigmatic example of platform capitalism, considering the changes produced in the capitalist productive exploitation model. Gil highlights the figure of the prosumer, being a type of employment that is usually anchored on personal exploitation and precariousness. Alan Quaglieri and Sheila Sánchez, for their part, analyze tourist accommodation as a collaborative economy phenomenon and propose a discussion of the solidity of the discourse around it produced by platform capitalism companies, although it is at the opposite end of what it would be an openly participatory and equitable economy, since they are structured around the monopoly of intermediaries. In this context, a panorama of political tension is generated around the phenomenon. Finally, we can mention the study about social movements carried out by Claudio Milano and José Mansilla, product of the social confrontation promoted by tourism activism against manifestations of neoliberalism, such as mass tourism through intensive exploitation. Thus, these movements discuss a phenomenon such as tourist saturation. The emergence of protesting social movements is generated that participate dually as they are based on a resistance towards tourism, as well as those where tourism can participate in the movement itself.

As we have described, Spanish cities have undertaken, in recent years, major revitalization projects in their cities that had a positive impact on the architectural state of areas previously abandoned by urban planning or simply fallen into decline. The positive effects of these efforts, however, are focused on increasing the stock of homes available for tourist exploitation. Tourist hotspots, therefore, begin to be subject to the pressures of real estate speculation, ultimately functioning as a wedge between proprietors and tenants (Valente et al., 2022). Access to housing is indeed becoming increasingly difficult for lifelong local neighbors. Arroyo-Alba's research (2019), which works on the case of the Lavapies neighborhood in Madrid, observes the effects of urban transformation and its consequences in the form of social problems. On the other hand, works such as those carried out by Birkbak et al. (2016) or by Goodwin (2019) focus on the decisions taken by the Barcelona administration in accordance with the effects of urban transformation and take it as a positive example of public policy around the issue of access to housing. However, authors such as Pareja and Simó (2014) or Blanco-Romero et al. (2017) are less optimistic regarding the regulatory decisions of the tourism sector made in that city, since they consider that they are rather insufficient when it comes to ensuring access to decent housing (Calo et al., 2021: 9). In the same sense, it is possible to highlight the study carried out by Vestri (2020) where the focus is placed, rather, on the legislative apparatus for the case of Seville.

Political paradigms or regimesthus turns out to be a transversal concern in the study of gentrification, and are frequently tackled by Spanish research. It is interesting, for example, to see the link that Vives-Miró (2011) makes between the neoliberal policies of deregulation promoted by the local administration and gentrification in Palma de Mallorca. His perception is consistent with that of González-Pérez (2019) for the same city. Studies that go in this sense aim to show the way in which local authorities have an influence in the advanced processes within the real estate market that are linked, in most cases, with the intention of adapting cities for traffic tourist (Crespi-Vallbona & Mascarilla-Miró, 2018; Díaz-Parra & Jover, 2019; Fernández-Tabales et al., 2020; Rodríguez-Barcón et al., 2020).

In recent years, studies that focus on the housing problem have been overwhelmed by the issue of so-called "collaborative economies", a type of organization of the real estate stock that appears before the large influx of tourist visitors and is based on renting flats in the short term. The appearance of accommodation platforms between individuals, such as the emblematic case of Airbnb, has invited researchers to observe what are the distinctive effects of the phenomenon (Sans & Domínguez, 2016; Yrigoy, 2017; Del Romero, 2018; Fierro & Salmón, 2018; González-Pérez, 2019; Bugalski, 2020; Etxezarreta-Etxarri et al., 2020; Yrigoy, 2020).

There are works in Spain that go beyond the study of the effects of the real estate market and urban policies related to tourism. Accelerated gentrification processes (especially transnational and which accompanied by tourism) shape the order of local economies, turning all efforts to cover the new type of demand created by the influx of a new type of population and individuals in the neighborhoods. The displacement of residents by a population with a different socioeconomic level, another nationality, and other interests—in short, another lifestyle—breaks traditional social relations and generates conflicts in displaced residents. In the Hispanic school of the study of gentrification there is great interest in its social and cultural effects, which many authors call a "loss of genuine urban identity" (Crespi-Vallbona

& Mascarilla-Miró, 2018; Gil & Sequera, 2018; García et al., 2019). In other words, tourist pressure in Spanish cities homogenizes the commercial network of previously traditional neighborhoods, thus displacing small businesses and giving rise to commercial typologies that are better valued both aesthetically and popularly by the new groups that arrive. In the case of tourism, this is consistent with a network focused on leisure and entertainment, especially at night (Calo et al., 2021: 12).

Of course, these dynamics generate resistance in the territories. Studies on these responses abound in the Spanish literature on the subject. The effects of tourism on the social foundations of cities generate growing feelings of discomfort and even rejection of the advent of tourist groups associated with a lifestyle disinterested in the preservation of a "good urban environment". Thus, there are investigations interested in being able to identify social movements and neighborhood networks specially formed to influence the pressures of gentrification. It is true that these investigations take an increasing role as tourism in cities expands and the groups against its effects become increasingly noisy and mediatic. The debates around "tourism-phobia" and "over-tourism" appear in the public realm as an essential question to think about the future of sustainable tourist cities (Cabrerizo et al., 2016; Milano, 2018; Egio-Rubio & Fernández-Toledo, 2020; Mundet & Pirillo-Ramos, 2020; Almeida-García et al., 2021).

In the following sections we will begin to delve into two debates that we consider relevant to think about the form that tourism policies will take in Spain in the post-pandemic era. To choose them, we made a reading of those topics that had the greatest relevance in recent years in the Spanish bibliography and we made an appreciation (based on what the authors have considered) of what the agents interested in the effects of tourist gentrification—local administrators, neighborhood networks, social movements, property owners, among others—should consider when thinking about more sustainable tourist cities.

17.4 The Debate Around the Approach to the Problem of Touristification

One of the biggest debates in the field of the study of tourism in Spain has to do with the relevance of the classical ways of understanding the phenomena of mobility and displacement of human masses. This means, the relevance of classical notions of gentrification within a context that, is generally understood, is different from that which occurred in other historical moments. The appearance of various innovations in the field of tourism – be they short-time rental platforms such as Airbnb or low-cost travels- had significant impacts on the social composition and dynamics of the real estate market within cities around the world. The Spanish case has to be situated within this global context, but presents, for several authors, different characteristics that require differential treatment (Cócola-Gant, 2016; Andújar Llosa et al., 2020; Calo et al., 2021).

This led many authors to conceptualize of new ways of thinking about the classic gentrification process introduced by Ruth Glass considering the particularities of the globalized world. And this, in turn, led studies on tourism in Spain to adopt as their axis three main processes that can be expressed in three different concepts: gentrification, transnational gentrification and touristification.

One of the most relevant approaches corresponds to that presented by, mainly, Cócola-Gant and Lopez-Gay (2020). These authors understood that the problem of displacement (or the expulsion of the local neighbor from the city) was not a mere consequence of tourism itself, but rather one driven by factors that mediate and accelerate the new rise of urban tourism as development strategy: real estate speculation, the withdrawal of the state from welfare provision (i.e.. the scarcity of social housing), and the deregulation of the housing market and rents agreements. Tourism appears in these perspectives as a multiplying factor of the exclusion that is generated when housing is above all an opportunity for profit. Their studies, carried out around displacement in the main neighborhoods of Barcelona, aim to identify the real relationship between the expansion of tourist activity and the dynamics of gentrification. The Gòtic, a central neighborhood associated with mass tourism, has undergone transnational gentrification in the last decade. In this case, despite the clear signs of a classical process—long-term divestment that generates housing rehabilitation for professional middle classes that displaces old middle-class residents—, the group that leads the gentrification process is a group of transnational migrants (especially from the global north).

This same logic can be found in other cities in Spain, such as Madrid or Seville.

The Spanish capital is a clear example of how a classic gentrification process can, with the drift of time, transform itself into a transnational gentrification. The semi-peripheral character of Madrid, an aspect that it shares with other southern European cities such as Lisbon or Barcelona itself, is one of the keys that allows us to understand the influx of migrants from countries of the European North and other countries of the Global North. The transformation of the real estate stock towards a short-term tourism-oriented market has a double effect: firstly, visitors see in the comparative low prices of accommodation the possibility of having a better and more authentic tourist experience in the place, different from that obtained with hotel accommodation; second, homeowners see this short-term real estate market as an alternative with higher profitability than the traditional long-term market (Urquiaga et al., 2020: 3096–3098). This encourages the arrival of immigrants with a high purchasing power and generates, in turn, a different socioeconomic profile from the previous one. Since 2013, it is possible to identify that a transnational gentrification is taking place in Madrid. Although the Spanish recession accelerated the loss of population in the historic center of Madrid, it also started a growth in the international immigration population, which grew by 5% between 2015 and 2018 (Urquiaga et al., 2020: 3103).

Seville has always been the preferred tourist destination in Andalusia and presents an interesting case that raises awareness about the differences between gentrification, transnational gentrification and touristification. This is due to the temporal compartmentalization of the different processes. Although these three processes are firmly interrelated, they can occur in different temporal spaces, also developing in particular ways. This conclusion is reached by Díaz-Parra and Jover (2019). In other words: they are differentiated in time. These authors observe, for the case of Seville, that gentrification began in the 1990s in the northern neighborhoods. These processes, in turn, correspond to those described by the classical notion of gentrification, that is, with the replacement of the working classes by middle and upper classes. In this sense, the municipal administration would have had an important preponderance, with programs such as the Seville Urban Development Plan, between 1995 and 1999, which meant an investment of up to 14 million euros in programs that included the restoration of emblematic buildings (Díaz-Parra & Jover, 2019: 7–8). The argument of these authors is that, without these primary regeneration processes in the historic centers, it would not have been possible for tourism to expand in Seville later on. In other words, the increase in a population of middle and upper classes in the city (and the influence of its lifestyle on the urban landscape) makes it more attractive to migrants from better-off countries. The entry of this population will produce a different process, a transnational gentrification.

Perhaps one of the most interesting findings of Díaz-Parra and Jover (2019) is the fact that, on occasions, gentrification (be it local or transnational) and touristification can collide in different directions. While gentrification pushes towards the transformation of neighborhoods in favor of the upper classes, touristification seeks to transform the urban scene into areas focused on tourist demand. The assumption in the latter is the fact that, in the tourist districts, few people would reside in the long term. Thus, disputes arise caused by displacement and by the use of public space in the face of inconveniences such as the increase in nocturnal noise or luminosity in late hours of the night (Díaz-Parra & Jover, 2019: 14). This is evident in recent times. Given the abrupt loss of local retail businesses in pursuit of the placement of modern bars and restaurants, the latter have grown by 230% since 1998 in the Alameda neighborhood. In turn, tourist accommodation increased significantly there: in 2018, the number of vacation homes in the neighborhood was 257, when 5 years earlier there were less than two dozen (Díaz-Parra & Jover, 2019: 12).

There are other approaches to the causes that lead gentrification that are interesting for their originality. We will highlight the research carried out by Yrigoy (2017) for the case of the proliferation of tourist rental apartments in Menorca. In this case, the author reflects on the possibility of the existence of a different type of gentrification: the so-called "popular gentrification". This reality is possible thanks to the emergence of platforms such as Airbnb, which allow the phenomenon of gentrification to be led not by large real estate companies, but by individuals. The truth is that several authors focus their eyes on the effects that Airbnb is having when it comes to changing the rules of the game. Etxezarreta-Etxarri et al. (2020) identify that a one standard deviation increase in Airbnb intensity is associated with an increase of 7,3% in rental prices.

We also highlight the study carried out in the historic center of the city of A Coruña by Rodríguez-Barcón et al. (2020). According to the authors, there are three different models of gentrification that occur in A Coruña classified by what motivation leads the process: (1) by previous business transformation processes (economic motivation); (2) positive social perception (socio-symbolic motivation); (3) strategic planning that seeks the revaluation of the architectural environment and the attraction of private capital (institutional motivation). This classification is a useful finding, as identifying the type of gentrification can help identify, in turn, the best ways to offset the inherent negative effects.

Now, what are those effects and how are they expressed in reality? Gentrification presents complex resistance and contestation apparatus in Spain. We will delve into these issues below.

17.5 The Social, Cultural, and Political Angle: Over-tourism, Loss of Identity and Tourism-Phobia

According to Fernández-Tabales et al. (2020), the Spanish public policies that aimed to solve the problems that appeared in urban centers because of the growth of tourism were, rather, typical of a sectoral approach: architectural renovation, construction of emblematic landmarks, beautification, public space equipment, the specialization of transport infrastructure for tourist mobility, and more recently the public investments for 'smart' services for visitors. However, when trying to track measures that have addressed the issue in a comprehensive way—linking it with its effects on social settings, local economies and housing provision—it is found that they have been, rather, scarce.

This inoperability is alarming, since gentrification processes have the characteristic of expanding and intensifying over time. The extensive deregulation that the real estate field presents in some cities (such as Palma de Mallorca or San Sebastián, where the use of platforms such as Airbnb prior to the pandemic was widespread) drive the loss of stock in long-term homes in pursuit of their use for short-term rentals for tourism. Not only does it hinder access to housing, displacing the most vulnerable population from urban centers, but this creates a mass of people inhabiting the city essentially made up of tourists, that is, people who are "in transit" and who are avid consumption of cultural, entertainment and leisure services.

The shortage of housing for residents and the overexploitation of tourist accommodation is the catalyst for urban spaces increasingly detached from their roots. That is: the increase in the price of housing and commercial land, local economies oriented to tourist demand, privatized urban spaces designed for constant exploitation, crowded environments of people at all hours, even (and especially) at night, and the precariousness and seasonality of employment associated with the service sector. All these are the consequences of over-tourism that are increasingly worrying the authors of Spanish literature. It is a phenomenon of overexploitation that not only depletes resources, but also wears out the very conditions necessary for tourism to take place.

Sequera and Janoschka (2015) focus on the different dimensions of symbolic gentrification, that is, how creativity, culture and retail can operate as gentrifying *dispositifs* that classify neighborhoods into different "products". For the authors, " the historic city center of Madrid is experiencing new civilities that exclude unwanted populations" (p. 387). This exclusion is closely linked to the "management" of public space and police control and punishment strategies.

In Lavapiés, the *dispositifs* are strongly related to culture, creativity and control of the public sphere, while in Triball it has more to do with the gentrification of fashion and retail trade that goes hand in hand with a commercialization, festivalization and trivialization of public space. This does not mean that Triball is exempt from the expulsion of the "unwanted" who are strategically evicted for a public space that is increasingly used for trade association activities. Some elements such as migration, counterculture and "authentic" taste, say the authors, are used as potential sources of "prosperity", important in times of economic crisis. This last question is of profound interest to understand the consequences that a crisis with the characteristics of a health emergency has in these Spanish cities that are increasingly dedicated to tourist monoculture. In these, a logic is developed that articulates the growing value of capital and investment through the creation of new use values: consumerism of multicultural, alternative, creative or bohemian symbols. Such policies, affirm Sequera and Janoschka (2015), harass the most vulnerable subjects in an increasingly unequal society, limiting access to public space and at the same time promoting social cleansing.

However, the authors acknowledge that the gentrification process is hampered by: (i) a disadvantaged non-European immigrant population giving rise to inter-ethnic solidarity networks; (ii) countercultural movements that take root in neighborhoods affected by social exclusion; (iii) the increase in struggles for the right to housing; and (iv) new residents who are not part of the expected profile of the neighborhood. In this sense, the examples from Madrid provided by Sequera and Janoschka (2015) provide a global understanding of the differences between gentrification in the 'Anglo-Saxon world' and the processes of accumulation of urban capital in Spain.

Milano (2018) understands that the tourism problem is not new. Since the 1970s and 1980s, various authors have already warned about the limits of tourism. In recent years, however, he admits, demonstrations began to be registered with greater intensity that put into question the economic model based on the growth of tourism. In other words, the most appropriate thing would be to speak of "old debates that have acquired new configurations" (Milano, 2018: 552).

In the same way, the author warns about the fact that saturation is not only produced in those places that register large flows of tourists. The perception of saturation always depends on the subjective parameters of the places in question. What is novel about the global post-financial crisis space, however, is the growing democratization of tourism. This causes the perception of the problem to be amplified and the malaise surrounding the tourist monoculture grows. Barcelona is an emblematic case, where, for the period 2016–2020, approximately 30 million annual visitors are estimated in a city with less than two million inhabitants (Ayuntamiento de Barcelona, 2016–2017).

The critical wave of recent years is due, above all, to social movements, the great politicization of the debate and its media resonance (Milano, 2018: 555). However, as found by Egio-Rubio and Fernández-Toledo (2020), in 2018 the national and regional media in Spain did not incorporate the problem of over-tourism and growing tourism-phobia in their agendas. They found that, in general, the issue is more important in the local media, evidencing a strong localization of a phenomenon that, as Milano (2018) points out, has origins and preponderances at a global level.

On the other hand, the result of the comparative study carried out by Almeida-García et al. (2021) about the relevance of tourism-phobia for the cases of the city of Malaga and Gdansk is interesting. The research finds that for residents, the annoyances generated by touristification are greater and more intense than tourism-phobia. In other words, in both study cases, the residents complain about the inadequate handling of the problem by public servants and the inefficient control measures. In this sense, the problem is not with tourists, but with local administrations that do not put regulations in place.

This leads us to ask what kind of decisions can be the best when it comes to providing solutions to the social unrest produced by over-tourism. Milano (2018) points out that few times the strategies of the administrations have aimed to improve the quality of life of the residents of tourist cities. These, in general, do not perceive in a direct or tangible way the benefits of tourist capital gains, which are usually aimed at tourists or landowners who speculate. The author makes a point against governments and recognizes that some of the reasons for not carrying out true controls have to do with a complicated interweaving of economic and political power behind tourism (Milano, 2018: 556). In turn, it points out that the strategies grouped under the name 5D—seasonally adjusted (desestacionalización), decongestive, decentralized, diversified and deluxe tourism—are, however, far from solving the issue, being temporary and time-consuming solutions to the "real" solution advocated by organizations of resistance and criticism such as the Assemblea de Barris pel DecreixementTurístic (ABDT): the decrease in tourism.

17.6 Conclusion and Challenges for the Post-pandemic Era

In 2019, the share of tourism in the Spanish economy was 12.4% of total GDP. In 2020, after the explosion of the pandemic, this share more than doubled: it stood at 5.5% of total GDP (INE de España, 2020). The decline is steep and represents a challenge for the country. The reparation of the Spanish tourist apparatus is necessary. But it cannot be done on just any term. It must be done with a view to long-term sustainable tourism.

The pandemic has opened new challenges to consider. In the first place, the health security of the tourist centers is key to ensuring sustainability. At least in the short term, large agglomerations represent a problem for public health and the spread of diseases and pollution of all kinds. The attention to these types of problems by the governments of the world should open a door for a real urban policy of decentralization and decongestion of spaces. This means diversifying the offer of cities to reduce the increasingly centralized influx in historic centers. This will require efforts by governments around more inclusive urban planning, incorporating the peripheral areas.

However, and as authors such as Milano (2018) or Cócola-Gant and Lopez-Gay (2020) argue, decentralizing and diversifying the tourist offer can mean a response only in the short term if it is not accompanied by regulatory measures that seek mitigate the effects of real estate speculation. Reducing the supply of accommodation in the short term and keeping the use of Airbnb at bay should be a policy implemented in tourist cities.

It is also important to consider the possible effects derived from the post-pandemic world. For example, an increasing number of people who work digitally remotely thanks to the advances and massification of home-office technologies. This will generate, increasingly, a large floating population not tied to specific places and characterized by having a high educational level. This will result in a higher demand for medium to short term rentals in the most popular cities. Transnational gentrification could intensify independently of touristification. This is a possibility to attend to.

Political decisions made in the future will have to consider the following points:

- Greater control of the real estate market and especially of short-term accommodation platforms in places where they lead the gentrification processes (Menorca and San Sebastian, for example). In other words, policies that regulate both the rise in rents and the over-demand for holiday apartments, as happens in cities where tourism is massive, such as Barcelona, Madrid or Seville.
- The protection of services and small and medium-sized businesses oriented not to a tourist demand, but to the neighbors. The regulation of nightlife and leisure and entertainment services goes hand in hand with this point.
- The diversification of the tourist offers not only within the cities, but in other parts of Spain, with the aim of balancing the effects of gentrification and being able to decentralize it.

Although the Great Crisis in 2008 banished interest in the role of crises within the social sciences, tourism studies have paid very little attention to moments of crisis. According to Murray et al. (2017), they tend to consider crises as "blank" moments. However, as these same authors recognize in the case of the Balearic Islands, booms in tourist centers may be the very fruit of crises. The crises and the political responses to them are crucial for a better understanding of the production of the tourist space. Consequently, the authors consider that focusing on the crisis can help build better critical narratives about the production of tourism space, because these arise from the same capitalist contradictions.

A first pattern detected by Murray et al. (2017) is that tourism has worked as a kind of "fix" to the capitalist crises in Spain and, paradoxically, tourism specialization has been increasing until it has become absolute in Spain. places like, for example, the Balearic Islands. These types of places have recovered quickly from previous crises, since they did not have to undertake long restructuring processes, but instead intensified the strategy of accumulation of tourism that turned out to be less expensive. A second pattern is that a constant expansion of the frontier of tourism products has been evidenced with each crisis. Although at first tourism production was concentrated in a few privileged areas, the crises have been overcome with the gradual incorporation of other areas.

On the other hand, while the new tourist spaces have presented growing profits, the old ones have suffered a process of devaluation and degradation. A fourth pattern that these authors have identified consists of the increase in tourist accommodation in times of crisis. This, combined with the fact that many small and medium-sized companies have gone bankrupt in the crisis, has paved the way for the concentration of capital, particularly in the hands of the large hotel chains. The penultimate pattern is the permanent tension between the policies that favor the expansion of the frontier of tourism products and the policies that restrict it. After the crisis and the rapid construction of illegal tourist beds in the first moments, the demands of social movements arise to limit tourist accommodation and protect natural spaces.

Finally, the successive crises have been resolved with labor policy reforms aimed at making working conditions more flexible and reducing labor costs. In turn, tourism production tends increasingly to be less labor intensive. This combination of factors leads to what Murray et al. (2017) consider a growing "accumulation by dispossession" of collective social work. All these considerations are key to thinking about possible egalitarian and fair solutions to the problems of the post-pandemic Spanish tourism industry.

The main take-out from the academic debate is that Spain needs a plan that considers tourism within the social and economic context of its main cities and the fine-grained geographies of social exclusion – especially housing exclusion – that gained ground in it during the last decade and are likely to have been amplified by the pandemic crisis. The sectorization and strong localization of tourism policies and urban planning make the coordination of efforts scarce: each district has its differences in how problems are solved. This is key because each district has its own edges to consider. But the truth is that sustainable tourism needs to consider global trends and, therefore, a clearer global orientation has to be integrated to urban policy discourses and funding schemes.

Acknowledgements This chapter has been partly funded through the R + D + i project ADAPTOUR (contract number PID2020-112525RB-I00, funding ref. MCIN/AEI/ 10.13039/ 501100011033).

References

Almeida-García, F., Cortés-Macías, R., & Parzych, K. (2021). Tourism impacts, tourism-phobia and gentrification in historic centers: The cases of Málaga (Spain) and Gdansk (Poland). *Sustainability, 13*(1). https://doi.org/10.3390/su13010408

Andújar Llosa, A., López Gay, A., & Salvati, L. (2020). Residential mobility, gentrification and neighborhood change in Spanish cities: A post crisis perspective. *Spatial Demography, 8*, 351–378. https://doi.org/10.1007/s40980-020-00069-0

Arroyo-Alba, M. (2019). Acceso a la vivienda en contextos de gentrificación y turistificación urbana. Tendencias Sociales. *Revista De Sociología, 4*, 79–98. https://doi.org/10.5944/ts.4.2019.25258

Ayuntamiento de Barcelona. (2016–2017). *Plan Estratégico de Turismo 2016 2020. Documento de diagnosis estratégica*. Dirección de Turismo, Gerencia de Empresa y Turismo.

Birkbak, A., Dredge, D., Elgaard-Jensen, T., Gymóthy, S., & Madsen, A. (2016). *The impact of regulatory approaches targeting collaborative economy in the tourism accommodation sector: Barcelona, Berlin, Amsterdam and Paris* (Impulse Paper No 9). Aalborg University, European Commission DG GROWTH.

Blanco-Romero, A., Blázquez-Salom, M., & Minguez-García, C. (2017). *Claves de la reestructuración turística de la ciudad. Naturaleza, territorio y ciudad en un mundo global* (págs. 1516–1524). UAM Ediciones. Obtenido de https://www.age-geografia.es/downloads/Naturaleza_Territorio_y_Ciudad_AGE2017.pdf

Bugalski, L. (2020). The undisrupted growth of the Airbnb phenomenon between 2014–2020. The touristification of european cities before the COVID-19 outbreak. *Sustainability, 12*. https://doi.org/10.3390/su12239841

Cabrerizo, C., Sequera, J., & Bachiller, P. G. (2016). Entre la turistificación y los espacios de resistencia en el centro de Madrid: Algunas claves para (re)pensar la ciudad turística. *Ecología Política, 52*, 78–82. Obtenido de https://www.jstor.org/stable/26333515

Calo, E., Rodríguez-Barcón, A., & Otero-Enríquez. (2021). Una revisión crítica sobre el análisis de la gentrificación turística en España. *ROTUR, Revista de Ocio y Turismo, 15*(1), 1–21. https://doi.org/10.17979/rotur.2021.15.1.7090

Cañada, E., & Murray, I. (2019). *Turistificación global. Perspectivas críticas en turismo*. Icaria.

Cócola-Gant, A. (2016). *Apartamentos turísticos, hoteles y desplazamiento de población*. Creative Commons.

Cócola-Gant, A., & Lopez-Gay, A. (2020). Transnational gentrification, tourism and the formation of 'foreign only' enclaves in Barcelona. *Urban Studies, 57*(15), 3025–3043. https://doi.org/10.1177/0042098020916111

Crespi-Vallbona, M., & Mascarilla-Miró, Ó. (2018). La transformación y gentrificación turística del espacio urbano. El caso de la Barceloneta (Barcelona). *Eure, 44*(133), 51–70. https://doi.org/10.4067/s0250-71612018000300051

Del Romero, L. (2018). Touristification, sharing economies and the new geography of urban conflicts. *Urban Science, 2*(4), 104. https://doi.org/10.3390/urbansci2040104

Díaz-Parra, I., & Jover, J. (2019). Gentrification, transnational gentrification and touristification in Seville, Spain. *Urban Studies, 57*(15), 1–16. https://doi.org/10.1177/0042098019857585

Egio-Rubio, C. J., & Fernández-Toledo, P. (2020). Coverage by the Spanish digital press of a geographic issue: Touristification. *Cuadernos de Turismo, 46*, 613–616.

Etxezarreta-Etxarri, A., Izagirre-Olaizola, J., Morandeira-Arca, J., & Carollo, I. M. (2020). Urban touristification in Spanish cities: consequences for the rental housing sector in San Sebastian. *Economic Research-Ekonomska Istraživanja, 33*(1), 1294–1310. https://doi.org/10.1080/1331677X.2020.1755878

Fernández-Tabales, A., Mínguez, C., & Piñeira, J. (2020). Vulnerabilidad y turistificación ¿quiénes son los perdedores del centro urbano? En G. Pons, A. Blanco-Romero, R. Navalón-García, L. Troitiño-Torralba, & M. Blázquez-Salom (Eds.), *Sostenibilidad turística: overtourism vs undertourism* (págs. 83–98). Illes Balears.

Fierro, Á., & Salmón, C. (2018). El impacto del turismo en el casco viejo de bilbao mediante los modelos economía colaborativa: una aproximación a través de una distribución binomial negativa. *Lurralde: Investigación y Espacio, 41*, 175–201. Obtenido de http://www.ingeba.org/lurralde/lurranet/lur41/41fierro.pdf

Fletcher, R. (2016). Cannibal tours revisited: The political ecology of tourism. En M. Blàzquez, M. Mir-Gual, I. Murray, y G. X. Pons (Eds.), *Turismo y crisis, turismo colaborativo yecoturismo*. XV Coloquio de Geografía del Turismo, el Ocio y la Recreación de la AGE. Obtenido de https://www.academia.edu/28966063/Cannibal_Tours_Revisited_The_Political_Ecology_of_Tourism

García, S., Barrera-Fernández, D., & Scalici, M. (2019). Touristification in historic cities. Reflections on Malaga. *Revista de Turismo Contemporáneo, 7*(1), 93–115. https://doi.org/10.21680/2357-8211.2019v7n1ID16169

Getz, D. (1986). Models in tourism planning. Towards integration of theory and practice. *Tourism Management, 7*, 21–32.

Gil, J., & Sequera, J. (2018). Expansión de la ciudad turística y nuevas resistencias. El caso de Airbnb en Madrid. *Empiria. Revista de Metodología de Ciencias sociales, 41*, 15–32. https://doi.org/10.5944/empiria.41.2018.22602

González-Pérez, J. M. (2019). The dispute over tourist cities. Tourism gentrification in the historic Centre of Palma (Majorca, Spain). *Tourism Geographies, 22*(1), 171–191. https://doi.org/10.1080/14616688.2019.1586986

Goodwin, H. (2019). Barcelona – Crowding out the locals: A model for tourism management? En R. Butler & R. Dodds (Eds.), *Overtourism: Issues, realities and solutions* (págs. 125–138). De Gruyter Oldenbourg. https://doi.org/10.1515/9783110607369-009

Guerra-Luzuriaga, P. (2020). Reactivación vs turistificación: los retos de las ciudades frente al turismo post pandemia. *INNOVA Research Journal, 5*(3.2), 134–150. https://doi.org/10.33890/innova.v5.n3.2.2020.1564

INE de España. (2020). *Cuenta Satélite del Turismo de España (CSTE). Revisión estadística 2019*. Instituto Nacional de Estadística. Obtenido de https://www.ine.es/prensa/cst_2020.pdf" https://www.ine.es/prensa/cst_2020.pdf

Ivars-Baidal, J. A. (2004). Regional tourism planning in Spain: Evolution and Perspectives. *Annals of Tourism Research, 31*(2), 313–333.

Milano, C. (2018). Overtourism, malestar social y turismofobia. Un debate controvertido. *Revista de Turismo y Patrimonio Cultural, 16*(3), 551–564. https://doi.org/10.25145/j.pasos.2018.16.041

Mundet, L., & Pirillo-Ramos, S. (2020). Tourism-phobia in Barcelona: Dismantling discursive strategies and power games in the construction of a sustainable tourist city. *Journal of Tourism and Cultural Change, 19*(1), 113–131. https://doi.org/10.1080/14766825.2020.1752224

Murray, I. (2015). *Capitalismo y turismo en España. Del "milagro económico" a la "gran crisis"*. Alba Sud Editorial.

Murray, I., Yrigoy, I., & Blázquez-Salom, M. (2017). El papel de las crisis en la producción, destrucción y reestructuración de los espacios turísticos. El caso de las Islas Baleares. *Investigaciones Turísticas*, 1–29.

Pareja, M., & Simó, M. (2014). Dinámicas en el entorno construido: renovación, gentrificación y turismo. El caso de la Barceloneta. *ACE: Architecture, City and Environment, 9*(26), 201–222. https://doi.org/10.5821/ace.9.26.3688

Rodríguez-Barcón, A., Calo, E., & Otero-Enríquez, R. (2020). Social Transformation and Urban Regeneration: Three interpretations on the phenomenon of gentrification in the historic center of ACoruña (Spain). En J. N. DeSena & J. Krase (Eds.), *Gentrification around the World, Volume II* (págs. 65–82). Palgrave Macmillan. https://doi.org/10.1007/978-3-030-41341-5_4

Sans, A., & Domínguez, A. (2016). 13. Unravelling Airbnb: Urban perspectives from Barcelona. En A. Russo & G. Richards (Eds.), *Reinventing the local in tourism: Producing, consuming and negotiating place* (págs. 209–228). Channel View Publications. https://doi.org/10.21832/9781845415709-015

Sequera, J., & Janoschka, M. (2015). Gentrification dispositifs in the historic centre of Madrid: a reconsideration of urban governmentality and state-led urban reconfiguration. En L. Lees, H. B. Shin, & E. López-Morales (Eds.), *Global gentrifications. Uneven development and displacement* (págs. 375–393). Policy Press.

Smith, N. (2012 [1996]). *La nueva frontera urbana. Ciudad revanchista y gentrificación*. Traficantes de sueños.

Urquiaga, A. A., Lorente-Riverola, I., & Sanchez, J. R. (2020). Platform-mediated short-term rentals and gentrification in Madrid. *Urban Studies, 57*(15), 3095–3115. https://doi.org/10.1177/0042098020918154

Valente, R., Russo, A. P., Vermeulen, S., & Milone, F. L. (2022). Tourism pressure as a driver of social inequalities: A BSEM estimation of housing instability in European urban areas. *European Urban and Regional Studies, 29*(3), 332–349.

Vestri, G. (2020). Acceso a la vivienda en la época de la «gentrificación» y de la «turisficación». El distrito casco antiguo de Sevilla como caso de estudio: legislaciones administrativas a prueba. *Dereito: revista xurídica da Universidade de Santiago de Compostela, 29*(Extra 1), 61–74. https://doi.org/10.15304/dereito.29.Ext.6554

Vives-Miró, S. (2011). Producing a "Successful City": Neoliberal Urbanism and Gentrification in the Tourist City—The Case of Palma (Majorca). *Urban Studies Research., 2011*, 1–13. https://doi.org/10.1155/2011/989676

Yrigoy, I. (2017). Airbnb en Menorca: ¿Una nueva forma de gentrificación turística? Localización de la vivienda turística, agentes e impactos sobre el alquiler residencial. *Scripta Nova. Revista Electrónica de Geografía y Ciencias Sociales, 21*. https://doi.org/10.1344/sn2017.21.18573

Yrigoy, I. (2020). El impacto del alquiler turístico sobre el alquiler residencial: abriendo (y cerrando) brechas de renta en el casco histórico de Palma. *Scripta Nova. Revista Electrónica de Geografía y Ciencias Sociales, 24*. https://doi.org/10.1344/sn2020.24.22643

Gustavo Javier Macías Mendoza has a bachelor's degree in tourism business administration from Eloy Alfaro Ray University of Manabi, Ecuador, and a master's degree in tourism management and consulting from international Iberoamerican University of Mexico. He is currently a PhD candidate in the Tourism and Leisure program at the Rovira i Virgili University in Tarragona, Spain. He has published two academic articles derived from his doctoral research, and has 20 years of experience in the hospitality field.

Antonio Paolo Russo is Professor of Urban Geography with the Department of Geography, Universitat Rovira i Virgili, Tarragona. He is the coordinator of the PhD program in Tourism and Leisure. Dr. Russo is author of 40+ publications in academic journals and books, on research topics ranging from tourism studies to urban geography and urban and regional planning. Currently, he leads the H2020 project 'SMARTDEST', tackling tourism mobilities and social exclusion, and coordinates the "Space Place Mobilities in Tourism" group in the ATLAS network.

Chapter 18
The Vicious Circle: Intersecting Leisure-Rooted Migrations and Ethnic-Based Segregation in the Mediterranean Spanish Cities

Jesús M. González-Pérez and Ismael Yrigoy

18.1 Introduction

"Cities have always been divided, (...) whether they are called divided, dual, polarised, quartered, or fragmented- This segregation process is intrinsic to urban development" (Marcuse, 2002:7). Ethnic or racial differentiation is one of the structuring elements of this segregation and, therefore, ethnic segregation has become a key field of inquiry when it comes to explaining the dynamics of urban environments across the globe. This type of segregation has been studied in contexts as disparate as the United States (Holloway et al., 2012; Massey et al., 2009), Latin America (Smets & Salman, 2016), Africa (Agyei-Mensah & Owusu, 2010) Oceania and Europe (Randolph, 2020).

There is, however, a need to strengthen research on how lifestyle migration, on the onc hand, and migration linked to tourist work, on the other, both shape the patterns of socio-spatial segregation in tourist cities.

Therefore, this chapter analyzes and compares how this dual migration process (migration for lifestyle linked to leisure and tourism and migration for work linked to tourism) shapes the processes of socio-spatial segregation in Palma, an illustrative example of the wider socio-spatial processes occurring in the touristic cities across

J. M. González-Pérez (✉)
Department of Geography, University of the Balearic Islands, Palma, Spain
e-mail: jesus.gonzalez@uib.es

I. Yrigoy
Department of Geography, University of Santiago de Compostela, Santiago, Spain

Department of Human Geography, Uppsala University, Uppsala, Sweden
e-mail: ismael.yrigoy@usc.es

A. Blanco-Romero, M. Blázquez-Salom (eds.), *Spanish Tourism Geographies*, Geographies of Tourism and Global Change,
https://doi.org/10.1007/978-3-031-39780-6_18

the Spanish Mediterranean. The chapter is structured in three sections. The first section contextualizes the study of ethnic segregation associated with urban processes in the tourist cities in Spain. The main contribution is to build bridges between the existing literature on lifestyle migration and on segregation. The second section presents the empirical contribution of the article: it demonstrates, based on the statistical exploitation of data from the Continuous Register of Inhabitants, the progressive concentration of low-income labor migrant population (Son Gotleu) and lifestyle migrations (Santa Catalina). The third section discusses how lifestyle and labor migrations and the associated socio-spatial segregation go hand in hand with displacement pressures, as illustrated by data on evictions and foreclosures. The fourth and final section concludes.

18.1.1 Placing the Role of Segregation in Leisure-Triggered Migrations

In the field of tourism studies, ethnic segregation has been investigated as an element linked to the construction of tourism megaprojects and which, ultimately, places physical barriers to the access of the local population to places commercialized by tourism -such as, for example, beaches- (Campuzano et al., 2014; López-López et al., 2006). Such an approach to segregation within tourism studies mainly refers to the exclusion mechanisms by which residents cannot gain access to specific spots following the construction of tourism resorts and the like. From the tradition of labor studies linked to tourism, the ethnic segmentation of jobs in different sectors linked to tourism has a prominent development, inspired mainly by Linda McDowell and the feminist school of labor studies. However, labor scholars have focused on the social aspects of segregation, leaving aside the spatial aspect of segregation.

Urban geography has been the most proactive in conducting studies on ethnic segregation, mainly in the context of North America and Latin America (Musterd, 2005; González-Pérez, 2021). The epistemological genealogy that explains the centrality of discussions on segregation goes back on the one hand to the factorial ecology studies of the Chicago school that theorized segregation "as an organic phenomenon and a transitory path towards integration" (Arbaci, 2019:21). These neo-Darwinist approaches of the 1960s and 1970s were progressively complemented and replaced by a combination of approaches that, drawing from epistemological currents as diverse as behavioral studies, neo-Weberism or neo-Marxism, articulated segregation studies around two key discussions. On the one hand, the discussion on the choices that different ethnic groups can make, leading to discussions such as the 'white flight' of ethnic Europeans from neighborhoods with the presence of migrants from the global south (Frey, 1979; Bråmå, 2006). On the other hand, the discussion on the limitations that certain ethnic groups suffer while residing in certain urban areas (Hanhörster, 2001). In Europe, ethnic segregation has been less studied, although there are important regional differences. While, in Southern Europe the incidence of the discussion on residential segregation is still an emerging field (with notable exceptions, such as Arbaci, 2019), studies on segregation have become central in the debates on urban evolution in Scandinavia (Bråmå, 2008; Malmberg et al., 2013). In Spain, the work of Leal (2002) has been one of the key references when explaining the phenomenon of ethnic segregation. In fact, studies on ethnic segregation in Spanish are mainly from the field of migration. Following the trail marked by Leal, the studies of Domínguez Mujica et al. (2008), Bayona i Carrasco and Gay (2011), Vono and Bayona (2011), Domínguez Mújica et al. (2010) and Parreño et al. (2022) have mainly focused on analyzing the spatial distribution of migrants from the global south. In other words, the aforementioned contributions refer mainly to social groups that are spatially segregated because of the restrictions they face in accessing the real estate market.

However, the insular and Mediterranean Spanish coastal areas have a double process of influx of migrants linked, in different ways, to tourism and leisure. There are on the one hand, migrants who arrive to coastal Spain to work within the tourism sector or sectors associated with tourism. On the other hand, there group of migrants from the Global North (particularly relevant in Spain are the cases of arrivals from Germany, the United Kingdom and Scandinavia), generally with purchasing power equal to or greater than the average local population, who partially move to fulfill a change in their lifestyle linked with an increasing leisure. This is a specific type of migration that Benson and O'Reilly (2009) defined as *lifestyle migration.* The combined effects of *both* kind of migrations to the issue of segregation is a key element that, as Huete et al. (2013), pointed out, needs to be further developed, particularly in the Spanish case.

Spain's specialization in tourism and real estate has historically been sustained by a double process of migration for tourism and for lifestyle. But both migrations have been transforming since their emergence at the dawn of Franco's developmentalism. On the one hand, labor migration has been globalizing, moving from a migratory dynamic framed within the Spanish rural exodus to a global South-North migration dynamic. On the other hand, migration by lifestyle has gone beyond the classic migration of pensioners to peripheral tourist urbanizations already anticipated by Gaviria (1974) or Jurdao (1979), to a new migration by lifestyle that is much more diverse in terms of age, family situation, origin, economic status and spatial location patterns. This lifestyle migration has a direct impact on urban segregation processes. In contrast to labor migration, which is closely linked to limitations in access to housing, lifestyle migration is linked to a process of residential segregation due not to a limitation, but to a choice.

The following sections explain the double process of how lifestyle migrations affect spatial segregation by choice, exemplified in the case of the Santa Catalina and Son Gotleu neighborhood in Palma, and how labor migrations shape -and is shaped- by spatial segregation.

18.2 Santa Catalina: Patterns of Segregation in Little Sweden

18.2.1 Introducing Santa Catalina: A History Between Marginality and Centrality

The neighborhood of Santa Catalina is located in the west of the city of Palma. It is a neighborhood expansion of the nineteenth century planned according to orthogonal schemes and originally of fishing and industrial origin (Ajuntament de Palma, 1994; Urgell, 2016; González-Pérez, 2017). At the beginning of the twentieth century, Santa Catalina was the most important neighborhood outside the city walls, both economically and demographically. In 1900 it had a population of 7406 inhabitants (12.1% of the municipal population of Palma), much higher than that of other neighborhoods outside the city walls at that time, such as the suburbs of El Molinar (1435 inhabitants) or La Soledad (1240 inhabitants) (Carbonero, 1991: 97–98).

This early seafaring and industrial vocation were progressively complemented by a tourist specialization given its proximity to key points of the early stages of the emergence of tourism. In fact, the location of Santa Catalina is adjacent to the neighborhood of El Terreno, the first neighborhood in Palma destined for international leisure since the early thirties. The neighborhood of Santa Catalina is also adjacent to the promenade of Palma, a space destined for nightlife, and the seaport of the city. This early tourist vocation of the neighborhood did not develop further from the 1960s onwards, due to the irruption of mass tourism in other spots across Mallorca. In fact, Santa Catalina stagnated demographically and declined economically, resulting pockets of urban degradation as happened in the neighboring neighborhood of Es Jonquet. This later neighborhood, bordering Santa Catalina, was the object of a Special Plan of Interior Reform in 1985, in the context of other similar plans within the historic city walls.

18.2.2 Socio-demographic Changes in Santa Catalina: Unveiling Patterns of Segregation

The population shifts in Santa Catalina illustrates a broader tendency occurring in touristic spots in coastal Spain: population from the Global South and Spanish nationals are shrinking because of the increase of population from the Global North.

By exploring the place of birth, foreigners by area of nativity, on five different dates: 1998 (change of migratory cycle in Spain and explosion of foreign migrations with the start of the real estate boom), 2004 (real estate bubble), 2010 (bursting of the bubble and economic crisis), 2015 (post-crisis) and 2021 (post-COVID stage and latest available data), we grasp the aforementioned shift.

Between 1998 and 2010 the census tract boundaries in Santa Catalina have changed. Consequently, any comparison between 1998 and 2004 with later years is merely indicative. Even the comparison between census tract averages is also indicative because the boundaries of all the census tracts studied between 1998–2004 and 2010–2015–2021 do not correspond exactly. Between 1998 and 2004 on the one hand, and from 2010 onwards, the census tracts have not been altered and therefore more reliable comparisons can be made.

In the first place, the stagnation and even a decreasing trend in the evolution of the total population in Santa Catalina stands out. Although in the period 1998–2004 there is a growing trend, from 2010 onwards, coinciding with the onset of the economic crisis, there has been a stagnation in the registered population in the sections studied for the neighborhood of Santa Catalina, from 7166 inhabitants in 2010 to 7157 inhabitants in 2021 (see Table 18.2-Annex). In this sense, the segregation processes linked to tourist migrations go hand in hand with a stagnation in the resident population.

Secondly, it is worth noting the increase in the non-Spanish population. This population has gone from representing an average of 4.25% in the sections that include only Santa Catalina in 1998 to 33% in 2021 (see Table 18.2-Annex). If the census sections that include both Santa Catalina and Es Jonquet are included, the increase is similar: from 1.95% in 1998 to 18.61% in 2021 (see Fig. 18.1). While the overall population is stagnant, the non-Spanish population has a pattern of growth. Indeed, the population of Balearic origin - descendants of the rural-urban migrations that took place on the island from the mid-nineteenth century to the mid-twentieth century – is on a downward trend. While in 1998 it represented approximately two thirds of the total population of the neighborhood, today it barely reaches 45%. This increase in the non-Spanish population denotes a pattern of segregation of the local population. On the one hand, such pattern could be interpreted because of residents being forced to emigrate due to the generalized price increase in many of the neighborhoods in the city, and that has in Santa Catalina one more exponent. And, on the other hand, there is a choice on the part of the original population of the

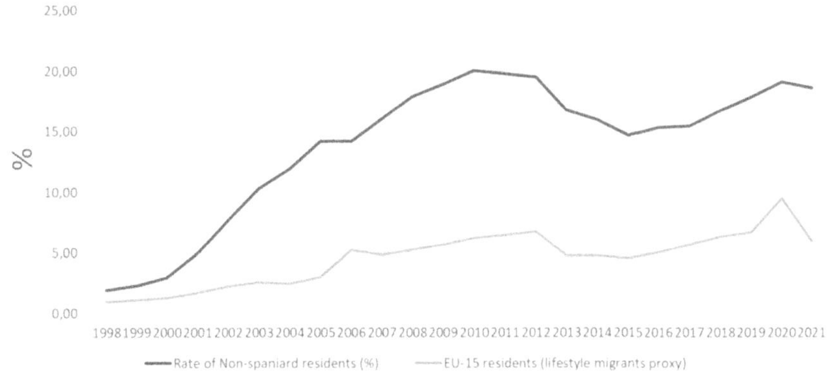

Fig. 18.1 Evolution of the rate of foreigners and migrants by lifestyle in Santa Catalina and Es Jonquet (1998–2021). (Source. Own elaboration from IBESTAT (2022))

neighborhood, because of what Valli (2015) called feelings of displacement. These feelings of displacement go hand in hand with a choice by residents, in principle not forced, to leave the neighborhood of their own free will. We refer here to the feelings of alienation from the neighborhood caused by the substitution of traditional businesses by tourist and/or migrant-oriented lifestyle businesses, the large-scale opening of nightlife and restaurant businesses, and the intensive use of public space by nightlife and restaurant businesses. Note here the similarity -and the differences- with the processes of abandonment by locals in impoverished neighborhoods. Both the fall in average purchasing power (linked to precarious tourism workers) and the rise in purchasing power (linked to lifestyle migrations) trigger feelings of alienation that have led segments of local residents to leave their neighborhoods. The symptoms of such alienation are partly similar (feeling of invisibility on the part of segments of the new residents) but partly different (feeling of insecurity in impoverished neighborhoods linked to tourism workers and feeling of rising prices in neighborhoods linked to lifestyle migrants).

Third, the evolution of the non-Spanish population between 1998 and 2021 in Santa Catalina illustrates three distinct moments in the spatial dialectic between immigration linked to tourism workers and lifestyle migrations. The period prior to the 2008 crisis (exemplified by the years 1998 and 2004) shows a process of conversion of the neighborhood into a center for immigrant workers linked to the service sector, particularly tourism. In fact, the population of Latin American origin increased during this period, while residents from European countries - used here as a proxy for lifestyle migrants - increased at a much slower rate (see Fig. 18.1), with some census tracts corresponding to Santa Catalina where migrants from European countries decreased both in absolute and relative terms (see Table 18.2). Then, the period after 2010 is characterized by a slowdown in migration processes: the total number and proportion of foreigners slightly decreases, both in the case of migrants linked to tourist work and lifestyle migrants. The third moment starts in 2015 when there is a conversion of the neighborhood into an epicenter of lifestyle migrations. Foreigners increase proportionally, and within the foreigners, those who increase the most are the citizens coming from the former EU-15 (see Fig. 18.1). This increase declined sharply between 2020–21, probably due to the disruptions linked to the pandemic (see Fig. 18.1). Changes between 2015 and 2020 point to a consolidation of this expansive phase of lifestyle migration. Given the current situation, it seems that Santa Catalina has become one of the bastions of lifestyle migrations in Palma, although these migrations are already expanding strongly to other parts of the city such as Son Espanyolet or Portitxol. The next movements in these lifestyle migrations may be precisely towards the neighborhoods bordering these, which, like an urban sprawl, are spreading throughout the city, including old mature tourist neighborhoods such as El Terreno (Table 18.2).

This lifestyle migration contributes to segregation processes in that census tracts with higher purchasing power tend to have higher housing prices, and these are the tracts with the highest presence of lifestyle migrants (see Fig. 18.2). And the areas with the highest purchasing power are the areas where there are higher prices and therefore greater restrictions on access to housing for social groups without high purchasing power.

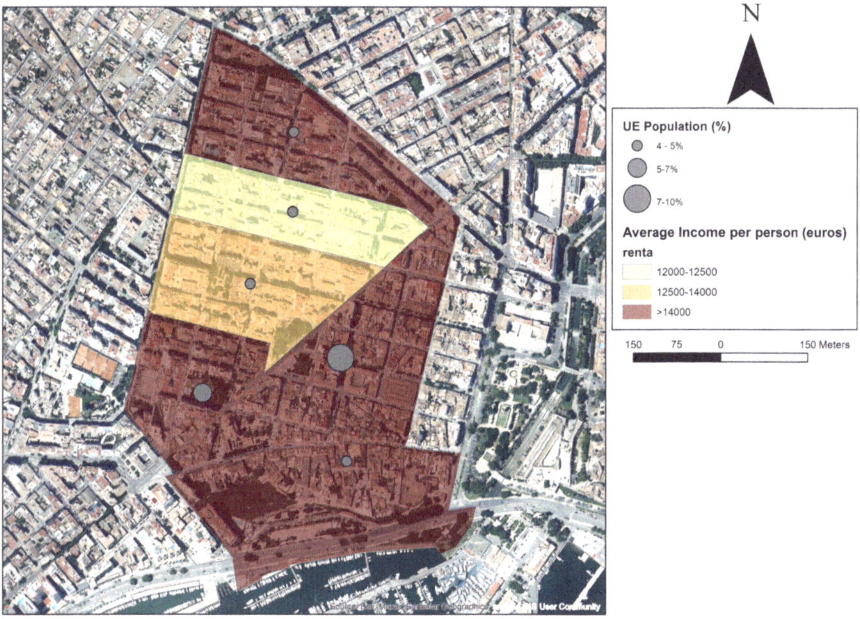

Fig. 18.2 Average income per person (2019) and location of EU-15 population (2021) by census tracts in the Santa Catalina neighborhood. (Source: Own elaboration from INE (2021) and IBESTAT (2022))

18.3 Son Gotleu: Low-Income Migration and Segregation

18.3.1 Introducing Son Gotleu: Migrations and Developmentalism *in the Periphery of the Spanish City*

The Spanish city of developmentalism has in the residential peripheries one of its most identifiable urban spaces (Capel, 2002). Built between the 1960s and 1980s, these are dense, monofunctional areas promoted largely by public initiative and fundamentally in the form of housing estates. From a socio-urban perspective, these are operations aimed at creating housing to accommodate the working class at the lowest possible cost, with peripheral locations, problems of integration into the urban fabric, deficient facilities and low construction quality (Capel, 2002). As a result of a planned segregation that isolates them from the rest of the city, these old peripheries were designed as containers for excluded and migrant populations, and particularly house industrial workers or low-skilled employees in the tertiary sector, especially in tourism in the case of the archipelagos in Spain (González-Pérez et al., 2022).

Son Gotleu neighborhood in Palma is a paradigmatic example of such dynamics. On the easternmost edge of Palma's Ensanche it continues to be peripheral and marginal from a social, perceptual and symbolic point of view.

The social housing promotions of developmentalism are the most recognizable and identifying landscape of the neighborhood. Dating back from the 1960s, such promotions are heirs to the principles of modern architecture of the 1920s and 1930s, where concepts such as the housing cell are included. However, the poor quality of the buildings, the small surface area (55 m^2) and their projection as residential containers for the low-income immigrant population led to their urbanization as a highly degraded area. The degradation of the common spaces is important both in the dwellings of 1960 and in those of 1964–1970 (González-Pérez, 2022).

Like so many other housing estates in the Spanish city, its origin is linked to the increase in population resulting from natural growth and the rural exodus within and between provinces that took place from the second half of the 1950s until well into the 1960s. Thus, from the beginning, it housed an immigrant working population, with low qualifications and income and, although there is no data on the subject, employed in tourism and construction activities (González-Pérez et al., 2022). This situation lasted at least until the end of the 1990s, when a process of change in the demographic composition began. With the emergence of labor-oriented international migrations, largely sponsored by the labor attraction produced by the real estate bubble, a process of ethnic diversification began (González-Pérez et al., 2022).

Six decades after its construction, Son Gotleu leads Palma's indicators of vulnerability and impoverishment. A study by the Ajuntament de Palma (Palma City Council) (2017 have placed Son Gotleu as the most vulnerable and with the most situations of social exclusion (Ajuntament de Palma, 2017). And, more recently and using other methodologies, González-Pérez and Piñeira-Mantiñán (2020) placed the five census sections of son Gotleu with the highest degree of vulnerability in Palma.

These indicators are supported by other types of data that confirm the problems and associated socio-urban processes. Son Gotleu is divided into five census sections. On the one hand, if we look at the economic level or income of the residents, four census sections of Son Gotleu are among the five with the lowest income of all the municipalities of the Balearic Islands according to the latest data of 2019. Section 4022 has the lowest per capita income (6153 Euros), and 4024 the second (6200 Euros). The section with the highest income in the neighborhood (4020) is the eighth lowest in the entire autonomous community.

On the other hand, the value of housing and the prices they acquire on the market, both for sale and rent, confirm the depreciation of the neighborhood. The average cadastral value of housing in Palma is 63,930 Euros (2017), whereas Son Gotleu has the second lowest average (30,977 Euros) in the city. Furthermore, according to data from the Association of Developers from the Balearic Islands (API), the average price per square meter of housing between party walls or multi-family sold between January 1, 2019, and June 25, 2022, in Palma ranges between 5157.07 Euros in Son Vida and 1028.25 Euros in Son Gotleu. The average rent in section 4022 is 5.4 Euros/m^2, the lowest in Palma. This is followed by section 4019 (5.6 Euros/m^2) and 4023 (5.7). These housing rental prices indicate the type and quality of thc housing offered and the economic level of the resident population. In short, an indicator that feeds back into the arrival of the poorest.

18.3.2 Ethnification: Immigration as the Basis of Its Social Composition

The birth and nationality data show the immigrant origin of Son Gotleu the continuous processes of displacement and substitution: when a family manages to rise in their social position, in many cases the children of the first Spanish immigrants, they leave the neighborhood and their place is taken by other immigrants who, in this case, are currently mostly of African origin. There is, therefore, a classic White flight process as conceptualized by Frey (1979).

As in Santa Catalina, we studied the evolution of the socio-demographic composition of Son Gotleu based on the analysis of four variables (place of birth, foreigners by area of nationality, most numerous foreigners and autonomous community with the highest number of foreigners born, except for the Balearic Islands), and on five different dates: 1998, 2004, 2010, 2015 and 2021.

In 1998, in the transit of the arrival of the great labor immigration flow to Spain in general and to Mallorca in particular, Son Gotleu preserves many of the characteristics that saw its birth three decades ago: importance of the population born in the same municipality (Palma),, scarce representation of population coming from other municipalities of the Balearic Islands and a noticeable presence of residents born in other autonomous communities. Among non-Balearic Spaniards, those born in Andalusia are in the majority in all sections and in all years analyzed. This demonstrates, on the one hand, the weight of peninsular immigration, especially of rural origin, to feed this insular urban development, as well as the scarce weight of the Majorcan rural exodus in the origin and in the urbanization process of Son Gotleu and, although further research would be necessary, in the majority of the housing estates of Palma. On the other hand, the resident Andalusians belong to the first generation of immigrants in the neighborhood. Regarding the rate of foreigners, it is very low (1.30%) and the majority are still EU-15 nationals. Another interesting feature is a certain social homogeneity among the five census sections. This model, inherited from its construction stage in the middle of the century, began to change at the end of the twentieth century.

In 2004, at the height of the real estate bubble, the main characteristic is the increase in the foreign-born at the expense of small reductions, more or less similar, in the other three groups (Palma, Balearic Islands, Spain). In 2004, the foreign-born rate rises to 8.68%. A progressive and unstoppable process of displacement and social substitution can be observed. A change where the arrival of the population from Latin America stands out, which dominates in all the census sections.

Six years later, in 2010, these trends intensified, with sharp falls in the rates of those born in Palma (−12.01%) and in other Autonomous Communities (−12.43%) above all, at the expense of an outstanding increase in foreigners (rate of 40.64%) but now fundamentally of African origin (almost 70% of these were born on this continent). The economic crisis of 2008 and the increase in unemployment rates contributed to the return to their countries of origin of many Latin Americans, but not Africans, who continue to enter the country and are located in degraded neighborhoods with cheaper housing (González-Pérez, 2022). An interesting change in this year is the progressive internal contrasts, with different or more pronounced dynamics in the three sections where the 1960 social housing developments are located. As degraded areas, it is where those born in Palma and even in other Autonomous Communities lose more representation at the expense of foreigners, especially Nigerians. In section 4024, 8 out of 10 foreigners are Africans, especially Nigerians, who are the most numerous in four of the five census sections of Son Gotleu.

Most of these processes are confirmed in 2015, with two fundamental changes that, to a lesser extent, also occurred in Santa Catalina: the rate of foreigners is reduced by more than ten points (30.20% in 2015) and, compared to 2010, the number of residents born in Palma increases while those born outside Spain are reduced. This is not only because there are proportionally fewer foreigners, but also because there is already a new generation of foreigners: the children of the first foreign immigrants who arrived in the last 20 years, who were already born in Palma. The dominance of Africans among immigrants is absolute, with the novelty of Moroccans as a new and growing nationality.

The latest data for 2021 again modify trends, a product of the enormous dynamism of the neighborhood and the disruption generated by covid: the number of those born in Palma has stabilized; those arriving from another Autonomous Community (a consequence of the first generation of immigrants) has decreased by about 20 points since 1998; those born abroad and the rate of foreigners has increased again (34.99%); Africans continue to be the largest group but Moroccans now outnumber Nigerians and Latin American immigration, especially from Colombia, has been reactivated (see annex Table 18.3 and Fig. 18.3).

These processes contribute to the social construction as an impoverished neighborhood but where, at the same time, important internal contrasts are detected. In these, immigration is once again the basis for spatial differences. Thus, there is a direct correlation between the three neighborhoods with the lowest income per

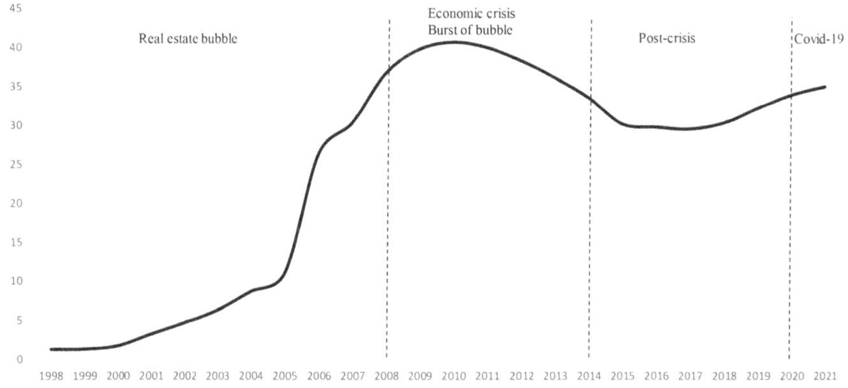

Fig. 18.3 Evolution of the rate of growth in number of foreigners in Son Gotleu (1998–2021). (Source: Own elaboration from IBESTAT (2022))

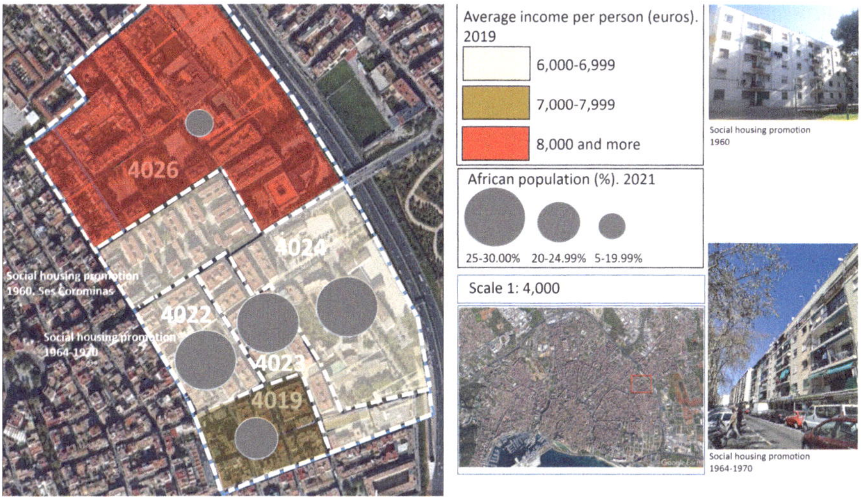

Fig. 18.4 Average income per person (2019) and percentage of African population (2021) by census tracts in the neighborhood of Son Gotleu. (Source: Own elaboration from INE (2022))

person, the highest percentage of African residents and the location of the 1960 social housing developments. An immigration of low-income population that contributes to feedback segregation and impoverishment (Fig. 18.4).

18.4 Evictions and Foreclosures: Displacements an Indicator of Segregation

Evictions and foreclosures can be used as indicators for the analysis of ethnic segregation. In Palma it has been shown how the rent component has caused gentrification and displacement, and how these two processes in turn entail social substitution (Vives et al., 2018a, b; González-Pérez et al., 2020). Although there is some precedent (González-Pérez, 2022), it is necessary to study how this social substitution has affected ethnic segregation depending on the nationality of foreigners. In this section we will study how the two types of eviction (rental evictions and foreclosures) have been used to displace former residents, in one case, to be replaced by foreigners of higher economic level coming from the global North (Santa Catalina-El Jonquet)[1], thus, a process under a

[1] In the analysis and mapping of Santa Catalina we include the neighboring El Jonquet, because despite having different morphological features, they function and, at times, are perceived as the same neighborhood.

gentrification objective, and in the other, as the displacements of the poorest in Son Gotleu are producing processes of regrouping of the poor, exemplified in the case of the African population.

Between 2008, the beginning of the economic crisis and the bursting of the housing bubble, and 2020, the last years obtained, in Palma there have been 8964 foreclosures and 18,515 rental evictions. This unbalanced distribution is respected in the two neighborhoods under study, although with a notable difference: Son Gotleu, as an impoverished neighborhood, has the highest foreclosure rates, approaching those of rental evictions. Almost 8 out of every 100 foreclosures in Palma have occurred in this small neighborhood of just over 10,000 inhabitants. The population of Son Gotleu represents 2.45% of the population of Palma but concentrates 5.35% of the evictions of the municipality in the period considered.

For the period 2003–2014 in Palma, it was shown that in gentrifying neighborhoods, rental evictions had a higher representation. In the impoverished ones, although both are important, foreclosures stand out (Vives et al., 2018a). This is confirmed in the two study neighborhoods in this more updated and longer time series. Now, our two cases, gentrification and impoverishment give a leading role to foreign population of different nationality and economic level. To a greater or lesser extent, and although they need further research, they participate in many of the dynamics of displacement and substitution. Bearing in mind that they have a similar population (7819 in Santa Catalina-El Jonquet and 10,284 inhabitants in Son Gotleu), the contrasts by neighborhood are important (Table 18.1).

El Jonquet (southernmost sector, in contact with the seafront) is a very elitist space, sparsely populated (662 inhabitants) and, therefore, with few evictions and concentrated in the limits with Santa Catalina. In the latter, the territorial distribution is more or less balanced. Meanwhile, in Son Gotleu these are a clear expression of the ethnic and economic segregation of the neighborhood. Those areas with more African population and lower income levels concentrate foreclosures and rental evictions. Displacements that increase impoverishment and segregation (Figs. 18.5 and 18.6).

In short, the accumulation of urban rents generalized through the crisis is the main mechanism driving evictions, which in turn becomes a fundamental mechanism for accelerating processes of gentrification and impoverishment. Processes that reproduce and intensify ethnic segregation.

Table 18.1 Population (2021), foreclosures and rental evictions in Santa Catalina-El Jonquet and Son Gotleu (2008–2020)

	Population	% Palma	Foreclosures	% Palma	Rent evictions	% Palma
Santa Catalina-El Jonquet	7819	1.86	104	1.16	423	2.28
Son Gotleu	10,284	2.45	710	7.92	760	4.10

Source: Prepared by the author based on Notices and Seizures Department (SCNE) data

Fig. 18.5 Foreclosures and rental evictions in Santa Catalina-El Jonquet and Son Gotleu (2008–2020). (Source: Own elaboration from Seizures Department (SCNE) data)

Fig. 18.6 Evictions in Santa Catalina-El Jonquet and Son Gotleu: distribution of foreclosures and rental evictions (2008–2020). (Source: Own elaboration from Seizures Department (SCNE) data)

Fig. 18.6 (continued)

18.5 Concluding Remarks

Tourist cities are a particularly relevant study territory for the comparative analysis of migrations (by typologies or motivations, or both) and the different segregation models (ethnic, economic). From the geography of tourism, the scientific literature has mainly focused on lifestyle migrants. From urban studies, more interest appears in the segregation in the city of the most vulnerable groups (foreigners from the global South, low-income population, and others). However, as we have studied in this chapter, there are significant interrelationships between these two phenomena: patterns of spatial segregation, on the one hand, and the origin and motivations for the movement of foreign migrants, on the other. Spanish Mediterranean cities, specialized in tourism-real state market binomial, are historically sustained by this double migratory flow. Income and housing are two of the indicators used to measure urban segregation, as well as to define processes and areas of gentrification and impoverishment in the city.

Palma is an example of a Mediterranean tourist city with a double migratory flow. That related to leisure and lifestyle, on the one hand, and labor, on the other. The location of these groups of different economic levels in the city generates socio-urban processes that generate ethnic segregation. This segregation is caused either by gentrification processes or by processes associated with impoverishment. The result is the construction of a segregated city (economic, social, residential, cultural...) largely as a consequence of an initial ethnic segregation associated, in one way or another, with the tourism and leisure component that economically and socially articulates the city of Palma.

The neighborhoods of Santa Catalina and Son Gotleu exemplify the importance of international migrations associated with tourism and leisure in the social construction and urban configuration of the city. In the former, the arrival of foreign groups is a recent phenomenon. The historical and patrimonial value of the neighborhood and its centrality serve as an attraction for a growing number of Europeans, mainly Swedes, who set up their main or secondary residence in one of the apartments in the neighborhood. Some 34% of foreigners are from the EU-15 in 2021. On the contrary, Son Gotleu is a neighborhood product of working-class labor migration, first peninsular Spanish and, because of substitution processes, mainly African at present. The latter represent 63% of the foreigners. Due to the relationship between the location of these groups and income distribution, as we have been able to demonstrate, we can affirm that the different origins and nationalities of the migrants are determining factors in the gentrification and impoverishment of the neighborhoods.

These processes of migration linked to leisure and tourism are closely linked to processes of displacement not only indirect, but direct in the form of evictions and foreclosures. Indeed, the revaluation caused by lifestyle migrations in Santa Catalina is causing evictions and foreclosures. These displacements are suffered by social groups other than lifestyle migrants (either local population or other migrant groups). On the contrary, the impoverishment linked to the precariousness of the workers residing in Son Gotleu leads as well, as a consequence, to displacement in the form of evictions and foreclosures. In any case, the interrelation of the city with leisure and tourism is resulting in direct forms of displacement that perpetuate segregation.

Acknowledgements Grants projects: "Housing and international mobility in cities of the Balearic Islands: the emergence of new forms of urban inequality" (RTI2018-093296-B-C22), and "Cities in transition. Urban fragmentation and new socio-spatial patterns of inequality in the post-pandemic context. The case of the urban area of Palma (Mallorca)" (PID2021-122410OB-C31), funded by MCIN/AEI/10.13039/501100011033 and by "ERDF A way of making Europe".

Appendix

Table 18.2 Population of Santa Catalina according to place of birth and nationality (1998–2021)

1998							
Census section		**3008**	**3007**	**3006**	**3005**	**3004**	**Santa Catalina**
	Total Population	2168	1157	1626	1007	745	6703
	Balearic Islands	63.24%	67.33%	66.85%	68.22%	63.49%	65.83%
	Other Autonomous Communities	35.52%	28.61%	29.09%	27.71%	31.68%	30.52%
	Foreign	4.24%	4.06%	4.05%	4.07%	4.83%	4.25%
Foreigners by nationality (%)	E.U.-15	52.27%	38.1%	55.55%	72.22%	23.81%	48.39%
	Latin America	34.09%	23.81%	29.63%	0%	52.38%	27.98%
	Africa	2.27%	23.81%	7.41%	22.22%	14.29%	14%
	Rest	11.37%	14.28%	7.41%	5.66%	9.52%	9.65%
2004							
Census section		**3008**	**3007**	**3006**	**3005**	**3004**	**Santa Catalina**
	Total Population	2256	1198	1614	1175	776	7019
	Balearic Islands	55.49%	58.76%	55.7%	56%	54.64%	56.12%
	Other Autonomous Communities	31.78%	25.04%	26.64%	28.51%	26.93%	27.78%
	Foreign	12.72%	16.19%	17.66%	13.78%	18.43%	15.76%
Foreigners by nationality (%)	E.U.-15	16.01%	16.13%	23.11%	20.55%	33%	21.76%
	Latin America	64.07%	67.74%	56.97%	68.38%	46.42%	60.71%
	Africa	11.68%	1.29%	8.36%	0	8.93%	6.05%
	Rest	8.24%	14.84%	11.56%	11.06%	11.65%	11.47%

(continued)

Table 18.2 (continued)

2010							
Census section		**2024**	**2023**	**2022**	**2021**	**2020**	**Santa Catalina**
	Total Population	2268	1163	1658	1240	837	7166
	Balearic Islands	49.78%	51.07%	53.37%	50.65%	34.65%	47.9%
	Other Autonomous Communities	28.97%	24.82%	24.97%	26.05%	22.22%	25.41%
	Foreign	21.25%	26.99%	21.65%	23.3%	31.18%	24.87%
Foreigners by nationality (%)	E.U.-15	27.87%	23.14%	33%	43.93%	40.17%	33.62%
	Latin America	6.01%	47.84%	38.3%	26.78%	6.11%	25.01%
	Africa	4.37%	0.39%	5%	5.43%	6.55%	4.35%
	Rest	28.25%	71.37%	23.5%	23.86%	47.17%	38.83%
2015							
Census section		**2024**	**2023**	**2022**	**2021**	**2020**	**Santa Catalina**
	Total Population	2256	1146	1642	1191	782	7017
	Balearic Islands	51.82%	51.48%	53.98%	50.97%	48.97%	51.44
	Other Autonomous Communities	27.62%	21.82%	24.26%	26.36%	24.17%	24.85
	Foreign	20.52	26.7%	21.75%	22.67%	26.85%	23.7
Foreigners by nationality (%)	E.U.-15	23.83%	18.8%	34.8%	36.68%	50.32%	32.86
	Latin America	42.96%	34.4%	23.2%	23.62%	18.3%	28.5
	Africa	1.17%	2.7%	18.4%	1%	3.26%	5.3
	Rest	32.04%	17.4%	23.6%	35.15%	28.12%	27.26
2021							
Census section		**2024**	**2023**	**2022**	**2021**	**2020**	**Santa Catalina**
	Total Population	2285	1121	1827	1147	777	7157
	Balearic Islands	49.72%	48.35%	39.03%	46.82%	45.05%	45.78%
	Other Autonomous Communities	25.47%	22.3%	13.03%	23.98%	20.59%	21.07%
	Foreign	24.81%	29.35%	47.95%	29.21%	34.36%	33.1
Foreigners by nationality (%)	E.U.-15	29.85%	28.38%	25.98%	40.7%	36.68%	32.32%
	Latin America	39.4%	38.28%	32.04%	19.49%	22.11%	30.26%
	Africa	5.07%	3.6%	7.85%	4.66%	2.01%	4.64
	Rest	25.68%	29.74%	26.28%	35.15%	49.2%	33.21

Source. Own elaboration based on INE (National Statistics Institute) Continuous Register of Inhabitants

Table 18.3 Population of Son Gotleu by census tracts according to place of birth and nationality (1998–2021)

1998							
		4019	**4022**	**4023**	**4024**	**4026**	**Son Gotleu**
Place of birth (%)	Palma	50.67	51.44	50.70	51.57	48.16	50.61
	Rest of Balearic Islands	6.00	7.30	7.41	12.07	12.66	9.20
	Other Autonomous Communities	40.63	37.47	38.73	33.26	34.71	36.82
	Foreign	2.71	3.79	3.16	3.10	4.47	3.37
Foreigners by nationality (%)	E.U.-15	20.69	47.06	56.25	40.91	54.55	54.84
	Latin America	13.79	27.27	27.27	38.70	39.28	29.55
	Africa	48.28	4.55	13.64	0	0	13.64
	Rest	17.24	21.12	2.84	20.39	6.17	1.97
More numerous	Autonomous Communities	Andalusia	Andalusia	Andalusia	Andalusia	Andalusia	Andalusia
	Foreigners	Senegal	No data	Italy	Italy	Bulgaria	Italy
2004							
		4019	**4022**	**4023**	**4024**	**4026**	**Son Gotleu**
Place of birth (%)	Palma	50.77	46.65	44.78	48.26	43.57	46.70
	Rest of Balearic Islands	4.05	5.74	6.91	12.99	10.80	7.93
	Other Autonomous Communities	36.17	35.71	35.81	32.73	30.78	34.24
	Foreign	9.01	11.89	12.49	6.02	14.85	11.13
Foreigners by nationality (%)	E.U.-15	8.26	9.63	10.20	39.29	18.94	14.74
	Latin America	67.89	58.29	59.86	50.00	54.19	58.13
	Africa	6.42	16.04	12.24	3.57	6.17	9.78
	Rest	17.43	16.04	17.69	7.14	20.70	17.36
More numerous	Autonomous Communities	Andalusia	Andalusia	Andalusia	Andalusia	Andalusia	Andalusia
	Foreigners	Colombia	Ecuador	Argentina	Ecuador	Argentina	Argentina

(continued)

Table 18.3 (continued)

2010							
		4019	**4022**	**4023**	**4024**	**4026**	**Son Gotleu**
Place of birth (%)	Palma	37.37	30.78	29.35	29.92	46.08	34.69
	Rest of Balearic Islands	3.44	2.55	2.49	2.80	4.25	3.11
	Other Autonomous Communities	21.59	18.86	19.35	22.62	25.71	21.81
	Foreign	37.59	47.81	48.81	44.66	23.96	40.39
Foreigners by nationality (%)	E.U.-15	3.53	1.83	2.47	3.13	8.46	3.29
	Latin America	15.99	15.98	13.92	11.02	44.53	17.07
	Africa	69.70	71.59	73.74	79.30	24.38	69.05
	Rest	10.78	10.6	9.87	6.55	22.63	10.64
More numerous	Autonomous Communities	Andalusia	Andalusia	Andalusia	Andalusia	Andalusia	Andalusia
	Foreigners	Nigeria	Nigeria	Nigeria	Nigeria	Ecuador	Nigeria
2015							
		4019	**4022**	**4023**	**4024**	**4026**	**Son Gotleu**
Place of birth (%)	Palma	42.20	40.89	35.37	37.93	49.95	41.37
	Rest of Balearic Islands	2.77	2.95	2.35	3.52	3.53	3.07
	Other Autonomous Communities	18.74	17.26	18.86	20.75	24.75	20.41
	Foreign	36.28	38.90	43.42	37.79	21.76	35.15
Foreigners by nationality (%)	E.U.-15	2.33	2.26	2.16	2.24	6.42	2.66
	Latin America	8.37	7.55	7.77	8.82	28.68	10.21
	Africa	72.79	76.42	75.97	79.83	38.49	72.82
	Rest	16.51	13.77	14.1	9.11	26.41	14.31

More numerous	Autonomous Communities	Andalusia	Andalusia	Andalusia	Andalusia	Andalusia	Andalusia
	Foreigners	Morocco	Nigeria	Nigeria	Nigeria	Morocco	Nigeria
2021							
		4019	**4022**	**4023**	**4024**	**4026**	**Son Gotleu**
Place of birth (%)	Palma	39.48	37.27	34.74	36.68	44.92	38.67
	Rest of Balearic Islands	2.71	1.75	1.56	3.52	3.06	2.58
	Other Autonomous Communities	13.55	13.03	12.16	15.75	21.77	15.56
	Foreign	44.26	47.95	51.54	44.05	30.25	43.19
Foreigners by nationality (%)	E.U.-15	0.72	0.92	1.21	1.12	4.60	1.50
	Latin America	11.39	7.76	9.77	11.50	43.10	14.45
	Africa	65.46	70.53	66.41	69.20	28.87	62.84
	Rest	22.43	20.79	22.61	18.18	23.43	21.21
More numerous	Autonomous Communities	Andalusia	Andalusia	Andalusia	Andalusia	Andalusia	Andalusia
	Foreigners	Morocco	Morocco	Morocco	Morocco	Colombia	Morocco

Source. Prepared by the authors based on INE Continuous Register of Inhabitants

References

Agyei-Mensah, S., & Owusu, G. (2010). Segregated by neighbourhoods? A portrait of ethnic diversity in the neighbourhoods of the Accra Metropolitan Area, Ghana. *Population, Space and Place, 16*(6), 499–516.

Ajuntament de Palma. (1994). *Els barris de Palma: revisió del Pla general d'ordenació urbana i d'actuacions urbanístiques*. Promomallorca.

Ajuntament de Palma. (2017). *Detecció de situacions de pobresa i/o exclusió social de la població de Palma*. Área de Benestar i Drets Socials, Ajuntament de Palma.

Arbaci, S. (2019). *Paradoxes of segregation: Housing systems, welfare regimes and ethnic residential change in Southern European cities*. John Wiley & Sons.

Bayona i Carrasco, J. B., & Gay, A. L. (2011). Concentración, segregación y movilidad residencial de los extranjeros en Barcelona. *Documents d'anàlisi geogràfica, 57*(3), 381–412.

Benson, M., & O'Reilly, K. (Eds.). (2009). *Lifestyle Migration. Expectations, Aspirations and Experiences*. Ashgate.

Bråmå, Å. (2006). 'White flight'? The production and reproduction of immigrant concentration areas in Swedish cities, 1990–2000. *Urban Studies, 43*(7), 1127–1146.

Bråmå, Å. (2008). Dynamics of ethnic residential segregation in Göteborg, Sweden, 1995–2000. *Population, Space and Place, 14*(2), 101–117.

Campuzano, E. P., Tello, C. A., & Everitt, J. C. (2014). Spatial segregation in a tourist city: the case of Puerto Vallarta, Mexico. *Journal of Latin American Geography, 13*, 87–112.

Capel, H. (2002). *La morfología de las ciudades (Sociedad, cultura y paisaje urbano)*. Ediciones del Serbal.

Carbonero, M. A. (1991). Estructura rural i indústria a Palma (1820–1930). In C. Manera & J. M. Petrus (coords) *Del taller a la fábrica. El procés d'industrialització a Mallorca* (pp. 91–100). Ajuntament de Palma.

Domínguez Mujica, J., García Coll, A., González-Pérez, J. M., & Sánchez Aguilera, D. (2008). La population étrangère en Espagne: vuelques éléments d'une géographie changeante. *Sud-Ouest Européen. Revue Géographique des Pyrénées et du Sud-Ouest, 26*, 71–88.

Domínguez Mujica, J., Parreño Castellano, J. M., & Díaz Hernández, R. (2010). *Inmigración y ciudad en España: Integración versus segregación socio-territoriales*. Scripta nova.

Frey, W. H. (1979). Central city white flight: Racial and nonracial causes. *American Sociological Review, 44*, 425–448.

Gaviria, M. (1974). *España a go-go. Turismo chárter y neocolonialismo del espacio*. Turner Madrid.

González-Pérez, J. M. (2017). A new colonisation of a Caribbean city: Urban regeneration policies as a strategy for tourism development and gentrification in Santo Domingo's Colonial City. In M. Gravari-Barbas & S. Guinand (Eds.), *Tourism and gentrification in contemporary metropolises* (pp. 25–51). Routledge.

González-Pérez, J. M. (2021). Racial/Ethnic Segregation and Urban Inequality in Kansas City, Missouri: A Divided City. *City & Community, 20*(4), 346–370. https://doi.org/10.1177/1535684121990799

González-Pérez, J. M. (2022). Evictions, Foreclosures, and Global Housing Speculation in Palma, Spain. *Land, 11*(2), 293. https://doi.org/10.3390/land11020293

González-Pérez, J. M., & Piñeira-Mantiñán, M. J. (2020). La ciudad desigual en Palma (Mallorca): geografía del confinamiento durante la pandemia de la COVID-19. *Boletín de la Asociación de Española de Geografía (BAGE)*, 87. https://doi.org/10.21138/bage.2998

González-Pérez, J. M., Vives, S., & Rullan, O. (2020). Evictions for unpaid rent in the Judicial District of Palma (Majorca, Spain): A metropolitan perspective. *Cities, 97*, 102466. https://doi.org/10.1016/j.cities.2019.102466

González-Pérez, J. M., Parreño-Castellano, J. M., & Sánchez-Aguilera, D. (2022). Fragmentación urbana en la ciudad española. Las viejas periferias urbanas del desarrollismo. In *Trabajos aportados al XVI Coloquio de Geografía Urbana* (pp. 181–193). AGE and Universidad de Málaga.

Hanhörster, H. (2001). Whose neighborhood is it? Ethnic diversity in urban spaces in Germany. *GeoJournal, 51*, 329–338.

Holloway, S. R., Wright, R., & Ellis, M. (2012). The Racially Fragmented City? Neighborhood Racial Segregation and Diversity Jointly Considered. *The Professional Geographer, 64*(1), 63–82.

Huete, R., Mantecón, A., & Estévez, J. (2013). Challenges in Lifestyle migration research: reflections and findings about the Spanish crisis. *Mobilities, 8*(3), 331–348.

IBESTAT. (2022). *Padrón continuo de habitantes.* Retrieved from: https://ibestat.caib.es/ibestat/estadistiques/poblacio/padro/2acef6cf-175a-4826-b71e-8302b13c1262. Accessed 29 Sept 2022

INE. (2022). *Atlas de distribución de renta de los hogares 2020.* Retrieved from: https://www.ine.es/dyngs/INEbase/es/operacion.htm?c=Estadistica_C&cid=1254736177088&menu=ultiDatos&idp=1254735976608. Accessed 13 Oct 2022.

Jurdao, F. (1979). *España en venta. Compra de suelo por extranjeros y colonización de campesinos en la Costa del Sol.* Ayuso.

Leal Maldonado, J. (2002). Segregación social y mercados de vivienda en las grandes ciudades. *RES, Revista Española de Sociología, 2*, 59–75.

López-López, Á., Cukier, J., & Sánchez-Crispín, Á. (2006). Segregation of tourist space in Los Cabos, Mexico. *Tourism Geographies, 8*(4), 359–379.

Malmberg, B., Andersson, E., & Östh, J. (2013). Segregation and urban unrest in Sweden. *Urban geography, 34*(7), 1031–1046.

Marcuse, P. (2002). The shifting meaning of the black ghetto in the United States. In P. Marcuse & R. van Kempen (Eds.), *Of states and cities. The partitioning of urban space* (pp. 109–142). Oxford University.

Massey, D. S., Rothwell, J., & Domina, T. (2009). The changing bases of segregation in the United States. *The Annals of the American Academy of Political and Social Science, 626*(1), 74–90.

Musterd, S. (2005). Social and Ethnic Segregation in Europe: Levels, Causes, and Effects. *Journal of Urban Affairs, 27*(3), 331–348.

Parreño, J. M., Piñeira, M. J., & González Pérez, J. M. (2022). *Fragmented City: International Mobility and Housing in Spain.* MPDI.

Randolph, B. (2020). Dimensions of urban segregation and the end of the Australian dream. In S. Muster (Ed.), *Handbook of urban segregation* (pp. 76–100). Elgar Online.

Smets, P., & Salman, T. (2016). The multi-layered-ness of urban segregation: On the simultaneous inclusion and exclusion in Latin American cities. *Habitat International, 54*, 80–87.

Urgell, L. (2016). *Canvis urbans i socials a les barriades de Santa Catalina i Es Jonquet. El procés de gentrificació.* Retrieved from: https://dspace.uib.es/xmlui/bitstream/handle/11201/2063/Llu+%C2%A1s%20Urgell%20Daviu_259091_assignsubmission_file_TFG.pdf?sequence=1. Accessed 25 Sept 2022.

Valli, C. (2015). A sense of displacement: Long-time residents' feelings of displacement in gentrifying Bushwick, New York. *International Journal of Urban and Regional Research, 39*(6), 1191–1208.

Vives, S., Rullan, O., & González-Pérez, J. M. (2018a). *Geografías de la desposesión de vivienda a través de la crisis. Los desahucios Marca Palma.* Icaria Editorial.

Vives, S., Rullan, O., & González-Pérez, J. M. (2018b, 591). Cartografías de los desplazamientos por desposesión de vivienda. Desahucios y ejecuciones hipotecarias en Palma a través de su geohistoria. *Scripta Nova, Revista Electrónica de Geografía y Ciencias Sociales, XXI.* https://doi.org/10.1344/sn2018.22.19774

Vono, D., & Bayona, J. (2011). Latinoamericanos en las grandes ciudades españolas. Concentración y segregación en el territorio. In J. Pujadas, A. Bayona, A. García Coll, & F. Gil Alonso (Eds.), *Población y Espacios Urbanos. XII Congreso de Población Española* (pp. 819–836). Universitat de Barcelona and AGE.

Jesús M. González-Pérez holds a PhD in Geography, and is currently a Full Professor at the University of the Balearic Islands (Spain). He is the President of the Spanish Geographical Association (AGE). He has stayed at about twenty universities in a Visiting Scholar capacity. In 2016, he was a Tinker Visiting Professor at Stanford University. He has published nearly 200 papers in numerous high impact journals and prestigious publishing houses (Routledge, Springer, Wiley, Peter Lang. . .), and has participated in over 30 research projects. His main research lines are urban inequality and fragmentation, social segregation and immigration, and tourist gentrification. His experience researching these phenomena goes beyond Europe, embracing the United States, Latin America and the Caribbean.

Ismael Yrigoy I am a geographer, currently a Ramón y Cajal researcher at the University of Santiago de Compostela and researcher at Uppsala University. I hold a BA and PhD in Geography from University of the Balearic Islands. I am currently a PI in a Project on Swedish lifestyle migrants funded by the Swedish Research Council on Sustainability. I am currently carrying out research on the political economy of real estate in Spain and the impacts of the irruption of short-term rentals. I have published my previous research in journals such as Antipode, International Journal of Urban and Regional Research, or Tourism Geographies amongst others.

Chapter 19
Touristification and Vulnerability in Urban Centres: Concepts and Analysis Approaches from the Geography of Spanish Tourism

Alfonso Fernández-Tabales, María José Piñeira-Mantiñán, and Carmen Mínguez

19.1 Introduction

Urban centres constitute a very important element of the image projected by cities. They are spaces in which great care is taken over the visual quality and they are easily identified because they are the areas of the highest volume of people and singular specialised and institutional buildings and a rich monumental heritage. This gives them greater visibility compared to other urban sectors (Zoido et al., 2013; Klinenberg, 2018). According to the sociologist L. Martín Santos[1] (1986), the centre of the city is where we can find its prestige, enjoy its monuments, where the space is intensified with the brightness of its lights. It is where urban identification takes place. This identity gradually dilutes as we move outwards. Far away from the centre, the urban identity gives way to an inorganic accumulation and the warehouse of frustrations.

[1] https://elpais.com/diario/1986/12/15/madrid/535033454_850215.html

A. Fernández-Tabales
Universidad de Sevilla, Seville, Spain
e-mail: aftabales@us.es

M. J. Piñeira-Mantiñán (✉)
Universidad de Santiago de Compostela, Santiago, Spain
e-mail: mariajose.pineira@usc.es

C. Mínguez
Universidad Complutense de Madrid, Madrid, Spain
e-mail: mcmingue@ucm.es

A. Blanco-Romero, M. Blázquez-Salom (eds.), *Spanish Tourism Geographies*, Geographies of Tourism and Global Change,
https://doi.org/10.1007/978-3-031-39780-6_19

The urban centre forms part of the collective memory and has a high symbolic value for the citizens. It is the focus of attention of the public administration when generating an attractive image and guaranteeing its dynamism. However, the pressure on the land of tertiary activities and the speculative processes occurring over the last few decades, fuelled by centrality, has given rise a dual process. On the one hand, the most dynamic sectors were affected by urban improvement processes, which, under different labels (regeneration, transformation, rehabilitation, recovery or requalification), brought about a landscape, social and functional change. On the other hand, the rest of the space was subjected to a process of degradation, with the deterioration of buildings, precarious living conditions, an ageing resident population, scarce resources and the emergence of pockets of marginalised population. As a result, the urban centres have become spaces of diversity, in which different population sectors coexist from a social, cultural and demographic point of view (Bruquetas et al., 2005; Alguacil Gómez, 2006).

Urban centres are vulnerable spaces, not only because they are home to a vulnerable population whose permanence is threatened by a lack of income, but also due to the pressure exerted by part of the citizens, who would prefer to eliminate these pockets of poverty which generate insecurity and give a "bad image"; and by the real estate promoters, sometimes supported by the government, who see in this space highly valuable plots of land and large capital gains in which to carry out urban renovation projects (Piñeira et al., 2020). This pressure has grown due to the gentrification and touristification processes which threaten the urban, social, economic and cultural characteristics of the urban centre.

Throughout this study, the objectives are to undertake a bibliographic review of the research on a Spanish and international level analysing touristification processes and how they have become a factor of urban vulnerability. Subsequently, their effects are addressed and finally, the initiatives that are being implemented by the government and society to mitigate their impact are discussed.

19.2 The Two Sides of Touristification: A Source of Income Vs. a Factor of Vulnerability

The importance of tourism as an engine of growth for the Spanish economy and as a strategic sector in times of crisis is a well-known and widely studied fact (Cuadrado-Roura & López Morales, 2015; Muñoz, 2016). According to the study "Aproximación al sistema de cuentas nacionales del turismo" (Estimating the national accounts of tourism) carried out by M. Figuerola (2019), tourism contributed 12.5% to GDP and generated 297,122 million euros in 2019, making it one of the sectors best perceived by the local governments for which it is a guaranteed source of income. At least this was the case until the COVID-19 health crisis. This crisis revealed the bitterest face of the vulnerability of the tourism activity, beyond aspects such as seasonality or the high concentration of the employee base (Caixabank, 2016).

But, what is tourism vulnerability? Many studies analyse vulnerability from the perspective of a threat, the exposure to uncertainty or risk (United Nations, 2003; Subirats & Martí-Costa, 2014; Méndez et al., 2015; Piñeira & Trillo, 2016; Hernández et al., 2018). However, authors such as Turner et al. (2003), define the concept in relation to tourism as the likelihood that a tourist destination will suffer damage due to its accidental exposure to risk. In our case, we extend and specify this concept, understanding that exposure to risk can refer to the tourist destination as a whole as a competitive product, to the quality of life of its local population or the state of conservation of its resources or attraction elements. Along these lines, Scheyvens and Momsen (2008) understand tourism vulnerability as the mixture of the economic dependence and environmental fragility of a tourist destination. This was later extended by Bramwell and Lane (2009), who indicate how the two most salient factors for the vulnerability of destinations are: a) the processes of economic recession with its social impacts, and b) climate change. These perspectives have been incorporated into more recent studies that take a comprehensive approach to the topic (Gallego, 2015). As this chapter addresses the problems inherent in urban centres, it will focus on the social impacts on them.

In this respect, Clark et al. (1998) characterised it in relation to the degree of exposure to risk and the capacity to address this risk, which included resistance (the capacity to absorb the impacts and continue to function) and resilience (the capacity to recover from losses after an impact). In this sense, it should be noted that before the health crisis, studies on tourism vulnerability referred to terrorist attacks (Liu & Pratt, 2017), natural disasters (Roselló et al., 2020), climate change (Scott et al., 2019), or the impact of the economic crisis (Murray, 2015; Méndez, 2022; Piñeira & González, 2018). COVID-19 obliged researchers to study new aspects in response to the emergence of previously unimaginable factors, such as the temporary blocking of travel structures, the confinement of potential tourists, the closure of borders, furloughing schemes, the limitation of capacity in restaurants and cultural spaces or the lack of resources to pay suppliers. In Spain, research has been conducted to analyse these factors and their socioeconomic repercussions, including the COVID pandemic Atlas in Spain, published by the National Geographic Institute (2021), the monographic issue of the AGE Bulletin (Lois et al., 2021) or the AGE-pandemia website.[2]

However, despite a fall of 69% in tourism GDP and the closure of 2021 with job losses (−7.1%), there is confidence in the capacity of resilience of the tourism sector. According to different authors, this resilience, together with adequate governance formulas that incorporate the public and private sectors in a coordinated way will enable its reactivation so that it may once again become a strategic sector for the economic recovery of the cities (Romagosa, 2020). Once this has been achieved, the negative impacts of tourism on the local populations will arise once again (Arias, 2018), together with the effects on urban, environmental, social and economic vulnerability.

[2] https://www.age-geografia.es/site/reflexiones-sobre-la-crisis-actual/

In the case of cities, tourism vulnerability has to be analysed together with urban vulnerability. Therefore, particular attention will have to be paid to those sectors whose population is affected by adverse circumstances and is in a situation of risk and/or fragility. This fragility sometimes increases in the urban centres due to regeneration policies which leads to, on the one hand, the expulsion of the poor population which is replaced by middle and high strata of society (gentrification); and on the other hand, the occupation of the space by allochthonous users. In this way, the urban centres have become spaces of consumption, business and leisure, in which tourism has acquired a central role in this process of gamification (Cabrerizo et al., 2016). However, at the same time, the urban centres run the risk of losing their traditional functions and the quality of life of the residents. They are sensitive spaces that require special attention (Egea et al., 2008; Arias, 2000), in which the spiral of obsolescence can determine the emergence or consolidation of situations of vulnerability as different environmental, economic, social, residential or sensory factors enter into play (Table 19.1). Their evolution and possible combination will depend on whether they are what R. Temes (2007, 2014) defined as mono-vulnerable urban areas in which one type of vulnerability is identified; poly-vulnerable areas, in which two types are identified (e.g., residential & economic; economic & social; social & residential); or integral vulnerable areas in which three types of vulnerability are identified simultaneously.

The European project SMARTDESK, directed by A. Russo,[3] analyses how the most vulnerable groups are those that suffer the worst effects of touristification (increase in the cost of living, scarcity and increased prices of property, congestion of services and public spaces, impoverishment of workers with temporary contracts, the transformation of the cultural identity of the urban centre, etc.). With tourism, the daily life city has become the back office of the show-room city (the commercialised city) (Harvey, 2012), far removed from becoming a sustainable city that improves the quality of life of all its citizens.

19.3 The Effects of Touristification and its Measurement in Urban Areas

As indicated in the previous section, which reviews the factors intervening in the subject of this study, the rapid expansion of mass tourism to urban spaces previously untouched by the phenomenon is one of the topics most addressed by the Social Sciences in recent years (García Hernández et al., 2017; Mordue, 2017).

There are several definitions of the concept of touristification. In this case we will use the one proposed by Blanco et al. (2021: 350), which states that it is the "Transformation process of a place in an eminently tourist space and its associated effects. This transformation, which operates on an attraction and destination level,

[3] https://smartdest.eu/

Table 19.1 Vulnerability factors in urban centres

Vulnerability	Triggering factor
Urban	Abandonment of the urban centres (principally the historical quarters) by the government until the 1980s (Campesino, 1984). Lack of comprehensive plans for the most vulnerable urban sectors (e.g., the improvement of the façades of buildings but not of the living conditions). Development of occasional sectoral interventions that accentuate the dualisation between dynamic and degraded (isolated) central urban sectors. Authenticated stagings, urbanalisation (Muñoz, 2008) and museumalisation in urban centres/historical quarters.
Economic	Increase in the value of the land. Increased competition for land uses. Generalisation of speculative processes on old plots (Bruquetas et al., 2005). Rent Act and updating of old rents (Fernández, 2018). Retail and services specialisation. Increase in tourism activity.
Social	Increase in social duality: elitisation-gentrification and pockets of poverty (Bruquetas et al., 2005). Increase in unemployment and labour precariousness. Increase in immigration (Parreño & Domínguez, 2013). Increasingly more anonymous neighbourhood context and the rupture of support networks (Egea et al., 2008). Lack of services that satisfy the basic needs of the resident population (Egea et al., 2008).
Residential	Increase in housing prices (Lois González et al., 2016). Increase in the homeless population (Sales, 2014). Reduction in the number of residential dwellings in the light of gentrification and buildings occupied by elite services. Scarce or no alternatives for housing for vulnerable groups: State-subsidised housing, social housing rents, purchase/rental promotions for young people.
Environmental	Increase in water stress. Higher electricity consumption. Increase in the volume of solid waste. Higher level of pollution (atmospheric, noise). Processes of artificialisation of land.

Source: Own elaboration

implies an adaptation of the place to the visitor, whether tourism is the dominant activity of this space or not".

Given the many and diverse implications of the phenomenon, the scientific treatment of its effects have also been numerous and sometimes confusing in terms of the conceptual aspect with few attempts to categorise its consequences or clearly differentiate causes, effects and collateral manifestations. Among these efforts and maybe the closest to the reality of the facts, is that proposed by De la Calle (2019), which differentiates five principal effects, to which we can add a sixth: 1. greater presence of visitors in the central areas of the city; 2. increase in the number of activities directly related to tourist consumption; 3. reorientation of an increasingly broad range of businesses towards foreign customers; 4. conversion of property into a new tourism good; 5. creation of a landscape or urban scene in which tourism elements are predominant (De la Calle, 2019: 15) and 6. loss of culture and cohesion of the neighbourhood (Gallagher, 2017). Despite these categorisation attempts, the majority of the studies combine several manifestations of the phenomenon. It is difficult to assign them to one category or another and the different visions are reviewed below.

It seems undeniable that the most relevant effect of this phenomenon is the alteration of the property market in the areas affected, with an intense increase in property prices (particularly in the rental market) which can lead to the replacement of the original resident population, displaced towards less central areas of the city. However, there is no sufficiently solid theoretical body or methodology on this topic. There are even conceptual differences when assigning the different phenomena. In the early years of this century, the concept of tourist gentrification arose in the study conducted by Gotham (2005) on New Orleans, which has subsequently and frequently been revisited (Sequera & Nofre, 2018; Cócola-Gant, 2018). However, there is a strand of opinion that considers that this expression should not be used, as the term gentrification, related to the replacement of the resident population with another equally resident population but with a higher level of income, is considered inappropriate to apply to tourism phenomena, in which the resident population is replaced by a floating population. Some approaches have attempted to resolve this terminological incoherence by alluding to the phenomenon as a new modality, "gentrification 4.0" (Sorando & Ardura, 2018). In any event, some authors have affirmed that "the academic tendency to give excessive elasticity to definitions enabled all of the types of development or neoliberal urban reurbanisation under the concept of gentrification, thereby losing the capacity to describe nuanced urban processes. Therefore, it is necessary to differentiate between gentrification, transnational gentrification and touristification" (Jover & Díaz-Parra, 2019: 3046). De la Calle is the author who has been most outright in indicating the conceptual risks of the incessant creation of new terms, sometimes unsubstantiated, when he states that: "The concept of tourist gentrification, undoubtedly successful in terms of media and academic marketing, could end up becoming a burden which limits the possibilities of studying tourism and its effects on the city" (De la Calle, 2019: 24). This author recommends against theoretical modelling, often arising in academic and institutional contexts and far removed from the areas of study, identifying the need to determine and analyse the differences and unique features of the local tourism-population relationship in each specific case.

In any event, we can refer to a long list of authors and researchers who have addressed this phenomenon from different approaches. On the one hand, there are those that have attempted to adopt a globalised view, assessing the impact in terms of how the common good of the pre-existing local population is affected (Requejo, 2016; Blanco et al., 2018; Lopez-Gay et al., 2021); or those that jointly analyse the role of the social actors and the public administrations (Comendador-Sánchez et al., 2021). On the other hand, there those who have studied the increase in the rents as a factor of change of part of the housing stock from the residential function to the tourism function, with the result of a displacement of tenants (Colomb & Novy, 2017). There is a long list of studies on this evolution in many cities around the world, such as the afore-mentioned study on New Orleans (Gotham, 2005), or those on Los Angeles (Lee, 2016), Boston (Merante & Horn, 2017), San Francisco and Valparaíso (Oppilard, 2017), Nueva York (Dudas et al., 2017), Madrid, Barcelona and Palma (García-Ayllón, 2018), Santiago de Compostela (López et al., 2019), Madrid (Velasco González et al., 2019), Málaga (Almeida et al., 2019), several Chinese cities (Wu, 2019), Lisboa (Cócola-Gant & Gago, 2019), Dublin (Clancy, 2020), or Hong Kong (Liang et al., 2021). The key issue resides in measuring the difference in profit obtained by the owner of a property between renting it as a residential tenancy with a long-term contract or as a tourism rental property with short stays and high rotation, through online platforms. It should be noted that the majority of the scientific literature indicated, together with reports elaborated by public institutions (Schneiderman, 2014), coincide in that the difference in profits clearly leans in favour of tourism rental, with the knock-on effect of reducing the number of properties in residential tenancy and an increase in its price. This constitutes a driver of the displacement of the population with a lower level of resources to other urban areas (Blanco & Blázquez, 2019). This aspect is analysed with significant methodological advances by García-López et al. (2020) for the case of Barcelona and for several European cities by Valente et al. (2022), who find a causal relationship between the increase in the number of tourist rental properties and the increase in the market prices of the properties of the most pressured urban areas.

Another of the effects repeatedly indicated is the rapid transformation of the daily life environment, which is inundated with a floating population whose conduct and habits can clash with those of the residents, particularly if they are of an older age, generating a perception of discomfort and insecurity (Gil & Sequera, 2018). This becomes more acute as the volume of visitors increases, producing massification effects in public spaces and congestion in public services and transport (Neuts, 2016; Cocola-Gant et al., 2020).

Partly related to the above, and as an aggravating factor, we can also refer to the disappearance of traditional shops and their replacement by others catering to the tourism demand (Gant, 2015; Blázquez-Salom et al., 2019). This phenomenon, called retail gentrification, increases the difficulties for the permanence of the pre-existing population to whom this retail fabric provided its services. This highlights its vulnerability to the disappearance of the residential function of the urban centre.

From the point of view of a sectoral analysis, some authors have analysed the effects of the expansion of tourist properties on other accommodation establishments, particularly hotels, with negative repercussions on the local population whose livelihoods are directly or indirectly related to them. These studies, often based on quantitative estimates, coincide in indicating that not only the generation of wealth and tax revenue, but also employment is considerably higher in proportional terms in hotel establishments than tourism rental properties (Zervas & Byers, 2016; Requejo, 2016; Fuentes & Navarrete, 2016). Although this problem is not related to the debate on the vulnerability of the population of urban centres, we should not ignore its impact on the local economy, which ultimately has a negative effect on the conditions of the most vulnerable segments (Piñeira et al., 2020).

Finally, a different approach which is worth mentioning is that of research specifically focused on the measurement of the effects related to the specific location of tourist properties (Camarillo-Naranjo et al., 2022). This approach is characterised by its distinct methodological nature, focused on obtaining maximum detail in the exact identification and location of these properties, through ICTs, and their causal spatial relationships with the location of tourism resources, public services, facilities or transport networks or with the distribution of land prices and properties. This approach is expanding and is essential in order to advance in a subject area which, until now has been predominantly based on the theoretical debate regarding the precise measurement of the phenomena. This approach undoubtedly opens fertile lines of research and, it should be noted that in some cases, a certain level of disproportion between the methodological effort and the results obtained, which have confirmed manifestations already known in the geography of tourism (for example, that the accommodation supply is located close to tourism resources). Without entering this debate, we can refer to some relevant contributions in more recent years, such as those by Schneiderman (2014) in the case of New York; Arias and Quaglieri (2016) for Barcelona; Lee (2016) for Los Angeles; Yrigoy (2016) for Palma de Mallorca; Gutiérrez et al. (2018) for Barcelona; Oppilard (2017) for San Francisco and Valparaíso; Ioannides et al. (2018) for Utrecht; García-Ayllón (2018) for Madrid, Barcelona and Palma; or Mínguez et al. (2019) for Seville.

19.4 The Social and Institutional Response

The strong impact caused by massification and touristification has led to a highly uneven institutional and social response. This has been studied by experts from different branches of the social sciences, such as Velasco González et al. (2019), Gil and Sequera (2018), Blanco et al. (2019), Russo et al. (2022) or Blázquez-Salom (in press), among others, generating an important corpus with which to compare initiatives responding to similar problems affecting an increasing number of cities.

Returning to the principal effects of current urban tourism, we can observe that neither the government nor civil society have responded with the same intensity to each of them; as the conversion of the residential property into a property for tourism use has attracted the most attention, even by the non-specialised media. The interest of the government initially focused on controlling the economic effects of this conversion. Therefore, the topic that has received most attention is the increase in land prices and the reduction in the area for residential use (Gravari-Barbas & Guinand, 2017; Cócola-Gant, 2020). Tax evasion and the competition with regulated accommodation establishments are also popular topics (Zervas & Byers, 2016; Xie & Kwok, 2017). We can observe that the common denominator is that the majority of the large international urban destinations have the need to regulate rental platforms in order to balance the interests of the visitors and residents/local businesses (Nieuwland & van Melik, 2020).

According to the first experts who analysed the initiatives developed until the mid-2010s, they can be grouped into three approaches, depending on the degree of rigidity: laissez-faire (which does not imply any regulation, rather the opposite), authorisation with certain restrictions and prohibition (Jefferson-Jones, 2014; Miller, 2014; Guttentag, 2015). The boom in properties for tourism use experienced over the last decade led to many measures consistent with its casuistry and legal framework. According to Von Briel and Dolnicar (2021), these measures can be classified into four types of responses, which many cities have adopted, following a chronological sequence. The first reaction is that of regulation and imposition; this is followed by the introduction of a registration system which enables the declaration and control of short-term rentals and, third, the modification of the regulation. Finally, in very few cases, such as Tokyo, there is collaboration between the government and the P2P accommodation platforms. The regulation of these types of platforms is very complicated (Espinosa, 2016), essentially due to the lack of instruments, which have had to be created *ad hoc* in search of a balance between the right to ownership and the right to access housing. But also due to the difficulty to determine whether the guests breach the rules or not (Gottlieb, 2013; Edelman & Geradin, 2016; Espinosa, 2016) and because the government does not have just one problem to deal with or a single root cause (Novy & Colomb, 2019). In short, we can observe that all of this leads to the application of restrictions that affect both the characteristics of the rented accommodation (Guttentag, 2015; Gottlieb, 2013; Miller, 2014; Jefferson-Jones, 2014) and the space in which it is located (Gurran & Phibbs, 2017). To do this, the governments have used existing urban planning instruments, such as the Plan de Ordenación Urbana (Urban Development Plan) and different municipal by-laws, although other specific instruments have been designed, such as the PEUAT of Barcelona or the Plan Especial de regulación del Uso Hospedaje (Special Plan for regulating accommodation use) (PEH) of Madrid.

There is a common trend which consists in replicating the agreed measures, although several authors indicate that each city should adopt its own, according to its characteristics and impacts (Guttentag, 2015; Oskam & Boswijk, 2016; Gurran & Phibbs, 2017). This is the case, for example, in the neighbourhood of Cabanyal-Canyamelar (Valencia), in which the use of the tourist properties is limited by a regulatory policy (Nieuwland & del Romero Renau, 2021); in Amsterdam, where it is prohibited in the centre; or in Berlin, which has been a pioneer in taking an exceptional measure to establish a limit to the rental prices of properties. This measure has been highly questioned, particularly by real estate and property owner associations and its legality and constitutional nature has been debated (Fischer-Lescano & Gutmann, 2020).

In parallel, a social response has occurred, characterised by being highly critical of the impact of tourism on the property market, the expulsion of the population and the loss of quality of life of the residents (Gil & Sequera, 2018). This urban activism created around tourism is channelled through platforms and, exceptionally, it has given rise to protests and a social reaction incorrectly called "tourismphobia" or anti-tourism (Blanco et al., 2019). In this way, different associations and platforms have emerged, such as ABTS (Assemblea de Barris per un Turisme Sostenible), Ciutat per qui l'habita in Palma, BiziLagunEkin in Donostia, Lavapiés, ¿dónde vas? in Madrid or Morar em Lisboa, in the Portuguese capital. All of them have a voice in websites, newspapers and social networks, enabling the exchange of information between them. Many of them have come together in the "Network of Southern European Cities against Touristification" which unites movements of fourteen European cities, ten in Spain, in order to pressure the public authorities to change their policies on tourism and regulate this activity in accordance with social, economic and environmental sustainability policies (Manifiesto de SETNet). This highlights the capacity of creating alternative networks (Gil & Sequera, 2018) with different forms of reaction, ranging from the most academic to the most artistic options, and, therefore, with the most media prominence.

The associations and platforms created as a result of touristification closely collaborate with other neighbourhood associations focused on other aspects, such as the conditions and access to housing, the protection of traditional retail and the defence of public spaces and services. In fact, they share critical approaches and similar objectives, such as the right to the city which is particularly threatened by tourism (Hernández Ramírez, 2018). In short, these new platforms continue the path initiated by other movements opposing gentrification trends, often launched by the government itself through its city embellishment programmes and assistance for improving the conditions of housing (Jover & Díaz-Parra, 2019). One salient fact is that these movements are not led by the most vulnerable population from a socio-economic point of view, but its protagonists are neighbours who once formed part of an initial gentrification process of the neighbourhood and who defend the living conditions of the neighbourhoods which have suffered a first gentrification process.

In parallel, another associative movement has also emerged which is less well-known but also the result of touristification, namely that of the owners of the tourist rental properties. These groups provide legal and taxation advisory services to the owners who wish to rent their properties and do business with them. Undoubtedly, this aspect is less well-known and, therefore, less studied.

In short, we should emphasise that the relevance acquired by tourism determines its firm inclusion in the political agenda of the local entities, bestowing the tourism activity with an economic, social and also political value, and also its inclusion in the agenda of the social movements (Milano et al., 2019). When the institutional response is compared to the social response, two fundamental issues can be identified. On the one hand, neither of the two directly defends the most vulnerable population whose situation of precariousness is intensified by the tourism pressure. On the other hand, the orientations of the institutional actions are more limited and have a more economic vision than that of the groups that contemplate broader and social aspects.

19.5 Conclusions

We have been able to determine the amount and diversity of studies carried out on the tourism activity and its impacts, specifically in Spain. Many studies indicate the benefits of tourism and how the political arena and the media have attempted to transfer the idea that tourism growth is beneficial, even though "spill over," for the whole of society, irrespective of income levels or position: so, although it is undeniable that some benefit more than others, ultimately all citizens will benefit. However, the increase in negative effects generated in many destinations has given rise to an alternative vision, according to which there are social groups for whom the cost-benefit balance of touristification is negative, understood as a loss in the quality of life and a deterioration of the access to public and basic goods.

An analysis of the scientific and institutional literature has enabled us to verify how public policies directed at mitigating the problems of urban centres have adopted, in general, a sectoral approach, with very few based on comprehensive approaches that combine social, economic and endowment measures. As a result, responding to the phenomenon of touristification, these public measures have not positively discriminated in favour of the most vulnerable social groups, which has led to their expulsion from the urban centres in parallel with the objective improvement of the conditions of the centres, paradoxically obtained thanks to public policies of urban regeneration.

Currently, the approaches to this subject matter have been disrupted by the effects of the COVID-19 pandemic. The sudden paralysis of tourist flows in 2020 and their slow and irregular recovery have significantly transformed the situation, whereby the scenario has shifted from a concern about the impact of touristification to a growing debate on the possible formulas to revive the tourism sector, in response to the economic and social problems derived from the pandemic. Although this second orientation, unanimous in the institutions and business worlds, is understandable in the light of the economic instability, it should be noted that the phenomena analysed can gain momentum again in a few years; with the same consequences, but with the aggravation that the deterioration derived from the crisis among the disadvantaged social groups can make them even more vulnerable than before. Therefore, a new future wave of touristification would have an impact on urban areas and groups whose vulnerability (capacity to assume an increase in the price of rent, level of organisational resistance, possibilities to construct an alternative discourse, etc.) would be worsened. It is too soon to make definitive conclusions on this scenario. However, we can, at least indicate that a fertile and necessary field of research and social action will be opened in the coming years.

Acknowledgments This work was supported by the Grant PGC2018-095992-B-I00, funded by FEDER and the State Research Agency through the Ministry of Science and Innovation (Spain); P20-00982, funded by the Regional Government of Andalusia (Spain); "The housing problem in Spanish fragmented metropolises. Permanence over time, new housing markets and alternatives from public policies". PID2019-108120RB-C31, State Program of R + D + i Oriented to the Challenges of Society, of the State Plan for Scientific and Technical Research and Innovation

2017-2020; and PID2020-114186RB-C21 funded by MCIN-AEI/ 10.13039/501100011033 "Urban tourism in changing times. Adaptation and resilience strategies in destination planning and management models (ADAPTATURciudad)", State Program of R + D + i Oriented to the Challenges of Society, of the State Plan for Scientific and Technical Research and Innovation 2017-2020.

References

Alguacil Gómez, J. (2006). Barrios desfavorecidos: diagnóstico de la situación española. En F. Vidal Fernández (dir.), *V Informe FUHEM de políticas sociales: La exclusión social y el estado del bienestar en España* (pp. 155–168). FUHEM.

Almeida, F., Cortés, R., & Balbuena, A. (2019). Tourism-phobia in historic centres: the case of Malaga. *Boletín de la Asociación de Geógrafos Españoles, 83*, 1–32. https://doi.org/10.21138/bage.2823

Arias, F. (dir.). (2000). *La desigualdad urbana en España*. Ministerio de Fomento. 191 pp.

Arias, A. (2018). Turismo y gentrificación. Apuntes desde Barcelona. In C. Donat, S. Porcel, & R. Gomà, *Gentrificació i dret a la ciutat* (Vol. 60, pp. 207–211). Papers, Regió Metropolinata de Barcelona.

Arias, A., & Quaglieri, A. (2016). Unravelling Airbnb: Urban perspectives from Barcelona. En A. P. Russo & G. Richards (Eds.), *Reinventing the local in tourism: Producing, consuming and negotiating place* (pp. 209–228). Channel View Publications.

Blanco, A., & Blázquez, M. (2019). Marchandisation touristique du logement et planification urbaine à Barcelone. *Sud-Ouest Européen, 46*, 9–22. https://doi.org/10.4000/soe.4568

Blanco Romero, A., Blázquez Salom, M., De la Calle Vaquero, M., Fernández Tabales, A., García Hernández, M., Lois González, R. C., Mínguez García, M. C., Navalón García, R., Navarro Jurado, E., & Troitiño Torralba, L. (2021). *Diccionario de turismo*. Ediciones Cátedra, 392 pp.

Blanco, A., Blázquez, M., & Canoves, G. (2018). Barcelona, housing rent bubble in a tourist city. Social responses and local policies. *Sustainability, 10*, 2043. https://doi.org/10.3390/su10062043

Blanco, A., Blàzquez, M., Morell, M., & Fletcher, R. (2019). Not tourism-phobia but urban-philia: Understanding stakeholders' perceptions of urban touristification. *Boletín de la Asociación de Geógrafos Españoles, 83*, 1–30. https://doi.org/10.21138/bage.2834

Blázquez-Salom, M. (in press). Clashing with corporate interests. Tourism governance in the Balearic Islands. In J. Saarinen (Ed.), *Handbook on tourism governance*. Edward Elgar Publishing.

Blázquez-Salom, M., Blanco, A., Guall, J., & Murray, I. (2019). Tourist gentrification of retail shops in Palma (Majorca). In C. Milano, J. Cheer, & M. Novelli (Eds.), *Overtourism: Excesses, discontents and measures in travel and tourism* (pp. 39–69). CABI Publishing.

Bramwell, B., & Lane, B. (2009). Economic cycles, times of change and sustainable tourism. *Journal of Sustainable Tourism, 17*(1), 1–4. https://doi.org/10.1080/09669580802608061

Bruquetas, M., Moreno, F. J., & Walliser, A. (2005). *La regeneración de barrios desfavorecidos* (68 pp). Fundación Alternativas.

Cabrerizo, C., Sequera, J., & Bachiller, P. (2016). Entre la turistificación y los espacios de resistencia en el centro de Madrid. Algunas claves para (re)pensar la ciudad turística. *Ecología Política, 52*, 78–82.

Caixabank. (2016). *La estacionalidad del turismo*. Informe mensual Focus. Avaible in: https://www.caixabankresearch.com/documents/10180/3228299/28%2BFocus%2B8%2BCAST.pdf

Camarillo-Naranjo, J. M., Vallejo-Villalta, I., Fernández-Tabales, A., & Santos-Pavón, E. (2022). Where is tourist housing actually located? New approaches and sources for detailed scale analysis. *European Planning Studies* (0965-4313/1469-5944) https://doi.org/10.1080/09654313.2021.2002825

Campesino, A. (1984). Los centros históricos, análisis de su problemática. *Norba: Revista de Geografía, 5*, 51–62.
Clancy, M. (2020). Tourism, financialization, and shortterm rentals: The political economy of Dublin's housing crisis. *Current Issues in Tourism*. https://doi.org/10.1080/13683500.2020.1786027
Clark, G., et al. (1998). Assessing the vulnerability of coastal communities to extreme storms: The case of Revere, MA.; USA. *Mitigation and Adaptation Strategies for Global Change, 3*, 59–82.
Cócola-Gant, A. (2018). Tourism gentrification. In L. Lees & M. Phillips (Eds.), *Handbook of gentrification studies* (pp. 281–293). Edward Elgar Publishing.
Cócola-Gant, A. (2020). Gentrificación turística. In E. Cañada & I. Murray (Eds.), *Turistificación global: perspectivas críticas en turismo* (pp. 291–308). Icaria.
Cócola-Gant, A., & Gago, A. (2019). Airbnb, buy-to-let investment and tourism-driven displacement: A case study in Lisbon. *Environment and Planning A: Economy and Space, 53*(7), 1671–1688. https://doi.org/10.1177/0308518X19869012
Cocola-Gant, A., Gago, A., & Jover, J. (2020). Tourism, gentrification and neighbourhood change: An analytical framework—Reflections from Southern European cities. In J. Oskam (Ed.), *The overtourism debate. NIMBY, nuisance, commodification* (pp. 121–135). Emerald.
Colomb, C., & Novy, J. (Eds.). (2017). *Protest and resistance in the tourist city* (344 pp). Routledge.
Comendador-Sánchez, A., Santos-Pavón, E., & Hernández-Ramírez, J. (2021). Players, interests and public policies in relation to tourist rental housing in Spain (2010–2019). Complexity and multiple conflicts. *Journal of Tourism Analysis: Revista de Análisis Turístico (JTA), 28*(1). https://doi.org/10.53596/jta.v28i1.373
Cuadrado-Roura, J. R., & López Morales, J. M. (2015). *El turismo, motor del crecimiento y de la recuperación de la economía española*. Universidad de Alcalá.
De la Calle, M. (2019). Turistificación de centros urbanos: clarificando el debate. *Boletín de la Asociación de Geógrafos Españoles, 83*, 1–40. https://doi.org/10.21138/bage.2829
Dudas, G., Vida, G., Kovalcsik, T., & Boros, L. (2017). A socio-economic analysis of Airbnb in New York City. *Regional Statistics, 7*, 135–151. https://doi.org/10.15196/RS07108
Edelman, B. G., & Geradin, D. (2016). Efficiencies and regulatory shortcuts: How should we regulate companies like Airbnb and Uber? *Stanford Technology Law Review, 19*(2), 293–328.
Egea, C., Nieto, J. A., Domínguez, J., & González, R. (2008). *Vulnerabilidad del tejido social de los barrios desfavorecidos de Andalucía. Análisis y potencialidades*. Centro de Estudios Andaluces.
Espinosa, T. P. (2016). The cost of sharing and the common law: How to address the negative externalities of homesharing. *Chapman Law Review, 19*(1), 597–627.
Fernández, G. (2018). *La "turistificación" de barrios y ciudades: definición y agentes intervinientes*. http://gumersindofernandez.com/blog/2018/05/25/la-turistificacion-de-barrios-y-ciudades-i-definicion-y-agentes-intervinientes/
Figuerola, M. (2019). *Aproximación al sistema de cuentas nacionales del turismo. Mesa de Turismo*. Universidad de Nebrija.
Fischer-Lescano, A., & Gutmann, A. (2020). Mietpreisregulierung im Föderalismus. Zur Verfassungsmäßigkeit des Berliner Wohnungsmietenbegrenzungsgesetzes. *KJ Kritische Justiz, 53*(1), 3–15. https://doi.org/10.5771/0023-4834-2020-1-3
Fuentes, R., & Navarrete, L. (2016). Tourists in hotels versus holiday homes: Economic impact and characterization. *Tourism Review International, 20*, 177–195. https://doi.org/10.3727/154427216X14791579617416
Gallagher, L. (2017). *The Airbnb story: How three ordinary guys disrupted an industry, made billions... and created plenty of controversy*. Houghton Mifflin Harcourt.
Gallego, I. (2015). *Vulnerabilidad de los destinos turísticos. Propuesta de un Sistema de indicadores para su gestión*. Universidad de Málaga.
Gant, A. (2015). *Tourism and commercial gentrification*. Avaible in: www.rc21.org/en/wpcontent/uploads/2014/12/E4-C%C3%B3cola-Gant.pdf

García Hernández, M., de la Calle Vaquero, M., & Yubero Bernabé, C. (2017). Cultural heritage and urban tourism: Historic city centres under pressure. *Sustaninability, 9*, 1346. https://doi.org/10.3390/su9081346

García-Ayllón, S. (2018). Urban transformations as an indicator of unsustainability in the P2P mass tourism phenomenon: The Airbnb case in Spain through three case studies. *Sustainability, 10*(8), 2933. https://doi.org/10.3390/su10082933

García-López, M. À., Jofre-Monseny, J., Martínez-Mazza, R., & Segú, M. (2020). Do short-term rental platforms affect housing markets? Evidence from Airbnb in Barcelona. *Journal of Urban Economics, 119*, 103278. https://doi.org/10.1016/j.jue.2020.103278

Gil, J., & Sequera, J. (2018). Expansión de la ciudad turística y nuevas resistencias. El caso de Airbnb en Madrid. *Empiria Revista de Metodología de Ciencias Sociales, 41*, 15–32. https://doi.org/10.5944/empiria.41.2018.22602

Gotham, K. F. (2005). Tourism gentrification: The case of new Orleans' Vieux Carre (French quarter). *Urban Studies, 42*(7), 1099–1121. https://doi.org/10.1080/00420980500120881

Gottlieb, C. (2013). Residential short-term rentals: Should local governments regulate the 'industry'? *Planning & Environmental Law, 65*(2), 4–9. https://doi.org/10.1080/15480755.2013.766496

Gravari-Barbas, M., & Guinand, S. (Eds.). (2017). *Tourism and gentrification in contemporary metropolises: International perspectives* (340 pp). Routledge.

Gurran, N., & Phibbs, P. (2017). When tourists move in: How should urban planners respond to Airbnb? *Journal of the American Planning Association, 83*(1), 80–92. https://doi.org/10.1080/01944363.2016.1249011

Gutiérrez, J., García-Palomares, J. C., Romanillos, G., & Salas-Olmedo, M. H. (2018). The eruption of Airbnb in tourist cities: Comparing spatial patterns of hotels and peer-to-peer accommodation in Barcelona. *Tourism Management, 62*, 278–291. https://doi.org/10.1016/j.tourman.2017.05.003

Guttentag, D. (2015). Airbnb: Disruptive innovation and the rise of an informal tourism accommodation sector. *Current Issues in Tourism, 18*(12), 1192–1217. https://doi.org/10.1080/13683500.2013.827159

Harvey, D. (2012). *Rebel cities: From the right to the city to the urban revolution* (208 pp). Verso.

Hernández Ramírez, J. (2018). La voracidad del turismo y el derecho a la ciudad. *Revista Andaluza de Antropología, 15*, 22–46. https://doi.org/10.12795/RAA.2018.15.02

Hernández, A., Rodríguez, R., & Rodríguez, I. (2018). *Barrios vulnerables de las grandes ciudades españolas 1991 | 2001 | 2011*. Ministerio de Fomento.

Instituto Geográfico Nacional. (2021). *La pandemia Covid-19 en España*. Centro Nacional de Información Geográfica.

Ioannides, D., Röslmaier, M., & Van der Zee, E. (2018). Airbnb as an instigator of 'tourism bubble' expansion in Utrecht's Lombok neighbourhood. *Tourism Geographies, 21*(5), 822–840. https://doi.org/10.1080/14616688.2018.1454505

Jefferson-Jones, J. (2014). Airbnb and the housing segment of the modern sharing economy: Are short-term rental restrictions an unconstitutional taking. *Hastings Constitutional Law Quarterly, 42*(3), 557–575. https://repository.uchastings.edu/hastings_constitutional_law_quaterly/vol42/iss3/3

Jover, J., & Díaz-Parra, I. (2019). Gentrification, transnational gentrification and touristification in Seville, Spain. *Urban Studies, 57*(15), 3044–3059. https://doi.org/10.1177/0042098019857585

Klinenberg, E. (2018). *Palaces for the people: How social infrastructure can help fight inequality, polarization, and the decline of civic life* (277 pp). Penguin.

Lee, D. (2016). How Airbnb short-term rentals exacerbate Los Angeles's affordable housing crisis: Analysis and policy recommendations. *Harvard Law Policy Review, 10*, 229–253.

Liang, C., Yeung, M. C. H., & Au, A. K. M. (2021). The impact of Airbnb on housing affordability: Evidence from Hong Kong. *Environment and Planning*. https://doi.org/10.1177/23998083211043123

Liu, A., & Pratt, S. (2017). Tourism's vulnerability and resilience to terrorism. *Tourism Management, 60*, 404–417. https://doi.org/10.1016/j.tourman.2017.01.001

Lois González, R., Piñeira Mantiñán, M. J., & Vives Miró, S. (2016). The urban bubble process in Spain: An interpretation from the theory of circuits of capital. *Journal of Urban and Regional Analysis, 8*(1), 5–20. https://doi.org/10.37043/JURA.2016.8.1.1

Lois, R. C., et al. (2021). Presentación del número monográfico: La Geografía frente a la COVID-19. Análisis territoriales y perspectivas multidisciplinares. *Boletín de la Asociación de Geógrafos Españoles, 91*(1). https://bage.age-geografia.es/ojs/index.php/bage/article/view/3230

López, L., Pazos Otón, M., & Piñeiro Antelo, M. A. (2019). ¿Existe overtourism en Santiago de Compostela? Contribuciones para un debate ya iniciado. *Boletín de la Asociación de Geógrafos Españoles, 83*. https://doi.org/10.21138/bage.2825

Lopez-Gay, A., Cocola-Gant, A., & Russo, A. P. (2021). Urban tourism and population change: Gentrification in the age of mobilities. *Population Space and Place, 27*(1), e2380. https://doi.org/10.1002/psp.2380

Méndez, R. (2022). Crisis pandémica y vulnerabilidad desigual de los destinos turísticos en España. In R. Martínez Cárdenas, L. F. Cabrales Barajas, M. Calle Vaquero, M. García Hernández, M.C. Mínguez García, & L. Troitiño Vinuesa (coords), *Leyendo el territorio. Homenaje a Miguel Ángel Troitiño* (pp. 546–555). Universidad de Guadalajara.

Méndez, R., Abad, L. D., & Echaves, C. (2015). *Atlas de la crisis. Impactos socioeconómicos y territorios vulnerables en España* (301 pp). Tirant lo Blanch.

Merante, M., & Horn, K. M. (2017). Is home sharing driving up rents? Evidence from Airbnb in Boston. *Journal of Housing Economics, 38*, 14–24. https://doi.org/10.1016/j.jhe.2017.08.002

Milano, C., Novelli, M., & Cheer, J. M. (2019). Overtourism and degrowth: A social movements perspective. *Journal of Sustainable Tourism, 27*(12), 1857–1875. https://doi.org/10.1080/09669582.2019.1650054

Miller, S. R. (2014). *Transferable sharing rights: A theoretical model for regulating Airbnb and the short-term rental market.* Avaible in: https://papers.ssrn.com/sol3/papers.cfm?abstract_id=2514178

Mínguez, C., Piñeira, M. J., & Fernández-Tabales, A. (2019). Social vulnerability and touristification of historic centers. *Sustainability, 11*(16), 4478. https://doi.org/10.3390/su11164478

Mordue, T. (2017). New urban tourism and new urban citizenship: Researching the creation and management of postmodern urban public space. *International Journal of Tourism Cities, 3*(4), 399–405. https://doi.org/10.1108/ijtc-04-2017-0025

Muñoz, F. (2008). *Urbanalización. Paisajes Comunes, Lugares Globales* (216 pp). Gustavo Gili.

Muñoz, R. (2016). El turismo como sector estratégico en las etapas de crisis y desarrollo de la economía española. *International Journal of Scientific Management and Tourism, 2*(4), 81–115.

Murray, I. (2015). *Capitalismo y turismo en España. Del "milagro económico" a la "gran crisis"* (424 pp). Alba Sud Editorial.

Neuts, B. (2016). An econometric approach to crowding in touristic city centres: Evaluating the utility effect on local residents. *Tourism Economics, 22*(5), 1055–1074. https://doi.org/10.5367/te.2015.0477

Nieuwland, S., & del Romero Renau, L. (2021). The impact and regulatory issues of Airbnb in a mid-sized city: Valencia (Spain). In *A modern guide to the urban sharing economy* (pp. 103–118). Edward Elgar Publishing.

Nieuwland, S., & van Melik, R. (2020). Regulating Airbnb: How cities deal with perceived negative externalities of short-term rentals. *Current Issues in Tourism, 23*(7), 811–825. https://doi.org/10.1080/13683500.2018.1504899

Novy, J., & Colomb, C. (2019). Urban tourism as a source of contention and social mobilisations: A critical review. *Tourism Planning & Development, 16*(4), 358–375. https://doi.org/10.1080/21568316.2019.1577293

Oppilard, F. (2017). From San Francisco's 'tech boom 2.00 to Valparaiso's UNESCO world heritage site: Resistance to tourism gentrification from a comparative political perspective. In C. Colomb & J. Novy (Eds.), *Protest and resistance in the tourist city* (pp. 129–151). Routledge.
Oskam, J., & Boswijk, A. (2016). Airbnb: The future of networked hospitality businesses. *Journal of Tourism Futures, 2*, 22–42. https://doi.org/10.1108/JTF-11-2015-0048
Parreño, J. M., & Domínguez, J. (2013). Crecimiento urbano-turístico, segregación residencial y políticas de intervención en el área urbana Gran Canaria Sur. In M. Valenzuela, D. A. Barrado, & J. Palacios (Eds.), *Las ciudades españolas en la encrucijada: Entre el "boom" inmobiliario y la crisis económica* (pp. 297–334). Real Sociedad Geográfica y Asociación de Geógrafos Españoles.
Piñeira, M. J., & González, J. M. (2018). Conflictos sociales y nueva gobernanza urbana en España: alternativas a la vulnerabilidad. In A. M. Autissier et al. (Eds.), *Urban dynamics. Conflicts, representations, appropriations and policies*. Peter Lang.
Piñeira, M. J., & Trillo, J. M. (2016). El efecto de la crisis sobre los barrios vulnerables: una aproximación al contexto español a través de un estudio de caso. En *Crisis, globalización y desequilibrios sociales y territoriales en España. Aportación Española al 33 Congreso de la Unión Geográfica Internacional.* Comité Español de la UGI.
Piñeira, M. J., Fernandez-Tabales, A., & Mínguez, C. (2020). Vulnerabilidad y turistificación: ¿quiénes son los perdedores del centro urbano? In *Sostenibilidad turística:"overtourism vs undertourism"* (pp. 83–98). Societatd' Història Natural de les Balears.
Requejo, J. (2016). Frente a la marabunta: Territorio y trabajo, bases del turismo responsable y sostenible. *Razones de Utopía, 2*, 30–47.
Romagosa, F. (2020). The COVID-19 crisis: Opportunities for sustainable and proximity tourism. *Tourism Geographies, 22*(3), 690–694. https://doi.org/10.1080/14616688.2020.1763447
Roselló, J., Becken, S., & Santana, M. (2020). The effects of natural disasters on international tourism: A global análisis. *Tourism Management, 79*, 104080. https://doi.org/10.1016/j.tourman.2020.104080
Russo, A. P., Soro, E., & Scarnato, A. (2022). Is another tourism possible? Shifting discourses in Barcelona's tourism politics. In *A research agenda for urban tourism* (pp. 257–280). Edward Elgar Publishing.
Sales, A. (2014). Crisis, empobrecimiento y personas sin hogar. *Dossier Catalunya Social Propostes des del Tercer Sector.* Avaible in: www.tercersector.cat/sites/default/files/dossier_crisis_empobrecimiento_y_personas_sin_hogar_2.pdf
Scheyvens, R., & Momsen, J. (2008). Tourism in small Island states: From vulnerability to strengths. *Journal of Sustainable Tourism, 16*(5), 491–410. https://doi.org/10.1080/09669580802159586
Schneiderman, E. T. (2014). *Airbnb in the city.* Office of the Attorney General of the State of New York. Avaible in: https://ag.ny.gov/pdfs/AIRBNB%20REPORT.pdf
Scott, D., Hall, M., & Gössling, S. (2019). Global tourism vulnerability to climate change. *Annals of Tourism Research, 77*, 49–61. https://doi.org/10.1016/j.annals.2019.05.007
Sequera, J., & Nofre, J. (2018). Debates shaken, not stirred. New debates on touristification and the limits of gentrification. *City, 22*(5–6), 843–855. https://doi.org/10.1080/13604813.2018.1548819
Sorando, D., & Ardura, A. (2018). Procesos y dinámicas de gentrificación en las ciudades españolas. *Papers. Regió Metropolitana de Barcelona, Territori, estratègies, planejament, 60*, 35–47.
Subirats, J., & Martí-Costa, M. (2014). *Ciudades, vulnerabilidades y crisis en España.* Fundación Pública Andaluza Centro de Estudios Andaluces, Sevilla 172 pp.
Temes, R. (2007). El tapiz de Penélope. *Transformaciones residenciales sobre tejidos sin valor patrimonial.* Tesis Doctoral inédita. Universitat Politécnica de Valencia.

Temes, R. (2014). Valoración de la vulnerabilidad integral en las áreas residenciales de Madrid. *EURE, 40*, 119–149. https://doi.org/10.4067/S0250-71612014000100006

Turner, B., et al. (2003). A framework for vulnerability analysis in sustainability science. *Proceedings of the National Academy of Sciences of the United States of America, 100*(14), 8074–8079. https://doi.org/10.1073/pnas.1231335100

United Nations. (2003). *Report on the world social situation*. United Nations.

Valente, R., Russo, A. P., Vermeulen, S., & Milone, F. L. (2022). Tourism pressure as a driver of social inequalities: A BSEM estimation of housing instability in European urban areas. *European Urban and Regional Studies, 29*, 332–349. https://doi.org/10.1177/09697764221078729

Velasco González, M., Herrero López, R., & López Sánchez, E. (2019). El orden del caos: la decisión del gobierno ante el problema del impacto del turismo en el centro urbano de Madrid. *Boletín de la Asociación de Geógrafos Españoles, 83*. https://doi.org/10.21138/bage.2832

Von Briel, D., & Dolnicar, S. (2021). The evolution of Airbnb regulations. In S. Dolnicar (Ed.), *Airbnb before, during and after COVID-19*. University of Queensland. https://doi.org/10.6084/m9.figshare.14195972

Wu, T. P. (2019). Does tourism affect house prices? Evidence from China. *Journal of China Tourism Research, 15*(4), 490–502. https://doi.org/10.1080/19388160.2019.1582449

Xie, K. L., & Kwok, L. (2017). The effects of Airbnb's price positioning on hotel performance. *International Journal of Hospitality Management, 67*, 174–184. https://doi.org/10.1016/j.ijhm.2017.08.011

Yrigoy, I. (2016). The impact of Airbnb in the urban arena: Towards a tourism-led gentrification? The case study of Palma old quarter. In M. Blázquez, M. Mir-Gual, I. Murray, & G. X. Pons (Eds.), *Turismo y Crisis. Turismo Colaborativo y Ecoturismo* (pp. 281–289). Societat d'Història Natural de les Balears.

Zervas, G., & Byers, J. (2016). The rise of the sharing economy: Estimating the impact of Airbnb on the hotel industry. *Journal of Marketing Research, 54*(5), 687–705. https://doi.org/10.1509/jmr.15.0204

Zoido, F., De la Vega, S., Morales, G., Mas, R., & Lois, R. (2013). *Diccionario de urbanismo*. Ed. Cátedra. 424 pp.

Alfonso Fernández-Tabales has focused his professional and academic career on the relationships between tourism activity and development processes at local and regional scale, and especially in the last several years on problems associated to increasing urbanisation and land use in tourism spaces. He obtained in the competitive process held on 7/11/2017 at the University of Seville the position of senior university professor. Currently Director of the PhD Programme in Geography at the University of Seville. Currently lead researcher of the Research Project titled "Spatial impacts of touristification in urban areas. New digital tools for diagnosis and management", Ministry of Science and Research.

Maria José Piñeira-Mantiñán is Associate Professor of the Department of Geography at the University of Santiago de Compostela. She has participated in more than 40 projects and public funding contracts. Among his recent research highlights those related to the processes of metropolitanization, new urban governance; social, economic and residential vulnerability, the impact of the real estate bubble in Spain, planning, urban landscape and smart cities. In addition, she has developed part of her research in prestigious foreign centers (Norway, Ecuador, Italy, France-Paris1, Brazil) and has published numerous books and articles in international journals. She is currently President of the IGU Urban Geography Commission and Vice-Dean of the Faculty of Geography and History.

Carmen Mínguez is an Associate Professor in Geography at the Complutense University of Madrid. Her research examines the relationships and the interdependencies between tourism and historic cities, specifically focusing on how to improve cultural heritage and functionally revitalize historic centers and monumental areas of cities. In recent years, Carmen has incorporated into her research the use of Big (Geo)data (social networks and photo-sharing services) and the Citizen Science approach for the study, planning and management of destinations. She is a member of the "Tourism, Heritage and Development" research group.

Chapter 20
Chambermaids: A Focus of Attention in Studies of Tourism Employment in Spain

Ernest Cañada

20.1 Introduction

In the field of social science in Spain, employment in the tourist sector has been analysed in an interdisciplinary way, largely from a sociological and anthropological point of view. For geographers specifically, it has not been a particularly important focus of study. This means that spatial aspects have not been taken into sufficient account in studies of tourism employment. Neither has much relevance been given to temporal perspectives. Hence, in the growing academic literature that can be found on the history of tourism in Spain, no particular emphasis has been given to employment (Palou, 2012; Vallejo & Larrinaga, 2018; Vallejo, 2022), with just a few noteworthy exceptions (Aroca, 2016, 2018), although it has been tackled in other disciplines (for instance, in political science or communication studies), albeit in an almost marginal way.

Since women account for a large part of the workforce in the tourist sector and much of this work is subject to growing precarity (Martínez-Gayo, 2019; Martínez-Gayo & Martínez Quintana, 2020; Martínez Quintana & Martínez-Gayo, 2019; Moreno & Cañada, 2014), the characteristics of the workforce and gender-based implications of its makeup have attracted attention. Three main fields of research can be highlighted: studies of female labour in rural tourism (Brunet & Alarcón, 2007; Canoves & Blanco-Romero, 2009; Flores & Barroso, 2011; Muñoz Semeco, 2012); the problems that women encounter in rising up the career ranks (vertical segregation or the glass ceiling) (García et al., 2018; Huete et al., 2016; Segovia-Pérez et al.,

E. Cañada (✉)
Universitat de les Illes Balears, Palma, Spain

Alba Sud, Barcelona, Spain
e-mail: ernest@alabasud.org

A. Blanco-Romero, M. Blázquez-Salom (eds.), *Spanish Tourism Geographies*, Geographies of Tourism and Global Change,
https://doi.org/10.1007/978-3-031-39780-6_20

2019; Sigüenza, 2011); and the dynamics behind the growing precarity of employment groups of a mainly female composition who have little hope of escape from this situation of insecurity (horizontal segregation or the sticky floor).

In recent years, studies of tourism employment in Spain have clearly turned the spotlight on one of these more insecure groups: chambermaids. The women who clean hotel rooms have passed from being invisible to society to coming under the academic beacon in studies of this kind, while also drawing substantial media attention. Consequently, not only has there been a clear bias toward hotels in studies of tourism employment, as forewarned years ago by Tom Baum (1993) in relation to the Anglo-Saxon world, but these studies have also concentrated very heavily on the housekeeping department and its workers.

The aim of this chapter is to systematize the knowledge that has been built up from an interdisciplinary perspective within the field of social science on the work of Spain's chambermaids. As a result, medical or legal analyses of this group have been excluded in order to restrict debate to the field of social science. The review starts in the 1970s and continues through to the latest studies. The main themes that these studies tackle are identified, together with the evolution of this employment group's analysis. Priorities for future research into tourism employment in Spain are also pinpointed, bearing in mind headway in international literature and salient debate at a social and political level. This review only takes into account articles published in scientific journals and books, not end-of-degree projects for bachelor's or master's degrees, doctoral dissertations or journalistic material or opinion articles, since they would be very hard to systematize due to their abundance.

20.2 Prevailing Approaches and Timeframes

Employment in the tourist sector has been widely associated with precarity. This link has been clearly highlighted in some international literature (Cañada, 2019a). Studies in Spain point to the existence of similar dynamics and this is the main focus that has predominated in research studies of work as a chambermaid, with most studies sharing a concern for dynamics that reflect situations of exploitation, inequality and discrimination, either due to the subjects they deal with or the approach they take. Hence, they mainly have very little to do with the human resource management approach that has prevailed traditionally in literature on tourism (Cañada & Murray, 2019).

In Spain's case, nonetheless, some studies have sought to come up with "technical solutions" to growing discontent among chambermaids. For instance, in the study by Díaz-Segura and Dorta-Afonso (2021), a series of management tools are proposed aimed at boosting employment satisfaction and reducing symptoms of staff burnout. These tools are directed specifically at chambermaids: a group characterized, according to the study, by recurring signs of cynicism and inefficiency. This type of approach, where priority is given to business management solutions, hinders a proper understanding of the employment conditions of this group of workers and

Table 20.1 Timeframe of studies of the work of chambermaids in Spain

Stage	Topics and studies
1st stage: Pioneering studies of a general nature	Analyses of how hotels operate from an employment perspective, paying specific attention to the housekeeping department within the framework of a more general analysis (Castellanos & Pedreño, 2006; Galán et al., 1977).
2nd stage: From growing precarity to a surge of discontent	The effects of the 2008 crisis and the 2012 labour reforms on the growing precarity of the conditions of chambermaids (Albarracín, 2010; Albarracín & Castellanos, 2013; Balbuena, 2021; Cañada, 2015). The consequences of outsourcing decisions (Cañada, 2016, 2018a; Megías Bas, 2019). The health impacts of the growing instability of this work (Garí & Cañada, 2020; Ferreira-Marante et al., 2017). Analyses of the work of chambermaids, based on a discussion of gender in tourism employment (Martínez-Gayo, 2021).
3rd stage: "The kellys" movement, a trailblazing experience	The historical process behind the creation of "the kellys": an independent movement, with no trade union links (Cañada, 2018b). Debate on relations between the "kellys" associations and trade unions, with contradictory interpretations (Barranco et al., 2022; Moral-Martin, 2020; Moral-Martin & Brunet, 2018; Moral-Martín & Icart, 2021; Godino & Molina, 2019; López-González & Medina-Vicent, 2020). The characterization of the movement's strategies in different spheres (Alcalde-González et al., 2021, 2022a, b; Fernández-Muñoz & Tomé-Caballero, 2020; Trellu, 2021; Valenzuela-Bustos et al., 2022).

Source: Own

the reasons for their dissatisfaction. Consequently, this review focuses on studies that strive to describe and analyse the employment and working conditions, labour relations and occupational health of these housekeeping staff rather than better management of this workforce by companies so as to suit their own corporate interests.

Studies of chambermaids in Spain can be divided into three main stages (shown in Table 20.1), largely conditioned by the emergence of an independent worker movement in 2015 called "the kellys",[1] with certain singularities when compared with the dynamics of this group of staff in other countries. More specifically, the following timeframes can be identified:

[1]Due to the coincidence between the name used colloquially to describe this employment group (derived from the term "la kelly" meaning "la que limpia" or "the woman who cleans") and the name of various different worker associations, in this article, the term "the kellys" is used to refer to numerous independent associations of chambermaids as opposed to just one of them.

- 1st stage: When pioneering studies of the work of chambermaids were conducted, starting in the 1970s and continuing through to the economic bonanza of the first decade of the twenty-first century. These analyses are few and far between in number, and the subject is tackled within the framework of a broader analysis of working conditions in hotels.
- 2nd stage: Focused on analysing the effects of the 2008 world financial crisis, the 2012 labour reforms and the resulting sharp rise in the precarity of chambermaids' conditions, leading to the first attempts by these workers in 2014 to organize themselves into what would become "the kellys";
- 3rd stage: Devoted to an analysis of "the kellys" movement and its different strategies, mainly from 2018 onward when it first started to be analysed in its capacity as a disruptive phenomenon that broke away from conventional forms of worker representation. This stage overlaps the previous one, although studies shifted their focus from the working conditions of chambermaids to the movement's development, its practices and its links with other forms of representation.

Outlined below are the main contributions of these studies during each of the above identified stages:

20.2.1 Stage 1: Pioneering studies

In Spain's case, some research studies allow us to reconstruct the evolution of this body of workers in recent years. These studies do not specifically address this group of housekeeping staff. Instead, the working and employment conditions of chambermaids are analysed within the broader framework of employment in the hotel sector at sun and sand destinations.

In 1977, anthropologist Antonio Mandly and a group of workers from the anarchist trade union, the Confederación Nacional del Trabajo (CNT), conducted a pioneering study of the organization of employment in hotels on the Costa del Sol (Galán et al., 1977). They wished to go beyond a mere analysis in the hope that these companies could be taken over and directly managed by their staff. As a result, the employment conditions and tasks that are carried out in each department, including the housekeeping department, are detailed with great precision. This was the boom period in tourism, with domestic migration from inland areas of Spain to its coastal fringes. This, in the hotel sector, gave rise to a very young set of workers. Although their work was hard–and this is how it is described in minute detail–, it was an escape from poverty, oppression and the moral conservatism that prevailed in many of the towns and villages from which they came. This was also a period of turbulence at a social and trade union level, seeking to recover democratic rights and labour entitlements. In the study, chambermaids are portrayed as a very young, fully female collective, with high hopes of improving their living conditions through their jobs.

Years later, in 2006, in the midst of the economic growth period of the time, a book was published by Mari Luz Castellanos and Andrés Pedreño that offers a fundamental insight into hotel work during the economic boom. Many Spanish chambermaids decided to give up this type of work during the period. Their rising age only exacerbated the effects of a particularly gruelling job, and employment opportunities in other jobs helps to explain this move. Hence, in the years prior to the big 2008 crisis, housekeeping departments started to employ high numbers of staff from other countries, mainly from Latin America, but also from Eastern Europe, the Philippines and, to a lesser extent, North Africa. Other chambermaids took advantage of the high demand for employment, changing their job with ease and improving their working conditions. Despite the economic growth of the period, employment in hotels and in housekeeping departments was still just as hard and intensive as before.

20.2.2 *Stage 2: The conditions that paved the way for a surge of unrest*

Shortly after the world economic crisis broke out in 2008, Daniel Albarracín conducted a study of the situation of chambermaids, based on a survey by the trade union Comisiones Obreras, with the subsequent collaboration of Mari Luz Castellanos in its analysis (Albarracín, 2010; Albarracín & Castellanos, 2013). The results highlight the growing decline at a social and employment level. The chambermaids, particularly those of Spanish origin, had aged and their work impacted on them even more severely than before. Although their working conditions had worsened, fear of losing their jobs made them accept increased workloads. It was no longer possible to switch to a different job due to the crisis and to mass unemployment, and all this went hand in hand with a radical drop in their individual and collective bargaining capacity with companies. The pressure not to lose their jobs was even higher: before, their pay might have been just part of the family income but now, in many homes, it was the only source of revenue. In the analysis of this workforce, signs of discontent with the sector's main trade unions could be observed and, to a certain extent, a growing sense of alienation from them.

Years later, between 2015 and 2018, my own research study highlighted how trapped many chambermaids felt, particularly middle-aged ones with long experience of the job, and, as the crisis advanced, and a feeling of having no escape was noted (Cañada, 2015). The situation became unsustainable, as illustrated by the pioneering mobilization of this group in ensuing years. Their worsening conditions were characterized by a steady reduction in fixed and fixed intermittent employment contracts and an increase in atypical formulae: temporary, part-time and, above all, subcontracted work, brought about by the 2012 labour reforms, in addition to a substantial increase in workloads (Cañada, 2016, 2018a).

The spreading use of subcontracted workers by housekeeping departments was one of the main strategies used to cut labour costs and to flexibilize labour in the Spanish hotel sector (Espino-Rodríguez & Ramírez-Fierro, 2018). It also meant that some fixed labour costs could be converted into variable ones, facilitating financialization processes in the sector (Yrigoy & Cañada, 2019). The impact of outsourcing decisions was a key focus of analysis in studies of the work of chambermaids (Balbuena, 2021; Cañada, 2016, 2018a; Megías Bas, 2019).

Declining working and employment conditions had a big impact on the health of this group of staff, both physically and psychologically, with a significant rise in medication and dependence on these drugs in order to put up with everyday conditions (Garí & Cañada, 2020; Ferreira-Marante et al., 2017).

These analyses of changing labour conditions in the hotel sector, with particularly severe changes in the case of chambermaids, were complemented by a study by Gema Martínez-Gayo (2021), who took a crosscutting gender-based perspective. Other aspects were explored besides changes in working and employment conditions and certain repercussions on the workers' health, such as difficulty in reconciling work and family life, the lack of free time, the scanty appreciation given to women's work, harassment in the workplace and sexual violence.

20.2.3 Stage 3: "The kellys" movement

Following the publication of an article in the newspaper *El País* (Cañada, 2014), the situation of chambermaids in Spain became known to the public. This article coincided with the launch of an international campaign to foster more decent conditions for chambermaids, promoted by the trade union UITA, leading in turn to resources like the book *Las que limpian los hotels (The Women Who Clean Hotels)* (Cañada, 2015), which made a very big impact in the media. Some chambermaids took advantage of the opportunity to launch their own platforms, initially for communication purposes and then as a tool in the mobilization of these workers. In 2014, the Facebook group "The kellys" was created. In 2015, it became an independent association, later giving rise to different regional "kellys" organizations.

This mobilization process and the rising visibility of Spain's chambermaids began to be analysed in academic circles from 2018. Fuelled by despair, lack of recognition and pent-up rage, these workers seized the opportunity of a trade union campaign to form their own independent movement, made up of hundreds of women from all over Spain, in a phenomenon known as "the rebellion of the kellys" that went far beyond the initial envisaged scope of the campaign (Cañada, 2018b).

From then on, different studies sought to analyse the characteristics of this unprecedented movement. One of the main fields of discussion revolved around the reasons for its emergence, particularly in the light of its independence, with no links with any of the main trade unions and differing degrees of antagonism or opposition to them depending on each local association and its different ideological stances. What was disconcerting was the sheer size of the movement and the echo in the media, and this was also an employment group that had traditionally been represented by trade unions. This set it apart from other groups of workers that had come into the public eye in recent years and had also organized themselves into independent movements, such as home helps or riders. Unlike chambermaids, they had not been represented by trade unions. In some research studies, a more in-depth analysis was made of the difficulty that workers in precarious jobs face in being represented by trade unions, together with possible strategies for facilitating certain processes of convergence and reunification (Barranco et al., 2022; Godino & Molina, 2019). In other cases, studies mainly focused on explaining why independent worker movements were needed, given the inefficiency of traditional trade unions in a context of growing precarity and rising female workforces in certain jobs (López-González & Medina-Vicent, 2020; Moral-Martin, 2020; Moral-Martin & Brunet, 2018; Moral-Martín & Icart, 2021).

Another research area sparked off by "the kellys" was the characterization of the movement's strategies through the identification of different action plans at both an organizational and communicative level, with several articles on the subject currently under review by scientific journals (Alcalde-González et al., 2021, 2022a, b; Fernández-Muñoz & Tomé-Caballero, 2020; Trellu, 2021; Valenzuela-Bustos et al., 2022).

20.3 A Comparison with International Literature

As is also the case in Spain, studies of tourism employment in international literature mainly concentrate on hotels and, within their different departments, on the housekeeping department in comparison with others, like food and beverages or the reception area. There are numerous possible reasons for this stronger focus on chambermaids in research at a social and healthcare level. In social science, greater attention to this group of workers is not only attributable to a keen interest in tourism employment but also to certain characteristics of this workforce. Because, numerically, this is a big, largely female employment group conspicuous for its high number of immigrant workers and low wages, this makes it an interesting field of study from various different perspectives and in differing disciplines. Furthermore, chambermaids make up a large part of a hotel's workforce and, traditionally, they have less trade union backing due to their makeup and the difficulty that many trade unions have in adopting a more egalitarian stance. This group of workers is hence a key target for companies wishing to cut costs and to introduce flexibilization measures, leading to a further decline in employment conditions while also drawing the attention of researchers in the field of social science.

Research into the work of chambermaids in Spain has some similarities with the international literature on the subject, although some specificities can be observed in the Spanish case due to the worker mobilization process. A systematic review of academic studies of this group of workers in the field of social science (Cañada, 2019b) showed that, in international literature, the subjects to which priority has been given are: a) the makeup of this body of workers, with a high proportion of women and immigrant groups; b) the segmentation of this workforce; c) the intensified dynamics of their work; d) occupational health; e) sexual harassment, and f) organizational processes at a trade union level. In the Spanish case, less attention has been devoted to studies of the makeup of this employment group and its internal dynamics. Neither has so much importance been lent to studies of the health effects of their working conditions, possibly because the high number of studies conducted in the United States are due to disputes between trade unions and mutual insurance companies over the cost of healthcare for these workers. In such scenarios, it is vital to gain a better understanding of the impact of a chambermaid's work on their health. In Spain, on the other hand, there has been broad debate on the associative organizational processes that have taken place in recent years. The other subjects largely coincide.

20.4 An Unfinished Research Agenda

As outlined above, there has been a sharp rise in research into the work of Spain's chambermaids due to the mobilization of this group of workers and to media coverage. Despite this flourishing stage in research into tourism employment in Spain, two limitations were found, which could be used to foster debate on the direction that future research in this field should take.

As in other countries, analyses of tourism employment have focused primarily on the hotel sector and, in particular, on housekeeping departments. Despite this research into the work of chambermaids as an employment body, there are still important gaps to be filled. More specifically, a more in-depth analysis must be made of their labour conditions and the makeup of this group, bearing in mind:

- Changes brought about by the COVID-19 pandemic, with evidence of extreme working poverty, plus the way in which tourism has been reactivated.
- The impact of technological and management changes by companies on the way in which work is organized in housekeeping departments.
- Changes in the makeup of this group of workers, with the incorporation of people from countries with even lower incomes than those of the former prevailing nationalities.

Some of the existing research studies have a certain tendency to exaggerate the achievements of "the kellys" movement. The same also applies in media coverage, in reactions by certain artistic circles and among left-wing activists wishing to use the movement to challenge Spain's leading trade unions, upholding it as a new associative model. Some of these analyses are biased by value judgements used to compare this movement with other models. The movement's idealization hinders a proper understanding of the employment dynamics of this group of workers and the real achievements of the mobilization process, which are by no means insignificant.

This is why, in research, greater attention must be paid to mobilization processes, both at a trade union level and through the "kellys" associations, given the growing precarity of employment and new campaigning and protest methods, without using these workers for political flag-waving purposes. A more in-depth comparative analysis must also be made of why chambermaids have managed to arouse the sympathy and support of broad sectors of society, the press and political groups while, in other types of highly unstable work with a predominance of female workers, like the care sector, this is not the case.

At the same time, one of the main challenges in the field of social science is to counterbalance the predominance that has been given to chambermaids in tourism studies in order to gain a better insight into the sector's employment dynamics. In the case of Spain, specific attention has only been devoted to a few groups of workers, aside from chambermaids: mainly tourist guides (Cañada, 2021; Leal & Medina, 2018; Navalón-García & Mínguez García, 2016, 2021) and, to a lesser extent, work associated with holiday rental properties (Cañada, 2020; Cañada & Izcara, 2021a, b; Gil, 2019). In both cases, the starting point has tended to be the rising precarity of this type of employment. Tourist guides have seen a sharp decline in their working conditions as a result of new online ways of organizing this kind of work, with the intrusion of free tours, the creation of intermediary platforms specializing in tourist experiences of all kinds, and pressure from big travel agencies to regulate the sector. This has given rise to the segmentation of the profession into tourist guides who try and maintain certain labour standards by working as self-employed professionals and others who started out in this insecure situation. As for work associated with holiday rentals, in the available studies, the growing precarity of those who host guests in their homes as prosumers is highlighted, together with the unstable, invisible dynamics of work conventionally associated with the whole of the value chain involved in these new forms of accommodation, with no trade union representation.

The research agenda must be opened up so as to broaden a vision that is currently too dependent on the dynamics of just a few employment collectives, in particular chambermaids. By focusing too heavily on this group of workers, other collectives might be neglected, thus failing to build up a picture that takes into account the diversity and complexity of employment in the tourist sector. The main priorities in this improved research agenda are:

- To fill the current gap in studies that characterize the employment and working conditions, labour relations and occupational health of workers in the tourist sector by opening up existing research to encompass other hotel departments and other employment groups from the tourist sector's multiple different fields of activity.
- To gain a better understanding of how tourism companies work, examining the causes of the constant pressure that comes with the devaluation of work, the mechanisms used for this purpose, and the consequences on the lives and jobs of these workers.

- To analyse changes in the characteristics of the work they do and the conditions in which it is done, in association with the main changes that the sector is undergoing (growing financialization, the role of platform capitalism, the intensification of work, etc.)
- To explore intersections in the makeup of certain employment groups based on class, gender and race. This would also entail opening up the research agenda to explore new issues from feminist and decolonial perspectives which might enrich understandings of tourism employment.
- To boost the number of analyses that focus on spatial and temporal aspects, since, until relatively recently, little importance was given to these key variables in understanding tourism unemployment.
- To systematize experiences of the organization of tourism activities through approaches other than those based on the reproduction of capital, so as to make headway in the development of post-capitalist formulae and to explore how they can be put into practice at an employment level.

20.5 Conclusions

In Spain, studies of tourism employment in the field of social science have mainly focused on chambermaids, as has also occurred in much of the available international literature. This workforce is almost entirely female with a high share of immigrant workers. Spanish studies can be found that offer an insight into the characteristics of their work, the makeup of the workforce and the evolution of this kind of employment from shortly after the tourist boom, between the 1960s and 70 s, through to today. In the beginning, this analysis of chambermaids formed part of a broader general characterization of employment in the hotel sector. Following the 2008 big financial crisis and its handling by the State in a way that clearly favoured employer interests, there was a sharp deterioration in the employment and working conditions of these workers. This led to more severe health problems and growing discontent: a situation that would finally be combatted through the creation of a novel independent worker movement that managed to unite different ideologies and practices. Initially, academic research mainly focused on exploring the changes brought about by this extreme dissatisfaction. As from 2018, efforts were made to try and explain the reasons for this highly singular movement and its characteristics. Despite the importance of these studies, research into tourism employment in Spain continues to suffer from big gaps and from a big imbalance. Analyses of chambermaids in Spain must be extended, without overlooking other employment groups from the tourist sector that have not received sufficient attention. It is also important to bear in mind that a political bias in the interpretation of some movements impedes an accurate vision of their development. Debate must be opened up so as to try and tailor research to meet the concerns and needs of a key sector in Spain's economy and in the generation of employment.

Acknowledgements I would like to thank Carla Izcara and Gema Martínez-Gayo, researchers from Alba Sud, for their review and discussion of this chapter. This chapter has been written with the support of the Grant RTI2018-094844-B-C31 funded by MCIN/AEI/10.13039/501100011033 and by "ESF Investing in your future".

References

Albarracín, D. (Ed.). (2010). *Relaciones laborales en los establecimientos hoteleros: los y las trabajadoras ante la crisis del modelo laboral y económico hotelero.* Gabinete Federal de Estudios FECOHT – CCOO.

Albarracín, D., & Castellanos, M. L. (2013). Las trabajadoras de los establecimientos hoteleros. Trayectorias en el túnel silencioso de la subordinación. *Sociología del Trabajo, 77*, 27–45.

Alcalde-González, V., Gálvez, A., & Valenzuela, A. (2021). No clean rooms, no hotel business: Subversion tactics in Las Kellys' struggle for dignity in hotel housekeeping. *Annals of Tourism Research, 91*, 103315. https://doi.org/10.1016/j.annals.2021.103315

Alcalde-González, V., Gálvez-Mozo, A., & Valenzuela-Bustos, A. (2022a). 'Las Kellys son las que limpian': Collective identity and social media in the mobilisation of room attendants in Spain. *New Media and Society*, in press. https://doi.org/10.1177/14614448221097891

Alcalde-González, V., Gálvez, A., & Valenzuela, A. (2022b). Diffusion of intersectionality across contemporary Spanish activism: The case of Las Kellys. *Social Movement Studies*, in press. https://doi.org/10.1080/14742837.2022.2142546

Aroca, M. (2016). El sindicalismo en la hostelería de Baleares: del franquismo a la democracia. *Historia, Trabajo y Sociedad, 7*, 77–103.

Aroca, M. (2018). *Sindicatos y turismo de masas en las Baleares. Del franquismo a la democracia.* Edicions Documenta Balear.

Balbuena, A. (2021). La realidad laboral de las camareras de piso. *TITUR Revista Iberoamericana de Turismo, 11*(1), 175–190.

Barranco, O., Alós, R., & Molina, O. (2022). Innovar para representar: trabajadores con dificultades de representación colectiva y estrategias sindicales en España. *Cuadernos de Relaciones Laborales, 40*(1), 115–137. https://doi.org/10.5209/crla.74452

Baum, T. (1993). Human resource issues in tourism: An introduction. En T. Baum (Eds.). *Human resource issues in international tourism* (pp. 3–21). Butterworth Heinemann.

Brunet, I., & Alarcón, A. (2007). Mujeres "emprendedoras" en el marco del turismo rural. *Revista de Gestión Pública y Privada, 12*, 53–74.

Cañada, E. (2014). Las que limpian los hoteles. *El País*, 30 de julio de 2014.

Cañada, E. (2015). *Las que limpian los hoteles. Historias ocultas de precariedad laboral.* Icaria Editorial.

Cañada, E. (2016). *Externalización del trabajo en hoteles. Impactos en los departamentos de pisos.* Alba Sud Editorial.

Cañada, E. (2018a). Too precarious to be inclusive. Hotel maid employment in Spain. *Tourism Geographies, 20*(4), 653–674. https://doi.org/10.1080/14616688.2018.1437765

Cañada, E. (2018b). La rebelión de las camareras de piso. En P. Ibarra, R. Gomà, R. González y S. Martí. *Movimientos sociales y derecho a la ciudad Historias de dignidad, resistencia y esperanza* (pp. 118–133). Icaria Editorial.

Cañada, E. (2019a). Trabajo turístico y precariedad. En E. Cañada & I. Murray (Eds.). *Turistificación global. Perspectivas críticas en turismo* (pp. 267–287). Icaria Editorial.

Cañada, E. (2019b). El trabajo de las camareras de piso: un estado de la cuestión. *Papers de Turisme, 62*, 67–84.

Cañada, E. 2020. Precarización laboral en la cadena de valor de las viviendas de uso turístico. En G. X. Pons, A. Blanco-Romero, R. Navalón-García, L. Troitiño-Torralba, & M. Blázquez-Salom, (Eds.). *Sostenibilidad Turística: overtourism vs undertourism* (pp. 333–346). Mon. Soc. Hist. Nat. Balears.
Cañada, E. (2021). *Guías de turismo. Una profesión en riesgo de precarización.* Alba Sud Editorial, Informes en contraste, 17.
Cañada, E., & Izcara, C. (2021a). *Precariedad laboral y viviendas de uso turístico. Alertas para una reactivación pospandemia.* Alba Sud Editorial, Informes en contraste, 16.
Cañada, E., & Izcara, C. (2021b). Vivienda turística y precaridad laboral: La invisibilidad del trabajo tradicional. En J. Sequera (Coord.). *Sé lo que hicisteis el último verano. La transformación del turismo urbano antes, en y después de la pandemia* (pp. 147–170). Bellaterra Edicions.
Cañada, E., & Murray, I. (2019). Introducción. Perspectivas críticas en turismo. En E. Cañada & I. Murray. *Turistificación global. Perspectivas críticas en turismo* (pp. 7–34). Icaria Editorial.
Canoves, G., & Blanco-Romero, A. (2009). Turismo, mercado de trabajo e inmigración en España. Un análisis de la situación en las Comunidades de Murcia, Canarias, Valencia y Andalucía. *Boletín de la Asociación de Geógrafos Españoles, 50*, 259–280.
Castellanos, M. L., & Pedreño, A. (2006). *Los nuevos braceros del ocio. Sonrisas, cuerpos flexibles e identidad de empresa en el sector turístico.* Miño y Dávila Editores.
Díaz-Segura, S. M., & Dorta-Afonso, D. (2021). Influencia de las prácticas de alto rendimiento de recursos humanos en el burnout y los resultados de las kellys. *Emprendimiento y Negocios Internacionales, 6*(2), 46–50.
Espino-Rodríguez, T. F., & Ramírez-Fierro, J. C. (2018). La intensidad de la externalización hotelera y sus beneficios y riesgos percibidos. *Cuadernos de Turismo, 41*, 161–186. https://doi.org/10.6018/turismo.41.326991
Fernández-Muñoz, C., & Tomé-Caballero, M. (2020). Empleo del storytelling y las narrativas en primera persona en la comunicación de las kellys como referente para las ongs. *Profesional de la Información, 29*(3). https://doi.org/10.3145/epi.2020.may.25
Ferreira-Marante, R., Rivas-Quarteni, N., & Viana-Moldes, I. (2017). Aproximación inicial al impacto del trabajo en las ocupaciones y en la salud de las camareras de piso desde una perspectiva de la justicia ocupacional. *TOG, 14*(26), 444–456.
Flores, D., & Barroso, M. d. l. O. (2011). La mujer en el turismo rural: un análisis comparativo de género en el Parque Natural de Sierra de Aracena y Picos de Aroche (comarca Noroccidental andaluza). *Ager, Revista de Estudios sobre Despoblación y Desarrollo Rural, 10*, 39–69.
Galán, J. J., Martín, A., Ruiz, J., & Mandly, A. (1977). *Costa del Sol. Retrato de unos colonizados.* Campo Abierto Ediciones.
García, S., Galante, F., & Poveda, V. (2018). El techo de cristal en la industria hotelera de Tenerife (Islas Canarias): el acceso de las mujeres al liderazgo en el sector. *Pasos Revista de Turismo y Patrimonio Cultural, 16*(4), 1105–1117. https://doi.org/10.25145/j.pasos.2018.16.076
Garí, A., & Cañada, E. (2020). Retos de la prevención de riesgos psicosociales frente a las nuevas formas del trabajo. El caso concreto de la externalización de los servicios de las camareras de piso. En *Iniciativa Interuniversitaria sobre el Futuro del Trabajo. Congreso de Sevilla, 7 y 8 de febrero de 2029. Volumen IV* (pp. 215–229). Madrid: Organización Internacional del Trabajo (OIT) y Ministerio de Trabajo y Economía Social.
Gil, J. (2019). Redistribución económica y precariedad. El caso de los anfitriones de Airbnb. *Recerca. Revista de Pensament i Anàlisi, 24*(1), 92–112. https://doi.org/10.6035/Recerca.2019.24.1.5
Godino, A., & Molina, O. (2019). *¿Quién debilita la negociación colectiva? Prácticas de externalización y regulación en España.* Recover Policy Paper – Quit Centre d'Estudis Sociològics sobre la Vida Quotidiana i el Treball.
Huete, R., Brotons, M., & Sigüenza, M. C. (2016). La desigualdad entre mujeres y hombres en el sector hostelero español. *Estudios y Perspectivas en Turismo, 25*(1), 73–87.

Leal, M. d. P., & Medina, F. X. (2018). Turismo y economía colaborativa: el caso de los recorridos gratuitos a pie en Barcelona. *Cuadernos de Turismo, 41*, 323–341. https://doi.org/10.6018/turismo.41.327051

López-González, J. L., & Medina-Vicent, M. (2020). Las Kellys y el turismo: de la invisibilidad del cuidado a la visibilidad política. *Digithum, 25*, 1–13. https://doi.org/10.7238/d.v0i25.3175

Martínez Quintana, V., & Martínez-Gayo, G. (2019). Empoderamiento y desigualdad en el trabajo turístico femenino. Una perspectiva comparada. *Papers de Turisme, 62*, 46–66.

Martínez-Gayo, G. (2019). ¿Empleos cinco estrellas? Reflexión sobre la precariedad laboral en la hostelería española. *Revista Internacional de Turismo, Empresa y Territorio (RITUREM), 3*(2), 1–15. https://doi.org/10.21071/riturem.v3i2.12283

Martínez-Gayo, G. (2021). ¿Conlleva la COVID-19 una mayor desigualdad de género? Pasado, presente y futuro de las trabajadoras turísticas en España. En E. Cañada & I. Murray (Eds.), *#TourismPostCOVID19. Turistificación confinada* (pp. 181–189). Alba Sud Editorial.

Martínez-Gayo, G., & Martínez Quintana, V. (2020). Precariedad laboral en el turismo español bajo la perspectiva de género. *Pasos. Revista de Turismo y Patrimonio Cultural, 18*(4), 649–665. https://doi.org/10.25145/j.pasos.2020.18.046

Megías Bas, A. (2019). La feminización de la pobreza a través de la externalización de servicios. Las camareras de piso. En V. Pérez de Guzmán, E. Bas Peña, & M. Machado Casas (coord.). *Gender Issues in Latin America and Spain: Multidisciplinary Perspectives* (pp. 217–229). Peter Lang, International Academic Publishers.

Moral-Martin, D. (2020). ¿Por qué se han organizado las camareras de piso? Algunas claves e interpretaciones desde la revitalización sindical. *Revista Española de Sociología, 29*(3), 97–115. https://doi.org/10.22325/fes/res.2020.59

Moral-Martin, D., & Brunet, I. (2018). La imagen del sindicato en el siglo XXI a la luz de su contestación por los Nuevos Movimientos Sindicales. *Sociología del Trabajo, 93*, 307–326. https://doi.org/10.5209/STRA.61833

Moral-Martín, D., & Icart, I. B. (2021). La necesidad de la revitalización sindical: una oportunidad para otras propuestas organizativas. *CIRIEC-España Revista de Economía Publica, Social y Cooperativa, 101*, 227–254. https://doi.org/10.7203/CIRIEC-E.101.16542

Muñoz Semeco, D. (2012). Las mujeres y el desarrollo del turismo en Castilla y León (España). *Geografía em Questão, 5*(2), 112–127. https://doi.org/10.48075/geoq.v5i2.4975

Navalón-García, R., & Mínguez, C. (2021). Tourist guides and free tours: A controversial relationship. *Tourist Studies, 21*(3), 486–506. https://doi.org/10.1177/1468797621110160

Navalón-García, R., & Mínguez García, M. D. C. (2016). El desarrollo del fenómeno de los tours gratuitos (*free tours*) en los destinos culturales. *El caso de la ciudad de Madrid: colaboración o negocio. Monografies de La Societat d'Historia Natural de Les Balears, 23*, 265–279.

Palou, S. (2012). *Barcelona, destinació turística. Un segle d'imatges i promoció pública.* Edicions Vitel.la.

Segovia-Pérez, M., Figueroa-Domecq, C., Fuentes-Moraleda, L., & Muñoz-Mazón, A. (2019). Incorporating a gender approach in the hospitality industry: Female executives' perceptions. *International Journal of Hospitality Management, 76*, 184–193. https://doi.org/10.1016/j.ijhm.2018.05.008

Sigüenza, M. C. (2011). La mujer en el subsector del alojamiento. El caso de la Costa Blanca (Alicante, España). *Investigaciones Turísticas, 2*, 102–119. https://doi.org/10.14198/INTURI2011.2.05

Trellu, J. (2021). De l'exploitation à l'empowerment: le cas des migrantes latino-américaines femmes de chambres dans l'hôtellerie en Espagne. *Documentos de Trabajo IELAT, 147*, 1–48.

Valenzuela-Bustos, A., Gálvez-Mozo, A., & Alcalde-Gonzalez, V. (2022). Invisible room attendants: Outsourcing as a dispositive of (in)visibility and the resistance of Las Kellys in Spain. *Work, Employment and Society*, in press. https://doi.org/10.1177/09500170221092353

Vallejo, R. (2022). *Historia del turismo en España, 1928–1962. Economía, política y administración pública.* Silex Ediciones.

Vallejo, R., & Larrinaga, C. (dirs.) (2018). *Los orígenes del turismo moderno en España. El nacimiento de un país turístico, 1900–1939*. Silex Ediciones.
Yrigoy, I., & Cañada, E. (2019). Fixing creditor-debtors' tensions through labor devaluation. Insights from the Spanish hotel market. *Geoforum, 98*, 180–188. https://doi.org/10.1016/j.geoforum.2018.11.012

Ernest Cañada is a post-doctoral researcher for the Margarita Salas aid programme, run by the Spanish Ministry for University Affairs, financed by Next Generation funds within the framework of the Recovery, Transformation & Resilience Plan managed by the University of the Balearic Islands. He is also a member of Alba Sud, an independent research centre specializing in critical approaches to tourism, which he coordinated between 2008 and 2021.

Chapter 21
Covid-19 Landing on Touristified and Unequal Spaces in Spain

Maria Antònia Martínez-Caldentey and Ivan Murray

21.1 Introduction

Since the outbreak of the 2008 crisis, academic and social debate on inequality has gained in importance, particularly with the publication of Thomas Piketty's book (2014) on capital in the twenty-first century. In parallel with spreading inequality as a result of the 2008 crisis, capital accumulation through tourism became one of the main solutions to the situation (Fletcher, 2011). Since Elizabeth Becker (2013) first cast light on the world tourism boom in her book *Overbooked: The Exploding Business of Travel and Tourism,* the issue of overtourism has gained in relevance in academic circles (Milano et al., 2019a). It is important to remember that debate on overtourism is very closely related to the intensification of urban conflicts brought about by tourism-related changes to cities. These changes have, on many occasions, occurred in contexts already affected by the negative consequences of the 2008 crisis, with this new phase of touristification aggravating existing processes of precarity, dispossession and inequality (Milano et al., 2019b). Nonetheless, in academic literature, a certain gap can be detected in studies of the link between tourism and inequality. Despite growing interest in analyses of inequality and notable headway in critical analyses of tourism (Cañada & Murray, 2019), critical research into tourism must be extended by exploring the explicit link between tourism production and spreading inequality.

The purpose of this paper is to analyse the connection between inequality and tourism in order to contribute to debate on inequality and the contradictions of tourism as a source of capital accumulation. It starts by exploring the concept of inequality and the different approaches that have been taken to its analysis. A brief

M. A. Martínez-Caldentey (✉) · I. Murray
Department of Geography, University of the Balearic Islands, Palma, Spain
e-mail: ma.martinez@uib.es; ivan.murray@uib.es

A. Blanco-Romero, M. Blázquez-Salom (eds.), *Spanish Tourism Geographies*, Geographies of Tourism and Global Change,
https://doi.org/10.1007/978-3-031-39780-6_21

review is then made of the way in which the issue of inequality and tourism has been tackled. This is followed by a presentation of two analyses of inequality and tourism in Spain and a study of the effect of the COVID-19 pandemic on tourism. The paper concludes by highlighting some possible areas for further research.

21.2 The Question of Inequality and Its Analysis

Inequality is an issue that can be linked to the theme of social justice. Like inequality, social justice can be approached from different theoretical perspectives. Broadly speaking, a conservative approach can be identified, based on the ideal of the *homo economicus* and the maintenance of the *status quo,* where current social relations are viewed as being fair in the sense that everyone has what they deserve (Hayek, 1973). In contrast with this, from a liberal standpoint, flaws in the social order are acknowledged to exist and so mechanisms are proposed in order to improve society, without altering the systems on which this social order is based (Rawls, 1971). Lastly, from a more radical perspective, injustice is seen as being a symptom of a capitalist, hierarchical, patriarchal system, and so social justice can be achieved by subverting the system (Fraser, 2008). According to Nancy Fraser (2008), justice transcends access to resources and their distribution, since it also includes the recognition of differences (gender, ethnic, religious etc.) and political representation. Hence, it surpasses Westphalian political frameworks in today's globalized world, given the inability of policies based on the principles of national sovereignty and citizenship (strictly construed within the context of nation-states) to offer a response to current social challenges, such as migrant or refugee problems. Nonetheless, with the post-structuralist turn that social science has taken, priority has come to be given to recognition of different aspects, leaving aside more material aspects of social justice that came under the umbrella of political economics. That is why Fraser (2008) advocates a dialectic approach to these three dimensions of justice: distribution, recognition and representation.

Inequality as an academic and political issue remained on the fringes of major debate until relatively recently. The 2008 financial crisis drew attention to a whole series of social conflicts masked by neoliberal globalization, which had facilitated the expansion of the so-called middle classes (Rodríguez, 2022). The contradictions of uneven geographical development came to a head with the crisis, and issues that had been marginalized in academic, political and social debate gained in importance (Harvey, 2014). One such issue was inequality, which Piketty (2014), in particular, brought to the fore.

Analyses of inequality differ depending on the school of thought on which they are based. For the dominant school of thought in economic science, inequality is viewed as being outside its scope of analysis. Hence, in accordance with what is known as positive economics, distributive aspects have been regarded as a political issue not to be tackled. In contrast with this, political economics does take them into account. From a neoclassical-conservative perspective, inequality is established as being intrinsic to humankind. It is also regarded as a driving force behind economic growth and the profits of those who receive the most trickle down to the rest of society, helping the latter to prosper (Galbraith, 2016). A second approach is the liberal-Keynesian one, according to which, generally speaking, inequality is viewed as a market flaw due to the trend toward monopolization, the asymmetry of information, the growing importance of rentier logic or the existence of externalities. With this approach, inequalities are resolved by making a series of adjustments (Stilwell, 2019). Lastly, from a radical standpoint, attention is mainly focused on analysing the structural characteristics of capital and class and power relations (Stilwell, 2019). According to Fraser and Jaeggi (2018), radical approaches should focus on more than just the effects of distributive inequalities, treating the structural mechanisms and institutional procedures that generate them as issues to be questioned and explored.

This is a general outline of the three main approaches, but a myriad of different ones can be found that might be classified as heterodox, including work by Thomas Piketty (2014, 2019), James K. Galbraith (2016), Richard Wilkinson and Kate Pickett (2010). While the first three authors mainly focus on inequality in terms of the distribution of income and wealth, Wilkinson and Pickett (2010) extend the framework for the analysis of inequality to include multiple dimensions, such as healthcare and education. Another key author is Amartya Sen (1992). He takes a humanist stance in the development of the capability approach, which underscores the importance of freedom of choice, social and individual heterogeneity, and the multidimensional nature of wellbeing. It must also be noted that many authors who deal with the issue of inequality have focused on analysing the impoverished popular classes, while only a few have turned the spotlight on the concentration of wealth and the wealthy (Dorling, 2019).

The incorporation of multidimensionality in inequality analyses was prompted by the contributions of feminist economics and also by the fight for civil rights. These inequalities are manifested in capitalist market economies and also in other economic areas where capital is not a direct vehicle (e.g. the care sector) (Pérez Orozco, 2014). In the field of political ecology, the framework for analysis has also been extended through studies of the uneven distribution of environmental benefits and damages, with environmental justice movements (Harvey, 1996) and the ecology of the poor deserving a special mention (Martínez-Alier, 2005).

In answer to the question why has there been this recent boom in studies of inequality, we find that, in many of them, inequality is highlighted as being a hindrance to economic growth, with four decades of neoliberal policies leading to an increase in inequality (Milanovic, 2016). According to these studies, if we are to move back onto the growth path, these inequalities must be reduced. These contributions bring to mind the theory formulated by Simon Kuznets (1955), where the relationship between economic growth and income takes an inverse U shape. Thus, in the early stages of economic growth, there is an increase in income inequality, but as the said growth continues, these inequalities smooth out. Spreading inequality has also been related to the advance of populism (Rodríguez-Pose, 2018).

In contrast with the above analyses, in the field of radical political economics, inequalities are seen to be a systemic condition of the capitalist social order and not a flaw in it. In fact, capital accumulation is founded on a contradiction between capital and labour in cooperation between both in the obtainment of income and on a clash between capital and labour in the distribution of this income. Paradoxically, capitalism would collapse if inequalities were eliminated, and yet if inequalities become too widespread, this also jeopardizes the reproduction of capital (Harvey, 2014).

With statistical advances and the availability of series of data for long periods, empirical headway has been made in studies of inequality. Most of the said studies focus on nation-states and they are based on an average statistical subject with no social attributes. Nonetheless, in studies of inequality and (in)justice, space matters (Harvey, 2019 [1973]; Soja, 2010). In the field of radical geography in particular, studies of uneven geographical development have been tackled on multiple scales. The geographies of capital produce spaces of accumulation, on the one hand, and spaces of dispossession, on the other. Hence, capital's uneven dynamics scar areas and, on these scars, we can analyse the complex fabric of inequality (Harvey, 2018 [1982]; Smith, 2010 [1984]). In a European context, within the framework of the 2008 crisis, different approaches to analyses of spatial inequalities at a regional level can be highlighted: first, studies of regional inequalities from a liberal-Keynesian perspective through quantitative methodologies (Rodríguez-Pose, 2018); second, critical analyses from the perspective of radical geography and the theory of uneven geographical development (Hadjimichalis, 2018); and, lastly, a combination of the theory of uneven geographical development and quantitative-spatial analyses (Martínez-Caldentey & Murray, 2019). Similarly, in the field of critical urban studies, an abundance of analyses of processes of segregation, dispossession and gentrification can be found (Darden & Wyly, 2010; Vives-Miró et al., 2018).

21.3 Approaches to Analyses of the Tourism-Inequality Nexus

Bodies like the World Tourism Organization have championed tourism as a mechanism in overcoming under-development and poverty in the Global South. Doubt was cast on this premise in the 1970s, with publications like *Tourism: Passport to Development?* (de Kadt, 1991 [1979]). Similarly, Evans and Timberlake (1980) found empirical evidence of the fact that multinational penetration and a shift toward the service sector in outlying countries was accompanied by an increase in inequality.

Critical studies of the political economics of tourism entered into a decline from the 1980s. This coincided with the academic institutionalization of tourism studies, neoliberal restructuring processes, and a predominant post-structuralist turn in social science. However, during the last decade, a new critical trend has emerged, partly due to increasing conflicts caused by growing touristification following the 2008 crisis (Cañada & Murray, 2019).

Although the issue of tourism and inequality has not been explicitly explored in any great profusion, the emergence of new forms or new types of tourism can be seen as a response to inequality. By way of an example, the following types of tourism can be cited: fair tourism, linked to the fair trade movement (Cleverdon & Kalisch, 2000); responsible tourism as a new form of travel ethics, characterized by a series of voluntary good practices (Rivera & Rodríguez, 2012); pro-poor tourism as a strategy for combatting poverty (Goodwin, 2009); social tourism, which has evolved from a model aimed at bringing tourism to the most disadvantaged to a broad concept based on wellbeing and the fight to overcome inequalities (Schenkel, 2019); inclusive tourism, which seeks to incorporate excluded or marginalized groups (Scheyvens & Biddulph, 2018); and accessible tourism, directed at people with different physical and mental needs (Buhalis & Darcy, 2010). Leaving aside the limitations and contradictions of all these types of tourism, their emergence highlights the existence of a framework of inequality around which tourism is deployed, in addition to showing that existing tourism really helps to reproduce inequality's systemic dynamics. The widespread idea that tourism helps to smooth out inequalities should therefore be questioned and explored.

Generally speaking, two approaches to exploring the link between tourism and inequality can be highlighted, based on almost opposing theoretical frameworks: on the one hand, the orthodox approach associated with neoclassical theories and, on the other, the heterodox or critical one, tied in with political economics. Largely speaking, the issue of inequality has not featured very prominently in tourism research, as demonstrated by the scarcity of studies on the subject, although the number of publications on the topic has risen since the book by Cole and Morgan (2010) came out. This scarcity is mainly due to the presumption that tourism spreads its tentacles in a beneficial way, with profits that are reaped by society as a whole and a multiplier effect on other sectors (Gascón & Cañada, 2017).

Among those studies that do explore the link between tourism and inequality, a wide variety focus on the demand. Bearing in mind that just 5% of the world population travels by plane and that most tourist travel is based on this form of transport, there is clearly still plenty of room for tourism consumption to grow. This is the idea behind the WTO initiative to uphold tourism as a human right (Gascón, 2019). Other initiatives also seek to open up access to tourism, such as accessible tourism or inclusive tourism.

From an orthodox point of view, in a reflection to a certain extent of the hypothesis behind the Kuznets curve, numerous studies defend the notion that the Kuznets curve serves to narrow the socioeconomic gap and to reduce social inequality. Raza and Shah (2017) confirm the hypothesis of the Kuznets tourism curve due to the positive relationship between tourism growth and the reduction in income inequality. However, a whole series of research studies reaches differing conclusions. Alam and Paramati (2016) explain how in the poor countries that they analysed tourism was associated with an increase in inequality. Most studies of income inequality and tourism have focused on impoverished countries, with few tackling rich and poor countries together. The study by Chi (2020) fills this gap, with interesting findings: in the case of wealthy countries, tourism has very little impact on inequality as they are more complex economies, whereas, in the case of poor countries, instead of an inverted U-shaped curve, the results define an N-shaped curve (an initial moment of tourism expansion when inequality grows, a second phase when tourism growth is accompanied by a reduction in inequality, a third phase when inequality shoots up again). One of the main criticisms of these studies is that they have been conducted at a nation-state level, overlooking the complex socio-spatial relations in each country. Studies on a smaller scale than a state level are few and far between, mainly due to a lack of data. There are some exceptions, such as the study by Lee (2009) of income inequality in the USA's most touristic counties, where the gap is higher than in other counties. The role that tourism has played in narrowing the disparity in regional income has also been analysed from orthodox perspectives in accordance with the convergence theory. Most of these studies, backed up by statistical data, conclude that tourism plays an important role in regional convergence in terms of income (Li et al., 2016; Lv, 2019; Soukiazis & Proença, 2008).

Critical analyses in the fields of political economics and political ecology, among others, have tackled the issue of inequality and tourism in very different ways from the above ones. By conceptualizing tourism as a specific form of capital accumulation, tourism is understood to reproduce and amplify the inequalities of the capitalist order. These studies have largely focused on analysing conflicts associated with tourism, with these conflicts being understood as the crystallization of processes of exclusion and dispossession derived from tourism. That is, they are distributive conflicts with tourism at their base. When inequality is analysed from these heterodox perspectives, priority is given to understanding the social conditions and structures that lead to these inequalities.

Merely by way of an example, certain aspects can be highlighted that have been tackled from a critical perspective. One aspect is the hypothesis that tourism's success in some parts of the planet is based on spatial inequalities, starting out from the fact that the main source countries for tourism are from the Global North, while leading tourist destinations are located in outlying or semi-outlying areas. In addition, complex corporate and financial transnational networks drain profits away from tourist areas to tourism control centres (Bianchi, 2018). Another big issue that is tackled from this perspective is conflicts associated with the dispossession or appropriation of certain resources, like water, for tourist use (Cañada, 2019). Thirdly, debate on urban conflicts tied in with overtourism reveals the growing urban inequalities that have been caused by the tourism commodification of cities following the 2008 crisis (González-Pérez, 2020). Special attention has been paid to labour-related issues, particularly after the 2008 crisis, when the intensification of tourism was accompanied by worse labour conditions in terms of workloads, wages, precarious employment contracts, outsourced work etc. (Cañada, 2015, 2016, 2018). Tourism-related labour issues have also been analysed from a gender perspective, revealing higher inequality in the case of women (Martínez-Gayo & Martínez Quintana, 2020). Sexual exploitation related to tourism can be interpreted as an extreme expression of the inequalities associated with the latter (Piscitelli, 2019). Lastly, one other field of research takes a critical look at what is known as alternative tourism and its impact on inequality. For instance, ecotourism was observed to have led to rising inequality among local communities (Fletcher, 2012; Ma et al., 2019) while, in the case of rural tourism, it has sparked off a whole series of conflicts due to the depletion of resources as a result of tourism activities, the progressive marginalization and diminishing value placed on rural activities in these areas, and the uneven distribution of tourism revenue (Gascón, 2019).

21.4 Methodology and Sources

In this study, an analysis is conducted to find out whether touristified Spanish municipalities differ from non-tourist ones in terms of their socioeconomic inequalities, while also measuring their resilience during the Covid-19 crisis. To examine the situation prior to the health crisis, particularly for the analysis of the link between tourism and inequality, a dataset was built of quantitative variables for 6284 municipalities in Spain in 2018, the last year when a complete dataset could be built. When municipalities are not featured, this is due to a lack of data. The chosen variables are some of the main causes of inequality, such as the proportion of foreign residents from impoverished countries or vice versa, or whether income is mainly from wages or welfare benefit payments. The level of specialization in tourism was captured through the accommodation capacity (tourism beds) and employment in tourism.

The methodology on which the study is based is a radical quantitative one. It consists of combining quantitative methods with a critical theoretical analysis. This involves posing research questions and searching for explanations using quantitative and spatial techniques, far removed from the positivist formulae associated with the status quo (Wyly, 2011). In this sense, this study explores the link between tourism and inequality by combining a statistical and map-based spatial analysis with the theory of radical geography.

One of the main novelties and strengths of this study is the construction of a dataset at a municipal level so that an analysis can be made of Spain's spatial complexities. Since this research study first began in 2020, other research studies in the field of geography have been developed that work on a municipal scale (Gago-García et al., 2021; Ojeda et al., 2021). Nonetheless, statistical analyses have been conditioned by the information available at a municipal level. Hence, due to a lack of public information about the amount of tourist accommodation at a municipal level in some autonomous communities, it was necessary to compile data from numerous different sources, either directly from regional tourism ministries or even by telephoning town councils. For the inequality analysis, the S80/S20 ratio was estimated. This measures the relationship between the income obtained by the 20% of the population with the highest income (in the fifth quintile) and that of the population with the lowest income (the first quintile). From the monthly data of workers affiliated to the Social Security, first, the mean yearly figure was calculated and then the Sargent-Florence coefficient of localization (CL) to show the specialization in tourism employment at a municipal level. For employment in the tourism sector, CNAE groups (National Classification of Economic Activities) 55 and 79 were considered, not including group 56 (food and beverage services) as this group includes catering services associated with tourism and others that are not. To calculate them, the following formula was applied:

$$CL = (Aij/Ai) / (Aj/At)$$

Aij: Workers affiliated to sector i (Tourism = CNAE 55 + CNAE 79) in municipality j.
Ai: Workers affiliated to sector i (Tourism) in Spain.
Aj: Total no. of affiliated workers in municipality j.
At: Total no. of affiliated workers in the whole of Spain.

Firstly, a multivariate factor analysis was conducted with the compiled data. The aim of this analysis is to reduce the data in order to create homogenous sets of interrelated variables so that factors can be extracted (Bosque & Moreno, 1994; Vilà et al., 2014). This factor analysis led to the obtainment of four factors with a KMO value of 0.713 and an explained variance of 76.48%. Subsequently, a hierarchical cluster analysis was conducted with the extracted factors. The cluster analysis was carried out using Ward's minimum variance method. The resulting dendrogram of the cluster analysis shows big discontinuities among the municipalities under study, indicating an optimum number of 10 clusters. The 10 clusters were then mapped using the ArcMAP programme in order to relate the analysed variables and to examine the issue of inequality and tourism at a municipal level.

Lastly, to analyse the social and employment impact of COVID-19, variations in the total number of workers affiliated to the Social Security were estimated between the year prior to the pandemic (2019), when there was a record number of tourist arrivals in Spain, and the two following years (2020, which coincided with the lockdown, and 2021 when, theoretically, tourism should have returned to normal after the pandemic). An analysis was made of changes in affiliation to the Social Security by the different CNAE groups (55, 56 and 79) during the said years. A decision was taken only to analyse municipalities with over five thousand inhabitants (1205 municipalities), not taking into account rural ones (construed as those with fewer than five thousand inhabitants) in order to avoid certain distortions detected in municipalities with a low or very low population. These results were depicted in map form to detect spatial patterns.

21.5 Inequality and Tourism in Spain Before COVID-19: A Complex Map

What is particularly striking is the lack of studies that analyse the relationship between inequality and tourism in Spain, bearing in mind the country's role as one of the world's leading specialists in tourism (Murray, 2015) and as one of the EU-27 member states with the highest inequality, with this inequality becoming particularly severe after the 2008 crisis (Serrano et al., 2016). The ten groups from the cluster analysis of the municipalities under study are characterized below (Table 21.1 and Fig. 21.1). Table 21.1 shows how the clusters were formed, depending on the weight of each of the variables. The weight of the clusters in Table 21.1 is based on a qualitative scale, where a value of 1 represents the highest incidence of the variable and 10 represents its most minor impact. The variable used to rank the clusters was the density of tourist beds per square kilometre as it was considered to be the best way of differentiating between tourism and non-tourism municipalities.

Hence, cluster 1 encompasses the places with the highest number of tourist beds per square kilometre and the highest percentage of the population born in the Global South. At the same time, it is also the cluster with the second highest specialization in tourist employment, and the one with the third highest level of inequality. Some examples of these hyper-touristified municipalities are Salou, Puerto de la Cruz, Benidorm, Ibiza and Fuengirola. The second cluster–the one with the second highest number of tourist beds–has the third highest specialization in tourism according to the number of affiliated workers and it ranks fifth in terms of the level of mean income and inequality. These municipalities are mainly located on the coast of mainland Spain and in the two archipelagos and they include Jávea (Valencia region), Marbella (Andalusia), Calvià (Balearic Islands) and Sant Bartolomé de Tirajana (Canary Islands). The third cluster is made up of the most densely populated municipalities, where numerous provincial capitals can be found (50% of the total), including big cities like Madrid, Barcelona, Seville and Palma, with a high number of tourist beds per square kilometre, but because they are complex urban areas, a high value was not achieved for the coefficient of specialization in tourism employment. These municipalities also have a very high mean income and level of inequality.

The fourth cluster stands out as not following the tourism pattern of the previous groups. Although it contains the municipalities with the highest specialization in tourism employment, this group ranks fourth in the density of its tourist accommodation. These municipalities have a low per capita income and also a low level of inequality, because a low mean income can be equated with a municipality with fewer inequalities. This group does not follow a uniform spatial pattern since these municipalities are scattered throughout Spain. By way of an example, they include Son Servera in northeast Mallorca, which includes the tourist area of Cala Millor; Pájara, the biggest municipality in Fuerteventura (the Canary Islands); and tourist municipalities in the Catalan Pyrenees (such as Naut Aran, with Baqueira Beret ski resort) and the Aragonese Pyrenees (such as Panticosa). The fifth cluster stands out for the medium density of its accommodation and its mid-level specialization in tourism employment, combined, on the one hand, with the lowest mean income and level of inequality and the highest proportion of income from welfare benefit payments. This group also has the highest proportion of people born in the Global North. The majority of the municipalities in this group are in the province of Alicante, such as Llíber, known as "little Europe" (Ortolà, 2021). They are magnets for pensioners from the Global North in search of a retirement home. Many of the foreigners are not resident in Spain, even though they are on the municipal census, and so they do not pay income tax in Spain but other taxes, such as property rates or

Table 21.1 Types of sources for the dataset at a municipal level, 2018

Variable	Source
Density of population.	Own, based on the continuous census by the Spanish Statistics Office: https://www.ine.es/dynt3/inebase/index.htm?padre=6235&capsel=6692
Proportion of people born in the Global North —except for those born in Spain—.	
Proportion of people born in the Global South.	
Mean yearly per capita income in euros.	Own, based on statistics for income tax payers. Classified by municipalities and taxable income bands. Spanish Tax Agency: https://sede.agenciatributaria.gob.es/AEAT/Contenidos_Comunes/La_Agencia_Tributaria/Estadisticas/Publicaciones/sites/irpfmunicipios_ccaa/2018/jrubikf6174bf3e5d92f32b246b97955c874fbf8ee4eb58.html#
S80/S20 ratio. Mean yearly income in euros for the fifth top quintile divided by the mean income for the first quintile.	
Percentage of income from wages in relation to the mean income.	Own, based on the distribution of household income atlas. Distribution by source of income. Spanish Statistics Office: https://www.ine.es/experimental/atlas/exp_atlas_tab.htm#
Percentage of income from welfare benefit payments in relation to the mean income.	
Beds in hotels and similar.	Own, based on data from the statistics offices of Spain's autonomous communities, regional tourism ministries and others.
Beds in rural tourist accommodation.	
Beds in tourist homes.	Measure of the number of tourist dwellings in Spain and their capacity. Spanish Statistics Office: https://www.ine.es/experimental/viv_turistica/exp_viv_turistica_tablas.htm
Density of tourist accommodation (total tourist beds/km^2).	Own, based on previously calculated tourist accommodation data, not including campsites.
Total workers affiliated to the social security by scheme on the last day of the month.	Secretary of State's Office for the Social Security and Pensions: https://www.seg-social.es/wps/portal/wss/internet/EstadisticasPresupuestosEstudios/Estadisticas/EST8/EST167/5b11b695-cf1c-4abe-8a12-ecd2d0d15271/2683/0985c546-dd0e-491b-b5e3-4e5e621c69b8
Number of workers affiliated to the Social Security (mean quarterly number) in group CNAE 55 (accommodation services).	Data from the Secretary of State's Office for Social Security and Pensions, available on request.
Number of workers affiliated to the Social Security (mean quarterly number) in group CNAE 79 (activities by travel agencies, tour operators, booking services and activities relating to the latter).	

Sources: Own based on the sources described in the table
CNAE is National Classification of Economic Activities

non-resident income tax. It should also be pointed out that a high degree of tax evasion occurs among the foreign population, since they are often recorded as non-residents when, in actual fact, they live in Spain. This probably explains why municipalities with a high proportion of foreigners from the Global North have such a low mean income (Ródenas & Martí, 2019).

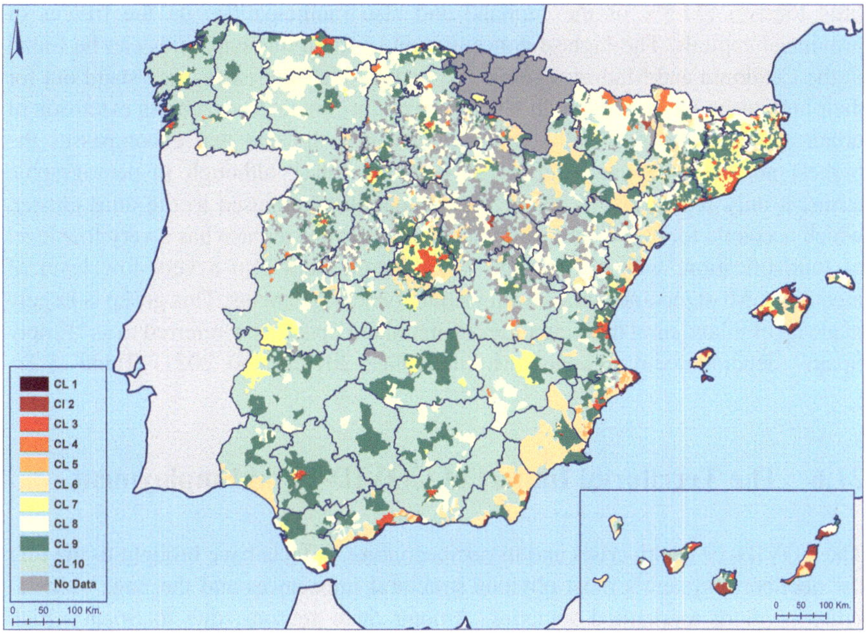

Fig. 21.1 Cluster map of Spanish municipalities in 2018. (Source: Own elaboration)

The sixth cluster has medium values for all the variables, even though these municipalities do not specialize very highly in tourism and they mainly lie on the fringes of mainland coastal Spain and in the Canary and Balearic Islands. The seventh cluster is made up of municipalities with a low density of tourist beds and the highest mean income and level of inequality. Most of them are concentrated around big cities, with 43.8% of the municipalities in the province of Barcelona forming part of this cluster and 21.3% of the ones in the Madrid region. In this sense, it can be confirmed that although big urban agglomerations are driving forces behind the accumulation of capital, the level of inequality rises as their size grows (Castells-Quintana, 2020). Another aspect to highlight is the fact that 55% of the municipalities in this group are in Catalonia, one of the regions with the highest level of inequality in Spain–one of the reasons why there has been political conflict over the independence of Catalonia (García-Agustín, 2021).

The eighth cluster stands out for its high specialization in tourism, although it mainly includes municipalities with a lower population density and a low density of tourist beds since they are large municipalities. However, this group also has a very low mean income and level of inequality and a very high proportion of income from welfare benefit payments. A large part of these municipalities are in northern mainland areas, particularly the Pyrenees and the Cantabria coast where there is a high proportion of rural tourism (Baraja-Rodríguez et al., 2020) (Fig. 21.1).

Among the municipalities with very little tourism, two groups can be distinguished. The first (cluster nine) is largely made up of provincial capitals with very little tourism (37.5% of the capitals) and also municipalities on the fringes of provincial capitals. The highest concentration of these municipalities can be found in the Catalonia and Madrid regions. As for their characteristics, they stand out for their high mean income and high level of inequality since they act as an extension of urban agglomerations. Finally, the last cluster is the one that encompasses the highest number of municipalities (55% of the total), although in demographic terms, it only represents 16.2% of the population as opposed to the third cluster, which accounts for 40% of the population. The last cluster also has a very low level of touristification, with a very low mean income and also a very low level of inequality. Most income comes from welfare benefit payments. This group is largely made up of inland rural municipalities on mainland Spain, now referred to as "Empty Spain" (depopulated rural areas) (Molina Ibáñez, 2019; Taibo, 2021) (Table 21.2).

21.6 The Territorial Impact of COVID-19 on Employment

The COVID-19 health crisis and its socioeconomic effects have brought to the fore the neoliberal system's most obvious structural imbalances and the huge vulnerability of more touristified societies. Among other reasons, due to Spain's high exposure to international tourism, it is one of the OCDE countries that has suffered the most from the pandemic (Cañada & Murray, 2021; Ayala et al., 2022), although the crisis has had an uneven impact geographically, with the worst affected areas being the ones most dependent on international tourism (Méndez, 2022). In Spain's archipelagos, the COVID-19 crisis brought their economies almost to a standstill, causing the most serious social emergency (Bauzà, 2020; Jiménez, 2020). Despite the sharp drop in the number of workers affiliated to the Social Security, there was not a huge rise in unemployment thanks to the cushion effect of furloughs, with tourism activities being the most hard hit (Ayala et al., 2022).

As already mentioned in the methodology, the analysis of the impact of the pandemic on employment was conducted for non-rural municipalities (with over five thousand people) in order to avoid outliers due to problems of over-representation in small municipalities. From the analysis of variance (ANOVA) between the difference in the total number of affiliated workers and affiliated workers in sectors related to tourism (employment activity codes CNAE 55, 56 and 79), results with a probability of 6.53^{-162} were obtained for 2019–2020 and 2.58^{-116} for 2019–2021, below the 0.05 significance level. These results indicate that the drop-in tourism employment (an average fall of 15.75% between 2019 and 2020, and 11% between 2019 and 2021) was statistically higher than the drop in general employment (an average fall of 3.3% between 2019 and 2020, and 0.5% between 2019 and 2021).

Employment in coastal municipalities–the most touristic ones–fell sharply between the years under study (Fig. 21.2). The two maps on the left show the general drop in affiliated workers. In the first map (A1), the biggest drops in worker affiliation can be seen to have occurred in 2020, affecting 85.65% of the

Table 21.2 Classification of the analysed clusters, depending on the weight of each of the analysed variables on a qualitative scale from 1–10

	Cluster									
Variable	CL1	CL2	CL3	CL4	CL5	CL6	CL7	CL8	CL9	CL10
Per capital income	3	5	2	7	10	6	1	8	4	9
%Welfare benefit payments	8	9	5	4	1	6	10	2	7	3
%Wages	3	5	2	6	10	7	1	8	4	9
S80/S20	5	6	2	8	10	4	1	9	3	7
Inhabitants/km^2	2	3	1	9	7	6	4	10	5	8
Tourist beds/km^2	1	2	3	4	5	6	7	8	9	10
%Global North	3	2	10	6	1	4	9	7	8	5
%Global South	1	2	3	9	6	4	5	10	7	8
CL Sargent-Florence	2	3	10	1	5	6	7	4	8	9

Characterization of clusters

CL1	Very high density of tourist accommodation, high proportion of residents from the Global South, high level of income and medium level of inequality.
CL2	Very high density of tourist accommodation, high proportion of foreign residents, medium levels of income and inequality.
CL3	Very high population density, high density of tourist accommodation, very high levels of income and inequality.
CL4	Very high specialization in tourism employment, high density of tourist accommodation, low level of income and very low level of inequality.
CL5	Medium density of tourist accommodation, very low levels of income and inequality, very high proportion of welfare benefit payments, very high foreign population from the Global North.
CL6	Medium density of tourist accommodation, high proportion of foreign residents, medium levels of income and inequality.
CL7	Very high levels of income and inequality, medium-low touristification.
CL8	High specialization in tourism employment, low density of tourist accommodation, low population density, very low levels of income and inequality.
CL9	Very little touristification, medium population density, high levels of income and inequality.
CL10	Very low touristification, very low population density, very low levels of income and inequality, high proportion of welfare benefit payments.

Source: Own elaboration

municipalities. The coastal regions show the biggest drops, although the darker shades are concentrated in the Balearic and Canary Islands. In the second map (A2), the drop-in worker affiliation between 2019 and 2021 is no longer so sharp–it decreases in 59% of the municipalities–, although the worst data is once again in tourist areas like Son Servera and Muro (Balearic Islands); Salou and Lloret de Mar (Catalonia); Candelaria (Canary Islands) and Benidorm (Valencia).

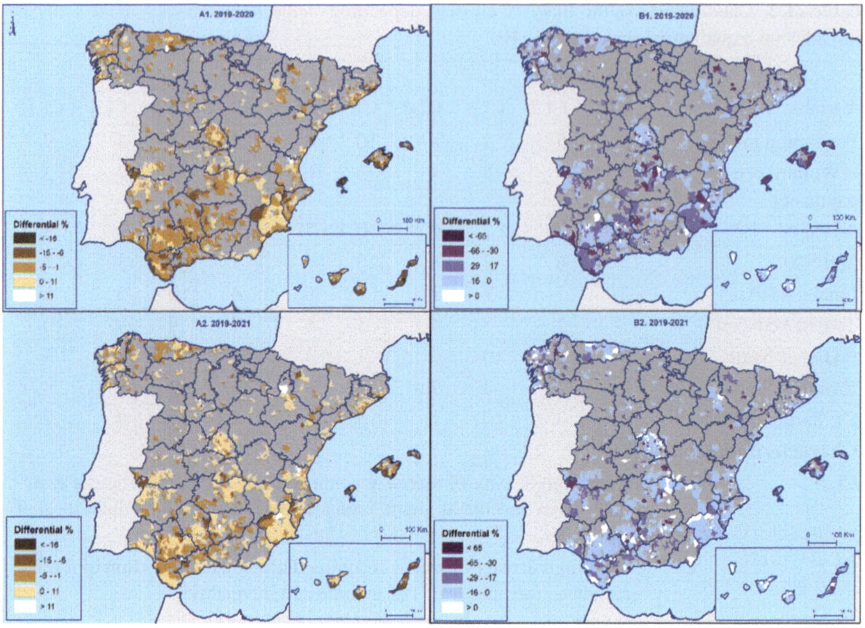

Fig. 21.2 Map of the difference between total worker affiliation (**a**) and worker affiliation in the tourist sector (**b**), 2019/2020–2019/2021 for municipalities with over 5000 inhabitants in Spain. (Source: Own elaboration)

The two maps on the right of Fig. 21.2 show the differences in worker affiliation for the tourist sector for the same time periods. The drop was much greater in the tourist sector, highlighting the extreme social vulnerability associated with tourist areas. The spatial patterns of the drop-in tourism employment overlap the pattern for the drop in total employment. As was the case with total employment, compared to 2019, the impact in 2020 was higher than that of 2021. The most heavily affected regions were the southern and eastern mainland ones close to the coast and the Balearic archipelago; that is, areas associated with high-season summer international tourism. In contrast, the Canary Islands were less strongly affected in 2020 because the high season was concentrated in the months of September through to March. The following year, however, with successive waves of the pandemic, employment in the Canaries was also hit, albeit less than in the Balearic Islands, where international tourism is far more seasonal. In short, from a social and employment perspective, the COVID-19 crisis has had more of an impact in tourist areas, particularly those associated with international tourism. If we consider that, in those areas, most of the population's income is from wages and that there is a high degree of inequality, the pandemic worsened the existing income inequality. In this type of employment crisis, an authentic social disaster might be expected. Nonetheless, the first studies on inequality and COVID-19 in Spain show that social protection policies, including furloughs, seem to have cushioned growing inequality (Aspachs et al., 2021).

21.7 Conclusions

In first place, this study attempts to focus academic debate on the link between inequality and tourism. In the field of tourism studies, little attention has been paid to the issue of inequality, given the prevailing assumption that tourism is a "passport to development". In second place, an analysis was made of the issue of inequality and tourism in Spain. One of the main novelties of the study is its analysis at a municipal level, combined with a radical quantitative methodology. Since Spain joined the EU and its industrial structures were dismantled as the tourism and real estate sectors were upheld as the road to capital accumulation (Murray, 2015), the commodification of tourism has spread throughout the whole of Spain. In this way, 85% of the municipalities have some kind of tourist accommodation and, in 2.7%, there are more tourist beds than inhabitants.

Spain is an example of tensions between inequality and tourism, since it heads the European ranks in both, although, geographically, the two phenomena are unevenly distributed. The country stands out for its huge territorial complexity, with some areas conspicuous for their high concentration of capital, high population density and high number of tourists and others with a low population density and low level of income. The highest level of income corresponds to big urban agglomerations and their functional areas, although this income is not evenly distributed. Since the 2008 crisis, these cities have become more touristified, while austerity policies have accentuated the level of inequality. The most touristic municipalities in coastal areas also have a high level of income, combined with a high degree of inequality. It is important to note that although Mediterranean municipalities with a high proportion of pensioners from the Global North seem to have very little inequality, this is a statistical misnomer because most of these pensioners do not pay tax in Spain. Lastly, in most depopulated rural areas of Spain, low incomes predominate, with most income coming from welfare benefit payments. It is also in these places that the lowest level of socioeconomic inequality was detected.

The outbreak of the COVID-19 pandemic halted the global circuit of capital accumulation and, after successive waves of the pandemic, it still failed to become fully reactivated. From an employment perspective, the pandemic had the highest impact on Spain's coastal areas and its archipelagos. This study highlights the high vulnerability of extreme dependence on tourism (Méndez, 2022), with far higher repercussions in places dependent on international tourism, although other municipalities in inland areas have even found themselves overwhelmed by the emergence of proximity tourism (Cañada & Izcara, 2021).

Lastly, to a certain extent, the social impact of the pandemic was cushioned by the deployment of a series of measures in the field of social protection. It must also be noted that those areas with less inequality in Spain and a lower level of income are the ones with a higher proportion of income from welfare benefit payments. In this sense, it can be claimed that public policies do impact on inequality. Hence, a universal living wage could play a major transformative role, particularly in coastal tourist areas where the climate crisis will have a very big effect.

Acknowledgements This research was supported by the Grant (RTI2018-094844-B-C31) funded by MCIN/AEI/ 10.13039/501100011033 and by "ERDF A way of making Europe".

References

Alam, M. S., & Paramati, S. R. (2016). The impact of tourism on income inequality in developing economies: Does Kuznets curve hypothesis exist? *Annals of Tourism Research, 61*, 111–126. https://doi.org/10.1016/j.annals.2016.09.008

Aspachs, O., Durante, R., Graziano, A., Mestres, J., Reynal-Querol, M., & Montalvo, J. G. (2021). Tracking the impact of COVID-19 on economic inequality at high frequency. *PLoS ONE, 16*(3), e0249121. https://doi.org/10.1371/journal.pone.0249121

Ayala, L., Laparra, M., & Rodríguez, G. (Eds.). (2022). *Evolución de la Cohesión Social y Consecuencas de la COVID-19 en España*. FOESSA.

Baraja-Rodríguez, E., Alario-Trigueros, M., & Molinero-Hernando, F. (2020). Turismo, paisaje y patrimonio en los territorios de la vid y el vino del Duero (España). In Serrano & Asins (Eds.), *Desafíos y oportunidades de un mundo en transición. Una interpretación desde la geografía* (pp. 429–439). Universitat de València.

Bauzà, J. (2020). La emergencia social alcanza a Mallorca con miles de familias sin lo mínimo para subsistir. *Diario de Mallorca*, 8/06/2020 https://www.diariodemallorca.es/portada/2020/05/17/mallorca-entra-fase-emergencia-social/1509728.html?utm_source=rss

Becker, E. (2013). *Overbooked. The exploding business of travel and tourism*. Simon & Schuster.

Bianchi, R. (2018). The political economy of tourism development: A critical review. *Annals of Tourism Research, 70*, 88–102. https://doi.org/10.1016/j.annals.2017.08.005

Bosque, J., & Moreno, A. (1994). *Prácticas de análisis exploratorio y multivariante de datos*. Oikos-tau.

Buhalis, D., & Darcy, S. (Eds.). (2010). *Accessible tourism: Concepts and issues*. Channel View Publications, 305 pp.

Cañada, E. (2015). *Las que limpian los hoteles. Historias ocultas de precariedad laboral*. Icaria.

Cañada, E. (2016). *Externalización del trabajo en hoteles. Impactos en los departamentos de pisos*. Alba Sud.

Cañada, E. (2018). Too precarious to be inclusive? Hotel maid employment in Spain. *Tourism Geographies, 20*(4), 653–674. https://doi.org/10.1080/14616688.2018.1437765

Cañada, E. (2019). Conflictos por el agua en Guanacaste, Costa Rica: Respuestas al desarrollo turístico. *Anuario de Estudios Centroamericanos, 45*, 1–23. https://doi.org/10.15517/AECA.V45I0.37666

Cañada, E., & Izcara, C. (2021). *Turismos de proximidad, un plural en disputa*. Icaria.

Cañada, E., & Murray, I. (2019). *Turistificación global. Perspectivas críticas en turismo*. Icaria.

Cañada, E., & Murray, I. (2021). *#TourismPostCOVID19. Lockdown touristification*. Alba Sud Editorial. https://www.albasud.org/detallespublicacion.php?id=103&lan=en

Castells-Quintana, D. (2020). Inequality and city size: An analysis for OECD functional urban areas. *Papers in Regional Science, 99*(4), 1045–1064. https://doi.org/10.1111/pirs.12520

Chi, J. (2020). Revisiting the tourism-inequality nexus: Evidence from a panel of developed and developing economies. *Current Issues in Tourism*. https://doi.org/10.1080/13683500.2020.1743243

Cleverdon, R., & Kalisch, A. (2000). Fair trade in tourism. *Tourism Research, 2*, 171–187.

Cole, S., & Morgan, N. (2010). *Tourism and inequality. Problems and prospects*. CABI.

Darden, J. T., & Wyly, E. (2010). Mapping the racial/ethnic topography of subprime inequality in urban America. *Urban Geography, 31*, 425–433. https://doi.org/10.2747/0272-3638.31.4.425

De Kadt, E. (1991 [1979]). *Tourism: Passport to development? – Perspectives on the social and cultural effects of tourism in developing countries*. Oxford University Press.

Dorling, D. (2019). *Inequality and the 1%*. Verso.

Evans, P. B., & Timberlake, M. (1980). Dependence, inequality, and the growth of the tertiary: A comparative analysis of less developed countries. *American Sociological Review, 45*(4), 531–552. https://doi.org/10.2307/2095006

Fletcher, R. (2011). Sustaining tourism, sustaining capitalism? The tourism industry's role in global capitalist expansion. *Tourism Geographies, 13*(3), 443–461. https://doi.org/10.1080/14616688.2011.570372

Fletcher, R. (2012). Using the master's tools? Neoliberal conservation and the evasion of inequality. *Development and Change, 43*(1), 295–317. https://doi.org/10.1111/j.1467-7660.2011.01751.x

Fraser, N. (2008). *Scales of justice*. Polity Press.

Fraser, N., & Jaeggi, R. (2018). *Capitalism. A conversation in critical theory*. Polity Press.

Gago-García, C., González-Relaño, R., Serrano, M., & Babinger, F. (2021). Impacto de la crisis de la COVID-19 en el empleo del sector turístico en España: Perspectivas territoriales y de género. *Boletín De La Asociación De Geógrafos Españoles*, (91). https://doi.org/10.21138/bage.3162

Galbraith, J. K. (2016). *Inequality. What everyone needs to know*. Oxford University Press.

García-Agustín, O. (2021). *Catalan Independence and the crisis of sovereignty*. Springer.

Gascón, J. (2019). Tourism as a right: A "frivolous claim" against degrowth? *Journal of Sustainable Tourism, 27*(12), 1825–1838. https://doi.org/10.1080/09669582.2019.1666858

Gascón, J., & Cañada. (2017). El mundo es finito, también para el turismo. *Del multiplicador turístico al conflicto redistributivo. Oikonomics, 7*, 28–34. https://doi.org/10.7238/o.n7.1705

González-Pérez, J. M. (2020). The dispute over tourist cities. Tourism gentrification in the historic Centre of Palma (Majorca, Spain). *Tourism Geographies, 22*(1), 171–191. https://doi.org/10.1080/14616688.2019.1586986

Goodwin, H. (2009). Reflections on 10 years of pro-poor tourism. *Journal of Policy Research in Tourism, Leisure and Events, 1*(1), 90–94. https://doi.org/10.1080/19407960802703565

Hadjimichalis, C. (2018). *Crisis spaces: Structures, struggles and solidarity in southern Europe*. Routledge.

Harvey, D. (1996). *Justice, nature and the geography of difference*. Blackwell.

Harvey, D. (2014). *Diecisiete contradicciones del capital y el fin del capitalismo*. Traficantes de Sueños.

Harvey, D. (2018 [1982]). *The limits to capital*. Verso.

Harvey, D. (2019 [2006]). *Spaces of global capitalism*. Verso.

Hayek, F. (1973). *Law, legislation and liberty*. Routledge & Kegan Paul.

Jiménez, J. (2020). Casi 40.000 hogares canarios no tienen ingresos: el cese de la economía sumergida acrecienta el drama y dispara la ayuda social. Eldiario.es. https://www.eldiario.es/canariasahora/sociedad/canarios-ingresos-economia-sumergida-acrecienta-renta-minima-Canarias_0_1017348895.html

Kuznets, S. (1955). Economic growth and income inequality. *American Economic Review, 45*(1), 1–28.

Lee, S. (2009). Income inequality in tourism services-dependent counties. *Current Issues in Tourism, 12*(1), 33–45. https://doi.org/10.1080/13683500802248001

Li, H., Chen, J. L., Li, G., & Goh, C. (2016). Tourism and regional income inequality: Evidence from China. *Annals of Tourism Research, 58*, 81–99. https://doi.org/10.1016/j.annals.2016.02.001

Lv, Z. (2019). Deepening or lessening? The effects of tourism on regional inequality. *Tourism Management, 72*, 23–26. https://doi.org/10.1016/j.tourman.2018.11.009

Ma, B., Cai, Z., Zheng, J., & Wen, Y. (2019). Conservation, ecotourism, poverty, and income inequality – A case study of nature reserves in Qinling, China. *World Development, 115*, 236–244. https://doi.org/10.1016/j.worlddev.2018.11.017
Martínez-Alier, J. (2005). *El ecologismo de los pobres*. Icaria.
Martínez-Caldentey, M. A., & Murray, I. (2019). Crisis y desarrollo geográfico desigual en la Unión Europea (2009–2015). *Revista de Geografía Norte Grande, 72*, 163–184. https://doi.org/10.4067/S0718-34022019000100163
Martínez-Gayo, G., & Martínez Quintana, V. (2020). Precariedad laboral en el turismo español bajo la perspectiva de género. *Pasos, 18*(4), 649–665.
Méndez, R. (2022). Turismo, pandemia y nuevos contrastes territoriales en España. Ikara. *Revista de Geografías Iberoamericanas*, (1). https://doi.org/10.18239/Ikara.3006
Milano, C., Cheer, J. M., & Novelli, M. (2019a). *Overtourism: Excesses, discontents and measures in travel and tourism*. CABI.
Milano, C., Novelli, M., & Cheer, J. M. (2019b). Overtourism and degrowth: A social movements perspective. *Journal of Sustainable Tourism, 27*(12), 1857–1875. https://doi.org/10.1080/09669582.2019.1650054
Milanovic, B. (2016). *Global inequality. A new approach for the age of globalization*. The Belknap Press.
Molina Ibáñez, M. (2019). Dimensión territorial de la despoblación. Aproximación al papel de las políticas públicas en un entorno europeo. En J. Farinós, J. Ojeda, & J. Trillo (Eds.), *España: Geografías para un Estado posmoderno* (pp. 153–170). Geocrítica.
Murray, I. (2015). *Capitalismo y Turismo en España. Del "Milagro Económico" a la "Gran Crisis"*. Alba Sud Editorial. Recuperado de www.albasud.org/publ/docs/68.ca.pdf
Ojeda, J., Paneque, P., Sánchez, E., & Pérez, J. P. (2021). Geografía de la renta de los hogares en España a nivel municipal: nuevos datos y nuevas posibilidades de geovisualización, exploración y análisis espacial en entornos cloud. *Investigaciones Geográficas, 76*, 09–30. https://doi.org/10.14198/INGEO.18993
Ortolà, B. (2021). Llíber, la pequeña Europa. Las Provincias 19/04/2021. https://www.lasprovincias.es/marina/lliber-pequena-europa20210419190116-nt.html
Pérez Orozco, A. (2014). *Subversión feminista de la economía. Aportes para un debate sobre el conflicto capital-vida*. Traficantes de Sueños.
Piketty, T. (2014). *El capital en el siglo XXI*. Fondo de Cultura Económica.
Piketty, T. (2019). *Capital e ideología*. Ediciones Deusto.
Piscitelli, A. (2019). ¿Turismo sexual? balance crítico de la producción del concepto. En E. Cañada y I. Murray (Eds.), *Turistificación global. Perspectivas críticas en turismo*. Icaria.
Rawls, J. (1971). *A theory of justice*. Belknap.
Raza, S. A., & Shah, N. (2017). Tourism growth and income inequality: Does Kuznets curve hypothesis exist in top tourist arrival countries. *Asia Pacific Journal of Tourism Research, 22*(8), 874–884. https://doi.org/10.1080/10941665.2017.1343742
Rivera, M., & Rodríguez, L. (2012). *Turismo responsable, sostenibilidad y desarrollo local comunitario*. Cátedra Intercultural, Universidad de Córdoba.
Ródenas, C., & Martí, M. (2019). *Informe Ineca: La situación social en la provincia de Alicante*. INECA. https://ineca-alicante.es/wp-content/uploads/2020/06/Estudio-Situaci%C3%B3n-Social-de-la-provincia-de-Alicante.pdf
Rodríguez, E. (2022). *El efecto clase media. Crítica y crisis de la paz social*. Traficantes de Sueños.
Rodríguez-Pose, A. (2018). The revenge of the places that don't matter (and what to do about it). *Cambridge Journal of Regions, Economy and Society, 11*(1), 189–209. https://doi.org/10.1093/cjres/rsx024
Schenkel, E. (2019). Turismo social: principals conceptos y enfoques teóricos. In E. Cañada & I. Murray (Eds.), *Turistificación global. Perspectivas críticas en turismo* (pp. 417–430). Icaria.
Scheyvens, R., & Biddulph, R. (2018). Inclusive tourism development. *Tourism Geographies, 20*(4), 589–609. https://doi.org/10.1080/14616688.2017.1381985

Sen, A. (1992). *Inequality re-examined*. Oxford University Press.
Serrano, P. J. G., Molero-Simarro, R., & Buendia, L. (2016). The impact of the 2008/9 crisis on inequality and poverty in southern Europe: The case of Spain. *The Journal of Australian Political Economy, 78*, 87–114.
Smith, N. (2010 [1984]). *Uneven development. Nature, capital and the production of space*. Verso.
Soja, E. (2010). *Seeking spatial justice*. University of Minnesota Press.
Soukiazis, E., & Proença, S. (2008). Tourism as an alternative source of regional growth in Portugal: A panel data analysis at NUTS II and III levels. *Portuguese Economic Journal, 7*(1), 43–61. https://doi.org/10.1007/s10258-007-0022-0
Stilwell, F. (2019). *The political economy of inequality*. Polity Press.
Taibo, C. (2021). *Iberia Vaciada. Despoblación, decrecimiento, colapso*. Libros de la Catarata.
Vilà, R., Rubio, M. J., Berlanga, V., & Torrado, M. (2014). Cómo aplicar un clúster jerárquico en SPSS. *REIRE. Revista d'Innovació i Recerca en Educació, 7*(1), 113–127.
Vives-Miró, S., Rullan, O., & González, J. M. (2018). *Geografies de la despossessió d'habitatge a través de la crisi. Els desnonaments Marca Palma*. Icaria.
Wilkinson, R., & Pickett, K. (2010). *The Spirit level. Why equality is better for everyone*. Penguin Books.
Wyly, E. (2011). Positively radical. *International Journal of Urban and Regional Research, 35*(5), 889–912. https://doi.org/10.1111/j.1468-2427.2011.01047.x

Maria Antònia Martínez-Caldentey Doctoral student of geography at the University of the Balearic Islands. Her main research is about inequality and tourism inside the economical geographies' studies from the perspective of critical geography and spatial analysis.

Ivan Murray Professor of geography at the University of the Balearic Islands. His research is framed from the perspective of critical geography, linking analysis from political economy, political ecology and ecological economics. His research focuses primarily on the spatial logics of tourism capitalism and the associated conflicts.

Chapter 22
Coastal Tourist Destinations in Spain: Growth, Social Reaction and Answers – Practices for a Post-growth Scenario

Enrique Navarro-Jurado, Yolanda Romero-Padilla, and José María Romero-Martínez

22.1 Introduction

In territories specialised in tourism, as is the case of the Spanish coast, conflicts and challenges are experienced as a result of tourism growth and expansion, which in Spain is associated with the real estate business (Barke, 2007; López Olivares et al., 2005; Naredo, 2010). The result is a model of urban-tourist growth based on economic interests that commodify the territory.

In the Spanish case, the model of the Franco dictatorship – in the 1960s – produced the so-called "years of development", based on home ownership. It then became a habit to re-qualify land by increasing the building capacity outside of the planning process. This led to "wild urbanism", which also facilitated second homes, together with the arrival of tourists to the Mediterranean coast and the islands. From 1978 onwards, in a first stage – with the arrival of democratic town councils after the dictatorship – there was an awareness among progressive professional elites to

E. Navarro-Jurado
Andalusian Institute for Research and Innovation in Tourism of the Universities of Granada, Malaga and Seville, University of Malaga, Málaga, Spain
e-mail: enavarro@uma.es

Y. Romero-Padilla (✉)
Business Organization, Faculty of Commerce and Tourism, Complutense University of Madrid, Madrid, Spain
e-mail: yromerop@ucm.es

J. M. Romero-Martínez
Department of Architectural and Engineering Graphic Expression, Higher Technical School of Architecture, University of Granada, Granada, Spain
e-mail: jmr2@ugr.es

A. Blanco-Romero, M. Blázquez-Salom (eds.), *Spanish Tourism Geographies*, Geographies of Tourism and Global Change,
https://doi.org/10.1007/978-3-031-39780-6_22

regain control over urban development. These elites gained access to local and regional administrations and other prestigious private entities, such as the architects' associations. From these entities, a revision of the ways of doing things began, which became more aware of the public and social condition of land and urban planning. However, with Spain's entry into the European Union and its development funds, from 1985 onwards, the Spanish model is an example of the spread of urban-territorial and tourist pathologies, which impose: (1) a territorial model with attractor spaces – population, capital and resources -, and supply and servitude spaces; (2) a diffuse and generic urban model (Koolhaas, 2004), which requires significant transport infrastructures to function; (3) a single construction model based on junk architecture (Koolhaas, 2008), which is abstracted from the environment, with a cladding independent of the structure and air conditioning (Naredo, 2010). This is the beginning of the neoliberal era. From 1996 onwards, the neoliberal model is fully consolidated, with the government of President Aznar, and the weak barriers imposed by planning are removed. This coincided with a wave of cheap and abundant liquidity flowing into the real estate-tourism sector, which aggravated the aforementioned pathologies. Two important business sectors – consolidated during the Franco regime – such as the real estate-construction and banking sectors, encouraged the spectacular urban-tourist expansion that took place at the time. The speculative real estate-financial growth formed the real estate bubble that developed in Spain between 1997 and 2007 (Naredo, 2010). At present, other global problems such as the aftermath of the economic-financial crisis of 2008, the health crisis of 2020 or the climate and energy crises are being added. The responses to these crises will determine the challenges and opportunities in the coming years.

Against this background, and focusing on the future of coastal tourism territories, there is an ongoing debate on the need for a paradigm shift, favouring alternative ideas, strategies and practices. Sustainable tourism has been one response, but it has been insufficient because it has practised weak sustainability, focused on compensating social and environmental conflicts with economic effects (Thiel Ellul & Navarro Jurado, 2018). Other research builds on concepts around ecological economics (strong sustainability), considers the exhaustion of the growth-based global tourism model (Fletcher, 2011). Some research explores proposals such as tourism degrowth (Andriotis, 2018), or prosperity without growth (Jackson, 2011), while others point to post-capitalist trends with already visible practices (Fletcher et al., 2021). These currents can be encompassed within a post-growth framework of analysis (Schulz & Bailey, 2014; O'Neill et al., 2018).

The aim of this research is to analyse, in Spanish coastal destinations: (1) problems and conflicts caused by urban-tourist growth policies and projects; (2) the social reaction they produce; and (3) the alternative responses that have arisen. The social reactions and responses analysed seek to cover geographical and temporal variety and types of practices. The purpose is to reflect on practices and procedures that may be applicable in a situation of post-growth, or eco-social transition.

22.2 Tourism Growth: Dysfunctionalities of the Growth Model in Tourist Destinations

Tourism as a global activity is linked to the evolution of the capitalist economy, whose worldwide expansion is carried out under the approach of unlimited growth and is based on the idea of modernity and progress. This growth process is characterised, among other elements, by (1) the extension of urban land, which progressively dilutes the boundaries with the rural environment (Harvey, 2013; López & Rodríguez, 2010); (2) the global economy based on processes of flexible and delocalised accumulation of capital; (3) the concentration of activities, people and capital in certain places, especially on the coasts (Antón Clavé & González Reverté, 2008; Romero-Padilla et al., 2016); (4) changes in consumption habits under the inertia of the technological and socio-demographic transformations of recent decades; and (5) the speed of its own functioning, accelerated in all its dimensions. This gives rise to a new reality, where tourism is one of the activities that best represents the global economy.

Since the 1970s there has been a good theoretical basis around tourism growth and its negative impacts (Doxey, 1975; Pizam, 1978; Butler, 1980; O'Reilly, 1986; Britton, 1991). The debate continued in the 1990s with research on sustainability (Saarinen, 2006), and in recent years has continued to be associated with the phenomenon of overtourism (WTO et al., 2018).

Spain, since the 1960s, has been marked by strategies to attract tourists and investment, and it has been impossible to separate its tourist development from urban growth (López Olivares et al., 2005). Spanish coastal areas have developed as urban-tourist growth machines (Navarro-Jurado et al., 2019), being the model that influences the structuring of the territory and its economic, social and cultural activity. The traditional discourse, on which tourism growth is based, has been associated with positive effects such as the creation of employment, the arrival of foreign investment, the increase in infrastructures and recreational facilities. However, this discourse has always ignored the negative impacts in all their dimensions (economic, social and environmental) and critical views (Auken, 2009; Malvárez et al., 2000; Britton, 1991). For example, employment is sustained by low-skilled jobs, low wages, precariousness, high dependence on external factors (Cañada, 2015; Exceltur, 2006). The distribution of wealth is questionable given the figures: tourist areas have some of the lowest household incomes in the country (Benidorm: €9450; Marbella: €8785; Malaga: €9955) (INE, 2020). This contrasts with the waste of public money on unnecessary and inadequate large-scale works and infrastructures (airports, motorways, conference centres, among others) which reached the figure of 200,000 million euros (half of it private initiative) between 1996 and 2016. In the middle years of the boom (2002–06), tourism-real estate waste in Spain reached 20% of GDP (Romero et al., 2018).

These dysfunctions have triggered social reactions, from the most pioneering, linked to environmental movements (Kousis, 2000), to the most recent, which are social reactions to the perception of over-tourism in some coastal cities (Milano et al., 2019).

22.3 Conflicts in Tourist Areas: Social Reactions

The history of social movements in tourist territories is wide and varied. Urban-tourist growth models have recurrently formed part of the geography of conflicts (Antón Clavé & González Reverté, 2008; Vilarino & Orea, 2013; Del Romero Renau, 2018). A common example has been the clash between the interests of agents promoting a particular private tourism-residential project versus environmental conservation interests (Kousis, 2000).

As a common phenomenon worldwide, social movements can act as important catalysts of tourism crises. They are now tending towards more informal and autonomous structures, with a more heterogeneous composition, discourses that assume transversal and global ideas (e.g. equality, welfare state, human rights...), but integrated and focused on more specific and local demands (Soja, 2008; Castells, 2012; Díaz Parra & Candón Mena, 2014; Talego & Hernández-Ramírez, 2017). At the same time, the use of technology and networks has become an indispensable tool for amplifying the message. This makes them both local and global movements: local because they are created in a specific context and space, and global because the use of the internet has enabled them to reach more people and to draw inspiration from the experiences of other movements (Castells, 2012). In their most recent evolution, technological changes and the economic-financial crisis of 2008 have had a decisive influence on the nature of these movements, the case of 15-M in Spain being a representative example.

The first social movements against mass tourism in Spain emerged in the mid-1990s, where local resistance arose (Boissevain, 1996) and reached a milestone with the proposed demolition of the Hotel El Algarrobico in Almería because this 21-storey white building in a protected natural area has become the symbol of the urban disaster of the Spanish coastline and its urban speculation.

Many of these movements are organised as collectives in "Defence of the Territory", and cases can be found throughout the Mediterranean destinations, from Catalonia to Andalusia, via the Balearic Islands. In cities such as Madrid, Valencia, Palma and Barcelona, there are also platforms that demand the right to the city against touristification; this is the case of the "Coordinadora de Asociaciones de Vecinos de Barrios Turísticos" (Coordinating Committee of Neighbourhood Associations of Tourist Neighbourhoods). These movements have paralysed projects and laws along the Spanish coastline, and are even promoting new legislative initiatives, as is the case of the social movements in the Mar Menor in Murcia, or the Grup Balear d'Ornitologia i Defensa de la Naturalesa (GOB), which will be analysed below. These movements are demonstrating in a practical way a way of thinking that rebels against the postulates of unlimited growth.

22.4 Postulates for a Change of Model: The Post-growth Approach

The concept of post-growth addresses an alternative scenario to the growth model, which advocates a socio-economic, ecological, democratic and just transition that guarantees the long-term viability of life on the planet. Schulz and Bailey (2014) review the main debates and contributions that -within the framework of Geography – have been developed on the limits of economic growth, its impacts and alternative currents; reviewing concepts related to ecological economics, such as degrowth (Georgescu-Roegen, 1995; Andriotis, 2018), or prosperity without growth (Jackson, 2011). These postulates invite us to rethink current policies and evaluate future proposals for viable economic development in the face of environmental challenges and human rights guarantees (Schulz & Bailey, 2014).

In the field of tourism, studies on carrying capacity and the establishment of limits in coastal destinations pioneered growth containment strategies (Lindbert et al., 1997). These studies have nurtured subsequent approaches to sustainability in tourism (Saarinen, 2006) and its transformation towards a circular economy based on local consumption and footprint reduction (Hall, 2009). More recently, the reopening of reflection from the degrowth approach (Bourdeau & Berthelot, 2008; Hall, 2009; Andriotis, 2018; Fletcher et al., 2019; Higgins-Desbiolles et al., 2019), opens up a way to consider other ways of thinking about the future of tourist destinations, through de-touristification (reduction of tourist activity), responsible re-touristification (breaking with the dynamics of growth), solidarity and autonomy (Blanco-Romero et al., 2019; Fletcher et al., 2019).

In practice, there is also a set of laws and projects that help containment, either proactively or in a protectionist/defensive manner. In the case of territorial planning, there are examples such as the 2011 Coastal Management Plan of Galicia based on the recognition of coastal territory with a proactive attitude of conservation, study and promotion of its character (it defines coastal territory as that from which the sea can be observed). There are also local experiences of containing urban sprawl and environmental protection. This is the case of the Island Plan of Menorca, the Valencia Metropolitan Territorial Action Plan (PATEVAL) of the Valencian Community, and the Plan for the Intervention of Tourist Areas (PIAT) of Mallorca (Blázquez-Salom et al., 2019).

In the case of specific projects, there are also notable examples such as the legalisation of the El Algarrobico Hotel with various ideas for its demolition (n'UNDO, 2012). And others that are more proactive, such as: (1) the ecological regeneration of the isthmus of O Grove and A Lanzada beach (Pontevedra), a project promoted by the national government, which deconstructs the infrastructure and reuses a light aircraft landing strip as a public car park for private vehicles on the beach, without having to do new work (García et al., 2010); (2) the deconstruction of the Club Med holiday development and restoration of the geological and landscape values of the Tudela-Culip site located in the Cap de Creus Natural Park in Gerona (TVE2, 2010).

According to Meadows' (1999, 2001) interpretation of Thomas Kuhn (1962), the process of paradigm change can be summed up as: having people with clear ideas in the new paradigm, people who put them forward, and people who carry them out. The review of research that proposes new alternative approaches to the unlimited growth of tourist destinations, the plans and projects proposed or executed for deconstruction and containment of urban-tourist expansion, as well as the defensive and proactive social movements, make up the necessary agents and instruments – on three levels: ideas, procedures and actions – so that a paradigm shift such as the socio-ecological transition or post-growth can come to fruition (Romero-Padilla et al., 2020).

22.5 Practices for a Post-growth Scenario on the Spanish Mediterranean Coastline

The Mediterranean coast, together with Madrid, has been the most dynamic area of Spain from a socio-economic point of view for at least 50 years. It covers an extensive coastal belt, from the province of Gerona (on the border with France) to the province of Huelva (on the border with Portugal), including the Balearic archipelago. It accounts for 21% of the country's total surface area and 42–43% of the population and businesses, respectively (Romero-Padilla, 2016).

The Mediterranean tourist destinations are the most important and longest-established in Spain: Costa Brava, Costa del Maresme-Garraf, Costa Dorada, Costa del Azahar, Costa de Valencia, Costa Blanca, Balearic Islands, La Manga (Costa Cálida), Costa de Almería, Costa Tropical, Costa del Sol and Costa de la Luz. These destinations account for 62% of hotel beds, 56% of travellers and 72% of overnight stays in the whole country.

The tourist model has been based on the classic "sun and beach" model, although over time it has become more complex (O'Reilly, 2009). It is a model based on urban expansion, with two trends standing out: the constant commitment to housing as a form of tourist accommodation, and the diversification of the product with the increase in the complementary offer (golf tourism, nautical tourism, congresses and events, etc.). The result of this process is a profound transformation of the territory that has left both positive aspects, such as social multiculturalism, tolerance or its creative capital (Romero-Padilla et al., 2016), and other problems of various kinds, environmental, social and economic, that distance it from the challenge of sustainability.

The demographic and economic attraction and urban and real estate speculation on the coast generate pressures on the coast, an ecologically fragile area which is a support and source of resources for human activity. The beachfront is urbanised with a high proportion of sealed land, and the urbanisation process has spread inland in most provinces (up to 20–30 km inland). The dynamic of beach regeneration has been broken, which means that every year millions of euros have to be invested in their regeneration, especially after storms, and their preparation for the tourist season; river courses have been modified (construction of reservoirs, diversions, canalisations...); aquifers have been overexploited and many wooded and natural areas have disappeared, except for some areas of special interest which have been expressly protected.

In the midst of this predominant growth dynamic, it is possible to find alternative – and reactive – practices. These practices are varied, ranging from those that have their origins in social reactions of opposition to projects that mortgage the territory and propose another possible future, to those that are based on the involvement of administrations on different scales. The following are five representative examples (Fig. 22.1).

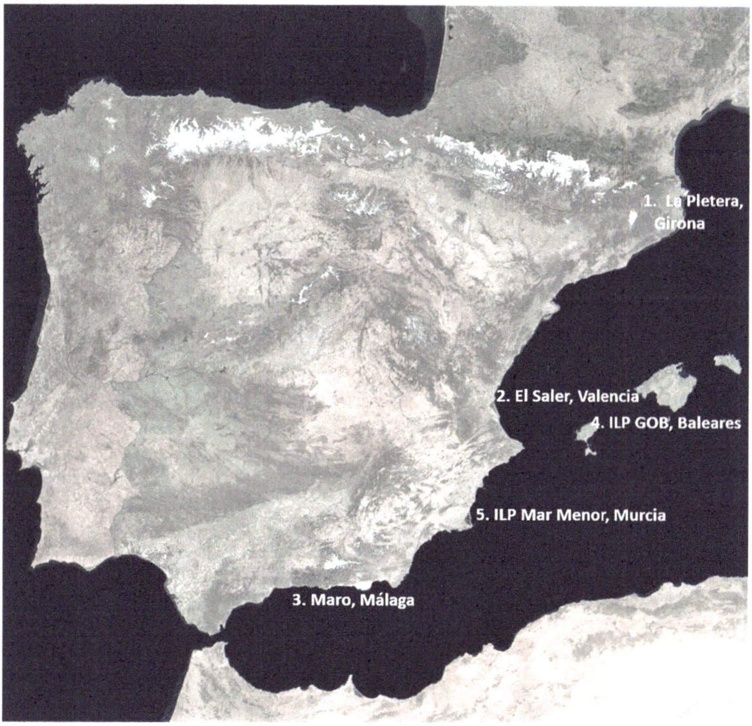

Fig. 22.1 Location of cases analysed on the Spanish coast. (Source: Own elaboration based on National Geographic Institute of Spain)

22.5.1 The Process of Deurbanisation and Recovery of the Marsh and Dune Area of La Pletera in Estartit (1999–2018), on the Costa Brava (Girona)

The Partial Plan for the urbanisation of La Pletera was approved in 1986 and the Urbanisation Project in 1988. Both were part of the revision of the General Urban Development Plan of 1983, which had received citizen demonstrations against it (Pié et al., 2018; Gamero et al., 2019). Subsequently, the marsh was filled and drained with construction debris, raising the height of the land. Environmental groups tried to stop the construction work. In 1989, the developer stopped the works and left the process unfinished with urbanised streets, raised plots, and the pedestrian promenade on the seafront. The stoppage was due to the lack of success of the housing development during the crisis of the nineties (Gamero et al., 2019).

The crisis forced a rethinking of the tourism model in a context in which demands for the environmental protection of the coast were gaining strength in the Empordà area. This change of model was reflected in the revision of the local planning of 1998–2001: La Pletera changed its classification and became undeveloped natural land (Quintana et al., 2018). In 2004, the urbanisation of La Pletera fell within the boundary of the public maritime-terrestrial domain: it was declared public, undevelopable and protected land.

In parallel, from 1999 onwards, Torroella de Montgrí Town Council implemented 3 projects in the framework of the LIFE programme (between 1999 and 2018), with a total investment of 5.3 million euros. The deurbanisation and environmental restoration carried out consisted of (Quintana, 2019): deconstructing the coastal urbanisation, parallel to the dune coastline; recovering the original functionality of the dune ecosystems; and recovering the functionality of the marshland ecosystems in the areas of the deurbanised streets (and their flora and fauna).

This action was selected in 2015 as a pilot destination to implement the European Indicator System for the Sustainable Management of Tourism Destinations and has received various international recognitions, such as the distinction of the European Charter for Sustainable Tourism or the 100 Top Green Destinations (Quintana et al., 2018).

22.5.2 La Dehesa de El Saler (Valencia): From the "El Saler per al poble" Movement to the Restoration of the Dune Area

The Devesa de El Saler is a sandy bar formed by dunes parallel to the coast and a system of water pools known as 'malladas'. It is one of the three ecosystems that make up the current Albufera de Valencia Natural Park, together with the freshwater lake connected to the sea by artificial channels, and the marshland that surrounds it, currently occupied almost entirely by rice fields that act as wetlands during crop flooding (Armengol et al., 2019). This area has several figures of environmental recognition and national and international protection: Natural Park (1986), Ramsar List of Wetlands of International Importance (1989), Site of Community Interest (1992), Special Protection Area for Birds (2009) and Natura 2000 Network.

Before the application of the protection figures, this space was subject to strong anthropic pressures, especially during the last century, which led to the disappearance of the dune ridges and the drying and silting up of the malladas (Armengol et al., 2019): (1) at the beginning of the twentieth century, for sanitary reasons, the malladas were perforated for drainage in order to eradicate the mosquito populations that transmitted the malaria disease; (2) in the 1960s, during the take-off of tourism under the dictatorship, the urbanisation project of real estate-tourist interest in La Devesa was approved, which included residential complexes with flat towers, single-family homes, hotels and recreational areas, sewerage system, car parks, secondary roads and promenade.

The urbanisation project was partially executed, due to a citizens' movement that was gaining strength under the emblem "El Saler per al poble" (University of Valencia and València City Council, 2017). This movement began with minority voices of biologists and environmentalists, which were joined by actions in the media such as the programme by Félix Rodríguez de la Fuente (RTVE, 1970) or the leadership of Consuelo Reyna in the newspaper Las Provincias (Cervera Sánchez, 2016). Along with environmentalist criticisms, other aspects were pointed out, such as the expropriation of communal property, the loss of a green lung of the city, and the economic damage to the municipal coffers. The actions gradually made their way into public opinion, and managed to gain the support of neighbourhood associations, institutions such as the College of Architects, and professionals. In the midst of the dictatorship, more than 15,000 signatures were collected against the development, but it was not until the arrival of the first democratic municipal government that it was definitively paralysed (University of Valencia and València City Council, 2017).

Since the late 1980s, the recovery of the ecological value of the area has been promoted with various projects to restore the coastal dunes and malladas (Armengol et al., 2019): (1) between 1988 and 1998, the local government restored part of the affected area; between 2007 and 2009, two Life Natura projects were developed (LIFE00 NAT/E/007339 and LIFE04 NAT/E/000044). With these projects, infrastructures (roads) were dismantled and part of the coastal dunes and interdune valleys were rehabilitated, sediments were removed from the malladas, some were restored to their original temporary function, with recolonisation of native plant species, and others were restored to their permanent function for use as reservoirs for endemic and endangered species.

22.5.3 Grupo Balear de Ornitología y Defensa de la Naturaleza (GOB): Fifty Years Caring for the Balearic Territory, from Advocacy Campaigns to Legislative Proposal

The Balearic Group for Ornithology and the Defence of Nature (GOB) is a non-profit association founded in 1973 in Palma de Mallorca as a birdwatching organisation. It gradually expanded its activities to include the protection of biodiversity, land use planning and education, playing an important role against the destruction of natural areas promoted by the advance of tourism and speculation. It is currently the main local NGO for nature conservation, declared of Public Utility, with sections in Mallorca, Menorca, Ibiza and Formentera and more than 5000 members (GOB, 2022a).

The GOB has promoted defence campaigns in different parts of the islands, such as Sa Dragonera, S'Albufera, Ses Salines d'Eivissa i Formentera, S'Albufera des Grao, Mondragó, Cabrera, Es Trenc, etc. It has contributed to the approval of the Law of Natural Spaces of 1991, the declaration of seven natural parks, one national park and one Biosphere Reserve, as well as the protection of some twenty enclaves along the Balearic coastline. Its activity has been recognised nationally and internationally. Since the 1980s it has managed La Trapa, a private nature reserve in Mallorca acquired by the GOB and open to the public, where they carry out conservation and restoration work (GOB, 2022a).

Among its latest actions is the campaign "Avui per demà" (Today for tomorrow), which promotes the *Popular Legislative Initiative Law for the Welfare of Present and Future Generations of the Balearic Islands (*ILP-LBGPF-Balearic Islands). This is a proactive action, based on the idea of intergenerational justice. The aim of this proposal is to "*ensure that the actions of the public institutions of the Balearic Islands guarantee the ecological, economic and social well-being of present and future generations*" (ILP-LBGPF-Balearic Islands). The legislative proposal includes the definition of eight principles: sustainable development, precaution, prevention, dignity of life, eco-dependence and interdependence, diversification, common good and non-regression; and establishes as a requirement for the processing of regulations, plans, projects and programmes, the preparation of an impact assessment report on the well-being of present and future generations (Borràs, 2021).

Currently, the legislative initiative has been processed by the Balearic Parliament and is in the signature collection phase. For the elaboration of the campaign and the legal texts, the GOB established a collaborative work with different entities such as La Pera Coop, Àgora Social and the Environmental Legal Clinic of the Rovira i Virgili University (URV), which allowed the participation of postgraduate law students from the URV and the University of Chile (GOB, 2022b).

22.5.4 Platform for the Mar Menor Popular Legislative Initiative

The Mar Menor is the largest saltwater lagoon in the Mediterranean. Its environmental importance and ecological value are recognised by protection figures and regulatory instruments promoted over the last twenty-five years, such as the following:

International: recognition as a RAMSAR Wetland (Wetlands of International Importance, especially as Waterfowl Habitat), ZEPIM (Specially Protected Area of Mediterranean Importance. Barcelona Convention); European: Natura 2000 Network, SPA (Special Area of Conservation for Birds) and SAC (Special Areas of Conservation); National: Regional Park of the Salinas de San Pedro and Protected Landscape *Open spaces and islands of the Mar Menor*; Wildlife Protection Area.

However, this recognition has been insufficient to protect the ecosystem of the Mar Menor as a whole and its basin. The urban-tourist pressure that surrounds the lagoon, and the pressure of intensive agricultural and livestock production that surrounds it, have been deteriorating its ecosystem. As a result, there have been serious episodes of mortality of its fauna due to anoxia and eutrophication, particularly serious in 2019 and 2021. However, these episodes have served to give visibility to the serious problem and to raise public awareness.

During the academic year 2019/2020, the Legal Clinic of the Faculty of Law of the University of Murcia began a study on the possibility of providing the Mar Menor with its own legal personality and rights, which concluded in May 2020. This was announced in the newspaper. From that moment on the Platform for the Popular Legislative Initiative Mar Menor-ILP was formed, for the granting of rights and legal personality to the entire ecosystem of the Mar Menor basin. This was won in July 2020 in the Town Hall of Los Alcázares. And it was not admitted for processing by the Regional Assembly of Murcia. That same month of July, the National ILP was initiated, which has been in the Congress of Deputies since 15 March 2022 for discussion and approval in the coming months.

The process has been extremely fast: just over two years. According to its promoters, the proposal for the Mar Menor was born from below, from the people. At first, both the PP and PSOE parties abstained from supporting the ILP. Now, everyone is looking favourably on the initiative. It has obtained 639,826 popular signatures, required for its acceptance by Congress. Moreover, it is the first European case of a recognition of rights to an ecosystem; and it has become a world-class reference for the UN. The importance of the initiative, apart from the bottom-up process – and of being recognised by the Public Administrations – lies in the fact that it grants recognition to an ecosystem, the capacity to exist, to regenerate and to evolve, and therefore, in the same way, to the relations of its agents among themselves and with their environment, as a whole with a life of its own.

The initiative, after having been accepted by the Spanish Congress of Deputies, will be presented at the UN General Assembly in New York, having been included in the Harmony with Nature Programme.

22.5.5 Platform 'Another Maro and Nerja Is Possible' Against the 'Plan Larios', Nerja: Costa del Sol

On 23 March 2020, the urban development agreement between the Town Council of Nerja and Sociedad Azucarera Larios Inmobiliaria, S.L. for the regularisation of previous agreements, land occupations and urban development of Maro was published in the Official Gazette of the Province. The agreement tries to compensate the debt acquired by the Town Hall for the occupation of land belonging to the Larios family, opening the development of the so-called 'Plan Larios': construction of a golf course (with proposal of declaration of tourist interest), residential urbanisation and building for tourist/hotel use.

The 'Plan Larios' is planned for an area of agricultural use and high cultural, natural and landscape value (it is one of the few undeveloped stretches of the Malaga coastline). As indicated in the agreement, the current PGOU classifies as no urban land, and protected by PGOU and POTAX (Plan de Ordenación Territorial de la Axarquía); some elements are also protected for its cultural interest. The agreement is not new, as it was attempted before 2015, as shown by the allegations presented by GENA-Ecologistas en Acción (2020). The citizen's response at the time led in 2017 to the creation of the platform *'Another Maro and Nerja is possible'* to confront the urban development project (Platform 'Another Maro and Nerja is possible', 2020).

With the change of government in the consistory, the process was paralysed, remaining a *zombie project* until 2020, when it was taken up again in the midst of the State of Alarm due to the health emergency, affecting the deadlines for public information and participation. The compensation established in the agreement includes modifications to the PGOU to allow a new urban development on land currently managed by Azucarera Larios Inmobiliaria S.L. The platform denounces, among other things, the inadequacy of the established compensation channels and the unjustified nature of the reclassification proposals, which leave the future of the municipality mortgaged (Yus Ramos, 2020): loss of the landscape as a cultural and natural value; as well as its value as a tourist resource; loss of the economic and productive diversification of the agricultural environment; construction of dispersed urbanisation – not compact – not connected to the existing urban nuclei, which will increase public spending and mortgage the provision of future services.

The publication in the BOP indicates the intention of the City Council to continue with the processing of the agreement that it has been promoting for more than five years.

The platform calls for (Platform 'Another Maro and Nerja is possible', 2020): (1) to unlink the debts contracted by the Ayto. (2) to audit the debt and draw up a payment plan that does not harm public interests or provoke social cuts; (3) to respect the protections contemplated in the current PGOU and the protection of other landscape, natural, cultural and architectural elements of the municipality, which are at risk, as well as the classification of General Systems of Open Areas and Social Use; (4) to give value to the heritage resources by incorporating facilities and trails for responsible tourism; (5) to ensure the maintenance of the land in healthy conditions and correct use, and to ensure the application of the Law of Rural Leases, avoiding situations of precariousness; (6) to create a structure of information and assistance to farmers on new crops, cooperatives, supply of peri-urban social economy vegetable gardens, as well as to promote the creation of a network of producers and consumers of organic products; and an area for the deposit and recycling of agricultural waste.

22.6 Conclusions

In Spain, the coast is the tourist area par excellence: the most important, in terms of volume, market recognition, employment, companies, territorial transformation, social change, environmental degradation, etc. The starting point of its developmentalist model based on urban-tourist expansion was born in the Franco dictatorship and continued during the democratic transition and integration into Europe. The idea that unlimited economic growth is viable and desirable is found in the global collective imagination, and is strengthened at the local level despite the conflicts it generates. This idea is reproduced whether or not it provides democratic guarantees.

The cases analysed present a variety of practices located along the Spanish Mediterranean coast. Those with the longest time span show how the first social reactions to urban-tourist growth occurred against projects inherited from the dictatorship. After an initial political resistance, the local administration ends up integrating these petitions and leads the revision of urban planning, the paralysation of projects, the declassification and protection of land, and the deurbanisation and ecological regeneration of the affected areas. Cases involving popular legislative initiatives represent a qualitative leap in practices born of social reactions. In the case of the Balearic Islands, the environmental justice that is being demanded goes beyond the present temporal scope. On the one hand, because an intergenerational objective is set and, on the other, because the scope is proposed for any type of future action. In the case of Murcia, it goes beyond the current legal conception of nature and proposes the recognition of an ecosystem as a subject of rights. This implies that nature ceases to be treated as an object, and its existence is comprehensively guaranteed through the preservation or regeneration of its structure, functions and vital ecosystemic processes. Finally, the Costa del Sol case is an example of the social reaction to projects promoted recently, but which have the same character as those promoted at the beginning of Spanish developmentalism.

The demands call for balanced tourism development, unlinked to real estate development and speculative construction. For these social movements, the promise of economic growth is no longer credible, and they are calling for projects with other aims, which preserve local identity and the diversity of economic sectors, which in the long term implies greater quality and sustainability of the territory and the destination.

The current situation of crisis – economic, health, energy and climate – is shaping a post-growth scenario. The outbreak of the pandemic in particular has been a turning point, generating a context in which eco-social transition policies are being promoted in Spain and, in general, in Europe. In the specific case of the tourism sector, alternative tourism practices have become visible that could fit better with the necessary global and local sustainability, as is the case of 'proximity tourism'. Debates advocating a reset of tourism have also been visible. In this context, through the cases shown in this chapter, it can be seen how social reactions to growth projects involving environmental destruction or social injustices can model a path of appropriate practices in this new scenario, either through examples of reactive actions of defence and protection, or through proactive examples of restoration or environmental justice. In this sense, social movements may constitute the agents of change necessary for the model transition. In future research, it will be of vital importance to continue to identify, analyse and categorise practices that enable the ecosocial transition to materialise.

Acknowledgements The research that gives rise to this article is funded by the research project entitled "Tourist saturation in Spanish coastal destinations. Strategies for tourism degrowth. An approach from the social dimension" (RTI2018-094844-B-C33) of the Ministry of Science, Innovation and Universities (Plan Nacional de I + D + i).

References

Andriotis, K. (2018). *Degrowth in tourism: Conceptual, theoretical and philosophical issues*. CABI. 232 pp.

Antón Clavé, S., & González Reverté, F. (Edit.). (2008). *A propósito del turismo*. UOC. 324 pp.

Armengol, X., Antón-Pardo, M., Ortells, R., & Olmo, C. (2019). The malladas of La Devesa de El Saler: The recovery of a set of coastal interdunal lagoons in the Albufera Natural Park. En X. Quintana, D. Boix, S. Gascón, & J. Sala (Coords.), *Management and restoration of Mediterranean coastal lagoons in Europe*. Ed. Càtedra d'Ecosistemes Litorals Mediterranis.

Barke, M. (2007). Second homes in Spain: An analysis of change at the provincial level, 1981–2001. *Geography, 92*, 195–207.

Blanco-Romero, A., Blàzquez-Salom, M., Morell, M., & Fletcher, R. (2019). Not tourism-phobia but urban-philia: understanding stakeholders' perceptions of urban touristification. *Boletín De La Asociación De Geógrafos Españoles* (83). https://doi.org/10.21138/bage.2834

Blázquez-Salom, M., Blanco-Romero, A., Vera, F., & Ivars, J. (2019). Territorial tourism planning in Spain: From boosterism to tourism degrowth? *Journal of Sustainable Tourism, 27*(12), 1764–1785.

Boissevain. (1996). But we live here. In Briguglio et al. (Eds.), *Sustainable tourism in Islands and Small States: Case studies* (pp. 220–239). Pinter.

Borràs, R. (2021). Contra el bla, bla, bla: ILP Avui per demà. Alba Sud, investigación y comunicación para el desarrollo. Available in: https://www.albasud.org/blog/es/1406/contra-el-bla-bla-bla-ilp-avui-per-dem-agrave

Bourdeau, P., & Berthelot, L. (2008, April 18–19). Tourisme et D_ecroissance: de la critique _a l'utopie? Paper presented at the First International Conference on Economic De-Growth for Ecological Sustainability and Social Equity, Paris.

Britton, S. (1991). Tourism, capital and place: Towards a critical geography of tourism. *Environmental and Planning D: Society and Space, 9*(4), 451–478.

Butler, R. (1980). The concept of a tourist area cycle of evolution: Implications for management of resources. *The Canadian Geographer/Le Geographe Canadien, 24*(1), 5–12.

Cañada, E. (2015). *Las que limpian los hoteles. Hidden stories of labour precariousness*. Edt. Icaria. 192 pp.

Castells, M. (2012). *Networks of indignation and hope: Social movements in the internet age*. Alianza. 294 pp.

Cervera Sánchez, A. M. (2016). María Consuelo Reyna. Periodista y empresaria en la Transición valenciana. *Historia y Comunicación Social, 21*(2), 399. Available in: https://core.ac.uk/download/pdf/81229903.pdf

Del Romero Renau, L. (2018). Touristification, sharing economies and the new geography of urban conflicts. *Urban Science, 2*(4), 1–17.

Díaz Parra, I., & Candón Mena, J. (2014). Geographical space and cyberspace in the 15M movement. *Scripta Nova. Electronic Journal of Geography and Social Sciences., XVIII*, 470.

Doxey, G. V. (1975). A causation theory of visitor/resident irritants: Methodology and research inferences. The impact of tourism. In *Proceedings of the travel research association 6th annual conference* (pp. 195–198). Travel Research Association.

Exceltur. (2006). *White paper on human resources*. Exceltur. 120 pp.

Fletcher, R. (2011). Sustaining tourism, sustaining capitalism? The tourism industry's role in global capitalist expansion. *Tourism Geographies, 13*(3), 443–461.

Fletcher, R., Murray Mas, I., Blanco-Romero, A., & Blázquez-Salom, M. (2019). Tourism and degrowth: An emerging agenda for research and praxis. *Journal of Sustainable Tourism, 27*(12), 1745–1763.

Fletcher, R., Blanco-Romero, A., Blázquez-Salom, M., Cañada, E., Murray, I., & Sekolova, F. (2021). Pathways to post-capitalist tourism. *Tourism Geographies*. https://doi.org/10.1080/14616688.2021.1965202

Gamero, J., Colomer, A., & Quintana, X. (2019). Re[making] the Landscape 2014–2018. Layman's report LIFE 13 NAT/ES/001001. Available in: http://lifepletera.com/wpcontent/uploads/2019/04/Life_Pletera_Laumans_report_2019_ENG.pdf

Georgescu-Roegen, N. (1995). *La décroissance. Entropie – Écologie – Économie*. Éditions Sang de la terre. 254 pp.

GOB. (2022a). *Què és el GOB?* Available in: https://www.gobmallorca.com/gob/sobre-nosaltres

GOB. (2022b). *Memòria 2021*. Available in: https://www.gobmallorca.com/images/stories/pdf/memoria2021.pdf

Hall, C. M. (2009). Degrowing tourism: Decroissance, sustainable consumption and steady-state tourism. *Anatolia, 20*(1), 46–61. https://doi.org/10.1080/13032917.2009.10518894

Harvey, D. (2013). *Rebel cities. From the right of the city to urban revolution*. Akal. 240 pp.

Higgins-Desbiolles, F., Carnicelli, S., Krolikowski, C., Wijesinghe, G., & Boluk, K. (2019). Degrowing tourism: Rethinking tourism. *Journal of Sustainable Tourism, 27*, 1926.

Jackson, T. (2011). *Prosperity without growth*. Icaria. 280 pp.

Koolhaas, R. (2004). "La ciudad genérica." En Martín Ramos, Á. (ed.): Lo urbano. : Edicions UPC, Universidad Politècnica de Catalunya, p. 73–82.

Koolhaas, R. (2008). *El espacio basura*. Gustavo Gili, SL. Colección GGmínima.

Kousis, M. (2000). Tourism and the environment. A social movements perspective. *Annals of Tourism Research, 27*(2), 468–489.

Kuhn, T. (1962). *The structure of scientific revolutions*. University of Chicago Press.

Lindbert, K., McCool, S., & Stankey, G. (1997). Rethinking carrying capacity. *Annals of Tourism Research, 24*(2), 461–465.

López Olivares, Antón Clavé, Navarro Jurado, Perilla Del Álamo, & Sastre Alberti. (2005). Changes and transformations in the current tourism model in Spain. *Annals of Tourism Research in Spanish, 7*(2), 423–446.

López, I., & Rodríguez, E. (2010). *End of cycle. Financing, territory and landowner society in the long wave of Hispanic capitalism (1959–2010)*. Traficantes de Sueños, 503 pp.
Malvárez, G., Pollard, J., & Dominguez, R. (2000). Origins, management and measurement of stress on the coast of Southern Spain. *Coastal Management, 28*(3), 215–234.
Milano, C., Novelli, M., & Cheer, J. M. (2019). Overtourism and degrowth: A social movements perspective. *Journal of Sustainable Tourism, 27*(12), 1857–1875.
Naredo, J. M. (2010). El modelo inmobiliario español y sus consecuencias. *Boletín CF+S, 44*, 13–27. Avalaible in: http://habitat.aq.upm.es/boletin/n44/ajnar.html
Navarro-Jurado, R.-P., Romero-Martínez, S.-M., Habegger, & Mora-Esteban. (2019). Growth machines and social movements in mature tourist destinations Costa del Sol- Málaga. *Journal of Sustainable Tourism, 27*(12), 1786–1803.
O'Reilly, A. M. (1986). Tourism carrying capacity: Concept and issues. *Tourism Management, 7*(4), 254–258.
O'Reilly, K. (2009). Hosts and guests, guests and hosts: British residential tourism in the Costa del Sol. In Pons, Crang, & Travlou (Eds.), *Cultures of mass tourism: Doing the mediterranean in the age of banal mobilities* (pp. 129–142). Ashgate Publishing.
Pié, Vilanova, Díaz, & Zhonero. (2018). "Salvem la Pletera!". Crònica d'un imposible. En Sala i Martí, P.; Puigbert, L.; Bretcha, G. (2018): Plecs de Paisatge: Reflexions 6. (Des)fer paisatges. Olot: Observatori del Paisatge de Catalunya. – 206 p. Available in: http://www.catpaisatge.net/cat/documentacio_plecs_ref_6.php
Pizam, A. (1978). Tourism's impacts: The social costs to the destination community as perceived by its residents. *Journal of Travel Research, 16*(4), 8–12.
Quintana, X., Capellà, J., Colomer, À., Hors, X., Marí, M., & Pié, R. (2018). *La restauració de la Pletera: crònica de la desurbanització. Llibre de la Festa Major de Torroella de Montgrí* (pp. 29–49). Available in: https://raco.cat/index.php/LlibreFestaMajor/article/view/341003
Romero, J., Brandis, D., Delgado Viñas, C., García Rodríguez, J. L., Gómez Moreno, M. L., Olcina, J., Rullán, O., Vera-Rebollo, J. F., & Vicente Rufí, J. (2018). Aproximación a la Geografía del despilfarro en España: balance de las últimas dos décadas. *Boletín de la Asociación de Geógrafos Españoles, 77*, 1–51.
Romero-Padilla, Y. (2016). *Metamorfosis de una invención. Turismo y clase creativa: el caso de la Costa del Sol en el mediterráneo español* (Doctoral dissertation, Universidad de Málaga).
Romero-Padilla, Y., Navarro-Jurado, E., & Malvárez, G. C. (2016). The potential of international coastal mass tourism destinations to generate creative capital. *Journal of Sustainable Tourism, 24*(4), 574–593.
Romero-Padilla, Y., Romero-Martínez, J. M., & Navarro-Jurado, E. (2020). Reflexiones desde el post-crecimiento: ideas, estrategias y tácticas para el turismo post-covid-19. In *Turismo post COVID-19: el turismo después de la pandemia global, análisis, perspectivas y vías de recuperación* (pp. 57–75). Ediciones Universidad de Salamanca. Available in: https://eusal.es/index.php/eusal/catalog/view/978-84-1311-461-3/5558/6275-1
RTVE. (1970). La Albufera de Valencia (Documental). Serie El planeta azul. Available in: https://www.rtve.es/play/videos/el-planeta-azul/vida-salvaje-albufera-valencia/3013598/
Saarinen, J. (2006). Traditions of sustainability in tourism studies. *Annals of Tourism Research, 33*(4), 1121–1140.
Schulz, C., & Bailey, I. (2014). The green economy and post-growth regimes: Opportunities and challenges for economic geography. *Geografiska Annaler: Series B, Human Geography., 96*(3), 277–291.
Soja, E. (2008). *Postmetropolis. Critical studies on cities and regions*. Traficantes de sueños. 590 pp.
Talego, F., & Hernández-Ramírez, J. (2017). New social movements re-enchant the world. *Quaderns-e, 22*, 35–49.

Thiel Ellul, D., & Navarro Jurado, E. (2018). *Measurement and analysis of sustainability: Synthetic indicators through multi-criteria methods and their relationship with tourism in the Andalusian coastline*. Editorial Síntesis. 266 pp.
University of Valencia and València City Council. (2017). EL SALER PER AL POBLE, ARA! El poder de la ciutadania en la transformacióresponsable del paisatge i del territorio. Available in: https://www.uv.es/uvweb/cultura/ca/presentacio-/exposicions-preparacio/saler-poble-ara-div-poder-ciutadaniatransformacio-responsable-paisatge-territori-/div-1285866236311/Activitat.html?id=1285989279915
Vilarino, A., & Orea, D. (2013). *Ordenación territorial*. Mundi-Prensa Libros. 544 pp.
World Tourism Organization; Centre of Expertise Leisure, Tourism & Hospitality; NHTV Breda University of Applied Sciences; & NHL Stenden University of Applied Sciences. (2018). *'Overtourism'? – Understanding and managing urban tourism growth beyond perceptions*. UNWTO. 12 pp.
Yus Ramos, R. (2020). Lo que pudo ser, lo que es, y lo que puede ser el territorio de Nerja y Maro. In Seminariotaller Covid-19 y efectos en el turismo: retos y oportunidades, 30 September and 1 October 2020. Nerja Museum. Organised by the Inter-University Research Institute for Tourism Intelligence and Innovation of the University of Malaga.

News and Report

Auken, M. (2009). *Report on the impact of extensive urbanisation in Spain and the application of EU law*. 18 pp. http://habitat.aq.upm.es/boletin/n40/auken.pdf
García, C., Aguirre, I., & Agrasar F. (2010). Project for the ecological regeneration of the isthmus of O Grove and A Lanzada beach (Pontevedra). http://citywiki.ugr.es/wiki/Monograficos_de_proyectos_14- 15/GRUPOS.
GENA-Ecologists in Action. (2020). Alegaciones al 'Convenio urbanístico entre el Ayto. de Nerja y Sociedad Azucarera Larios Inmobiliaria, S.L. para la regularización de anteriores convenios, ocupaciones de suelo y desarrollo urbanístico de Maro'. Disponible en: https://f3431ce0-d274-4822-8ba5-10b9428a3f6e.filesusr.com/ugd/8a9436_9891e9f5c8a243859a73380c727a6fbd.pdf
INE. (2020). *Urban indicators*. Edition 2020. Available at: https://www.ine.es/prensa/ua_2020.pdf
n'UNDO. (2012). http://archivo-es.greenpeace.org/espana/es/news/2012/June/nUndo-arquitectos-y-Greenpeace-demuestran-que-la-demolicion-del-Algarrobico-puede-ser-sostenible-y-generar-cerca-de-400-puestos-de-trabajo/
O'Neill, et al. (2018). *Europe, it's time to end the growth dependency*. Available at: https://degrowth.org/2018/09/06/post-growth-open-letter/
Platform 'Another Maro and Nerja is possible'. (2020). Manifesto No to 'Plan Larios' in Maro.
Quintana, X. (2019). Ecological criteria in the restoration of the marshes of La Pletera in Torroella de Montgrí-l'Estartit (Girona). Conference at the Escuela Técnica Superior de Arquitectura, 10-12-19. University of Granada.

Dr. Enrique Navarro-Jurado PhD in Geography, Director of University Institute for Research in Tourism Innovation at University of Malaga. Principal Investigator of R + D + i projects and participated in more than 30 research projects (Brazil, Dominican R., Cuba, Mexico, Argentina) and in technology transfer contracts. With more than 40 publications in high impact journals. Member of several boards of trustees, scientific associations and journal editor. Advisor to the Blue Plan (UNEP-UN) and various regional and local plans. Research interests: Planning and management of tourism destinations; Limits to growth and tourist carrying capacity; Sustainability indicators; Multi-criteria analysis and its application to sustainability and climate change.

Dr. Yolanda Romero-Padilla PhD. in Tourism. Professor of Business Management at Complutense University of Madrid, member of the Dean's Office at the Faculty of Commerce and Tourism. Her research is linked to sustainable management of tourist destinations and organisations, focusing on social and urban transformations in Spanish Mediterranean destinations.

Dr. José María Romero-Martínez PhD Architect, Professor of Architectural Projects at University of Granada, Spain. He has received awards such as the Europan III Prize (Belgium, 1993) and the Malaga Architecture Prize for new buildings (1993 and 2003) and urban planning (2005). His research is linked to architecture, territory, sustainability and collaborative processes. He is currently researching the theories and practices of degrowth and post-growth applied to the disciplines of architecture and urban planning, particularly in coastal areas.

Chapter 23
Package Holidays and Charter Airlines Companies in Spain (2004–2021): The End of a Fordist Pair?

David Ramos-Pérez

23.1 Introduction

In literature in the field of geography, the relationship between air transport and tourism has not been analysed in any depth, even though most authors acknowledge the existence of strong mutual ties between both activities (Bieger & Wittmer, 2006; Forsyth, 2006; Lumsdon & Page, 2004). Almost a decade ago, Duval (2013) pointed out that this relationship was still an incipient field of research in geography, while Landré and Peeters (2011) went one step further, referring to the "big gap" in knowledge. Prior to these authors, over 25 years ago in the field of tourism geography, Ioannides (1995) highlighted the need to turn the spotlight on the supply in order to gain a better insight into the structure and organization of the tourism production system. Within this supply, he pointed to airline companies, tour operators and hotel chains as being the stakeholders most worthy of attention. More recently, due to debate on the lost centrality of transport geography in human geography (Hanson, 2003 & Goetz, 2006), several authors proposed a work agenda that would re-engage both of them, highlighting the role that tourism can play in this task (Shaw & Sidaway, 2011).

D. Ramos-Pérez (✉)
Department of Geography, University of Salamanca, Salamanca, Spain

Department of Geography, University of Porto, Porto, Portugal
e-mail: a13004@usal.es

A. Blanco-Romero, M. Blázquez-Salom (eds.), *Spanish Tourism Geographies*, Geographies of Tourism and Global Change,
https://doi.org/10.1007/978-3-031-39780-6_23

When we turn our attention to former charter companies,[1] the gaps in our knowledge are even greater. While it is true that studies by Gillmor (1973, 1996), Spill (1976), Vanderheyden (1979) and Pearce (1987a, 1987b, 1987c) offer a series of important insights into the subject, they refer to the 1970s and 80 s: the golden age of charter airline companies. Among these studies, mention must be made of the contributions of New Zealand geographer David Pearce, since he allowed certain important changes to be identified in the organization of charter networks and in the diversification of the origin airports in some European outbound countries. Nonetheless, it was not until the middle of the first decade of the twenty-first century that Belgian geographers Dobruszkes et al. (2006) again turned their attention to the issue. This apparent lack of academic interest in the subject–a paradoxical situation in a scenario in which a genuine surge in air transport studies in the field of geography can be observed–can be explained by significant difficulties in obtaining breakdowns of charter airline data back then. Taylor (2016) recently emphasized the need not to lose sight of new outbound tourism markets in Central and Eastern Europe when analyses of leisure airlines are made in order to avoid biased conclusions. Lastly, Ramos-Pérez and Dosbruszkes (2019) analysed the impact of the irruption of low-cost airlines on the spatial configuration of the flight networks of European leisure airlines, based on two case studies, demonstrating that competition by low cost airlines had not affected the dominant position of former charter companies on short and mid-haul routes, hence seeming to discount the hypothesis that is often put forward with regard to their specialization in long-haul tourist routes.

In order to help fill in some gaps in existing knowledge of the relationship between air transport and tourism in Spain, this chapter analyses the demand that is channelled through former charter airlines (now operating as leisure carriers) on flight routes between Spain and European outbound markets between 2004 and 2021. The business model followed by these airlines specializes in passenger transport to sun and sand destinations of Fordist origins, located in the extended Mediterranean basin[2] (Fig. 23.1). They are airlines that were identified in official statistics as non-scheduled traffic. Although this association no longer makes sense following the liberalization of EU air transport, it continues to be widely used, leading to the systemic under-representation of their market share, a bias that this study aims to amend.

[1]The term charter airline was widely used in Europe before liberalisation to designate airlines specialised in transporting passengers from the outbound markets of central and northern Europe to the tourist destinations of the Mediterranean basin. As most flights were not sold directly by the airline to passengers but were included in package holidays offered by tour operators, the passengers they carried were considered non-scheduled traffic. As such, the flight was considered a non-scheduled service. Today, however, within the liberalised EU market, most flights to these destinations are operated as scheduled services, albeit often on a seasonal basis. A small proportion of these flights are still operated as non-scheduled services, especially on routes with low demand.

[2]This is understood to mean the space between the Iberian archipelagos in the Atlantic (the Azores, Madeira and the Canaries) and the coasts of the Black and Red Seas (Brito, 2009).

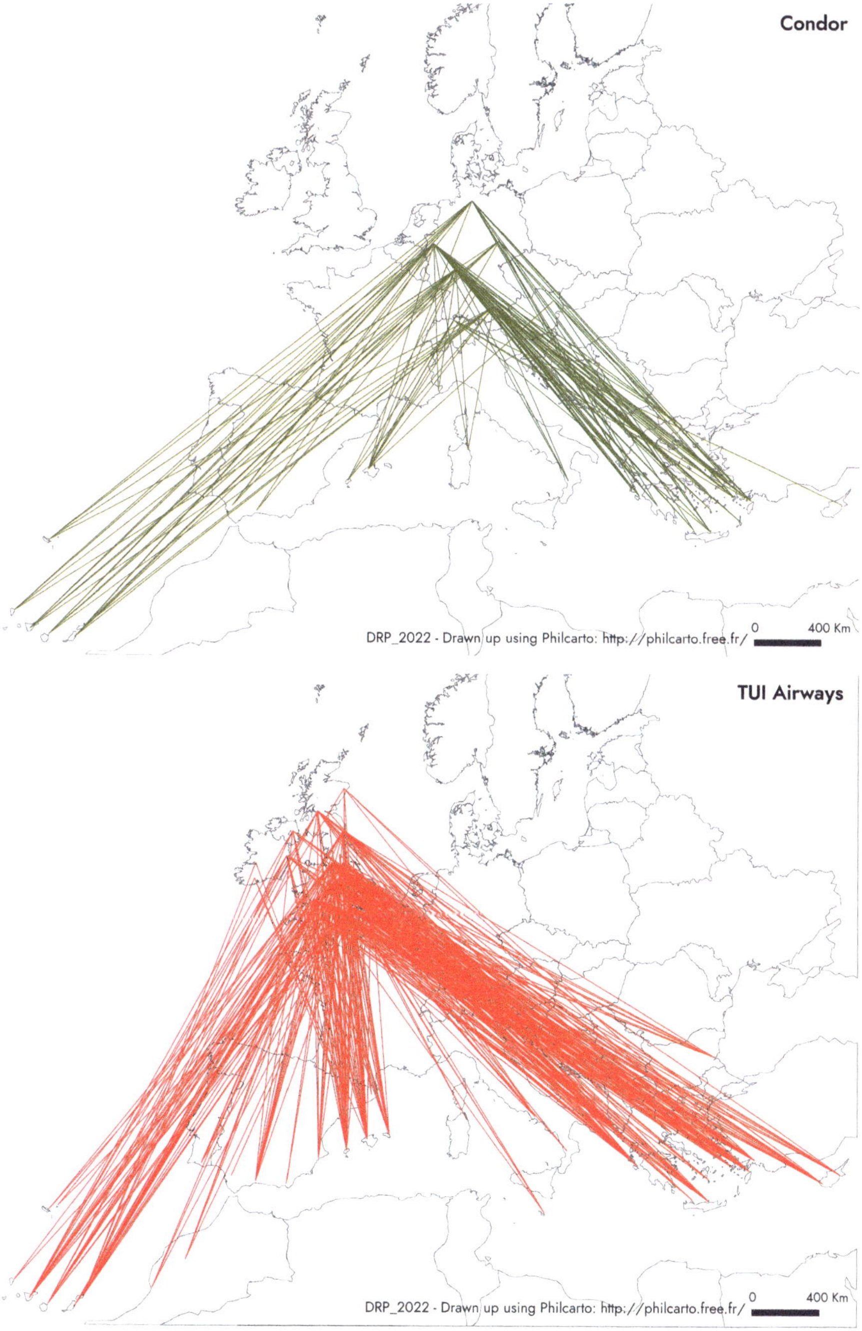

Fig. 23.1 Short and mid-haul routes operated by Condor & TUI Airways (June 2022). (Source: Fligthradar24 (flights programmed from June 6th to 12th 2022))

By rectifying it, a greater understanding will be gained of competition between leisure airlines and low-cost carriers, one of the processes triggered by the liberalization of the EU air transport market for which there is less empirical knowledge. Indeed, the disappearance of charter airlines from short and mid-haul European routes has been widely forecast in academic literature since the early twenty-first century. It was said that the liberalization of EU air traffic and the decline in package holidays would lead to their replacement by low-cost operators better able to cope with the flexibility needed to fit in with the more individualistic practices of post-Fordist tourists. Although the bankruptcy of tour operator Thomas Cook in 2019 and some associated airlines would seem to confirm these predictions, it is important not to overlook the growth in package holidays over the last decade or the survival of a wide range of leisure airlines.

Given the objectives of this book, the chapter opens with a review of academic literature in the field of Spanish geography that focuses on relations between tourism and air transport. This is followed by a discussion of substantial methodological importance regarding conceptual misunderstandings of the term "charter", a fact that has led to the systematic under-representation of traffic carried by leisure airlines. Next, an assessment is made of the market share of these airlines at Spanish airports where they played a dominant role prior to liberalization, using statistics by AENA that allow air traffic to be broken down by company and airport for the period under study. Following on from this, an outline is given of the impact of the Covid-19 pandemic on the strategies that have been taken by the airlines that channel tourist traffic to the airports under consideration during the last 2 years. The chapter then concludes with a brief presentation of the main conclusions.

23.2 The Tourism/Air Transport Pair in Spain

In Spanish geography, only limited attention has been paid to studies of air transport to Spanish tourist destinations despite the subject's relevance in the birth and development of these destinations and despite the over thirty pages that Cano García (1980) devoted to analysing the relations between charter companies and the growth of tourism in this country in his seminal study of the geography of air transport in Spain. In addition to highlighting the close links between these airlines and the dominant tour operators of the period, he also analysed the spatial and temporal distribution of passengers on charter carriers. For this purpose, he took advantage of different official statistics dating back to 1958, using them to offer an invaluable insight into the subject, although his study does lack a breakdown by routes in order to identify certain spatial-temporal patterns in the growth of the phenomenon, as Pearce later showed (1987a, 1987b).

Almost two decades would pass until Serrano Martínez (1999) alluded to the important role of tourist traffic in a study of Spain's airport network, focused on the evolution of the makeup of traffic at the said airports, its uneven distribution and changes in its hierarchical organization between 1973 and 1998. However, the author limited himself to using tourism as an explanatory variable for the high number of international passengers recorded by some airports, without entering into more weighty discussions. More recently, after calculating the coefficient of connectivity for Spain's airports in 1970 and 2008, Díez Pisonero (2012) once again pointed to tourism as being the main explanatory factor for the high connectivity values achieved by certain island and coastal airports, and this was in a study that only took into account scheduled flights, hence minimizing the connectivity of the said airports in 1970. The primary and multiple linkage analysis tested by Seguí Pons and Martínez Reynés (1994) clearly shows this dissymmetry between scheduled and non-scheduled connectivity in the case of Palma (Mallorca) in 1989, a characteristic of tourist airports that persisted until the liberalization of EU air transport. Among the different studies that highlight the close links between tourist activities and traffic at certain airports, the most important conceptual contribution can be found in an unpublished thesis by Martinez Reynés (1997), since it analyses the relations between Mediterranean Europe's tourism models and air transport in the early 1990s, highlighting the latter's high dependence on what the author describes as "regions specializing in international hotel tourism, with a high or very high role in tourism", coinciding with mature tourist destinations in the Mediterranean area.

Following the liberalization of EU air traffic, new studies were sparked off by the emergence of low-cost airlines, given the interest aroused by the waning market share of charter airlines at airports where they had played a dominant role for over three decades. Using Mallorca as a case study, Seguí Pons and Martínez Reynés (2010, 2012 and 2013) related the growing share held by low-cost airlines to the emergence of a more heterogeneous, individualistic post-Fordist tourist who fitted in better with the flexibility afforded by low-cost companies than with the standard package holidays and charter flights typical of Fordism. Likewise, they drew attention to the environmental impact of the growth in low-cost traffic, both in terms of greenhouse gas emissions and the transformation of the landscape and land uses. Gago García et al. (2011) noted that, in the Canary Islands, a relative delay could be observed in the replacement of charter airlines with low-cost operators, particularly in comparison with Mediterranean tourist destinations. If distance was a handicap in the growth of Fordist tourism in the Canaries until the widespread use of jet planes (Hernández Luis et al., 2011), that same distance would be responsible for a delay in the irruption of low-cost operators in the archipelago, since the flight times needed to reach the Canaries from European outbound markets did not initially fit in with their operating strategy, which would explain the certain delay in the spread of post-Fordist tourist practices there. Saladié et al. (2014) and Clavé et al. (2015) explored changes in tourist profiles and found that, in the case of Reus, the existence of a low-cost airline on a route increased the possibility of the Costa Dorada being chosen as a destination by tourists living in the catchment area of East Midlands Airport.

One more recent field of research has focused on changes in the spatial organization of tourist destinations possibly related to the new profiles of travellers using low-cost carriers. Ivars Baidal and Menor Muñoz (2008) and Vera Rebollo and Ivars Baidal (2009) showed that in the provinces of Alicante and Murcia, the consolidation of low-cost airlines at their respective airports has ended up by affecting urban and spatial planning policies. The growing preference for villa and apartment accommodation and the boom in residential tourism –in this last case, thanks to cheaper tickets and improvements in connectivity– have led to increasing specialization on the part of the real estate sector at these destinations, spurring on the approval of numerous residential tourism complexes outside traditional tourist areas. Bauzà van Slingerlandt (2009) reaches the same conclusions for Mallorca, noting that the doubling of Palma Airport's capacity in the second half of the 1990s played a key role in the transformation of the whole island, leading to a process of time-space compression in which "all space is urban tourism space and all time is urban tourism time". In this sense, Serrano Martínez et al. (2016) show that, in Murcia, planning processes influenced by these tourism and real estate expectations led to the construction of a new airport by the regional government 25 kilometres from the existing state-owned one, which, paradoxically, had received a large amount of investment for the extension and modernization of its passenger terminal, which is currently closed. Alicante, Murcia and Mallorca are just three examples of the dynamics that have spread to most Spanish coastal tourist areas.

In addition to these studies of Spain, others must be mentioned that conduct a worldwide analysis. For instance, Gago-García et al. (2019) identify the role that tourism played in the increased air connectivity of the world's 134 cities with the highest number of international tourists in 2006 and 2016, highlighting the irruption on the scene of heritage and cultural destinations, which until recently ranked second in comparison with sun and sand destinations or global cities. Lastly, Ramos-Pérez (2019) proposes a theoretical framework for the interpretation of relations between air transport and tourism in an approach far removed from technological determinism, incorporating the spatial logic of capitalism, in particular its tendency to expand geographically in order to take advantage of differences in income (and prices) in different parts of the planet. Air transport is seen as an important facilitator in this strategy by swiftly putting outbound markets into contact with tourist destinations, making it the perfect tool in fostering time-space compression and in speeding up the turnover of capital; that is, in the globalization of tourism.

23.3 From Charter Companies to Leisure Airlines: Some Conceptual Clarifications

In Europe, the tourist-related boom in air mobility is clearly linked to the *Multilateral Agreement on Commercial Rights of Non-Scheduled Air Services in Europe*, approved in Paris in 1956. At the time, the *Paris Agreement*, as it is known, which was signed by 19 member countries[3] of the European Civil Aviation Conference (ECAC), allowed for a certain liberalization of the architecture of international civil

[3]The signatory countries of the Paris Agreement were Germany, Austria, Belgium, Denmark, Spain, Finland, France, Ireland, Iceland, Italy, Luxembourg, Norway, the Netherlands, Portugal, the UK, Sweden, Switzerland, Turkey and Greece, although the latter never came to formally ratify it.

aviation, as defined by the 1944 Chicago Convention. Bilateral air service agreements were the mechanisms stipulated in the said convention for the establishment of flight connections between States and, in them, certain factors relating to international flight routes were established, from the city pairs for flight operations to the frequency of flights, maximum seating capacity, number of companies that each State could authorize, and the tariffs to be applied. Flights operated in accordance with these parameters, mainly comprising flight connections between the signatory countries' foremost cities, were given the name *scheduled* flights. The remaining ones were deemed to be *non-scheduled*, although they were barely dealt with in the convention (article 5) since it was thought that there would be few of them.

Hence, the connections between the outbound markets and tourist destinations that were of interest to the travel agencies and tour operators of the time could only be offered through non-scheduled flights. Although these non-scheduled flights could be operated in a more flexible way than scheduled ones, they had to be authorized by the States that were involved. The *Paris Agreement* eliminated the need for this authorization, replacing it with administrative notice provided that the envisaged flights involved signatory countries, were operated by the airlines of the said countries, were "international flights for remuneration or hire, other than scheduled international air services", and that they coincided with one of the six activities contemplated in article 2 of the Agreement, which included "the transport of passengers between regions which have no reasonably direct connection by scheduled air services" (Spanish Official Gazette, 1961). On those non-scheduled flights, direct sales of seats were forbidden, a prerogative limited to scheduled flights on the routes established in bilateral agreements.

The *Paris Agreement* contributed to the birth of European charter companies; that is, airlines specializing in renting their capacity to travel agencies and tour operators. These, in turn, created *packages*, where potential tourists were offered transport, accommodation and meals at reduced rates at a certain destination for a limited time. This gave rise to a business model that has lasted through to today by applying Fordist principles to the tourist industry (Lyth, 2009), with the massification of tourism in the Mediterranean as a clear corollary. This sheds better light on how it was that these companies came to account for over 40% of the international air traffic between member states of the ECAC in the late 1970s (ECMT, 1980), with a business model (charter airlines) that came to be almost fully identified with a certain type of air service (non-scheduled services). This was basically a valid equivalence until the third package of measures to liberalize EU air transport came into effect in 1992, from which point the distinction between scheduled and non-scheduled flights was no longer of any importance in the programming of international air services within the EU.

Since the early twenty-first century, numerous doubts have been raised by certain authors about the future of charter operators (Buck & Lei, 2004; O'Connell & Bouquet, 2015; Williams, 2001). They suggest that charter companies will not be able to withstand competition from the low-cost carriers that have appeared following the liberalization of short and mid-haul routes, particularly if the demand by independent travellers grows at the expense of those who opt for package holidays.

Doganis (2010) qualifies this by saying that charter companies will have a future on certain mid-haul routes; for instance, flight connections with a high volume of passenger traffic between big cities in main European outbound tourism markets and the Mediterranean's top tourist destinations or seasonal connections with little traffic from the same outbound markets to small island destinations and/or relatively isolated outlying ones.

However, the analysis on which these arguments is based focuses almost exclusively on the recent trend in the UK outbound market (Taylor, 2016), and solely data for non-scheduled traffic is used to demonstrate the declining role of charter carriers on short and mid-haul routes. By continuing to identify non-scheduled traffic with charter airlines, these studies overlook the fact that most operations by these airlines have been accounted for as scheduled traffic since the EU liberalization process, when they started to sell some of their seats directly. Evidently, this leads to a notable bias in assessments of their market share (Ramos-Pérez & Dosbruszkes, 2019). As a result of this misunderstanding, academic studies continue to equate current traffic by charter companies with figures for non-scheduled traffic, conditioning the conclusions that are drawn, as can be seen in work by Castillo-Manzano et al. (2017) on Spanish airports.

The trend toward the individualization of travel is a reality, driven not so much by a change in tourist habits or preferences as Urry (1990) upholds, but, above all, by society's increasing *Mcdonaldization,* leading to predictable, efficient, pre-calculated, controlled tourist travel outside the package holiday circuit (Ritzer & Liska, 1997). At the same time, it is also true that package holidays are far from extinct. Indeed, a clear increase in package holidays since 2009 can even be noted, as can be confirmed from an analysis of mature outbound markets like the United Kingdom (Fig. 23.2) or well-established tourist destinations such as Spain (Fig. 23.3). These packages are no longer solely associated with a plane trip on a charter airline because both low-cost and network carriers are actively involved in this business. Even so, some airlines in Europe mainly continue to carry package holiday passengers and their network of flight connections (Fig. 23.1) is clearly conditioned by agreements signed with the tour operators who purchase seating capacity on their flights (Doganis, 2010).

These former charter companies, which are increasingly commonly referred to now as *leisure airlines* or *holiday airlines* (DLR, 2008), have undergone certain changes in order to deal with the aforementioned pressure of competition, with the creation of a hybrid model that combines elements of a low-cost airline with others inherited from their traditional connections with tour operators (Papatheodorou & Lei, 2006). Thus, although there are still airlines that focus exclusively on selling their seating capacity to tour operators and their traffic continues to be regarded as non-scheduled, most former charter airlines now carry a mix of independent passengers and ones on package holidays sold by one or several tour operators in varying proportions. One clear example is Condor, which, in 2007, sold 40% of its seating capacity to Thomas Cook, the tour operator with which it was associated, while 24% was purchased by other tour operators and the remaining 36% was sold directly to consumers (Doganis, 2010).

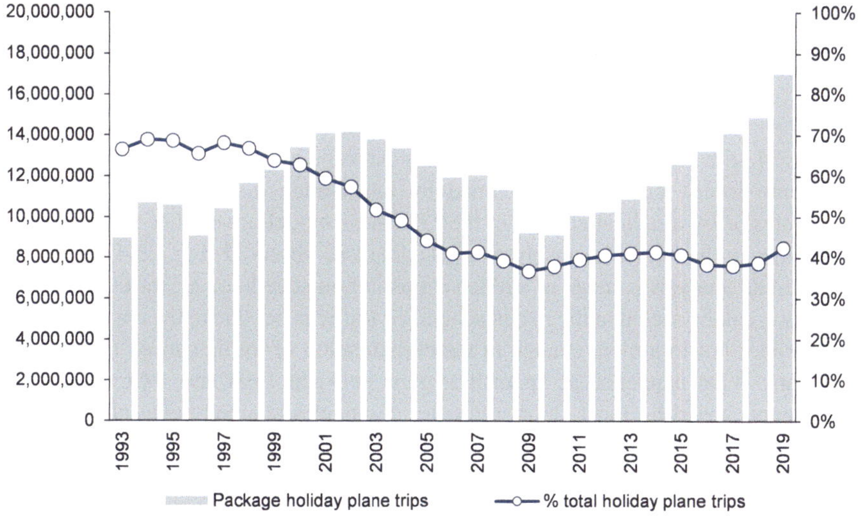

Fig. 23.2 Tourist trips by Britons on package holidays to European destinations (1993–2019). (Source: Based on the UK International Passenger Survey. Available at https://www.ons.gov.uk/peoplepopulationandcommunity/leisureandtourism/datasets/travelpac)

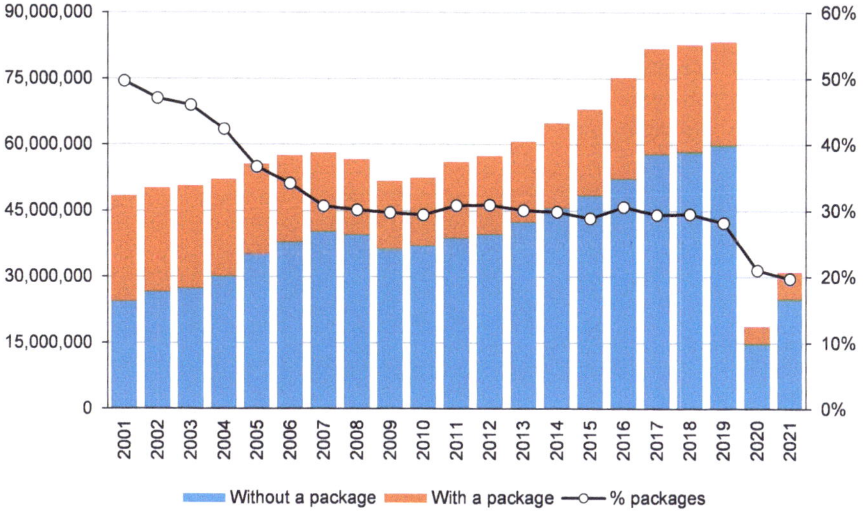

Fig. 23.3 Foreign tourist arrivals to Spain by type (2002–2021). (Source: Based on Statistics for Border Tourist Movements (FRONTUR). Available at https://www.ine.es/uc/JtaQBzp8)

23.4 How to Identify Leisure Airlines in Official Spanish Air Traffic Statistics

The drop in the relative importance of non-scheduled air traffic at Spanish airports as from 1994 –the year after the last liberalizing package of measures for EU air traffic came into force– cannot be explained by the sudden disappearance of charter companies or by a radical change in their business model, given that an immediate sharp drop in this traffic in absolute terms cannot be observed (Fig. 23.4).

When an analysis is made of sudden drops in non-scheduled air traffic at certain airports, it can be seen that the rise in scheduled traffic is not due to the appearance of new airlines but to a mere change in the classification of traffic carried by existing operators. What happened at Spanish airports between 1994 and 1996 is that the passengers carried by most German charter companies stopped being classified as *non-scheduled traffic* and they were instead deemed to be *scheduled traffic* in AENA's statistics. The same was not true, however, of airlines based in the UK, Belgium, the Netherlands and Sweden, whose passengers continued to be labelled as *non-scheduled.* Although, prior to liberalization, German charter companies already sold some of their seats directly to travellers in a practice known as *seat only,* some airlines from the aforementioned countries also started to proceed in a similar way, such as the Dutch company Transavia. To add to the confusion, traffic by German charter companies continued to be deemed non-scheduled traffic at La Palma and Jerez airports until December 1998, from which point it was considered to be scheduled, which explains why scheduled traffic came to predominate at both airports during the year 2000 (Table 23.1). All this shows that, from the said dates, classification problems with *scheduled* and *non-scheduled* services in official

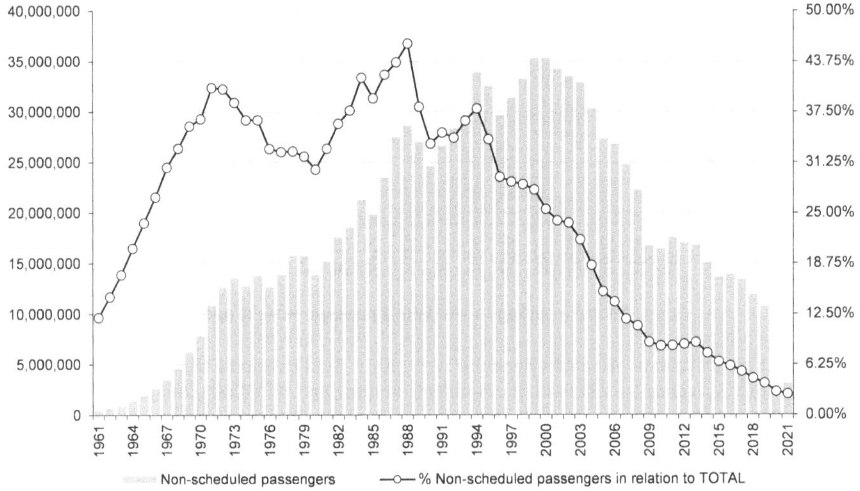

Fig. 23.4 The evolution of non-scheduled international air traffic at Spanish airports (1961–2021). (Source: Based on official air traffic statistics by the Spanish Directorate General for Civil Aviation)

Table 23.1 Distribution of international traffic at La Palma Airport (SPC) by company and type of service in 1996 and 2006

1996			2006		
	Airlines	Passengers		Airlines	Passengers
NON-SCHEDULED	**LTU**	**93,109**	SCHEDULED	**Condor**	**107,382**
	Condor	**79,203**		**LTU**	**21,089**
	Trans European Airways	12,834		**Air Berlin**	**31,288**
	Transavia Airlines	11,879		Hapag-Lloyd	18,843
	Air Holland	8,462	NON-SCHEDULED	Transavia Airlines	29,855
	Air Berlin	**8,298**		Martinair Holland	13,414
	Germania	6,105		Thomas Cook Airlines Belgium	6,938
	Deutsche BA	4,509		Jetairfly (TUI Airlines Belgium)	3,785
	Spanair	2,554		Hamburg International	1,789
	Aero Lloyd	1,226		Blue Panorama	1,300
	Edelweiss Air	918		Itali Airlines	1,049
	Centennial	792		Otras	816
	TOTAL	**229,889**		**TOTAL**	**237,548**

Source: Based on AENA's statistical server AENA https://www.aena.es/es/estadisticas
The airlines whose traffic was switched from non-scheduled to scheduled between 1996 and 2006 by AENA are shown in bold.

statistics make it hard to gain a proper insight into the trend that charter airlines have followed or their real share of traffic at the airports where they operate.

To eliminate the bias that is caused by equating charter airlines with non-scheduled ones in statistics, we need to find out the volume of traffic carried by charter airlines, following the liberalization of EU air traffic that has officially been classified as scheduled. Reconstructing the series using data for traffic broken down by company is the only way of approaching the problem, a methodology previously used for Faro Airport in Portugal (Ramos-Pérez & Almeida, 2015). Only in this way can charter traffic be subtracted from the number of scheduled passengers and then added to the figure for non-scheduled passengers, assuming that this last figure corresponds to operations solely linked to the sale of package holidays.

As from at least 1996, AENA –a company with State participation and the successor of a state-owned company of the same name in charge of managing Spanish airports of general interest since 1991– draws up statistics for the traffic at each of its terminals, broken down to a degree that is unusual in European contexts. This allows us to find out the traffic generated by each airline at the different airports in the network run by AENA, differentiating between scheduled and non-scheduled services and arrivals and departures. For the period 2004–2022, this information is available on an online statistical website and, from last year onward, mass downloads of data can be made. As for previous years, we were able to consult AENA's Statistical Yearbook in Access format for the years 1996 and 1998. So far it has been impossible to gather data for 1997 or for the period 1999–2003.

For the study, we selected fifteen airports which we shall refer to as *tourist airports*. In 1994, 2 years after the third liberalizing package, these airports continued to specialize in non-scheduled international traffic, both in absolute and relative terms. By crossing the two chosen variables (the number of non-scheduled international passengers and the percentage of these passengers in relation to the total traffic at the airport), these fifteen airports were identified (Fig. 23.5), located in both Spanish archipelagos and on the mainland Mediterranean coast (Fig. 23.6). Although the level of specialization is lower at La Palma and Jerez than the other thirteen airports, it is important to remember that international charter flights only started operating at La Palma in 1987 and that although non-scheduled traffic has been recorded at Jerez Airport since the early 1980s, charter companies only came to acquire any relevance from 1991 onward.

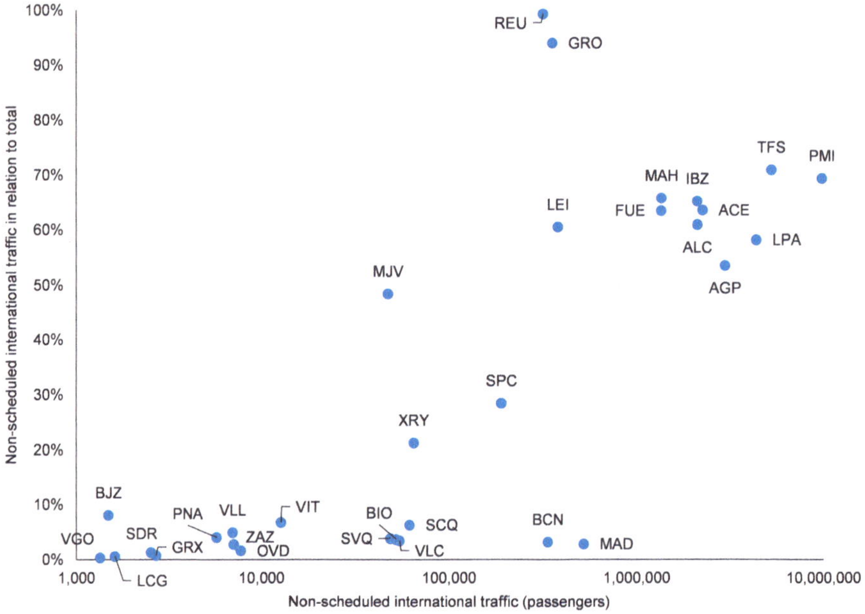

Fig. 23.5 Spanish airports specializing in non-scheduled international traffic in 1994. (Source: Own, based on official air traffic data by the Spanish Directorate General for Civil Aviation)

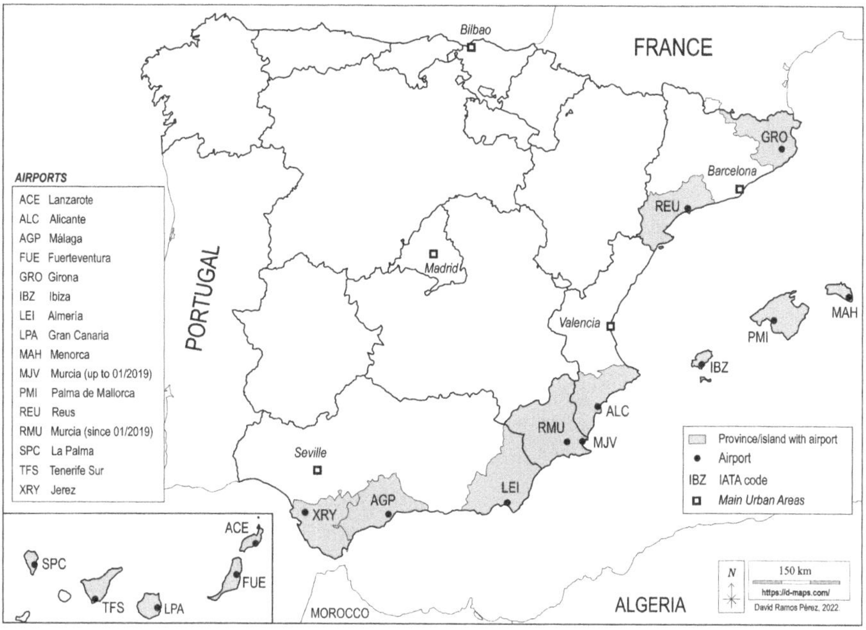

Fig. 23.6 Location of the selected airports. (Note: Own design, based on: https://d-maps.com/)

To identify the *leisure airlines* whose traffic has been classified as scheduled at these fifteen airports, the inventory of European charter companies drawn up in 1990 by the Institut du Transport Aérien (Belet & Colomb, 1991) was used, together with the list of airlines included in the *Non-Scheduled Flight Studies* for each IATA season, drawn up by the Spanish Directorate General for Civil Aviation between 1989 and 2000 (DGAC, 1989–1999). From that date on, the charter airline ranking drawn up yearly by the magazine *Airline Business* was used to track mergers, bankruptcies and newly established charter companies through to today.

We are nonetheless aware that it is no easy task to identify a company with a specific business model in a context like the current one where a trend toward their hybridization can be noted (Daft & Albers, 2015). This might cast doubt on the representativeness of the companies that we deem to be *leisure airlines*. That is why, in our selection, European *leisure airlines* were also considered to be companies whose traffic must revolve around a series of airports clearly associated with sun and sand tourism of Fordist origins, located in the extended Mediterranean area (Fig. 23.1). That is, they must demonstrate specialization in traffic that replicates the characteristics of charter companies up until 1992. This can clearly be seen in the Spanish case, where the scheduled traffic between 2004 and 2021 operated by the leading airlines that we have identified as leisure airlines was almost exclusively concentrated at the fifteen selected airports (Table 23.2). Air Berlin and Transavia are the only exceptions due to their stronger hybrid mix. In the first case, this resulted in a network carrier integrated in a global alliance, while, in the second, its formula is

Table 23.2 Leading leisure airlines by scheduled international traffic to Spanish airports (2004–2021)

Airline	Country	Total traffic Spain	Traffic 15 airports	% 15 airports
Germania	Germany	3,818,768	3,804,954	99.6%
TUI Airlines Nederland	Netherlands	2,233,169	2,220,260	99.4%
Thomas Cook Airlines (1)	UK	14,247,624	14,129,224	99.2%
Thomas Cook Belgium	Belgium	1,843,454	1,817,676	98.6%
Condor	Germany	52,410,649	51,575,738	98.4%
TUI Airlines Belgium	Belgium	15,495,194	15,142,619	97.7%
Neos	Italy	2,577,753	2,471,000	95.9%
Jet2.com	UK	52,505,540	50,191,927	95.6%
TUIfly (2)	Germany	51,221,966	48,868,228	95.4%
Edelweiss	Switzerland	3,418,670	3,260,348	95.4%
Monarch Airlines	UK	40,595,706	38,004,429	93.6%
LTU	Germany	5,494,909	5,060,132	92.1%
Globespan	UK	4,807,746	4,230,943	88.0%
TUI Airways (3)	UK	3,056,251	2,573,569	84.2%
Niki	Austria	17,540,498	14,736,021	84.0%
Sterling Airlines	Denmark	4,437,831	3,625,451	81.7%
Smartwings	Czech Republic	3,123,027	2,509,028	80.3%
Air Berlin	Germany	126,094,554	92,414,532	73.3%
Transavia Holland	Netherlands	33,858,677	23,652,156	69.9%
Transavia France	France	7,597,738	2,408,071	31.7%
	TOTAL 20 airlines	**446,379,724**	**382,696,306**	**85.7%**

Source: Drawn up using AENA's statistical server https://www.aena.es/es/estadisticas. Shown in darker grey are the airlines operating in 2021. NOTES: (1) Includes traffic by My Travel Airways; (2) Includes traffic by Hapag-Lloyd and Hapag-Lloyd Express; (3) Includes traffic by Thomson, Britannia and First Choice.

similar to that of a low-cost airline, with a French branch with more diversified destinations. Even so, they can still be regarded as leisure airlines due to their past history as charter companies and their close links with different German and Dutch tour operators. In addition to identifying the leisure airlines, we also proceeded to classify the remaining operators with a share of scheduled traffic at these airports, in accordance with the different business models habitually acknowledged in academic literature (DLR, 2008): full service network carriers (FSNC), regional carriers and low-cost carriers, (LCC).

Lastly, doubt might also be cast on the idea of equating non-scheduled traffic at the selected airports with that of leisure airlines. However, from an analysis of the makeup of non-scheduled international traffic at all the Spanish airports for the period 2004–2021, this traffic can be seen to be headed by companies whose operational strategy focuses on the same tourist flows typical of charter airlines (Table 23.3).

Table 23.3 Leading European airlines by non-scheduled international traffic to Spanish airports (2004–2021)

Airline	Country	Traffic	%
TUI Airways (1)	UK	82,338,417	26.4%
Thomas Cook Airlines (2)	UK	33,788,291	10.8%
Sunclass (3)	Denmark	17,584,768	5.6%
Transavia Holland	Netherlands	9,948,257	3.2%
TUIFly Nordic (4)	Sweden	7,320,443	2.3%
Smartwings	Czech Republic	5,330,980	1.7%
SAS	Sweden/Denmark	4,874,010	1.6%
Monarch Airlines	UK	3,999,182	1.3%
Jet Time	Denmark	3,655,522	1.2%
Thomas Cook Belgium	Belgium	3,356,988	1.1%
Primera Air	Denmark	3,130,483	1.0%
TUI Airlines Nederland	Netherlands	3,044,314	1.0%
Enter Air	Poland	2,949,787	0.9%
Germania	Germany	2,639,654	0.8%
Norwegian (5)	Norway	2,600,983	0.8%
TUI Airlines Belgium	Belgium	2,558,794	0.8%
Nova Airlines	Sweden	2,544,800	0.8%
Small Planet Airlines (5)	Lithuania	2,106,059	0.7%
Hamburg International	Germany	1,970,759	0.6%
Air Mediterranée	France	1,827,827	0.6%
Others	--	*114,058,085*	*36.6%*
	TOTAL	**311,628,403**	**100.0**

Source: Based on AENA's statistical server https://www.aena.es/es/estadisticas. Shown in darker grey are the airlines operating in 2021. NOTES: (1) Includes traffic by Thomson, Britannia and First Choice; (2) Includes traffic by MyTravel Airways; (3) Includes traffic by Thomas Cook Scandinavia and MyTravel; (4) Includes traffic by Britannia AB; (5) Includes the different subsidiaries of the parent companies.

Hence, it is possible to identify leisure airlines whose traffic was considered to be non-scheduled at some airports through to a certain date. They include Transavia; some British companies classified by AENA as non-scheduled (TUI Airways, Thomas Cook Airlines) even though they sold some of their seats directly; independent Scandinavian airlines (Jet Time) and ones associated with leading European tour operators (Thomas Cook Scandinavia, TUI Fly Nordic) that continue to link their flights to the purchase of a package holiday; Dutch and Belgium companies in a similar situation to the previous ones (Jetairfly and TUI Nederland); and, more recently, companies from countries in Central and Eastern Europe (Enter Air, Travel Service, Small Planet) with a charter-airline-type strategy similar to the dominant one prior to liberalization, targeted at new issuing markets like those of their countries of origin (Taylor, 2016), without overlooking mature markets like the French one (Bugnot, 2012), where they engage in a form of social dumping due to their low wage costs. There are also the former Scandinavian (SAS) and Finnish flag carriers (Finnair), highlighting the peculiarities of some outbound markets, where these airlines have always maintained close links with tour operators.

23.5 Leisure Airlines in Spain: A Decline but Not a Collapse

For decades, non-scheduled passenger traffic has acted as a fundamental variable in understanding the evolution of traffic at Spanish airports. The available statistics show a growth in this traffic in parallel with the Spanish economy's specialization in tourism, promoted by the Franco dictatorship (Pack, 2009), with non-scheduled traffic accounting for 40.5% of passenger traffic to the country's airports in 1971 (Fig. 23.2). After the economic crisis in the second half of the 1970s, this figure dropped, with a new recovery in the 1980s leading to non-scheduled traffic's peak share of Spain's total air traffic in 1990: 46% of the total. These non-scheduled passengers were mainly tourists travelling through tour operators on charter airlines to destinations on the Spanish coast where accommodation was concentrated. Hence, the fifteen selected airports accounted for over 93% of the non-scheduled air traffic to all Spanish airports between 1961 and 2021.

Although in absolute terms, the growth in non-scheduled traffic would continue through to 1999, the date when it reached its historical peak, the same did not apply in relative terms, with a reduction in its importance that became clearly visible from 1996. This general trend was replicated by all the selected airports, where a first drop in non-scheduled traffic can be observed from 1993, just a year after the beginning of the liberalization of EU air traffic. This decrease in non-scheduled traffic has continued through to today, with non-scheduled operations accounting for 2.6% of the total traffic to Spanish airports in 2021.

If we reconstruct the series for the fifteen selected airports, based on the approach described in the methodology, separating leisure airline traffic from scheduled traffic and adding it to non-scheduled traffic, we find that the drop experienced by these airlines seems not to have been so sharp. In fact, in 2019, they continued to account for over 35% of the international traffic at this group of airports, a figure that drops to 10.6% if we only consider the official data for non-scheduled traffic. Furthermore, during the growth period in the tourism demand that took place in the midst of the economic and financial crisis, the traffic by these carriers underwent a notable rise in absolute terms, helping to slow down the loss of their market share (Fig. 23.7).

Although there are clear differences between the airports, these global figures discount the relatively widespread notion in academic literature that leisure airlines have disappeared. Likewise, they also show that low-cost operators are far from having achieved a total hegemony at these airports, coinciding with the conclusions obtained for Faro Airport in Portugal, where an identical methodology was applied (Ramos-Pérez & Almeida, 2015). Last but not least, the figures also show that traditional package holidays, whose makeup has become increasingly flexible thanks to the introduction of formulae like *dynamic packaging*, are now far from solely associated with charter companies since they currently involve a wider variety of airlines.

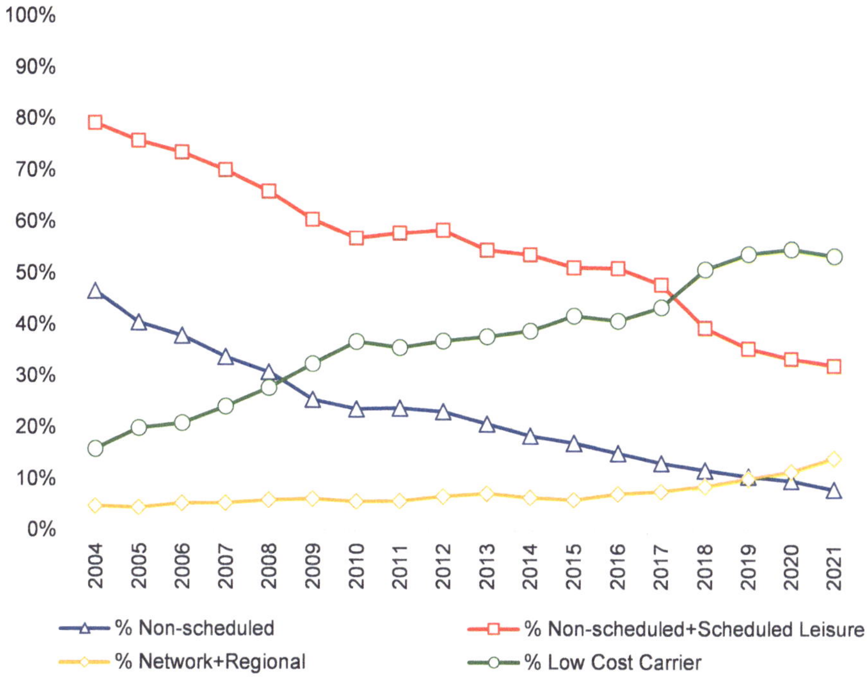

Fig. 23.7 International traffic at the selected tourist airports by airline business model (2004–2021). (Source: Based on AENA's statistical server https://www.aena.es/es/estadisticas)

As for differences among the airports, from Fig. 23.8, where the total international traffic is crossed with the relative importance of the sum of non-scheduled and leisure traffic, a certain correlation between both variables can be noted if we except the cases of Girona and Murcia, since the coefficient of determination (R^2) then rises from 0.237 to 0.659. That is, leisure companies have a higher market share at airports with less traffic, while at airports with a higher volume of passengers, they have largely been replaced by low-cost airlines. This partly confirms the hypothesis by Doganis (2010) that these companies can continue to play an important role at destinations with a low seasonal demand which one or two weekly flights at most can cover, since this kind of service is hard to integrate in the flight programmes of low-cost airlines. However, changes can be expected in forthcoming years because Ryanair, in particular, and Easyjet, to a lesser extent, tend to expand, incorporating routes with fewer than one daily flight (de Wit & Zuidberg, 2012).

The cases of Tenerife Sur and Gran Canaria seem to contradict this logic although, in their case, the importance of distance as an explanatory factor should not be overlooked because connections from Europe's main outbound markets involve a flight of over 4 hours, which complicates the management of low-cost fleets, whose business model is focused on shorter routes where benefits can be obtained from more frequent aircraft turnarounds (*ibid*, 2012). Indeed, Ryanair started operating in Gran Canaria in late 2009,

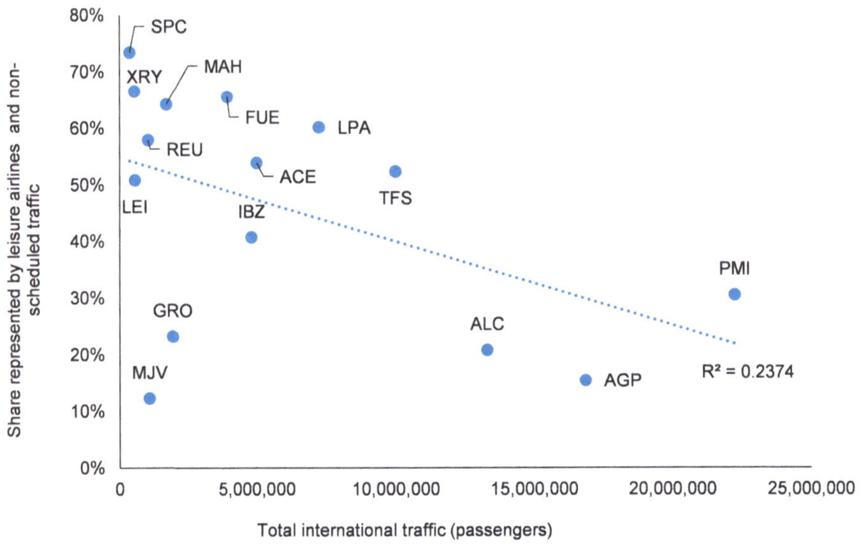

Fig. 23.8 Leisure companies at tourist airports in 2019. (Source: Based on AENA's statistical server https://www.aena.es/es/estadisticas)

2 years after it had first done so in Mallorca and 7 years after its first operations in Girona. As for Girona and Murcia, they are two special cases due to the early strong launch of operations there by Ryanair, which made it hard for leisure airlines to survive at both airports. After a progressive drop in flights by Ryanair at Girona, starting in 2011 and intensifying in 2014, leisure companies gradually returned to the airport, thus explaining that market share of 23% in 2019 when, in 2013, it barely reached 4%. This also demonstrates the lower dependence of leisure airlines on public subsidies from the local and regional authorities, aid which was so fundamental in Ryanair's operations at Girona Airport (Bel, 2009; Ramos-Pérez, 2016).

23.6 The Impact of Covid-19: Strategies by Airlines to Adapt to the Pandemic

The mobility restrictions imposed in different European countries from March 2020 in order to slow down the spread of Covid-19 has caused a significant drop in the air passenger demand over the last 2 years. However, this drop in traffic has not been a uniform one across all segments of demand. While business travel and long-haul travel underwent a genuine collapse, tourism travel to coastal destinations was the first to start up again from the summer of 2020. Although it has been pointed out that low-cost companies had better strengths to cope with this scenario, due to their lower dependence on the first two types of passengers (Suau-Sanchez et al., 2020), the most significant change that can be observed at the Spanish tourist airports analysed here is the stronger presence of network carriers, whose market share grew by four percentage points between 2019 and 2021 (Fig. 23.7).

Faced with a loss in traffic and falling returns, the former flag carriers reorganized their flight networks, taking advantage of the high number of aircraft that were not being used, with a big increase in routes and seats to Mediterranean tourist destinations. Their interest in these destinations was not totally new, since traditionally they have been an important niche market for operators like SAS or Finnair. Furthermore, from the last decade onward, other companies had started to boost their services to these airports, such as Lufthansa and Swiss. As from the summer of 2020, however, this presence rose substantially, with the incorporation of airlines that had not operated at tourist airports or only marginally so, such as Air France, KLM and LOT. The predominance of passenger traffic to/from these fifteen airports in network carriers' total traffic to/from Spain in 2021 highlights the importance of this strategy (Table 23.4) and its continuance beyond the summer and winter of 2020.

The boom in network carriers contrasts with the slowdown in operations by low-cost companies (a drop of 0.3 percentage points between 2019 and 2021) and the continued decline of leisure airlines (a drop of 3.4 percentage points between both dates, Fig. 23.7). Although in 2020, low-cost companies' market share reached a peak, this progressive rise stopped in 2021, with a certain incapacity on the part of these operators to take more advantage of a potentially favourable scenario for them,

Table 23.4 Percentage of passengers to/from tourist airports in the scheduled traffic of European network carriers operating in Spain (2019 versus 2021)

Airline	Country	Pax 2010 (%)	Pax 2019 (%)	Pax 2021 (%)	Difference 21-19 (p.p.)
Swiss	Switzerland	22.1	34.1	59.2	25.1
LOT	Poland	4.6	0.0	23.9	23.9
Lufthansa	Germany	6.9	21.6	45.0	23.5
Finnair	Finland	40.5	50.6	73.6	23.0
Alitalia	Italy	7.2	18.3	40.8	22.5
British Airways	UK	17.3	48.2	70.2	22.0
Air France	France	0.2	3.2	23.6	20.4
KLM	Netherlands	0.0	8.5	24.7	16.2
Brussels Airlines	Belgium	25.4	48.2	62.2	14.0
TAP	Portugal	1.9	13.6	22.8	9.2
Luxair	Luxembourg	72.0	73.6	81.8	8.1
Aer Lingus	Ireland	67.4	71.1	78.4	7.3
SAS	Sweden/ Denmark	57.9	97.7	89.1	-8.7
Austrian Airlines	Austria	23.4	99.9	57.2	-42.7

Source: Based on AENA's statistical sever: https://www.aena.es/es/estadisticas

given the characteristics of the fleet of aircraft they operate. The aggressive strategy by network carriers may have influenced these dynamics, together with the recovery of leisure airlines in 2021. The bid by leisure airlines to operate triangular routes, where a single operation combines two destinations in order to reduce costs and to increase the occupancy of the aircraft and ensuing profit margins, has allowed them to recover destinations whose demand does not currently justify a direct flight from certain issuing markets. The growing proportion of transit passengers shown in the statistics for Minorca, Jerez, La Palma, Alicante, Gran Canaria and Tenerife Sur Airports in 2021 is one outcome of the increase in these triangular flights.

23.7 Conclusions

The makeup of traffic at the airports that acted as driving forces behind the growth of Fordist tourism to the Spanish coast is so diverse and complex that the over-simplified approach that is sometimes taken to this issue does not apply. It cannot be concluded that former charter airlines have simply been replaced by low-cost operators. At the fifteen selected airports, there is a mix of airlines based on different business models, with a varying market share depending on the characteristics of each airport.

Former charter companies, now transformed into leisure airlines, are still far from extinct, even though a big drop of 47.1 percentage points can be observed in their relative importance since 2004. It is too soon to know whether this trend will culminate in the disappearance of all these operators, since leisure airlines showed an unusual strength between 2010 and 2012 during the peak period of the economic and financial crisis. This situation also has to be influenced by the fact that package holidays are more attractive in a context of greater job and income instability since they afford greater control over holiday budgets, given the uncertainties associated with individual purchases of travel, accommodation and food.

Given all the above and bearing in mind the worsening socioeconomic scenario after the Covid-19 pandemic, the stability of these operators might be seen as a very plausible likelihood. However, the data for Spain for 2020 and 2021 does not validate this hypothesis, with the detection of a drop in the number of tourists who buy a package holiday and a reduction in the market share of leisure airlines at the airports under study. We will have to wait until 2022 and 2023, when most of the population of the outbound markets have been vaccinated and mobility restrictions have been removed, to confirm whether it is a temporary drop motivated by a search for accommodation where social interaction is kept to a minimum or whether, on the other hand, this is a stable trend.

It can also be confirmed that leisure airlines are no longer the only ones to benefit from recoveries in sales of package holidays, since the almost exclusive link between charter companies and package holidays that was so typical of Fordist tourism is diminishing. Even in situations in which leisure airlines are completely dependent on tour operators, as is the case of the Scandinavian market, these tour operators work with a variety of airlines in order to flexibilize the design of their package holidays; hence the increasing presence of former flag carriers and network carriers at our selection of airports from 2015 in a process that has clearly been boosted during the last 2 years by the drop in business and long-haul travel and the reorganization of their services, with the redirection of flights toward destinations like the ones analysed here.

The obtained results help to characterize the timeframe of charter companies' replacement by low-cost ones, depending on the type of outbound market, the distance, the size of the existing demand and the extent of liberalisation in the case of non-EU Mediterranean third countries. In order to hone the conclusions of this study and confirm whether the results can be generalized, its geographical scope must be extended beyond Spain and an analysis must be conducted at a flight route level in order to draw more significant conclusions on competition among airlines at this type of tourist airports.

Lastly, the trends observed in the Spanish case confirm the hypothesis put forward by Lobbenberg (1995), shortly after the liberalization of the EU market, when he contended that routes formerly operated almost exclusively by charter companies were being taken over by a mix of airlines. This mix ranged from traditional charter companies selling most or all their seats to tour operators to public flag carriers. Between both extremes, there were some airlines that opted for mixed models, where the same aircraft carried package holiday and independent passengers. Subsequently, low-cost airlines joined them, which increased the diversity of the airlines operating these routes. With the hybridization and convergence of business models that has being going on for years, the differences between one set of airlines and another are currently far less evident than they were at the end of the last century.

Acknowledgements This research is included in the project "Tourist saturation in Spanish coastal destinations. Strategies for tourism degrowth", funded by the Ministry of Science, Innovation and Universities (reference RTI2018-094844-B-C31; period 2019-2022).

References

Bauzà Van Slingerlandt, A. (2009). El Aeropuerto de Palma de Mallorca. La puerta que abre el proceso de compresión espacio-temporal de la isla. En I. Caravaca, V. Fernández, R. Silva (dirs). *Ciudades, culturas y fronteras en un mundo en cambio* (Actas del IX Coloquio de Geografía Urbana) (pp. 259–276). Consejería de Obras Públicas y Transportes.

Bel, G. (2009). How to compete for a place in the world with a hand tied behind your back: The case of air transport services in Girona. *Tourism Management, 30*(4), 522–529.

Belet, D., & Colomb, L. (1991). *European charter airlines: Strategies for the 90s*. ITA.

Bieger, T., & Wittmer, A. (2006). Air transport and tourism. Perspectives and challenges for destinations, airlines and governments. *Journal of Air Transport Management, 12*, 40–46.

Brito, S. P. (2009). *Território e Turismo no Algarve*. Edições Colibri.

Buck, S., & Lei, Z. (2004). Charter airlines: Have they a future? *Tourism and Hospitality Research, 5*(1), 72–78.

Bugnot, F. (2012, 25 de mayo). Ces compagnies charter de l'Est qui profitent du ciel français. *L'Echo Touristique*. http://www.lechotouristique.com/article/ces-compagnies-charter-de-l-est-qui-profitent-du-ciel-francais,51337

Cano García, G. (1980). *El transporte aéreo en España*. Ariel.

Castillo-Manzano, J., Castro-Nuño, M., López-Valpuesta, L., & Pedregal, D. (2017). Measuring the LCC effect on charter airlines in the Spanish airport system. *Journal of Air Transport Management, 65*, 110–117.

Clavé, S. A., Saladié, Ó., Cortés-Jiménez, I., Fernandez-Young, A., & Young, R. (2015). How different are tourists who decide to travel to a mature destination because of the existence of a low-cost carrier route? *Journal of Air Transport Management* (42(C), pp. 213–218). https://doi.org/10.1016/j.jairtraman.2014.11.001

Daft, J., & Albers, S. (2015). An empirical analysis of airline business model convergence. *Journal of Air Transport Management, 46*, 3–11.

De Wit, J., & Zuidberg, J. (2012). The growth limits of the low cost carrier model. *Journal of Air Transport Management, 21*, 17–23.

DGAC (Dirección General de Aviación Civil). (1989–1999). *Estudio sobre vuelos no regulares* (temporadas de invierno y verano). Ministerio de Transportes, Turismo y Comunicaciones.

Díez Pisonero, R. (2012). La incidencia del turismo en la evolución de la conectividad aérea española (1970–2008). *Cuadernos de Turismo, 29*, 137–159. https://revistas.um.es/turismo/article/view/153821

DLR. (2008). *Airline Business Models*. Deutsches Zentrum für Luft- und Raumfahrt.

Dobruszkes, F., Schepens, V., & Decroly, J.-M. (2006). Éléments pour une géographie de l'offre charter européenne face à la concurrence des compagnies low-cost. In X. Bernier & C. Gauchon (Eds.), *Transport et tourisme* (pp. 65–76). Collection Edytem.

Doganis, R. (2010). *Flying off course. Airline economics and marketing* (4th ed.). Routledge.

Duval, D. T. (2013). Critical issues in air transport and tourism. *Tourism Geographies, 15*(3), 494–510.

ECMT. (1980). *Transport and the activity of the conference* (General report on activity, 26th annual report, 1979). European Conference of Ministers of Transport.

Forsyth, P. (2006). Tourism benefits and aviation policy. *Journal of Air Transport Management, 12*, 3–13.

Gago García, C., Diez Pisonero, R., & Córdoba y Ordóñez, J. A. (2011). Low cost y actividad turística. Adaptaciones recientes en enclaves de alta especialización turística. En Hernández Luiz, J.A. (Ed.). *Servicios, globalización y territorio* (Actas del V Congreso Geografía de los servicios, Tomo 2) (pp. 469–494). Anroart ediciones.

Gago-García, C., Díez-Pisonero, R., & Córdoba-Ordóñez, J. (2019). Transporte Aéreo y Conectividad: Evidencias de la Emergencia de Nodos Turísticos en el Sistema Urbano Mundial. *Revista de Estudios Andaluces, 37*, 161–183. https://doi.org/10.12795/rea.2019.i37.08

Gillmor, D. A. (1973). Irish holidays abroad: The growth and destinations of chartered inclusive tours. *Irish Geography, 6*(5), 618–625.

Gillmor, D. A. (1996). Evolving air-charter tourism patterns: Change in outbound traffic from the Republic of Ireland. *Tourism Management, 17*(1), 9–16.

Goetz, A. (2006). Transport geography: Reflecting on a subdiscipline and identifying future research trajectories: The insularity issue in transport geography. *Journal of Transport Geography, 14*(3), 230–231.

Hanson, S. (2003). Transportation: Hooked on speed, eyeing sustainability. En E. Sheppard & T. Barnes (eds) *A companion to economic geography* (pp. 468–483). Blackwell.

Hernández Luis, J. Á., Armengol Martín, M., González Morales, A., & Sobral García, S. (2011). El hándicap del transporte aéreo para el desarrollo inicial del turismo de masas en las Islas Canarias. *Cuadernos de Turismo, 28*, 75–91. https://revistas.um.es/turismo/article/view/147221

Ioannides, D. (1995). Strengthening the lies between tourism and economic geography: A theoretical agenda. *The Professional Geographer, 47*(1), 49–60.

Ivars Baidal, J. A., & Menor Muñoz, M. (2008). El impacto de las compañías de bajo coste en la actividad turística del área de influencia del aeropuerto de Alicante. *Estudios Turísticos, 175–176*, 89–104.

Landré, M., & Peeters, P. (2011). Editorial: Transport and tourism: A weak symbiosis. An introduction to the special issue. *European Journal of Transport and Infrastructure Research, 11*(3), 276–280.

Lobbenberg, A. (1995). Strategic responses of charter airlines to single market integration. *Journal of Air Transport Management, 2*(2), 67–80.

Lumsdon, L., & Page, S. J. (2004). *Tourism and transport: Issues and agenda for the new millennium*. Elsevier.

Lyth, P. J. (2009). Flying visits. The growth of British air package Tours, 1945–1975. En Segreto, L., Manera, L. y Pohl, M. (Eds.) *Europe at the seaside. The economic history of mass tourism in the Mediterranean* (pp. 11–31). Berghahn Books.

Martinez Reynés, M. R. (1997). *Turismo y transporte aéreo en Mallorca. Análisis jerárquico y funcional* (Tesis doctoral inédita). Universidad de las Islas Baleares. https://www.educacion.gob.es/teseo/mostrarRef.do?ref=191052

O'Connell, J., & Bouquet, A. (2015). Dynamic packaging spells the end of European charter airlines. *Journal of Vacational Marketing, 21*(2), 175–189.

Pack, S. (2009). *La invasión pacífica. Los turistas y la España de Franco*. Turner.

Papatheodorou, A., & Lei, Z. (2006). Leisure travel in Europe and airline business models: A study of regional airports in Great Britain. *Journal of Air Transport Management, 12*(1), 47–52.

Pearce, D. G. (1987a). *Tourism today. A geographical analysis*. Longman.

Pearce, D. G. (1987b). Mediterranean charters – A comparative geographic perspective. *Tourism Management, 8*(4), 291–305.

Pearce, D. G. (1987c). Spatial patterns of package tourism in Europe. *Annals of Tourism Research, 14*(2), 183–201.

Ramos-Pérez, D. (2016). State aid to airlines in Spain: An assessment of regional and local government support from 1996 to 2014. *Transport Policy, 49*, 137–147.

Ramos-Pérez, D. (2019). El transporte aéreo y la globalización del turismo. En E. Cañada Mullor & I. Murray Mas (Eds.). *Turistificación global: perspectivas críticas en turismo* (pp. 241–266). Icaria.

Ramos-Pérez, D., & Almeida, C. (2015). Transporte aéreo y turismo en el Algarve (Portugal): más allá de la dualidad chárter-bajo coste. En Espinosa Seguí, A. (coord.), *El papel de los servicios en la construcción del territorio: redes y actores* (pp. 543–558). Ediciones Universidad de Alicante.

Ramos-Pérez, D., & Dosbruszkes, F. (2019). The end of European charter airlines: Myths and realities. In A. Graham & F. Dobruszkes (Eds.), *Air transport : A tourism perspective* (pp. 143–162). Elsevier.

Ritzer, G., & Liska, A. (1997). McDisneyization and post-tourism: Complementary perspectives on contemporary tourism. In C. Rojek & J. Urry (Eds.), *Touring cultures: Modern transformations in leisure and travel* (pp. 96–109). Routledge.

Saladié, Ó., Anton Clavé, S., Cortés-Jiménez, I., Fernandez Young, A., & Young, R. (2014). La influencia de las rutas de vuelos de bajo coste en la elección del destino turístico. *Cuadernos de Turismo, 34*, 287–312.

Seguí Pons, J. M., & Martínez Reynés, M. R. (1994). El aeropuerto de Palma en la red de tráfico aéreo de pasajeros. Un análisis de relaciones. *Boletín de la Asociación Española de Geografía, 19*, 67–88.

Seguí Pons, J. M., & Martinez Reynés, M. R. (2010). Movilidad y sostenibilidad en el transporte aéreo: las compañías de bajo coste en los destinos turísticos de España y Baleares. *Revista Transporte y Territorio, 2*, 68–98. https://doi.org/10.34096/rtt.i2.234

Seguí Pons, J. M., & Martínez Reynés, M. R. (2012). Los tráficos internacionales y las compañías de bajo coste en las Islas Baleares. Movilidad versus sostenibilitad. *Territoris, 8*, 241–264. https://dialnet.unirioja.es/servlet/articulo?codigo=3911336

Seguí Pons, J. M., & Martínez Reynés, M. R. (2013). New mobilities and old problems. Tourist flows and low-cost Airlines in the Balearic Islands and Spain. *Current Issues of Tourism Research, 3*(2), 4–20. https://citr.up.krakow.pl/article/view/1643
Serrano Martínez, J. M. (1999). Tráfico aéreo de pasajeros, turismo y red aeroportuaria en España a finales del siglo XX. Algunas consideraciones. *Cuadernos de Turismo, 4*, 73–88. https://revistas.um.es/turismo/article/view/22891
Serrano Martínez, J. M., Espejo Marín, C., & García Marín, R. (2016). Accesibilidad aérea en la región de Murcia: una encrucijada compleja desde una perspectiva turística. *Papers de Turisme, 59*, 62–85.
Shaw, J., & Sidaway, J. (2011). Making links: On (re)engaging with transport and transport geography. *Progress in Human Geography, 35*(4), 502–520.
Spanish Official Gazette. (1961). Instrumento de ratificación del acuerdo multilateral relativo a derechos comerciales de los servicios aéreos no regulares europeos. *Boletín Oficial del Estado*, 117 de 17 de mayo, 7.419–7.421. https://www.boe.es/boe/dias/1961/05/17/pdfs/A07419-07421.pdf
Spill, J.-M. (1976). Les charters en Méditeranée. *Annales de Géographie, 468*, 206–224.
Suau-Sanchez, P., Voltes-Dorta, A., & Cugueró-Escofet, N. (2020). An early assessment of the impact of COVID-19 on air transport: Just another crisis or the end of aviation as we know it? *Journal of Transport Geography, 86*, 102749. https://doi.org/10.1016/j.jtrangeo.2020.102749
Taylor, Z. (2016). Air charter leisure traffic and organised tourism in Poland: Are charters passé? *Moravian Geographical Reports, 24*(4), 15–25.
Urry, J. (1990). *The tourist gaze: Leisure and travel in contemporary societies*. Sage.
Vanderheyden, A. (1979). L'offre touristique de quelques tours operators belges. *Travaux Géographiques de Liége, 167*, 63–78.
Vera Rebollo, J. F., & Ivars Baidal, J. A. (2009). Spread of low-cost carriers: Tourism and regional policy effects in Spain. *Regional Studies, 43*(4), 559–570.
Williams, G. (2001). Will Europe's charter carriers be replaced by "no-frills" scheduled airlines? *Journal of Air Transport Management, 7*(5), 277–286.

David Ramos-Pérez was born in La Palma (Canary Islands, Spain). He is Associate Professor of Human Geography at the University of Salamanca (Spain). His research work has focused on the geography of air transport, where his contributions on the spatial repercussions of the process of liberalisation of the EU market stand out. He has also been interested in the analysis of tourism, with some contributions on the spatial diffusion of holiday housing on the island of La Palma. He is currently a Visiting Professor at the Department of Geography of the University of Porto (Portugal), where he is broadening his training in the fields of urban and regional planning.

Chapter 24
Accessible Tourism in Spain: How Are Smart Cities Performing?

Rosario Navalón-García, Ana Clara Rucci, and Raquel Huete

24.1 Introduction and Theoretical Background

Approximately 68% of the world's population will live in cities by the year 2050 (UN, 2019), the population aged 65 and over is growing faster than all other age groups globally (UN, 2019), and more than a billion people, around 15% of the world's population, live with some type of disability (WHO, 2020). All of this has led public bodies to develop an agenda in order to address these issues. In this respect, the United Nations 2030 Agenda for Sustainable Development Goals (SDGs) has generated a challenge for the transformation of cities so as to make them more inclusive and resilient. Particularly, goal 10 refers to "reducing inequality within and among countries" and goal 11 to "making cities and human settlements inclusive, safe, resilient and sustainable", based on reducing inequalities in groups at risk of exclusion, such as the elderly or people with disabilities (United Nations Development Programme, 2021).

There is no doubt about the relevance of accessible tourism in Spain and it is gradually being incorporated as an issue of interest in tourism research. Generally speaking, accessible tourism constitutes a new approach that can contribute to deseasonalizing destinations, improving the quality and experience for tourists and

R. Navalón-García (✉)
Department of Regional Geographical Analysis, University of Alicante, Alicante, Spain
e-mail: r.navalon@ua.es

A. C. Rucci
University of Alicante, Alicante, Spain
e-mail: anaclara.rucci@econo.unlp.edu.ar

R. Huete
Department of Sociology and researcher, University Institute for Tourism Research of the University of Alicante, Alicante, Spain
e-mail: r.huete@ua.es

A. Blanco-Romero, M. Blázquez-Salom (eds.), *Spanish Tourism Geographies*, Geographies of Tourism and Global Change,
https://doi.org/10.1007/978-3-031-39780-6_24

favoring the competitiveness of tourist destinations belonging mainly to the sun and beach segment (Domínguez Vila et al., 2015; Fraiz Brea et al., 2008; González Velasco, 2008). It also complements other types of supply related to nature tourism (López Abril et al., 2014) or urban and cultural tourism (Molina Hoyo & Cànoves Valiente, 2010; Olivera Poll, 2006).

The different approaches to the concept are based on the Baker Report (1989) and the definition of tourism proposed by official bodies (European Commission, EUROPARC, OMT, and authors such as Fernández Villarán, 2007), giving rise to other related or complementary concepts, such as accessible tourism (Fernández, 2009), inclusive tourism (Brinckmann & Wildgen, 2003), or social tourism (Clemente-Soler et al., 2018; Molina Hoyo & Cànoves Valiente, 2010), among others, such as tourism for the elderly, or senior tourism (Lardies-Bosque et al., 2015; Raccanello & Cuamatzin, 2019), or accessible social tourism, coined by Garcés y Ferri (2012), (cited by Clemente-Soler et al., 2018). In addition to these terms, there are many references that relate this activity to quality of life and universal design (Espinosa Ruiz & Bonmatí Lledó, 2013), or smart mobility (Carrasco Campos et al., 2013) and universal accessibility as a relevant strategic factor (González Velasco, 2008; Hernández-Galán & Borau Jordán, 2017), due to its capacity to deseasonalize, increase the volume of visitors and improve the quality of the services (Domínguez Vila et al., 2015) and, therefore, competitiveness.

Until recently, studies on accessible tourism in Spain were, overall, conducted from an economic scientific or sociological perspective. There are many reference documents that have been published by entities and institutions related to the development and promotion of accessibility (Ministry of Social Affairs, CERMI, PREDIF, ENAT, Fundación ONCE, EUROPARC-España, 2007). However, from the beginning of the 2000s, the field of geography also began to produce research studies related to the tourism activity with accessibility, which are increasing in number and becoming more diverse in terms of the topics addressed. In view of the bibliographic evidence, we can say that accessibility and accessible tourism have been associated with the geography of tourism fundamentally from a destination planning perspective. There are studies that seek to incorporate accessibility on different scales and to various products, including the predominant sun and beach model, but there are also cases of accessibility being applied to protected natural spaces, urban destinations and exhibition spaces or interpretation centers. Within the diversity, we can observe a certain thematic clustering, with studies based on the analysis of the concept and regulatory framework within which accessible tourism is developed (Clemente-Soler et al., 2021; Millán Escriche, 2010). Others focus on the size and adaptation of the tourism supply on different scales (local, regional and national) or are based on compared proposals (Fernández Alles, 2007; Jurado Almonte, 2014; Molina Hoyo & Cànoves Valiente, 2010). There are also studies that analyze the adaptation of specific products, such as enotourism (Molina Hoyo, 2008), nature tourism (Leco Berrocal et al., 2011; López Abril et al., 2014; Santana-Santana et al., 2022), urban centers and museums (Molina Hoyo, 2008) or beach areas (Yepes et al., 2000; Fernández-Allès & Moral Moral, 2011) and the study of their future potential (Fraiz Brea et al., 2008). Although the majority focus on identifying obstacles and specific isolated aspects, without considering the accessibility chain (Hernández-Galán & Borau Jordán, 2017), there are also studies that examine the definition of the accessibility chain of accessible tourism and the definition of the indicators that enable a destination to be classified as accessible to all people (Cristóbal et al., 2017; Rubio Gil, 2000; Santana-Santana et al., 2022; Tejada Cruz & Fernández-Bermejo, 2017).

The implementation of accessibility in the territory is addressed from at least four dimensions: (1) First, space-time accessibility, which implies the displacement of people (Miralles, 2002); urban accessibility with the use of GIS to analyze and design routes (Hernández-Galán & Borau Jordán, 2017; Zimmermann-Janschitz, 2018) and the design of tourism applications (Françoso et al., 2013; Santana-Santana et al., 2020). (2) Also interesting is the study of social issues associated with disability and the territorial implications, which, in turn, connect the matter with the rights and care of people with different disabilities, but also with social and senior tourism (Alén González et al., 2010; Raccanello & Cuamatzin, 2019). (3) We can also identify another approach taken in studies that analyze tangible issues that constitute specific obstacles to access. For example, transport, equipment, information, or different tourism activities, which affect the tourism experience of people with disabilities, generating social discrimination (Huete García, 2013). In this respect, adaptations to enable inclusion (Alonso, 2007; Hernández-Galán, 2011; Tejada Cruz & Fernández-Bermejo, 2017) often associated to the disciplines of architecture and design (Alonso, 2002) are also studied. (4) Finally, we can identify the perspective of the economic effects and impacts of accessible tourism in the places and destinations in which it is developed (Fraiz Brea et al., 2008). In short, we can say that accessibility is a topic that has been gradually associated with the geography of tourism in Spain from a destination planning point of view. In general terms, it is connected with the diagnosis of accessibility of territories and resources and with the identification of the most appropriate strategic responses to the needs of a demand with special requirements.

Therefore, we can observe that the development of accessible tourism based on different lines of action, together with a smart tourism specialization (Santana-Santana et al., 2022) improves the competitiveness and positioning of an area, with a direct effect on the image projected and perceived by residents and visitors. Cities and tourist destinations are fundamental elements in global, integrated policies "for everyone" aimed at achieving environmental, social and economic sustainability (Koch & Ahmad, 2018). In these policies, as we have indicated, accessibility and governance in planning constitute fundamental aspects, as do transparency and a transversal approach to territorial smartness in the current technological and intensely competitive territorial context (Ivars-Baidal & Vera Rebollo, 2019; Luque Gil et al., 2015). Thus, it is appropriate to consider the connection between accessibility and territorial smartness. On the one hand, tourism is considered a right for everyone (McCabe, 2020). Achieving universally designed tourism environments would require the elimination of mobility, communication and other barriers to enable persons with disabilities to travel independently, equality and dignity (Buhalis & Darcy, 2011; Darcy & Dickson, 2009; McKercher & Darcy, 2018). On the other hand, smart tourist destinations have emerged as new models of governance for tourism development that consider innovation and technology as means to achieve the sustainability of the activity and to improve the quality of life of citizens (Rucci, 2021; Rucci et al., 2021).

The concept has a broad meaning that is not only associated with the field of technology, but also encompasses people and communities, information, infrastructure and social equity (Albino et al., 2015; Okafor et al., 2022). A smart destination implies that citizens, visitors and businesses benefit from more efficient traditional services and networks through mainly digital solutions, offering greater interactivity and safety in public spaces (European Commission, 2021). In this context, destinations are adjusting to these new technologies, not only in order to satisfy new traveler's needs (Gretzel et al., 2015) and to help tourists to anticipate their needs or to make real-time decisions (Boes et al., 2016) but also to positively influence the development of cities (United Nations Conference on Trade and Development, 2020). In this way, applying smartness across the whole tourism value chain could correct current problems related to sustainability, including wealth distribution, accessibility or overtourism, among others (Ramón-Rodríguez & Such-Devesa, 2018; Perles-Ribes et al. 2020a, b). According to Lai et al. (2020) a smart city should meet the needs of all consumer groups, offering, in addition, accessible technology, ensuring a society in which everyone can fully participate (Buhalis et al., 2019; Zhang & Cole, 2016). However, participation in tourism activities continues to exclude certain social groups, mainly people with disabilities (Figueiredo et al., 2012). A recent study that measures the progress of smart tourism destinations from the Region of Valencia, Spain, reveals that accessibility is the most complicated dimension for destinations to become smart, showing that accessibility indicators have the worst performance, leaving more room for improvement (Ivars-Baidal et al., 2021a). Ensuring access to travel and tourism opportunities for people with disabilities, as well as for the entire population requires considerable knowledge, effort and commitment, together with an inclusive design that is essential for all citizens. The Inclusive Smart City considers "all" citizens, recognizing their diversity, eliminating not only physical but also digital barriers (Hollands, 2008).

Within this framework, this chapter links the issues of accessibility and smartness in tourism for cases in Spain, taking the study carried out by Ivars-Baidal et al. (2021b) as a point of departure. It analyzes the planning instruments applied in smart strategies in 69 Spanish cities, focusing on how sustainability indicators are considered under the smart paradigm. The plans for both smart cities and smart destinations in Spain (Perles et al., 2020b) lay the foundations for the identification and collection of data related to initiatives, good practices and awards for accessible tourism applied to those cities. The cities and destinations that have been awarded have improved the mobility, communication or information for people with disabilities. In this sense, the objective of this chapter is not to provide empirical evidence on the relation or correlation of accessibility "attributes" or "assets" of cities and destinations. It is a first approach to the discussion on the existence and relationship between accessibility and smart cities/destinations plans, as conditions that can benefit and help cities to win an accessibility prize or to implement a smart destination plan (considering that accessibility constitutes a dimension of the smart model).

The abundance of public policies and literature research in this field justifies the consideration of Spain as a case of study. According to the World Economic Forum (WEF, 2019), Spain is regarded as the most competitive country in terms of tourism and has the highest rate of population ageing in the world after Japan (OECD, 2017). Likewise, tourists from countries such as the United Kingdom, France and Germany are the main international tourists visiting Spain and those considered to be the main demand for accessible tourism in the European Union (INE, 2019; Miller, 2014).

In terms of accessible tourism, Spain is considered as one of the countries that has made the most progress in terms of accessible tourism policies and has been taken as a case study by several authors all over the world (Domínguez Vila et al., 2015, 2020; Rucci et al., 2021; Rucci & Porto, 2022). Spain has also carried out specific actions such as the Queen Letizia Awards for Universal Accessibility of Municipalities.

With respect to digital transformation, Spain is considered the European country that has made the greatest efforts in terms of the digital transformation of tourism destinations (Herrera Priano, 2018) and is the only country with standards and certification in Smart Tourism destinations (UNE 178501 and 178,502). Moreover, since 2012, the Ministry of Industry, Trade and Tourism of Spain has developed a National Tourism Plan based on the transformation of destinations, considering concepts such as innovation, digitization, sustainability, accessibility, among others.

The chapter is structured as follows. A first section presents the conceptual bases of accessibility and destination smartness as an opportunity in the current context of tourism competition. The following section describes the different accessible tourism initiatives, good practices and awards, identifying the cities of Spain and the effectiveness of the decisions taken to build accessible cities and destinations based on a contrasted evaluation by expert juries. The third section analyzes the list of smart city and destination plans based on the study by Ivars-Baidal et al. (2021b) in which, among other indicators, the different accessibility planning tools described in the previous section and their relationship with the granting of accessibility awards are also discussed. Finally, a section for discussion and conclusions sheds some light on the field of research.

24.2 Accessible Tourism Initiatives, Good Practices and Awards in Spain

Public and private institutions have sought to reward the materialization of accessibility in cities and tourism destinations as we can observe in Table 24.1. On an international level, in 2019 the UNWTO launched the call to recognize those destinations "that make an important effort so that their destination can be enjoyed

Table 24.1 Accessible Tourism's awards and good practices

Award	Organization	Application range
Accessible Tourism Destinations	United Nations – World Tourism Organization & ONCE Foundation	International
Good Practice	Design for all Foundation	International
European Destinations of Excellence (EDEN)	European Commission	Europe
Access City Award	European Commission	Europe
European Capitals of Smart Tourism	European Commission	Europe
Queen Letizia Awards	Organizations from Spain	Spain
CERMI	Spanish Committee of Representatives of Persons with Disabilities	Spain

Source: Own elaboration

by any tourist, regardless of their physical, sensory or cognitive abilities" as Accessible Tourism Destinations (UNWTO, 2019). Another award is the "Good Accessibility Practices", by the Design for all Foundation (a non-profit organization internationally recognized for its trajectory in the field). On the European level, we can refer to awards such as "EDEN – European Destinations of Excellence" and "Access City Award", both organized by the European Commission. At the Ibero-American level, the "Queen Letizia Awards on Universal Accessibility of Municipalities" is prominent both in Spain and in Latin American countries; and the "CERMI.ES Awards" is organized by the Spanish Committee of Representatives of Persons with Disabilities (Ravlic, 2020).

In general terms, the awards for good practices recognize a specific situation, having no impact on its process or offering any advice, with calls launched every 1 or 2 years. In most cases they are open to all applications, they require the submission of a form and a report that justifies and evidences the achievements in accessibility. The decision of the winner is determined by a jury made up of experts and representatives of different organizations, who base their assessment on certain pre-established criteria. Each of these recognitions has its own characteristics and requirements, which can lead to a prize with or without a financial reward.

A brief description of each award:

24.2.1 Accessible Tourist Destinations – UNWTO

In 2019, the World Tourism Organization (UNWTO), together with the ONCE Foundation, awarded the International Recognition "Accessible Tourism Destination" (ATD) for the first time.

Its recipients are the public entities that manage tourist destinations, and its scope has a global perspective, considering the different links in the value chain of accessibility in tourism, whose seven main areas are: Planning, prior information and reservations; Arrival/departure transportation to the destination; Accommodation; Restoration; Urban and interurban transport in the destination; Tourism resources; and Public tourism administrations. Participants must provide a Justifying Report on the Accessibility of the Tourist Destination that is represented. After studying all the applications, a joint assessment is issued, selecting a single Tourist Destination as an example of accessibility for each year, in addition to the special mentions by type of destination. The results were announced in the framework of the 23rd UNWTO General Assembly, held in September 2019 in Saint Petersburg and Portugal was chosen as the best "Accessible Tourist Destination".

In addition to the award for ATD, a special distinction was given to "Urban Destination" and a mention to "Emerging Destination". The Urban Destination distinction was awarded to Barcelona (Spain) given that this Spanish city is "really advanced in terms of accessibility, presenting good conditions in its resources, and accommodation and restaurant establishments, especially in transport". The mention as an Emerging Destination went to the city of Thrissur (India) due to it being "in full development of accessibility, with a great involvement of public entities in this aspect".

24.2.2 Good Practice – Design for all Foundation

Design for all Foundation is an international non-profit entity founded in 2001, in Spain. The 11th edition of its international awards was held in 2021 with the aim of recognizing all the efforts, large or small, that have been carried out by companies, entities, administrations and professionals from around the world to demonstrate that any application of Design for All/ Universal Design contributes to improving the quality of life of all people. Unlike all the other awards mentioned in this section, it is the only one that charges an application fee for those who are not members of the Network.

It grants the Good Practices award to all those applications that meet the criteria of excellence. However, each year an international jury selects the 5 "Best Practices" which receive the Design for All Foundation International Award, standing out for their impact and marking the way forward in the future for best practices in Design for All /Universal Design.

Nominations may be submitted in one of the following two categories: (1) Environments, products and services that people can already enjoy; (2) Projects, proposals, initiatives, methodologies and studies. And within each of the categories it covers, in a non-exclusive manner, the following areas: domestic, industrial, graphics, ICTs, urban and natural spaces, buildings, transport and mobility, services, etc.

In the category "Environments, products and services in use", Vilamuseu Network of Museums and Monuments of La Vila Joiosa City Council has won the prize two times, in 2018 and in 2019.

24.2.3 EDEN – European Destinations of Excellence

This initiative was launched in 2006 and is awarded every 2 years. It aims to promote sustainable tourism development models throughout the European Union (EU). In the case of accessible tourism, only one call was made in 2013.

The award recipients are emerging, little-known destinations located in EU countries. Every 2 years, the European Commission chooses a theme in close cooperation with the national tourism authorities. Eligible destinations must show that an economically viable tourism offer has been developed based on that year's EDEN theme. For each participating country or selected destination, a winner is selected, along with four other mentions. In 2013, the Natural Park of the Mountains and Canyons of Guara in the province of Huesca was the selected destination in Spain.

24.2.4 Access City Award

Since 2011, the European Commission has convened the "Access City Award" (ACA) for all cities in the European Union with more than 50,000 inhabitants. The prize is awarded to the city that has clearly and sustainably improved accessibility in fundamental aspects of life in the city, and that has concrete plans for further improvements. The aim of the Prize is to inspire other cities, which may face similar challenges, and to promote good practice across Europe. Likewise, cities in second and third place are awarded, and initiatives in infrastructure, technology or heritage for which they have applied are highlighted.

These prizes and the Queen Letizia award are the only ones that grant an economic contribution to the winners.[1]

Cities of Spain has won different prizes throughout the years:

First prize: Ávila (2011) and Breda (2019)
Second prize: Castellón de la Plana (2020)
Third prize: Barcelona (2022)

Special mention: Terrassa (2012), Pamplona and Bilbao (2013), Burgos (2014), Málaga (2014), Logroño (2015), Arona (2015), Lugo (2017), Vigo (2019), and Palma (2022).

[1] More information about Access City Award official website, available at https://ec.europa.eu/social/main.jsp?catId=1141

24.2.5 Queen Letizia Awards "Universal Accessibility of Municipalities"

On a national level in Spain, an example of awards with extensive experience and recognition are the former Queen Sofía Awards, which from 2015 were renamed the Queen Letizia Awards, on the occasion of the head of state succession and the acceptance by the Queen of the Honorary Presidency of the Royal Board of Trustees on Disability of Spain. This organization, together with the ACS Foundation, annually convene these awards in Spain in five categories: (1) Promotion of Labor Insertion; (2) Universal Accessibility of Municipalities; (3) Technologies of Accessibility; (4) Rehabilitation and Integration, and (5) Inclusive Culture.

The Spanish Agency for International Development Cooperation (AECID) grants the "Universal Accessibility of Municipalities" award of great importance, in view of the positive impact of these awards have on Spanish municipalities. For this reason, since 2009 the Queen Letizia Awards have been extended to incorporate applications from Latin American Municipalities (Table 24.2).

24.2.6 European Capitals of Smart Tourism

This is a reward created by the European Commission in 2018, which recognizes European cities, giving them the opportunity to share their practices as smart tourism destinations in one of four categories: accessibility, digitalization, sustainability, creativity and cultural heritage. Each city that wants to be taken in consideration is asked to write four essays to present its best practices and measures implemented under each of the above-mentioned categories. A group of experts compares and selects two winners as "European Capital of Smart Tourism" and another four cities are recognized with European Smart Tourism Awards for their outstanding achievements in the individual categories of the competition (Rucci et al., 2021).

Unlike other awards, this one represents an inspiration for exemplary and innovative solutions in sustainable and accessible tourism development as well as innovative and inclusive digital tools to enhance the visitor's experience.

Málaga and Valencia (Spain) has won both prizes in "best practices" and "accessibility" categories, however, Valencia was awarded as European Capital of Smart Tourism in 2022.[2]

[2] More information about European Capitals of Smart Tourism official website, available at https://smart-tourism-capital.ec.europa.eu/index_en

Table 24.2 Queen Letizia Awards "Universal Accessibility of Municipalities" for period 2005–2021

Year	Less than 10,000 inhabitants	Between 10,001 and 100,000 inhabitants	More than 100,001 inhabitants
2005	Breña Alta (Santa Cruz de Tenerife)	Quart de Poblet (Valencia)	Málaga
2006	Ribadesella (Asturias)	Lugo	Santiago de Compostela
	Fontanar (Guadalajara)		
2007	Vegadeo (Asturias)	Aldaia (Valencia)	Jeréz de la Frontera (Cádiz)
2008	Blanca (Murcia)	Ávila	Elche (Alicante)
2009	Hacinas (Burgos)	Arona (Santa Cruz de Tenerife)	Santander
2010	Alcaracejos (Córdoba)	Basauri (Vizcaya)	
2011	Buenavista del Norte (Tenerife)	Torrent (Valencia)	Valencia Vitoria
2012	Carranque (Toledo)	Cáceres	Burgos Valladolid
2013	Ayuntamiento de Frigiliana	Ayuntam. de Linares (Jaén)	Ayuntamiento de Logroño
2014	Cubillos del Sil (León)	Palencia	Badajoz
2015	Almodóvar del campo (Ciudad Real)	Pozuelo de Alarcón (Madrid) Plasencia (Cáceres)	Lérida
2016	Sabiñánigo (Huesca)	Ávila Carreño-Candás (Asturias)	Terrasa (Barcelona)
2017	*Closed*		
2018	Medina de Rioseco (Valladolid)	Antequera (Málaga)	Murcia
2019	*Information not available*		
2020	*Information not available*		
2021			Mérida

Source: Own elaboration based on the Queen Letizia Awards official website. https://www.rpdiscapacidad.gob.es/actividades/premios.htm

24.2.7 CERMI.ES: Spanish Committee of Representatives of Persons with Disabilities Awards

The Spanish Committee of Representatives of Persons with Disabilities Awards (CERMI.ES) are a recognition and an expression of gratitude for the solidarity shown by Spanish society with the objectives defended by this organized representation of people with disabilities and their families. They are intended for: people; projects; experiences; Spanish companies, entities or administrations that propose initiatives in favor of the rights, community participation, inclusion and well-being of people with disabilities and their families. The CERMI.ES were created in 2002 and in the category of Best Regional and/or Local Action the prize have been awarded, chronologically, to:

In 2020, the Ministry of Social Policy of Galicia (ex-aequo) was awarded for the scope and vigor of its policies for the inclusion of people with disabilities deployed in dialogue and co-governance with Galician civil society; and, the Institute for People with Disabilities of the Barcelona City Council (ex-aequo) through the Diagnosis of accessibility of the city of Barcelona, it reissued the historical commitment of the Catalan capital with the values of universal accessibility. In 2019, the award went to the Government of Extremadura for its quick political decision to preserve the economic sustainability in the third sector of disability in the region. In 2018, to Ceuta for the work of the Technical Office of Accessibility. In 2016, to the Government of Aragón for the legal regulation of the social services management modality based on a non-profit civic initiative, innovative management formula. In 2015, to the Regional Center for Child Assessment and Coordination of Madrid because it is a pioneering social resource conceived to offer comprehensive, specialized and free care to girls and boys with disabilities or with support needs. In 2014, to the City Hall of Pozuelo de Alarcón, Madrid, for its permanent consideration towards disability as an important element in political action and in the management of municipal public services, within a framework of active collaboration with the representative entities of the sector. In 2013, to Logroño City Hall because in a time of crisis in which most of the Administrations sacrificed inclusion policies, it reactivated public action in terms of disability within a framework of active collaboration with the representative entities of the sector. In 2012, to the City Hall of Elche, Alicante, for successfully developing inclusive local policies for people with disabilities and their families in recent years, promoting the participation, commitment and active citizenship of this population group. In 2011, to the City Hall of Santanderfor the Plan for the Care and Promotion of Persons with Disabilities. In 2010, to the City Hall of Málaga for its Málaga Accessible City initiative, which aims to extend, with growing success, universal accessibility and design for accessible to all spaces for civic relations and coexistence in this city of Andalusia. In2009, to the Badajoz Provincial Council for undertaking, in close collaboration with the associative fabric of disability in the territory, accessibility plans in all the municipalities that make up the province, systematically extending the paradigm of universal accessibility and reaching rural areas where the difficulty to exercise the rights of people with disabilities increases considerably. In 2008, to the City Hall of Ávila for its commitment to work in favor of universal accessibility, considering it a transversal axis of its municipal policy as a whole. In 2007, to the City Hall of Pontevedra for its intention and continued action to "turn the Galician city into an accessible, open and free space for all people". In 2006, to the Regional Government of Madrid for the project "Independent Living and Promotion of Social Participation", of the Ministry of Family and Social Affairs, for which thirty-five people from Madrid with very severe disabilities received the help of a personal assistant. In 2005, to the Ministry of Public Administrations for having substantially reformed, through a Royal Decree, the access of people with disabilities to the public service sector, expanding the number of positions reserved for people with disabilities within the Administration to five percent. Cumulative and preference when choosing a destination. And, in 2004, to the City Hall of Jerez de la Frontera for the approval of the "Municipal Plan for the accessibility and social integration of people with disabilities 2004–2007", which provided a total of 177 measures in urban accessibility, professional training, employment, social services, health, housing, leisure and sports, as well as the creation of the Municipal Office for Attention to Disability.

24.3 Accessibility Standards in Spain

In Spain, for some time another way of bringing about conditions of accessibility, has been the establishment of standards, particularly those referring to accessibility. The first dates back to the year 2001 (Accessibility in building and urbanism. Design general criteria) and the last to 2018 (Management system of smart tourist destinations, which includes accessibility requirements). Related to accessibility, there are other technical standards linked to cultural heritage, smart cities, public transport, building construction, beaches and signs.

A special mention goes to the International Standard on Accessible Tourism (ISO 21902: 2021 Tourism and related services – Accessible tourism for all – Requirements and recommendations). It was created by the ONCE Foundation for the Cooperation and Social Inclusion of People with Disabilities in Spain, the UNWTO and the Spanish Association for Standardization (UNE). This standard establishes the requirements and recommendations for accessible tourism and is aimed at both service providers and tour operators, as well as public administrations and tourism destinations. The ISO-21902 standard was published on July 12, 2021, establishing that the historical absence of homogeneous measures on accessibility in tourism destinations and that the results from the implementation of this standard will only be seen in the coming years.

24.4 Analysis of Smart Plans and Accessibility Tools in Spain

Table 24.3 presents the cities analyzed in Ivars-Baidal et al. (2021b) linked with the accessibility tools mentioned in the previous section. Smartness plans are divided into two approaches: smart city, which is referred in Table 24.3 as (a) Smart Plans and (b) EDUSI; and smart destinations, referred in Table 24.3 as (c) RED.ES and (d) Smart Plan / STD. While accessibility tools are divided into three levels of application: international, referred as (e) Accessible Tourism Destination awarded by UNWTO and (f) Good Practice awarded by Design for all Foundation; at the European region, referred in Table 24.3 as (g) EDEN awarded by Design for all Foundation, (h) Access City Award and (i) European Capitals of Smart Tourism, the three of them awarded by European Commission; and, in Spain, referred in

Table 24.3 Smartness and accessibility tools applied in 69 cities of Spain

	Smartness				Accessibility						
	City		Destination		International		Europe			Spain	
City	a	b	c	d	e	f	g	h	i	j	k
Albacete	2013	2016									
Alcalá de Henares		2016									
Algeciras	2018	2016									
Alicante		2016	2016								
Almería		2017									
Aranjuez		2016	2018								
Ávila	2016	2018						2011		2008 2016	2008
Badajoz		2016	2015							2014	2009
Barcelona	2016	2016			2019			2022			2020
Benalmádena		2018	2015								
Benidorm	2019	2018	2018								
Bilbao	2018	2016	2018					2013			
Burgos	2015									2012	
Cáceres		2016	2016							2012	
Cartagena	2016	2020									
Castelldefels	2018										
Castellón de la Plana	2017	2016		2017				2020			
Ciudad Real		2016	2015								
Córdoba		2016	2018	X							
Coruña	20,122,020	2017	2016	20,172,020							
Cuenca	2015	2016									
El Hierro			2016								
El Puerto de Santa María	2014	2016	2018								

(continued)

Table 24.3 (continued)

	Smartness				Accessibility						
	City		Destination		International		Europe			Spain	
City	a	b	c	d	e	f	g	h	i	j	k
Elche		2017		20,162,018						2008	2012
Estepona		2016	2015								
Fuengirola		2018	2015								
Gijón	2013	2018	2016								
Granada	2018	2016	2015	2018							
Guadalajara		2017	2015								
Huelva	2014	2016	2015								
Huesca		2018	2018				2013				
Jaca			2018	X			2013				
Jaén		2018								2013	
Las Palmas (S)	2014	2016	2016					2022			
León	20,182,027	2016 2022		20,182,027							
Lloret de Mar			2018								
Logroño	20,162,020	2018		20,182,022				2017		2013	2013
Lorca	2016	2018									
Lugo		2016	2016					2017		2006	
Madrid	2014			2016							2006 2020
Málaga	2018		2015						2019	2005	2010
Marbella	2013		2015								
Mérida		2017								2021	
Motril		2016									
Murcia	2016	2017	2016	20,172,020						2018	
Orihuela	2017	2016									

Oviedo		2016	2014 2019	2022							
Palencia	2010	2016	2016							2014	
Palma de Mallorca	2013		2018								
Pamplona	2013		2018					2013			
Paterna		2016									
Ponferrada			2016								
Roquetas de Mar		2017	2018								
Salamanca		2016	2018								
San Bartolomé de Tirajana		2018	2018								
San Cristobal de la Laguna	2015	2018									
Santa Cruz de Tenerife		2016	2018								
Santander	2015	2016	2016	20,152,020						2009	2011
Santiago de Compostela	2015	2016		20,172,022						2006	
Segovia			2016								
Sevilla	2014	2016	2015								
Tarragona			2018	2022							
Toledo		2017	2015								
Valencia	2013		2016						2022	2011	
Valladolid		2018	2016	20,162,019						2012	
Vigo	2015	2016						2019			
Vila Joiosa (La)						2018					
Vitoria-Gasteiz	2017			20,182,020						2011	
Zaragoza	2012		2018	20,172,021							

Source: Own elaboration

Note: a – Smart Plan; b – EDUSI; c – RED.ES; d – Smart Plan / STD; e – Accessible Tourism Destination (UNWTO); f – Good Practice; g – EDEN; h – Access City Award; i – European Capitals of Smart Tourism; j – Queen Letizia Awards; k – CERMI.ES

Table 24.3 as (j) Queen Letizia Awards and (k) CERMI.ES. The year of the implementation of the smart plan or the awarding of the accessibility prize is also shown in Table 24.3.

According to the main results of the analysis of smart plans by Ivars-Baidal et al. (2021b:8):

> *The plans, which are in most cases linked to the availability of public funds, reveal a qualitative and quantitative leap forward in the obtaining and analysis of data by cities. In contrast, the results from the questionnaire addressed to the technical officers of smart cities and destinations show a rather limited implementation of these tools. Most projects are still under execution and their effective application has been postponed due to legal, organizational or technical problems (including interoperability issues).*

This study focuses on selected cities that are part of the Smart Cities Network (RECI). Of the total of 69 cities that have advanced in smart destination plans, 37.68% have obtained awards related to accessibility. This may lead us to believe that these documents with a territorial intelligence approach raise reflections and actions to improve the accessibility chain and that their implementation is the trigger that leads to achieving these awards. However, in the detailed analysis of the respective documents, both with the Smart City approach, or EDUSI (Integrated Sustainable Urban Development Strategy), and the Smart Tourism Destination, only six cities can be found to implement approaches to improve people's accessibility in a concrete way, which highlight the link between these conditions and the prizes granted a posteriori. These cities are: Barcelona, Las Palmas, Málaga, Mérida, Murcia and Valencia. It is worth mentioning some specific cases: Barcelona has obtained important achievements in terms of both smartness and accessibility. The Catalan capital held the Olympic and Paralympic Games in 1992, creating a legacy of accessibility conditions that places Barcelona in a superb position compared to other cities. Last year, Barcelona won prizes from CERMI, Access City Award and special mentions at the international level: Accessible Tourism Destination by UNWTO. Another example could be Málaga or Valencia. Both of these cities have been making great efforts in accessibility over the last few years. They have both won the European Capital of Smart Tourism award which seems to reflect the combination of both smartness and accessibility improvements.

Overall, in the period of analysis of this study, from 2012 to 2022, for the 69 cities analyzed, 41 awards or prizes related to accessibility proposals have been won, but the relationship of these with the proposal of plans and projects related to territorial smartness is more than doubtful. Not surprisingly, 73.17% of these awards were obtained prior to the approval of the aforementioned smart plans, so it seems clear that the decisions on accessibility improvements in the cities analyzed are not the result of a smart approach, as they were taken before the study plans were drafted. If there is a relationship between accessibility actions and these plans, which only occurred a posteriori in 26.8% of the cases, it seems clear that those based on a Smart City approach seem to have a greater relationship, as 72.7% of the awards obtained after the plans were drawn up are in cities with Integrated Sustainable Urban Development Strategy plans (called EDUSI) or Smart City plans (63.63%), with Red.es plans accounting for 45.4% and Smart tourism destination for only 18.18% of the cases. This may suggest that the determination of plans with a clearer tourism-orientated focus on aspects more related to technology, awareness raising and even sustainability, but that the accessibility aspects of destinations do not form an essential part of the proposals. Thus, it does not seem that there is a simple causal link between territorial intelligence plans and progress in improving the accessibility of cities and destinations, and that the social, legislative and even urban planning or economic context can provide valid explanatory arguments.

24.5 Conclusions

This study argues that it is interesting and intelligent to focus on improving accessibility for visitors and residents. Countries and destinations are beginning to incorporate improvements in accessibility, as we can observe in the review of the international awards won by Spanish cities. When comparing the smart planning approach and urban improvements in accessibility, it has been found that there is no correlation between the smart approach and *a posteriori* improvements in accessibility, with percentages that show the scarce incidence, as in only six cases improvements in accessibility could be related to the proposals of the plans, subsequently 23.08%. However, the focus of this study is not to attempt to analyze whether cities began first with smart plans or accessibility initiatives. It should also be noted that only smart plans and accessibility tools have been considered. However, there are others such as legislation, specific plans, and technological developments, which have allowed the materialization of these issues in cities. For this reason, this study does not intend to carry out a causal analysis but rather to propose a first discussion on the relationship between the themes in cities in Spain.

In this study, it is understood that accessibility tools and intelligence plans are meant to improve the conditions of more inclusive and resilient cities, as established in the Agenda 2030 SDG. However, there are many other factors in cities that influence the better or worse conditions in achieving smartness, such as topography or whether the city is the provincial capital or not. This study leads future research for studying the legacy of cities that "became" smart. What did they do beforehand to achieve smartness? What previous conditions are needed to "become" smart in an easier way? Do accessibility tools play a key role in achieving accessibility? Are cities that have been awarded with accessibility prizes better prepared than other cities to become smart? Further research is needed.

Finally, as there are studies that indicate that tourists find great differences in the imagined accessibility (before traveling to the destination) and what is perceived when they visit the place (within the destination) (Buhalis & Darcy, 2011; Cheng & Chen, 2015; Lättman et al., 2018), international standards are a good tool for preventing these differences. However, in this study accessibility standards are not considered due to the lack of information about public and private organizations that have implemented or certified some of them. This information could be a good indicator of accessibility or even of smartness (considering accessibility within it) distributed by tourism destinations.

References

Albino, V., Berardi, U., & Dangelico, R. M. (2015). Smart Cities: Definitions, Dimensions, Performance, and Initiatives. *Journal of Urban Technology, 22*(1), 3–21. https://doi.org/10.1080/10630732.2014.942092

Alén González, M. E., Domínguez Vila, T., & Fraíz García, A. (2010). El turismo senior como segmento de mercado emergente. *Cuadernos de Turismo, 26*, 9–24.

Alonso, F. (2002). *Libro Verde de la accesibilidad en España*. Imserso-Acceplan. http://www.ceapat.es/InterPresent1/groups/imserso/documents/binario/libroverdedelaaccesibilidadene.pdf

Alonso, F. (2007). Algo más que suprimir barreras: conceptos y argumentos para una accesibilidad universal en Trans. *Revista de traductología, 11*, 15–30.

Baker, M. (1989). *Tourism for All: A report of the working party chaired*. English Tourist Board.

Boes, K., Gretzel, U., Zhong, L., Koo, C., Buhalis, D., & Inversini, A. (2016). Smart tourism destinations: Ecosystems for tourism destination competitiveness. *International Journal of Tourism Cities, 2*(2), 108–124.

Brinckmann, W. E., & Wildgen, J. S. (2003). Desafíos para los estudiosos del turismo: la construcción de la sociedad inclusiva y del turismo accesible. *Cuadernos de Turismo, 11*, 41–58. https://revistas.um.es/turismo/article/view/19621

Buhalis, D., & Darcy, S. (Eds.). (2011). *Accessible tourism: Concepts and issues*. Channel View Publications.

Buhalis, D., Harwood, T., Bogicevic, V., Viglia, G., Beldona, S., & Hofacker, C. (2019). Technological disruptions in services: lessons from tourism and hospitality. *Journal of Service Management, 30*(4), 484–506. https://doi.org/10.1108/JOSM-12-2018-0398

Carrasco Campos, A., Martínez Fernández, L. C., & Moreno Mínguez, A. (2013). Revisión crítica de la medición del bienestar desde una perspectiva interdisciplinar: hacia una propuesta de indicadores subjetivos y espaciales. *Prisma Social, 11*, 91–122.

Cheng, Y. H., & Chen, S. Y. (2015). Perceived accessibility, mobility, and connectivity of public transportation systems. *Transportation Research Part A: Policy and Practice, 77*, 386–403. https://doi.org/10.1016/j.tra.2015.05.003

Clemente-Soler, J. A., Bote, M., & Sánchez, P. (2018). El turismo social accesible como nuevo modelo turístico. *Cuadernos de Turismo, 41*, 139–159.

Clemente-Soler, J. A., Rodríguez, D., & Martínez- Martínez, A. L. (2021). Evolución de la accesibilidad en el turismo europeo: retos y desafíos. Challenges of the evolution of accesibility in European tourism. *Gran Tour: Revista de Investigaciones Turísticas, 23*, 25–49.

Cristóbal, E., Martín-Fuentes, E., & Daries-Ramón, N. (2017). *Turismo accesible: una oportunidad de diferenciación para los destinos*. https://www.researchgate.net/publication/324018227_Turismo_accesible_una_oportunidad_de_diferenciacion_para_los_destinos

Darcy, S., & Dickson, T. (2009). A Whole-of-Life Approach to Tourism: The Case for Accessible Tourism Experiences. *Journal of Hospitality and Tourism Management, 16*(1), 32–44.

Domínguez Vila, T., Darcy, S., & Gonzalez Alen, E. (2015). Competing for the disability tourism market. A comparative exploration of the factors of accessible tourism competitiveness in Spain and Australia. *Tourism Management, 47*, 261–272. https://doi.org/10.1016/j.tourman.2014.10.008

Domínguez Vila, T., Alén González, E., & Darcy, S. (2020). Accessibility of tourism websites: the level of the countries' commitment. *Universal Access in the Information Society, 19*, 331–346. https://doi.org/10.1007/s10209-019-00643-4

Espinosa Ruiz, A., & Bonmatí Lledó, C. (2013). *Manual de accesibilidad e inclusión en museos y lugares del patrimonio cultural y natural*. Ediciones TREA.

EUROPARC-España. (2007). *Catálogo de buenas prácticas en materia de accesibilidad en espacios naturales protegidos*. Fundación Fernando González Bernáldez. https://www.redeuroparc.org/system/files/shared/manual_5.pdf

European Commission. (2021). "European Capitals of Smart Tourism. Best Practices", European Capitals of Smart Tourism. https://smart-tourism-capital.ec.europa.eu/bestpractices_en.

Fernández, M. (2009). Turismo accesible y turismo para todos en España: antecedentes, estado de la cuestión y futuras líneas de investigación. *Estudios turísticos, 180*, 141–153.

Fernández Alles, M. T. (2007). *Turismo accesible: análisis de la accesibilidad hotelera en la provincia de Cádiz*. Tesis doctoral. Facultad de Ciencias Económicas y Empresariales. Universidad de Cádiz. https://rodin.uca.es/bitstream/handle/10498/15665/mtfernandezalles.pdf?sequence=1

Fernández Villarán, A. (2007). El acceso al turismo para las personas con discapacidad. *Desafíos y compromisos del turismo: hacia una visión más humana, Congreso UNIJES*. Universidad de Deusto. 47–49.

Fernández-Allès, M. T., & Moral Moral, M. (2011). La accesibilidad en las playas españolas como factor estratégico de calidad de la oferta turística. En *XV Congreso Anual de la Academia Europea de Dirección y Economía de la Empresa* (The markets of tomorrow: basis for their analysis today).

Figueiredo, E., Eusebio, C., & Kastenholz, E. (2012). How diverse are tourists with disabilities? A pilot study on accessible leisure tourism experiences in Portugal. *International Journal of Tourism Research., 14*, 531–550. https://doi.org/10.1002/jtr.1913

Fraiz Brea, J. A., Alén González, M., & Domíngez Vila, T. (2008). La accesibilidad como oportunidad de mercado en el management de destinos turísticos. *Revista de Análisis Turístico, 5*, 30–45.

Françoso, M. T., Cortijo Costa, D., Mello Valin, M., & Rodrigues Amarante, R. (2013). Free software for development of Web GIS in tourism accessibility. *Applied Mechanics and Materials, 256–259*, 2953–2956. https://doi.org/10.4028/www.scientific.net/AMM.256-259.2953

Garcés, J. y Ferri, M. (2012). *Establecimiento de los criterios de accesibilidad universal a los recursos turísticos*. Proyecto: Investigación para la mejora del conoci miento en turismo social accesible para todos en España (SOCIALTOURISM). proyecto financiado por el Ministerio de Economía y Competitividad (ref.:CS0201015425).

González Velasco, D. J. (2008). *El mercado potencial del turismo accesible para el sector turístico español*. Ministerio de Industria, Turismo y Comercio. http://www.accesturismo.com/assets/img/publicaciones02/Mercado_Potencial_Turismo_Accesible.pdf

Gretzel, U., Koo, C., Sigala, M., & Xiang, Z. (2015). Special issue on smart tourism: Convergence of information technologies, experiences and theories. *Electronic Markets, 25*(3), 175–177.

Hernández-Galán, J. (2011). *Manual de accesibilidad para técnicos municipales*. Fundación ONCE y Vía Libre. https://biblioteca.fundaciononce.es/publicaciones/coleccionespropias/coleccion-accesibilidad/manual-de-accesibilidad-para-tecnicos

Hernández-Galán, J., & Borau Jordán, J. L. (2017). *Guía técnica de accesibilidad en espacios naturales*. Organismo Autónomo Parques Nacionales y Fundación ONCE. https://biblioteca.fundaciononce.es/sites/default/files/publicaciones/documentos/guia_impresa_2017-marcadores-01_revisado.pdf

Herrera Priano, F. (2018). Ciudades inteligentes en España: ¿se está produciendo realmente un cambio? *Equipamiento y servicios municipales, 183*, 40–45.

Hollands, R. (2008). Will the real Smart City please stand up? Creative, progressive or just entrepreneurial? *City: Analysis of Urban Trends, Culture, Theory, Policy, Action, 12*, 303–320. https://doi.org/10.1080/13604810802479126

Huete García, A. (2013). La exclusión de la población con discapacidad en España. Estudio específico a partir de la Encuesta Social Europea. *Revista Española de Discapacidad, 1*(2), 7–24. https://doi.org/10.5569/2340

Ivars-Baidal, J. A., & Vera Rebollo, J. F. (2019). Planificación turística en España. De los paradigmas tradicionales a los nuevos enfoques: planificación turística inteligente. *Boletín de la Asociación de Geógrafos Españoles, 82*, 1–31. https://doi.org/10.21138/bage.2765

Ivars-Baidal, J. A., Celdrán-Bernabeu, M. A., Femenia-Serra, F., Perles-Ribes, J. F., & Giner-Sánchez, D. (2021a). Measuring the progress of smart destinations: The use of indicators as a management tool. *Journal of Destination Marketing & Management, 19*(1), 100531. https://doi.org/10.1016/j.jdmm.2020.100531

Ivars-Baidal, J. A., Vera-Rebollo, J. F., Perles-Ribes, J. F., Femenia-Serra, F., & Celdrán-Bernabeu, M. A. (2021b). Sustainable tourism indicators: What's new within the smart city/destination approach? *Journal of Sustainable Tourism, 31*, 1556–1582. https://doi.org/10.1080/09669582.2021.1876075

Jurado Almonte, J. M. (2014). El turismo accesible en Andalucía y Portugal. *Cuadernos de Turismo, 33*, 121–150.

Koch, F., & Ahmad, S. (2018). How to measure progress towards an inclusive, safe, resilient and sustainable city? Reflections on applying the indicators of sustainable development goal 11 in Germany and India. In S. Kabisch, F. Koch, E. Gawel, A. Haase, S. Knapp, K. Krellenberg, J. Nivala, et al. (Eds.), *Urban transformations: Sustainable urban development through resource efficiency, quality of life and resilience* (pp. 77–90). Springer International Publishing.

Lai, C. S., Jia, Y., Dong, Z., Wang, D., Tao, Y., Lai, Q. H., Wong, R. T. K., et al. (2020). A Review of Technical Standards for Smart Cities. *Clean Technologies, 2*(3), 290–310. https://doi.org/10.3390/cleantechnol2030019

Lardies-Bosque, R., Fernández-Mayoralas, G., Rojo-Pérez, F., Forjaz, M. J., Martínez Martin, P., & Rojo-Abuin, J. M. (2015). Leisure activities and quality of life among older adults in Spain. In F. Maggino (Ed.), *A new research agenda for improvements in quality of life* (Social Indicators Research Series) (Vol. 57, pp. 163–189). Springer.

Lättman, K., Olsson, L. E., & Friman, M. (2018). A new approach to accessibility–Examining perceived accessibility in contrast to objectively measured accessibility in daily travel. *Research in Transportation Economics, 69*, 501–511. https://doi.org/10.1016/j.retrec.2018.06.002

Leco Berrocal, F., Pérez Díaz, A., & Mateos Rodríguez, A. B. (2011). Accesibilidad en espacios protegidos: las áreas de uso público de la Reserva de la Biosfera de Monfragüe. *Geographicalia, 59–60*, 213–225.

López Abril, M., Melero Bellmunt, J., & Hernández Rofa, J. (2014). *Educación Ambiental sin barreras: algunas reflexiones y experiencias de educación ambiental para todos*. Centro Nacional de Educación Ambiental. https://www.miteco.gob.es/es/ceneam/articulos-de-opinion/2014-10-lopez-melero-hernandez_tcm30-163432.pdf

Luque Gil, A. M., Zayas Fernández, B., & Caro Herrero, J. L. (2015). Los Destinos Turísticos Inteligentes en el marco de la Inteligencia Territorial: conflictos y oportunidades. *Investigaciones Turísticas, 10*, 1–25.

McCabe, S. (2020). Tourism for all?' Considering social tourism: a perspective paper. *Tourism Review, 75*(1), 61–64. https://doi.org/10.1108/TR-06-2019-0264

McKercher, B., & Darcy, S. (2018). Re-conceptualizing barriers to travel by people with disabilities. *Tourism Management Perspectives, 26*, 59–66. https://doi.org/10.1016/j.tmp.2018.01.003

Millán Escriche, M. (2010). Turismo accesible/turismo para todos, un derecho ante la discapacidad. *Gran Tour: Revista de Investigaciones Turísticas, 2*, 101–126.

Miller, G. (2014). *Economic Impact and Travel Patterns of Accessible Tourism in Europe Service*. Contract SI2. ACPROCE052481700, Commission européenne, DG Entreprises et industrie. http://tinyurl.com/lzc6962. Accessed 13 Nov 2020.

Miralles, C. (2002). *Ciudad y transporte. El binomio perfecto*. Ariel Geografía.

Molina Hoyo, M. (2008). *Turisme accessible i enoturisme: Una proposta d'introducció de l'accessibilitat al Museu de les Cultures del Vi de Catalunya*. Universidad Autónoma de Barcelona. https://docplayer.es/71970817-Turisme-accessible-i-enoturisme.html

Molina Hoyo, M., & Cànoves Valiente, G. (2010). Turismo accesible, turismo para todos: la situación en Cataluña y España. *Cuadernos de Turismo, 25*, 25–44.

National Statistics Institute [INE]. (2019). *España en cifras 2019*. Publication. https://www.ine.es/prodyser/espa_cifras/2019/48/

Okafor, C. C., Aigbavboa, C., & Thwala, W. D. (2022). A bibliometric evaluation and critical review of the smart city concept – making a case for social equity. *Journal of Science and Technology Policy Management, Emerald Publishing Limited, 14*, 487–510. https://doi.org/10.1108/JSTPM-06-2020-0098

Olivera Poll, A. (2006). Discapacidad, accesibilidad y espacio excluyente. Una perspectiva desde la Geografía Social Urbana. *Treballs de la Societat Catalana de Geografia, 61–62*, 326–343.

Organización para la Cooperación y el Desarrollo Económico [OECD]. (2017). *Preventing ageing unequally*. http://www.oecd.org/employment/preventing-ageing-unequally9789264279087-en.htm

Perles-Ribes, J. F., Ramón-Rodríguez, A. B., Moreno-Izquierdo, L., & Such-Devesa, M. J. (2020a). Machine learning techniques as a tool for predicting overtourism: The case of Spain. *International Journal of Tourism Research, 22*(6), 825–838. https://doi.org/10.1002/jtr.2383

Perles-Ribes, J. F., Moreno-Izquierdo, L., Torregrosa-Marí, M. T., & Such-Devesa, M. J. (2020b). The relationship between satisfaction and tourism expenditure in 'sun and beach' destinations: a structural equation modelling approach. *Current Issues in Tourism, 24*, 1–15. https://doi.org/10.1080/13683500.2020.1849046

Raccanello, K., & Cuamatzin, Y. (2019). Demanda de turismo accesible para adultos mayores: discapacidad y barreras. Elderly's accessible tourism demand: impairment and barriers. *Gran Tour: Revista de Investigaciones Turísticas, 19*, 54–78.

Ramón-Rodríguez, A. B., & Such-Devesa, M. J. (2018). Is hotel efficiency necessary for tourism destination competitiveness? An integrated approach. *Tourism Economics, 24*(1), 3–26. https://doi.org/10.5367/te.2016.0555

Ravlic, I. (2020). *Propuesta y elaboración de directrices de gestión en accesibilidad para municipios turísticos: una estrategia superadora y complementaria en el camino de la calidad turística para todas las personas (Master Thesis)*. Buenos Aires, Universidad Nacional de San Martín, Argentina.

Rubio Gil, A. (2000). Paradojas en el paraíso: la accesibilidad de los servicios de naturaleza turística en España. Por un turismo de calidad y para todos. *Estudios turísticos, 143*, 65–91.

Rucci, A. C. (2021). *Accessible tourism and competitiveness: assessing different dimensions of accessibility in destinations [PdD thesis]*. University of Alicante, Spain. http://rua.ua.es/dspace/handle/10045/121569

Rucci, A. C., & Porto, N. (2022). Accessibility in tourist sites in Spain: does it really matter when choosing a destination? *European Journal of Tourism Research, 31*, 3108. https://doi.org/10.54055/ejtr.v31i.2165

Rucci, A. C., Moreno-Izquierdo, L., Perles-Ribes, J. F., & Porto, N. (2021). Smart or partly smart? Accessibility and innovation policies to assess smartness and competitiveness of destinations. *Current Issues in Tourism*. https://doi.org/10.1080/13683500.2021.1914005

Santana-Santana, S. B., Peña-Alonso, C., & Pérez-Chacón Espino, E. (2020). Assessing physical accessibility conditions to tourist attractions. The case of Maspalomas Costa Canaria urban area (Gran Canaria, Spain). *Applied Geography, 125*, 102327. https://doi.org/10.1016/j.apgeog.2020.102327

Santana-Santana, S. B., Correa Silva, M. P., Cantillana Reyes, J. C., & Pérez-Chacón Espino, E. (2022). La interpretación del paisaje en la práctica del turismo inclusivo: recursos adaptados para personas con discapacidad visual. *Cuadernos de Turismo, 49*, 225–260. https://doi.org/10.6018/turismo.521901

Tejada Cruz, A., & Fernández-Bermejo, M. (2017). El espacio público accesible como elemento de integración social. Aplicación en zonas costeras e islas. Études caribéennes, 36. https://doi.org/10.4000/etudescaribeennes.10752

United Nations [UN]. (2019). *Perspectivas de la población mundial 2019*. Document of the Population Division of the Department of Economic and Social Affairs. https://population.un.org/wpp/

United Nations Conference on Trade and Development. (2020). *Impact of the COVID-19 Pandemic on Global FDI and GVCs. Updated Analysis*. Global Investment Trend Monitor, Vol. 35, UNITED NATIONS, Ginebra. https://unctad.org/system/files/official-document/wir2020_en.pdf

United Nations Development Programme. (2021). *The SDGS in Action*. https://bit.ly/3DDqgaZ

United Nations World Tourism Organization [UNWTO]. (2019). *Accessible Tourism Destination, 2019 Edition*. https://www.unwto.org/es/accessibility
World Economic Forum [WEF]. (2019). The Travel & Tourism Competitiveness 2019. *Report*. https://www.weforum.org/reports/the-travel-tourism-competitiveness-report-2019
World Health Organization [WHO]. (2020). *WHO Coronavirus Disease (COVID-19) Dashboard*. Retrieved from: https://covid19.who.int/. Accessed 29 Mar 2022.
Yepes, V., Cardona, A., & Vallés, A. (2000). Diseño y gestión de playas turísticas accesibles. *Equipamientos y Servicios Municipales, 88*, 9–14.
Zhang, Y., & Cole, S. T. (2016). Dimensions of lodging guest satisfaction among guests with mobility challenges: A mixed-method analysis of web-based texts. *Tourism Management, 53*, 13–27.
Zimmermann-Janschitz, S. (2018). Geographic Information Systems in the context of disabilities. *Journal of Accessibility and Design for All, 8*(2), 161–192. https://doi.org/10.17411/jacces.v8i2.171

Rosario Navalón-García Senior Lecturer in the Department of Regional Geographical Analysis at the University of Alicante. Her research focuses on the territorial planning of tourism, the relationship between the different elements that make up the tourism system, and the analysis of the potential of cultural heritage and its tourist activation. She is also interested in the concept of accessibility and its application in destinations and tourist attraction resources, as well as the professionalization of the tourist guide as a tourist dynamizer.

Ana Clara Rucci PhD in Tourism (University of Alicante, Spain). Professor, researcher and head of the Tourism Research Institute at Universidad Nacional de La Plata, Argentina. Her field of research is accessible tourism, competitiveness and smart tourism destinations. She is the head of the Accessible Tourism Argentinian Association (Asociación Argentina por el Turismo Accesible).

Raquel Huete Senior Lecturer in the Department of Sociology and researcher at the University Institute for Tourism Research of the University of Alicante. Prior to joining academia, she has worked in several companies related to the tourism sector, foundations and social research consultancies. She also served as Director-General for Tourism at the Valencia Region Government. Her current interests are focused on lifestyle mobilities, social sustainability of tourism and accessible tourism.

The manufacturer's authorised representative in the EU is Springer Nature Customer Service Centre GmbH, Europaplatz 3, 69115 Heidelberg, Germany. If you have any concerns regarding our products, please contact ProductSafety@springernature.com

Printed and bound by CPI Group (UK) Ltd, Croydon, CR0 4YY
01/10/2026
02225892-0017